# WITH AND WITHOUT GALTON

## About the Publisher

**Open Book Publishers** is a not-for-profit, scholar-led Open Access academic press and we are dedicated to revolutionising academic publishing, breaking down the barriers of high prices and restricted circulation so that outstanding academic books are available for everyone to read and share.

If you believe that knowledge should be available to everyone, you can support our work with a monthly pledge or a one-off donation and become part of the OBP community:

https://www.openbookpublish/pledge

## About the Author

**Nikolai Krementsov** is a Professor at the Institute for the History and Philosophy of Science and Technology, University of Toronto (Canada). He has published several monographs and numerous articles on various facets of the history of science, medicine, and literature in Russia and the Soviet Union. His latest publications include *A Martian Stranded on Earth: Alexander Bogdanov, Blood Transfusions, and Proletarian Science* (2011), *Revolutionary Experiments: The Quest for Immortality in the Bolshevik Science and Fiction* (2014), and *The Lysenko Controversy as a Global Phenomenon* (2017), 2 vols. (co-edited with William deJong-Lambert).

# With and Without Galton

## Vasilii Florinskii and the Fate of Eugenics in Russia

*Nikolai Krementsov*

https://www.openbookpublishers.com

© 2018 Nikolai Krementsov

This work is licensed under a Creative Commons Attribution 4.0 International license (CC BY 4.0). This license allows you to share, copy, distribute and transmit the text; to adapt the text and to make commercial use of the text providing attribution is made to the authors (but not in any way that suggests that they endorse you or your use of the work). Attribution should include the following information:

Nikolai Krementsov, *With and Without Galton: Vasilii Florinskii and the Fate of Eugenics in Russia*. Cambridge, UK: Open Book Publishers, 2018. http://dx.doi.org/10.11647/OBP.0144

Copyright and permissions for the reuse of many of the images included in this publication differ from the above. Copyright and permissions information for images is provided separately in the List of Illustrations.

Every effort has been made to identify and contact copyright holders and any omission or error will be corrected if notification is made to the publisher.

In order to access detailed and updated information on the license, please visit https://www.openbookpublishers.com/product/800#copyright

Further details about CC BY licenses are available at https://creativecommons.org/licenses/by/4.0/

All external links were active 23/7/2018 unless otherwise stated and have been archived via the Internet Archive Wayback Machine at https://archive.org/web

Updated digital material and resources associated with this volume are available at https://www.openbookpublishers.com/product/800#resources

EISBN Paperback: 978-1-78374-511-1
ISBN Hardback: 978-1-78374-512-8
ISBN Digital (PDF): 978-1-78374-513-5
ISBN Digital ebook (epub): 978-1-78374-514-2
ISBN Digital ebook (mobi): 978-1-78374-515-9
DOI: 10.11647/OBP.0144

Cover image: Portrait of Vasilii Florinskii by Vasilii Cheremin (1997). Photo courtesy of V. Puzyrev. Cover design by Corin Throsby.

All paper used by Open Book Publishers is SFI (Sustainable Forestry Initiative), PEFC (Programme for the Endorsement of Forest Certification Schemes) and Forest Stewardship Council(r)(FSC(r) certified.

Printed in the United Kingdom, United States, and Australia by Lightning Source for Open Book Publishers (Cambridge, UK)

*To A.-E. and E. F.*

*"Pro captu lectoris habent sua fata libelli."*

Terentianus Maurus, c. 2nd century CE

# Contents

| | |
|---|---|
| *Preface* | xi |
| *List of Abbreviations* | xiii |
| *List of Illustrations* | xvii |
| *Note on Names, Transliterations, and Translations* | xxi |
| *Acknowledgments* | xxiii |
| The Faces of Eugenics: Local Mirrors and Global Reflections | 1 |
| **I. "HYGIENIC" AND "RATIONAL" MARRIAGE** | **23** |
|    1. The Author: Vasilii Florinskii | 25 |
|    2. The Publisher: Grigorii Blagosvetlov | 73 |
|    3. The Book: Darwinism and Social Hygiene | 125 |
|    4. The Hereafter: Words and Deeds | 183 |
| **II. "BOURGEOIS" AND "PROLETARIAN" EUGENICS** | **237** |
|    5. Rebirth: Eugenics and Marxism | 239 |
|    6. Resonance: Euphenics, Medical Genetics, and *Rassenhygiene* | 293 |
|    7. Afterlife: Medical Genetics and "Racial" Eugenics | 351 |
|    8. Science of the Future: With and Without Galton | 409 |
| Apologia: The Historian's Craft | 461 |
| Notes | 495 |
| Index | 655 |

# Preface

This book is an outgrowth of a project I have been pursuing on and off for nearly thirty years: a fully-fledged history of eugenics in Russia. I became interested in the subject at the very beginning of my career as a historian of science in the mid-1980s. Indeed, it was in one of my first public talks, delivered to a May 1989 conference on the social history of Russian science in Leningrad, that I first ventured into this peculiar history. But for the next two decades, numerous other topics captivated my attention and overshadowed this particular interest. Plus, during this very time several scholars in the Soviet Union, the United States, and elsewhere, most notably, Mark B. Adams in Philadelphia, Vasilii V. Babkov in Moscow, and Mikhail B. Konashev in St. Petersburg were publishing extensively on various aspects of that history, and I felt that I did not need to pursue it any further.

Nevertheless, I kept reading on the history of eugenics worldwide and kept collecting whatever materials pertinent to the history of eugenics in Russia I would stumble upon in archives and libraries while going after other subjects. Eventually, I came to realize that the history of eugenics I wanted to write would differ substantially from the works produced by many others who have taken up the subject during the intervening years. So, I decided to actually do it. In 2010-2014, a research grant from the Social Sciences and Humanities Research Council of Canada enabled me to launch the project in earnest and to spend several months each year hunting for relevant materials in Russian archives, libraries, and museums.

At first, I approached my subject in a rather conventional way. I studied the institutions, publications, patrons, methods, agendas, and practices of eugenics; followed individuals and various disciplinary and professional groups involved with its development; and explored its representations in contemporary journalism, literature, cinema, and theater in Russia. The stacks of copies, notes, drafts, and plans for projected chapters steadily grew and soon threatened to completely overflow my modest home office and to turn my little project into a multi-volume edition of unmanageable length and undetermined duration. I began writing up (and occasionally publish) pieces and bits of a complex story that was slowly emerging out of the mountain of materials I have collected over the years.

I would perhaps still be searching through the numerous nooks and crannies of this mountain, if it were not for a Guggenheim Fellowship that gave me a twelve-month leave in the 2015-2016 academic year. The wonderful freedom (from teaching, committees, and other delights of university life) afforded me a peace of mind to rethink my project and reshape it into something very different from what I had at first envisioned. Instead of writing a history of eugenics in Russia by going systematically through its institutional, intellectual, cultural, personal, disciplinary, political, ideological, and many other dimensions, I decided to experiment and to approach my subject from a decidedly different angle.

In my explorations of various episodes in the history of eugenics in Russia, I noticed a recurrence of one particular book, Vasilii Florinskii's *Human Perfection and Degeneration*. Originally published as a series of essays in 1865, it came out in book format less than a year later. For sixty years it lay dormant and apparently unread, but in 1926 it was reprinted and actively discussed. Yet, just a few years later, any references to its existence disappeared and resurfaced again only in the early 1970s. A new edition of the book came out a quarter of a century later, in 1995, and then — just as I was in Moscow doing my archival research — in 2012, it was republished once more. This seemed rather peculiar. Why would an obscure mid-nineteenth-century book be repeatedly revived and forgotten and revived again during nearly 150 years? Intrigued, I began going through my materials with a fine-tooth comb looking for clues. Soon I realized that the life story of this book — of its author, contents, publishers, editors, commentators, and readers — offers a unique lens to cast the history of eugenics in Russia in an unusual and very revealing light. In addition, a novel (for me) way to write a history of science by means of a "biography of a book" promised to be quite challenging and exciting. I hope the results do convey at least some of that excitement.

St. Petersburg, 8 December 2017

# List of Abbreviations

| | |
|---|---|
| AMN | *Akademiia medistinskikh nauk*, Academy of Medical Sciences |
| AN | *Akademiia nauk*, Academy of Sciences |
| APS | American Philosophical Society |
| ARAN | *Arkhiv Rossiiskoi Akademii nauk*, Archive of the Russian Academy of Sciences |
| ARGB | *Archiv für Rassen- und Gesellschafts-Biologie* [journal] |
| ASMiOG | *Arkhiv sudebnoi meditsiny i obshchestvennoi gigieny*, Archive of Legal Medicine and Social Hygiene [journal] |
| BAN | *Biblioteka Akademii nauk*, The Library of the Academy of Sciences |
| BME | *Bol'shaia meditsinskaia entsikolpediia*, Great Medical Encyclopedia |
| BSE | *Bol'shaia sovetskaia entsiklopediia*, Great Soviet Encyclopedia |
| DZP | *Dnevnik zagranichnogo puteshestviia*, Diary of [My] Foreign Trip [Florinskii's unpublished memoir] |
| EES | Eugenics Education Society |
| ERO | Eugenics Record Office, United States |
| GARF | *Gosudarstvennyi arkhiv Rossiiskoi Federatsii*, State Archive of the Russian Federation |
| GASO | *Gosudarstvennyi arkhiv sverdlovskoi oblasti*, State Archive of the Sverdlovsk Region |
| GIS | *Gigiena i sanitaria*, Hygiene and Sanitation [journal] |
| GIZ | *Gosudarstvennoe izdatel'stvo*, State Publishing House |
| IEB | *Institut eksperimental'noi biologii*, Institute of Experimental Biology |
| IFK | *Institut fizicheskoi kul'tury*, Institute of Physical Culture |

| | |
|---|---|
| IIET | *Institut istorii estestvoznaniia i tekhniki,* Institute for the History of Natural Science and Technology |
| IMG | *Institut meditsinskoi genetiki,* Institute of Medical Genetics |
| IMSA | *Imperatorskaia mediko-khirurgicheskaia akademiia,* Imperial Medical-Surgical Academy |
| INION | *Institut nauchnoi informatsii po obshchestvennym naukam,* Institute of Scientific Information on Social Sciences |
| ITU | *Izvestiia tomskogo universiteta,* Herald of Tomsk University [journal] |
| JHB | *Journal of the History of Biology* |
| L. | Leningrad |
| M. | Moscow |
| MBI | *Mediko-biologicheskii institut,* Medical-Biological Institute |
| MV | *Meditsinskii vestnik,* Medical Herald [journal] |
| Narkompros | *Narodnyi Komissariat prosveshcheniia,* People's Commissariat of Enlightenment |
| Narkomzdrav | *Narodnyi Komissariat zdravookhraneniia,* People's Commissariat of Health Protection |
| Narkomzem | *Narodnyi Komissariat zemledeliia,* People's Commissariat of Agriculture |
| NEP | *Novaia ekonomicheskaia politika,* New Economic Policy |
| NKVD | *Narodnyi Komissariat vnutrennikh del,* People's Commissariat of Internal Affairs |
| NMRT | *Natsional'nyi muzei Respubliki Tatarstan,* National Museum of the Tatarstan Republic |
| OMM | *Okhrana materinstva i mladenchestva,* Protection of maternity and infancy |
| OZDP | *Okhrana zdorov'ia detei i podrostkov,* Protection of children's and adolescents' health |
| PZM | *Pod znamenem marksizma,* Under the Banner of Marxism [journal] |
| PZORV | *Protokoly zasedanii Obshchestva russkikh vrachei v S. Peterburge,* Protocols of the Meetings of the St. Petersburg Society of Russian Physicians [journal] |
| RAZh | *Russkii antropologicheskii zhurnal,* Russian Anthropological Journal |

| | |
|---|---|
| RES | *Russkoe evgenicheskoe obshchestvo*, Russian Eugenics Society |
| REZh | *Russkii evgenicheskii zhurnal*, Russian Eugenics Journal |
| RGALI | *Rossiiskii gosudarstvennyi arkhiv literatury i iskusstva*, Russian State Archive of Literature and Arts |
| RGASPI | *Rossiiskii gosudarstvennyi arkhiv sotsial'no-politicheskoi istorii*, Russian State Archive of Socio-Political History |
| RGIA | *Rossiiskii gosudarstvennyi istoricheskii arkhiv*, Russian State Historical Archive |
| RGVIA | *Rossiiskii gosudarstvennyi voenno-istoricheskii arkhiv*, Russian State Military Historical Archive |
| RO IRLI | *Rukopisnyi otdel Instituta russkoi literatury*, Manuscript Department of the Institute of Russian Literature (St. Petersburg) |
| RO RNB | *Rukopisnyi otdel Rossiiskoi natsional'noi biblioteki*, Manuscript Department of the Russian National Library (St. Petersburg) |
| RS | *Russkoe slovo*, Russian Word [journal] |
| RSFSR | *Rossiiskaia Sovetskaia Federativnaia Sotsialisticheskaia Respublika*, Russian Soviet Federated Socialist Republic |
| SF | Science fiction |
| SNK | *Sovet narodnykh komissarov*, Council of People's Commissars |
| SP | *Sovremennaia psikhiatriia*, Modern Psychiatry [journal] |
| SPb. | St. Petersburg |
| TsGAM | *Tsentral'nyi gosudarstvennyi arkhiv goroda Moskvy*, Central State Archive of the City of Moscow |
| USSR | *Soiuz Sovetskikh Sotsialisticheskikh Respublik*, Union of Soviet Socialist Republics |
| VARNITSO | *Vsesoiuznaia assotsiatsiia rabotnikov nauki i tekhniki dlia sodeistviia sotsialisticheskomu stroitel'stvu*, All-Union Association of the workers in science and technology for assisting socialist construction [journal] |
| VASKhNIL | *Vsesoiuznaia akademiia sel'sko-khoziaistvennykh nauk im. V. I. Lenina*, Lenin All-Union Academy of Agricultural Sciences |
| VIET | *Voprosy istorii estestvoznaniia i tekhniki*, Issues in the History of Science and Technology [journal] |

| | |
|---|---|
| VKA | *Vestnik Kommunisticheskoi akademii*, Herald of the Communist Academy [journal] |
| VMA | *Voenno-meditsinskaia akademiia*, Military-Medical Academy |
| VMZh | *Voenno-meditsinskii zhurnal*, Military-Medical Journal |
| VOGiS | *Vsesoiuznoe obshchestvo genetikov i selektsionerov*, All-Union Society of Geneticists and Breeders |
| ZV | *Zapiski i vospominaniia*, Notes and Recollections [Florinskii's unpublished memoir] |

# List of Illustrations

| | | |
|---|---|---|
| 1-1 | A postcard featuring Perm Seminary, c.1900. Courtesy of RNB. | 33 |
| 1-2 | A postcard featuring the main building of the Imperial Medical-Surgical Academy (IMSA), c.1900. Courtesy of RNB. | 43 |
| 1-3 | A page from Vasilii Florinskii's diary, 1858. Courtesy of NMRT. | 52 |
| 1-4 | Vasilii Florinskii in Paris, c.1862. Photo by Sergei Levitsky. Courtesy of NMRT. | 61 |
| 1-5 | The Florinskii family, c.1862. Photographer unknown. Courtesy of NMRT. | 65 |
| 1-6 | Maria Florinskaia, née Fufaevskaia, c.1863. Photographer unknown. Courtesy of NMRT. | 70 |
| 1-7 | Solomko house on Malaia Italianskaia Street (currently Zhukovskogo Street) in St. Petersburg. Photo by the author, 2016. | 71 |
| 2-1 | *Russkoe slovo*, 1859, 1, title page. Courtesy of RNB. | 76 |
| 2-2 | Grigorii Blagosvetlov, c.1860s. A lithograph from G. E. Blagosvetlov, *Sochineniia* (SPb.: E. A. Blagosvetlova, 1882). Courtesy of BAN. | 80 |
| 2-3 | Cartoon of the Russian journalistic scene entitled "A concerto in C-dur(nom) tone" by Petr Borel', from *Zanoza*, 3 March 1863, supplement. Courtesy of RO IRLI. | 86 |
| 2-4 | Cartoon of Blagosvetlov entitled "A good-natured editor whose left hand does not know what the right hand does," from *Iskra*, 23 December 1864, p. 640. Courtesy of RNB. | 113 |
| 3-1 | Vasilii Pukirev's painting *The Unequal Marriage*, 1862, oil on canvas, 173 x 136.5 cm. From https://commons.wikimedia.org/wiki/File:Pukirev_ner_brak.jpg. Tretyakov Gallery, Moscow. | 155 |
| 3-2 | Table for calculating the degrees of kinship, from S. P. Grigorovskii, *O rodstve i svoistve* (SPb.: Trud, 1903), 6th ed., p. 25. Courtesy of RNB. | 171 |

| | | |
|---|---|---|
| 4-1 | F. [V. M.] Florinskii, *Usovershenstvovanie i vyrozhdenie chelovecheskogo roda* (SPb.: printed by Riumin & Co., 1866), title page. Courtesy of RNB. | 191 |
| 4-2 | F. Gal'ton, *Nasledstvennost' talanta, ee zakony i posledstviia* (SPb.: Znanie, 1875), title page. Courtesy of RNB. | 200 |
| 4-3 | Maria Florinskaia with her children, c.1870. Photo by Ludvik Cluver, St. Petersburg. Courtesy of NMRT. | 203 |
| 4-4 | Tomsk University, c.1888, from *Pervyi universitet v Sibiri* (Tomsk: Sibirskii vestnik, 1889), insert. Courtesy of RNB. | 217 |
| 4-5 | Map of Western-Siberian Educational District, c.1888, from *Pamiatnaia knizhka Zapadno-Sibirskogo uchebnogo okruga* (Tomsk, 1897), 5th edn., insert. Courtesy of RNB. | 219 |
| 4-6 | Vasilii Florinskii, c.1888, from *Pervyi universitet v Sibiri* (Tomsk: Sibirskii vestnik, 1889), insert. Courtesy of RNB. | 221 |
| 4-7 | Maria Florinskaia, c.1877. Photo by Iulii Shteinberg, St. Petersburg. Courtesy of NMRT. | 223 |
| 4-8 | Advertisement for Florinskii's book and Darwin's *Descent of Man*, from *Delo*, 1872, 9, backmatter. Courtesy of RNB. | 230 |
| 5-1 | Mikhail Volotskoi at the time of his graduation from high school, c.1912. Photographer unknown. Courtesy of TsGAM. | 256 |
| 5-2 | Mikhail Volotskoi at the time of his graduation from Moscow University, c.1918. Photographer unknown. Courtesy of TsGAM. | 259 |
| 5-3 | The opening of an exhibition at the State Museum of Social Hygiene, 11 July 1919. Photographer unknown. Courtesy of the Museum of the History of Medicine at the Sechenov First Moscow Medical University. | 261 |
| 5-4 | Nikolai Kol'tsov with his students, c.1913. Photographer unknown. Courtesy of ARAN. | 262 |
| 5-5 | Iurii Filipchenko with his students, 1923. Photographer unknown. Courtesy of S. Fokin. | 268 |
| 5-6 | "Bio-social eugenics, its scientific foundation, conditions of development, and methods," from M. V. Volotskoi, *Sistema evgeniki kak biosotsial'noi distsipliny* (M.: Izd. Timiriazevskogo instituta, 1928), insert. Courtesy of INION. | 281 |

| | | |
|---|---|---|
| 5-7 | "Eugenics tree." From Harry H. Laughlin, *The Second International Exhibition of Eugenics held September 22 to October 22, 1921, in Connection with the Second International Congress of Eugenics in the American Museum of Natural History, New York* (Baltimore: Williams & Wilkins, 1923), p. 15. | 281 |
| 5-8 | V. M. Florinskii, *Usovershenstvovanie i vyrozhdenie chelovecheskogo roda* (Vologda: Severnyi pechatnik, 1926), title page. Courtesy of RNB. | 287 |
| 5-9 | Female fashion models, from *Vestnik mody*, 1923, 5, insert. Courtesy of BAN. | 289 |
| 6-1 | Illustration for the article "Marital Choice," from *Gigiena i zdorov'e rabochei i krest'ianskoi sem'i*, 1930, 17-18, p. 12. Courtesy of BAN. | 302 |
| 6-2 | An advertisement for the GIZ series "Classics of Natural Science," from I. I. Mechnikov, *Lektsii o sravnitel'noi patologii vospaleniia* (M.: GIZ, 1923), backmatter. Courtesy of RNB. | 305 |
| 6-3 | Mikhail Volotskoi, c.1926. Photographer unknown. Courtesy of N. Bogdanov. | 312 |
| 6-4 | Alexander Serebrovskii at the Fifth International Genetics Congress in Berlin, September 1927. Photographer unknown. Courtesy of ARAN. | 314 |
| 6-5 | Solomon Levit at Herman J. Muller's lab in Texas, 1931. Photographer unknown. Courtesy of the Lilly Library, Indiana University, Bloomington. | 336 |
| 6-6 | Cartoons depicting genetics as a "fascist," "imperialist" science, by Boris Efimov, from *Flash*, 1949, 11, pp. 14-16. Courtesy of BAN. | 348 |
| 7-1 | Iurii Filipchenko with his students, 1925. Photographer unknown. Courtesy of N. Medvedev. | 358 |
| 7-2 | Gregor Mendel's bust in front of Ivan Pavlov's Laboratory for the Experimental Genetics of Higher Nervous Activity in Koltushi, Russia. Photo by the author, 2016. | 360 |
| 7-3 | *Soviet Eugenics*, 1991, 2, cover page. Courtesy of BAN. | 377 |
| 7-4 | A bronze medal, commemorating Vasilii Florinskii by A. Shamaev, 1990, the author's collection. | 380 |

7-5 Evgenii Iastrebov's "family tree", from E. Iastrebov, ed., *Sto neizvestnykh pisem russkikh uchenykh i gosudarstvennykh deiatelei Vasiliiu Markovichu Florinskomu* (Tomsk: Izd-vo Tomskogo Universiteta, 1995), p. 6. Courtesy of BAN. 383

7-6 Valerii Puzyrev, c.2000. Photographer unknown. Courtesy of V. Puzyrev. 387

7-7 V. M. Florinskii, *Usovershenstvovanie i vyrozhdenie chelovecheskogo roda* (Tomsk: Izd-vo Tomskogo universiteta, 1995), title page. Courtesy of V. Puzyrev. 388

7-8 Vladimir Avdeev, a portrait by Roman Iashin, c.1998, oil on canvas, 120 x 100 cm; from http://racology.ru/galereya. Courtesy of V. Avdeev and R. Iashin. 393

7-9 Vladimir Avdeev with Jared Taylor, Moscow, 2016. Photographer unknown. From http://racology.ru/galereya. Courtesy of V. Avdeev. 395

7-10 V. B. Avdeev, ed., *Russkaia evgenika* (M.: Belye Al'vy, 2012), title page. The author's collection. 397

8-1 Vasilii Florinskii's portrait, oil on canvas, 72 x 88 cm, 1997, by Vasilii Cheremin (1926-2002). Courtesy of V. Puzyrev. 423

8-2 *Russkii evgenicheskii zhurnal*, 1922, 1, title page. Courtesy of BAN. 428

# Note on Names, Transliterations, and Translations

This book deals with more than a century and a half of Russian history, especially the history of medicine, science, education, and journalism. During this period the very name of the country (to say nothing of its borders or its political, economic, administrative, and social organization) changed radically several times. To avoid an unnecessary confusion and cluttering of the text, throughout the book I often use its colloquial name, Russia, to refer to the Russian Empire (before 1917), the Russian Soviet Federated Socialist Republic (1917-1922), the Union of Soviet Socialist Republics (1922-1991), and the Russian Federation (after 1991), even though strictly speaking it is not historically accurate. Furthermore, numerous Russian cities, regions, and municipalities also repeatedly changed their names. To give but one example, St. Petersburg became Petrograd in 1914, Leningrad in 1924, and regained its original name in 1991. Throughout the text I use that name for a particular locale, which was in use at the time I describe, occasionally noting its current name and administrative subordination. Similarly, many institutions described in this book have also been renamed multiple times. Thus the St. Petersburg Imperial Academy of Sciences became the Russian Academy of Sciences in 1917, the USSR Academy of Sciences in 1925, and once more the Russian Academy of Sciences in 1991. Again, to avoid an unnecessary clattering of the text, I use the generic name "Academy of Sciences" throughout, unless its proper full name is called for by a particular context.

In rendering various Russian names and words in the Latin alphabet, I use the Library of Congress's transliteration system, except for the commonly adopted spellings of well-known names, such as, for example,

"St. Petersburg," "Alexander," "Leon Trotsky," and "Fedor Dostoevsky," instead of "Sankt-Peterburg," "Aleksandr," "Lev Trotskii," and "Fedor Dostoevskii," respectively. Except for the names of the country's two major newspapers, *Pravda* and *Izvestiia*, and a popular-science magazine, *Priroda* (*Nature*), I translated into English the titles of various periodicals in the text, but preserved their Russian names in the references.

Although some of the original Russian (and, occasionally, French and German) sources I cite are available in English translations, all of the translations in the book are my own. Indeed, a correct translation of various Russian texts became a major challenge in writing this book and an essential part of my analysis of its subject. So much so, that I felt compelled to write an extensive essay on translations as a requisite part of the historian's craft and to include it in the "Apologia" appended to this book.

# Acknowledgments

As is always the case with any scholarly work, this book bears the mark of numerous individuals who helped me conceive, research, write, and, finally, publish it. First and foremost, I am greatly indebted to Mark B. Adams, who encouraged and nurtured my interest in the history of eugenics over the nearly thirty years of our friendship. Our numerous conversations (in person, by phone, and via countless emails) deeply influenced my thinking about and writing up this book.

A major part of any historical study is gathering all possible information about its subjects: individuals, institutions, ideas, times, and places. Given the multitude of themes, characters, and locales examined in this book, I would have never been able to do it all by myself. I owe the greatest debt of gratitude to the unsung heroes of historical profession — archivists and librarians. This book benefited tremendously from the generous assistance offered by the staff members of numerous Russian and North American archives, libraries, and museums, including the Archives of the Russian Academy of Sciences, the State Archive of the Russian Federation, the Russian State Archive of Socio-Political History, the Russian State Archive of Literature and Arts, the Russian State Historical Archive, the Russian State Military-Historical Archive, the State Archive of the Sverdlovsk Region, the Manuscript Department of the Institute of Russian Literature, the Manuscript Department of the Russian National Library, the Library of the Russian Academy of Sciences, the Russian National Library, the Fundamental Library of the Military-Medical Academy, the Rare Book Collection of the Scientific Library of Moscow University, the Russian State Library, the Museum of the History of Medicine of the Sechenov First Moscow Medical University, the National Museum of the Tatarstan Republic (Kazan), the US National Library of Medicine (Bethesda, MD), the Libraries of the University of Toronto, the Library of the American Philosophical Society (Philadelphia, PA), and the Lilly Library of Indiana University (Bloomington, IN).

The help of many colleagues, including Aleksandra Bekasova, Anastasia Fedotova, Julia Laius, Galina Savina, Nataliia Semenova, and Marina Sorokina, in my search for necessary documents, illustrations, and publications was indispensable. Nikolai Bogdanov, Sergei Fokin, Mikhail Konashev, Susan G. Solomon, and Daniel P. Todes generously shared with me their own precious archival finds and their expertise in various areas of Russian history covered in the present volume. Valerii Puzyrev helped obtain copies of several rare publications, as well as a "Florinskii Medal" issued by the Military-Medical Academy in 1990, and shared with me his recollections about his involvement with publishing Florinskii's book in 1995. Dmitrii Kozlov and Artem Borisov provided much appreciated assistance in copying materials in various archives and libraries in Moscow and St. Petersburg.

My students at the University of Toronto willingly served as "experimental subjects" for testing some of the major ideas elaborated in this book. In 2013 and 2015, graduate students at the Institute for the History and Philosophy of Science and Technology joined me in exploring the historical developments of eugenics at a research seminar on "The Rise of Eugenics: A Comparative History." Our sometimes animated discussions sustained my interest in and excitement about the project, and also helped hone and refine my arguments.

Numerous colleagues, including Francesco Cassata, Michael D. Gordin, Diane B. Paul, Valerii Puzyrev, Susan G. Solomon, and Marina Sorokina patiently heard and/or read pieces and bits of this book at various stages in its development and offered helpful criticisms and suggestions. I am particularly indebted to Mark B. Adams, Anne-Emanuelle Birn, Nathaniel Comfort, Daniel P. Todes, and John Waller, each of whom read the entire manuscript, for their thoughtful comments that helped improve the book in many significant ways.

Several institutions provided financial support for my work for which I am very grateful. Funding for research leading to this book came from the Social Sciences and Humanities Research Council of Canada and from the John Simon Guggenheim Memorial Foundation, while its publication was facilitated by a generous grant from the University of Toronto's Victoria College.

The editorial team of Open Book Publishers made the production of this book nearly painless.

My family stoically endured my long absences during frequent research trips to Russia and fostered my obsession with this project in every possible way — I would not have done it without their support.

Naturally, I alone bear the responsibility for any mistakes and misinterpretations.

# The Faces of Eugenics: Local Mirrors and Global Reflections

> "We cannot even guess,
> How our word will echo…"
>
> Fedor Tiutchev, 27 February 1869

The two men whose names appear in the title of this book — British polymath Francis Galton and Russian gynecologist Vasilii Florinskii — were contemporaries. Although they never met and likely never even heard of one another, in the history of science their names appear to be closely linked. In 1865, each published in an influential monthly a scholarly piece that years later their followers in Britain and in Russia, respectively, would hail as the foundation of a new "science of improving human stock," which Galton named eugenics. Numerous scholars have examined the development of Galton's version of this new "science" in his homeland and beyond. By tracing the punctuated life story of Florinskii's 1865 treatise on *Human Perfection and Degeneration* — reprinted in 1866, 1926, 1995, and 2012 — this book analyses the history of eugenics in his homeland and explores its implications for the understanding of eugenics as a transnational phenomenon.

In the last few decades, the history of eugenics has turned into a virtual industry, with the number of publications on its various facets rapidly multiplying with each passing year.[1] Yet in all of this vast literature, what exactly was (and still is) covered by this composite name that Galton had constructed from the Greek eu-genes (εὐ-γενής) — well-born — remains ambiguous. Numerous students of its history have examined eugenics' institutions and practices, research and policies, prophets and victims, mobilization campaigns and intellectual roots, political resonance and patronage patterns, stated goals and tacit ideals, public perceptions and state legislation, cultural representations and ideological underpinnings, as well as its local variations and international trends. They have portrayed eugenics as a scientific discipline, a creed, a social movement, a biological doctrine, an ideology, a variant of social medicine, a pseudoscience, an array of policies, and so on. They have linked the development of eugenics to the major ideological and political currents of the late nineteenth and early twentieth centuries, including racism, nationalism, socialism, statism, anarchism, fascism, feminism, neo-Malthusianism, progressivism, scientism, technocratism, social Darwinism, elitism, and even spiritualism. They have explored the relationship of eugenics to a number of disciplines, specialties, and professions in public health, natural and social sciences, agriculture, education, medicine, and jurisprudence, as well as to various religious doctrines.

Such a diversity of views and approaches notwithstanding, most scholars tacitly agree on tracing the origins of eugenics to Galton and, following his early disciples, on hailing him as its "founding father."[2] After all, he had given *it* the name. This conventional genealogy, however, is, for the most part, an artifact deriving from interconnected historical and historiographical incidents. Historically, the consistent efforts to unite into a more or less coherent whole the various individuals and groups interested in what Galton christened eugenics took place in the English-speaking countries: the First International Eugenics Congress had been held in Britain in 1912 and the next two in the United States in 1921 and 1932. Historiographically, the studies of eugenics had begun, and for many years were focused largely on its development, in Britain and the United States, often skirting similar developments in the rest of the world.[3] The rise of English as the lingua franca of science and scholarship in the second half of the twentieth century further enhanced this "anglicizing" of both the global history and the genealogy of eugenics.

Already at the First International Eugenics Congress convened in July 1912 in London, however, German physician Alfred Ploetz successfully challenged this linear genealogy. Ploetz effectively pressed the London congress into acknowledging the German "priority" in "endeavouring to co-ordinate eugenic work in various countries."⁴ Indeed, nearly a decade earlier, Ploetz had coined an alternative name for what Galton had termed eugenics — *Rassenhygiene* (race or racial hygiene). He had also founded an international Society for Racial Hygiene (*Gesellschaft für Rassenhygiene*), created its first specialized periodical, *Archiv für Rassen- und Gesellschafts-Biologie* (1904), and convened its first international conference.⁵

Furthermore, the London congress's proceedings vividly demonstrated that what Galton had named "national eugenics" in Britain also had its "national" analogues in Belgium, Denmark, France, Germany, Italy, Norway, Switzerland, and the United States. Indeed, Karl Pearson, Galton's right-hand man in developing British eugenics, forcefully asserted a year prior to the congress that "Eugenics must ... be essentially national, and eugenics as a practical policy will vary widely according as you deal with Frenchmen or Japanese, with Englishmen or Jews."⁶ These "national" variants developed in different institutional, intellectual, and social contexts and under different names, such as, for instance, *Rassenhygiene* and *Fortpflanzungshygiene* in Germany, humaniculture, euthenics, and stirpiculture in the United States,⁷ and *pédotechnie, puériculture,* and *eugénnetique* in Belgium and France. The growing number of historical studies on various "national" eugenics and "proto-eugenics," along with "socialist," "Jewish," "proletarian," "Latin," "Baltic," "East-Central European," "liberal," "reform," "state," "private," and many other incarnations and permutations of eugenics, have undermined the traditional genealogy even further, complicating the issue of what eugenics was (and is) far beyond the questions of onomastics and "founding fathers."⁸

The protean nature of eugenics manifested in the burgeoning number of historical studies is often magnified, ironically, by the *ahistorical* treatment of their common subject by some of its students, who tend to see it in static, rather than dynamic terms. But even Galton's understanding of what eugenics was did not remain unchanged. It underwent considerable modifications from his first 1865 musings on "hereditary talent and character" — through the 1883 *Inquiry into Human Faculty,* in which he first introduced the name — to the early 1900s when he endowed the first research institution (the Eugenics

Record Office, soon renamed the Francis Galton Eugenics Laboratory, at the University of London) and initiated the formation of the British Eugenics Education Society. Galton's disciples and detractors used the name in multiple, sometimes radically different ways, while many others propagated very similar ideas under a variety of other names. Advanced by numerous actors in different times and places, "eugenics," then, meant many different things to many different audiences.

What can explain the protean nature of eugenics, its appeal to diverse audiences, and its multiple local histories? Why did (and do) some observers attach the label "eugenics" to such vastly different phenomena as infanticide in Ancient Sparta and genetic engineering, cloning and Plato's or H. G. Wells's meditations on the "breeding" of ruling elites, taboos surrounding incest and Peter the Great's decree on the prohibition of "fool marriages," sperm banks and the birth-control or anti-abortion movements, the Holocaust and regulation of consanguineous marriages, prenatal diagnostics and sterilization laws?

One might suggest that what all these (and many other) phenomena habitually described as "eugenics" share is explicit or implicit intervention into *human reproduction* and the implications of such intervention for the *future*.[9] The scientific foundations, ideological underpinnings, social and political goals, projected targets and scope, appointed agents and agencies, moral justifications, and proposed instruments of such intervention have varied substantially in different times and places. Yet, the intervention itself and its anticipated future consequences have remained at the core of "eugenic" concepts propounded everywhere, whether named eugenics, *Rassenhygiene*, *eugénnetique*, aristogenics, homiculture, or something else entirely.

A close reading of the literature on the history of eugenics suggests that in each of these concepts, their authors and adherents fused together certain *ideas, values, concerns,* and *actions* regarding human reproduction, heredity, variability, development, and evolution characteristic of their own particular society and their own era, which informed and justified a particular vision of that society's future. For the purposes of a historical analysis, it might be useful to disentangle the four major components of such amalgams. The first component is an array of intertwined ideas and conceptions about human reproduction, heredity, variability, individual and social development, and evolution, often blended together under the shorthand of "human nature." The second is a system of norms, beliefs, values, mores, customs, traditions, and ideals that assigns individuals and groups specific roles and places in the structures of a given society and Nature writ large. The third is a constellation of perceived social concerns (both hopes and fears) associated with the issues of human reproduction, heredity, diversity, development, and evolution. And the fourth is a set of practices, policies, measures, and actions directed at "controlling" human reproduction, heredity, development, and evolution, which purport to address these concerns and aim at either conserving or altering "human nature," along with existing societal norms and structures. Each component of this quadriga exhibited numerous "geographical variations" and underwent manifold transformations, often unrelated to those of the other three elements.

In certain times and places, however, the four components coalesced in the writings and actions of concrete historical figures. These individuals produced a peculiar *amalgam* of distinct ideas, specific values, characteristic concerns, and particular practices that targeted "human nature" to address some perceived social problems, and thus shape the future of their own societies and humanity as a whole. It is this fusion that explains the attractiveness of eugenics to virtually every professional, occupational, and disciplinary group interested in human reproduction, heredity, variability, individual and social development, and evolution, be they psychiatrists, geneticists, civil servants, pediatricians, anthropologists, social reformers, gynecologists, educators, public health specialists, feminists, legislators, or preachers. And it is a particular vision of the future embedded in this fusion that made eugenics a favorite trope of science-fiction (SF) writers from H. G. Wells to Robert A. Heinlein, Orson Scott Card, and Margaret Atwood.

Seen in this light, eugenics, as articulated by Galton, is just such an amalgam quite specific to late Victorian England. It fused together Galton's (anthropological, statistical, biological, Darwinian, and so on) notions of human reproduction, heredity, variability, development, and evolution with his — upper-middle-class, racist, imperialist, sexist, bourgeois, atheistic, etc. — value system. It proposed a series of policies and actions (including state regulations of marriage, stipends to talented youth, and propaganda of eugenic ideas) that, he hoped, could uplift the "human faculty" of the British nation and alleviate such "social ills" as criminality, feeble-mindedness, pauperism, and differential fecundity, which troubled his contemporary society, thus assuring its future survival and progress. And it (eventually) found support and elicited criticism from individuals representing a quite specific array of disciplines, professions, ideologies, and occupations.

One could easily imagine a different amalgam — forged out of different (or the same) notions of human reproduction, heredity, variability, development, and evolution, imbued with a different (or the same) set of norms, values, and ideals, addressing different (or the same) social concerns, and calling for a different (or the same) sort of actions. One could as easily imagine countless possible permutations such an amalgam could attain in different times and places, different visions of the future it could embody, and a different assortment of individuals and groups who would either support or criticize it. This book examines the history of one such amalgam created simultaneously with, but independently from, Galton's in a setting very different from Victorian England — Imperial Russia.

## Eugenics and Eugamics

In June and August 1865, in London, the capital of the "perpetually sun-lit" British empire, the influential monthly *Macmillan's Magazine* carried Galton's article, titled "Hereditary Talent and Character."[10] This short twenty-page piece laid the first stone in the foundation of what nearly twenty years later its author would name "eugenics" and define as "the science of improving [human] stock" devoted to "the investigation of the conditions under which men of a high type are produced."[11] Based on the statistical analysis of "blood relations" among "British men," Galton suggested that "if a twentieth part of the cost and pains were spent in measures for the improvement of the human race that is spent on the improvement of the breed of horses and cattle, what a galaxy of genius might we not create!"[12] Galton's grand idea was quite simple: "by selecting men and women of rare and similar talent, and mating them together, generation after generation, an extraordinarily gifted race might be developed."[13]

At almost exactly the same time, thousands of kilometres from London, in the capital of the "perpetually-frozen" Russian Empire, St. Petersburg, the August 1865 issue of the leading "literary-political" journal *Russian Word* opened with a nearly sixty-page-long essay, titled "Human Perfection and Degeneration."[14] Three more essays appeared under the same general title in the journal's October, November, and December issues.[15] In late August 1866, the journal's publisher released the essays in book format.[16] This 200-page treatise was not penned by one of *Russian Word*'s renowned regular contributors, which included such leading intellectuals of the day as Dmitrii Pisarev, Nikolai Shelgunov, Petr Tkachev, Varfolomei Zaitsev, and Grigorii Blagosvetlov, the journal's editor and publisher. It was authored by Vasilii Florinskii (1834-1899), a little known (outside of narrow professional circles) adjunct professor of gynecology and obstetrics at the Imperial Medical-Surgical Academy (IMSA), the country's premier medical school.

Florinskii's stated goal was to acquaint the journal's readers with an "as yet unexamined, one can even say untouched, subject," namely the "general conditions of human perfection and degeneration." The treatise's brief introduction charted its main lines of inquiry and its two-pronged approach. The first half was to focus on "the variability and perfection of the human type in general." It was to include special sections on "heredity as the main cause of human variability and perfection" and on "conditions conducive to stock perfection" such as "taste and demand for certain qualities, influence of the external conditions of life, rational marriage, and sex life." The second half was to discuss "the degeneration of the human type in general" and "conditions that could facilitate such degeneration." Among such conditions Florinskii listed "incest and the lack of stock renewal, inequality [of partners] in marriage, [and] the influence of drunkenness, debauchery, diseases, poverty, and slavery."

Like Galton, Florinskii drew parallels with animal breeding:

> Much attention is paid to, and whole doctrines exist about, the betterment of stocks in cattle, sheep, horses, dogs, even chickens, pigeons, and so on, and the goal is actually being achieved. Systematically cultivated breeds of animals astonish us by their perfection; whilst man in the successive generations breeds diseases and physical weakness rather than perfection.

But in contrast to Galton's ultimate goal of breeding "an extraordinarily gifted race" and producing "men of a high type," Florinskii's was to perfect "the human type in general" and to stave off its possible degeneration. Accordingly, instead of Galton's selective mating of "men and women of rare and similar talent," Florinskii proposed removing existing barriers to "mixed" marriages between men and women of different ancestry, confessions, ethnicities, talents, physiques, social standings, and so on. Were Florinskii to search for a moniker for his major idea of "hygienic" or "rational" marriage, he would probably have coined the word *eugamics* — well-married (from the Greek gámos [γάμος] — marriage), not *eugenics* — well-born, as did Galton.[17]

The fact that Florinskii's "eugamic" treatise was published nearly simultaneously with Galton's first "eugenic" article is not a mere coincidence. As Galton himself once said, "Great discoveries have often been made simultaneously by workers ignorant of each other's labours. This shows that they have derived their inspiration from a common but hidden source, as no mere chance would account for simultaneous discovery."[18] Such a common and very obvious source for the two approaches to the "improvement of the human stock" was Charles Darwin's evolutionary concept formulated just six years earlier in his book *On the Origin of Species* (whose very publication, ironically, had been prompted by the "simultaneous discovery" of Darwin's main idea of natural selection by his compatriot Alfred Russel Wallace). It was Darwin's detailed examination of such fundamental factors of species evolution (both in nature and under domestication) as *variability*, *heredity*, and *selection* that inspired the futuristic vision of a *directed human evolution* accomplished by manipulating human reproduction and embodied in both Galton's "eugenics" and Florinskii's "eugamics." Darwin's analysis of various forms of selection (natural, sexual, and artificial) as the mechanism of such basic evolutionary processes as species adaptation, divergence, and extinction provided a solid grounding to this vision.

The contemporaneity of Galton's and Florinskii's ideas not only suggests the shared intellectual impulse, but also points to their common mid-nineteenth-century scientific contexts, most important, the emergence of science as a particular social institution. This development included the entwined processes of the formation of a professional workforce (the scientist), the institutionalization of separate disciplines, the increasing efforts to find patrons to support scientific endeavors, the popularization of science among the educated public, and the rising appreciation of science as the engine of social progress, all of which unfolded during the middle decades of the nineteenth century.[19] In this respect, the rapid growth of two particular disciplines — physical anthropology and social hygiene — both of which were institutionalized during the 1860s in Russia and Britain, provided much fodder for, and played an especially prominent role in, the formulation of both Galton's and Florinskii's concepts.

These common intellectual and social contexts indicate that, based on the analogy with animal breeding, the central idea underpinning both Galton's eugenics and Florinskii's eugamics — the possibility to direct human future evolution to a desired end by manipulating human reproduction — likely have found similar "national" expressions in other locales. After all, the same processes of science professionalization and institutionalization were taking place around the same time in soon-to-be-united Germany and Italy, the Austro-Hungarian Empire, France, and the United States.[20] Consider, for instance, the following statement:

> If the same amount of knowledge and care, which has been taken to improve the domestic animals ... had been bestowed upon the human species in the last century, there would not have been so many moral patients for the lunatic asylum, or for our prisons, at present. That the human species are as susceptible of improvement as the domestic animal, who can deny?

This excerpt echoes almost verbatim both Galton's and Florinskii's pronouncements. But it appeared two years earlier, in a revised 1863 edition of *New Domestic Physician*, a popular medical manual published by John C. Gunn in Cincinnati, Ohio.[21]

Notwithstanding the striking resemblance between Galton's and Florinskii's approaches to their common subject, certain significant differences between them point to profound dissimilarities between the cultural, social, political, and economic terrains of their respective homelands in the aftermath of the 1853-1856 Crimean War that had pitted the two empires against each other. These dissimilarities were clearly reflected in different social concerns, which held the attention of contemporary British and Russian societies and to which Galton's and Florinskii's concepts responded. To give just one example, unlike in Britain, in mid-nineteenth-century Russia marriage remained an exclusive domain of the church. The notion of "civic marriage" was only beginning to make inroads into the country's social conscience. The Russian Orthodox Church had a number of very strict regulations regarding marriage, including the prohibition of "mixed" marriages (between individuals of different religious confessions) and "kin" marriages (between relatives in both blood and "spirit" — such as God-mothers and God-fathers — to the fourth degree), as well as the nearly unsurmountable barriers to getting a divorce or a marriage annulment.[22] In the heady atmosphere of the Great Reforms, which had been initiated in post-Crimean Russia by the young Emperor Alexander II and dramatically reshaped almost every facet of the country's life, Florinskii's essays clearly responded to and challenged what historian Gregory L. Freeze has described as the "marital order of a rigidity unknown elsewhere in Europe."[23] Although Florinskii presented "kin" marriages as a main source of degeneration, thus, in a way endorsing the church's prohibition, he saw "mixed" marriages — between individuals of different social standings, religious confessions, talents, and physique — as a major instrument for averting degeneration and advancing human perfection.

Much as happened to Galton's 1865 article, at the time of its publication Florinskii's treatise went virtually unnoticed. Not a single review of his book appeared in Russian medical, scientific, or "literary-learned" periodicals. But unlike Galton, who would spend the rest of his life and a large portion of his personal fortune on developing his eugenic concept by investigating both the "laws of inheritance" that underpin it and the possible ways it could be implemented in the actual life of his homeland, Florinskii never returned to the subject of "rational" marriage in the course of his long and distinguished career. He did absolutely nothing to promote it among his colleagues or the general public. As a result, his treatise was soon completely forgotten.

Yet, in 1926, exactly sixty years after its first publication as a book, Mikhail Volotskoi (1893-1944), an anthropologist and a founding member of the Russian Eugenics Society established in Moscow six years prior, reissued Florinskii's treatise, hailing its author as a "precursor" to

Galton and his eugenic ideas.[24] Volotskoi's reprint gave Florinskii's book a new lease on life. This time it found an attentive audience and proved influential in shaping the debates among the proponents of eugenics in the Union of Soviet Socialist Republics (USSR) that emerged on the ruins of the former Russian Empire. Volotskoi found in Florinskii's book an inspiring model for creating a "proletarian," "socialist," "bio-social" eugenics, instead of what he came to see as a "bourgeois" eugenics created by Galton and propagated by his fellow members of the Russian Eugenics Society. The resurrection of Florinskii's ideas, however, proved short-lived. In 1930 in the Soviet Union eugenics was condemned as a "bourgeois," "fascist" science and Florinskii and his treatise again slipped into oblivion.

But in the early 1970s, Florinskii's name resurfaced in some Soviet publications on the history of human genetics, where his *Human Perfection and Degeneration* was again viewed through the prism of Galtonian eugenics. This time, however, it was hailed as the foundational work not of eugenics, but of medical genetics.[25] In 1995, almost seventy years after its previous publication, Florinskii's treatise was reprinted again by Valerii Puzyrev, director of the Tomsk Institute of Medical Genetics. In his foreword to the new edition, Puzyrev reiterated the idea that the book was a foundational work of *both* eugenics and medical genetics.[26] Finally, in 2012, Florinskii's tract was reissued once more, this time as part of a reader on "Russian eugenics" published by Vladimir Avdeev, a self-styled expert on a "new science of raciology," who claimed that its author had founded a particular "Russian," "racial" eugenics.[27]

For nearly a century, practically all of the commentators on Florinskii's treatise have followed the simplistic trope of seeing this work and its author as mere "precursors" to or "contemporaries" of Galton and eugenics.[28] The focus of this book is different. It looks at Florinskii's treatise through the lens of its own time and reads it within its own multiple (personal, social, scientific, national, medical, philosophical, etc.) contexts. *Human Perfection and Degeneration* raises a number of intriguing questions about its author, its contents and aims, the timings and venues of its different editions, its intended audiences, and its reception. This book, then, is not a biography of Vasilii Florinskii, but rather a "biography" of his treatise.

## The Biography of a Book

A book manuscript is usually the product of a single individual, no matter how many others its author acknowledges as being helpful and instrumental in its creation. A published work, however, is the result of the joint labors of its author, editor(s), and publisher(s) — and a host of other people involved in its actual production, including typographers, translators, censors, illustrators, and so on. Furthermore, upon its first publication, a book becomes a "thing in itself" and attains a life of its own, often completely independent from the individuals who have first brought it to life. Its distribution depends on booksellers, advertisers, and librarians, while its reception is often shaped by newspaper columnists, magazine critics, textbook authors, anthology editors, and so on. The biography of a book, then, requires a thorough analysis of the motivations, inspirations, goals, and efforts not only of its author, but also of the book's editors, publishers, censors, sellers, translators, readers, keepers, reviewers, and commentators.[29]

The biography of a book has long won a respectable place in historical writings as a popular genre in both book and literary studies[30] and, in the last few decades, has made significant inroads into the history of science and medicine.[31] In addition to the life stories of certain books as cultural artefacts,[32] most works in this genre fall into (and occasionally combine) two broad categories. One focuses on the issues of the "evolution" of a specific text during its author's lifetime and beyond, documenting the author's changing worldviews, ideas, and skills and, occasionally, addressing those of the text's editors, translators, and publishers.[33] Another deals primarily with the readership and reception/impact of a particular text,[34] often tracing its various editions and translations through time and/or space, as volumes in the popular English-language series "A book that shook the world" and the no less popular Russian-language series "Fates of books" readily attest.

As exciting and interesting as writing the biography of a book is in and of itself, my goals in writing the biography of Florinskii's treatise extend beyond chronicling its birth and punctuated life, tracing its textual and para-textual changes, and assessing its readership and reception. I use it as a convenient lens through which to look at the unusual — as compared to the development of eugenics in other locales during the same time period — historical trajectory of eugenics in Russia and to examine the multiple factors that account for the particular fate of Russian eugenics.

Indeed, even though eugenic ideas had begun to filter into Russia's professional and public discourse shortly after Florinskii's death in 1899, it was only after the 1917 Bolshevik Revolution that eugenics became an established scientific discipline, inspired a grassroots following, and exerted considerable influence on various social policies and cultural productions. Among the many countries that featured well-organized eugenic movements during the interwar period, none seemed to provide a less likely locale for concerns with the "racial degeneration" or the increasing fertility of "lower classes," which at the time commanded the attention of Galton's numerous followers, than Bolshevik Russia. Why and how could a "proletarian state," which claimed to build a classless society and loudly denounced racism and nationalism, become a hotbed of eugenic debates, support eugenic research and institutions, and adopt eugenics-inspired policies? Why, after a decade of rapid development and growing popularity, was eugenics condemned in the Soviet Union in 1930, long before any other country in the world adopted a similar stance towards the eugenic programme of "bettering humankind"? And why did eugenics reappear in Russia in the late 1960s and early 1970s, then in the 1990s, and again in recent years? The biography of Florinskii's treatise offers telling clues to answer many of these questions.

Compared to the ever-growing and variegated literature on the history of eugenics in other countries (especially Britain, France, Germany, Scandinavia, and the United States), the history of eugenics in Russia (particularly during the imperial and late-Soviet periods) has attracted relatively little scholarly attention. Thanks to the pioneering studies of Mark B. Adams published more than a quarter of a century ago, the institutional and intellectual developments of eugenics as *a science* of human heredity have been outlined, largely in relation to the growth of genetics during the early Soviet period.[35] This work has enabled a preliminary analysis of the similarities and differences of Soviet developments to experiences in other countries,[36] along with the role western eugenics and genetics communities (especially in Germany and the United States) played in shaping eugenics in Russia.[37] At the same time, the history of Russian eugenics as *an ideology* — a particular normative, value-laden way of thinking about human reproduction, heredity, diversity, development, and evolution — remains essentially uncharted territory.[38] One of the largest holes in our knowledge is the history of eugenics as *a policy*: the influence eugenics and eugenicists in Russia exerted on actual policy-making and implementation in a variety of fields, from social hygiene to family planning and from abortion to ethnic policies.[39] Although previous scholarship has demonstrated strong links between genetics and eugenics, the representation and influence of other fields and disciplines — from medicine and jurisprudence to demography and pedagogy — in the Russian eugenic movement requires additional research. Similarly, public and professional attitudes to eugenics as a science, a policy, and an ideology in Russia still await careful investigation.[40] Despite the fact that during the last 25 years a large corpus of new archival and printed materials has become available, a comprehensive history of eugenics — as a specific amalgam of ideas, values, concerns, and actions regarding human reproduction, heredity, development, and evolution — remains to be written. Far from providing such a comprehensive history, the biography of Florinskii's treatise, nevertheless, offers an illuminating glimpse of the complicated trajectory of eugenics in Russia, along with numerous individuals and forces that shaped it.

This book is divided into two parts. Part I details the origins, contents, and initial reception of Florinskii's tract during its author's lifetime. I begin with an examination of the life-story of Vasilii Florinskii up to the late spring of 1865 when he started working on his essays. The first chapter, "The Author," presents a multitude of individuals, events, ideas, places, institutions, and ideals, which all together and each separately molded the future author of *Human Perfection and Degeneration*. Born to the clergy and educated at primary and secondary theological schools in the Urals and Siberia, Florinskii wanted to continue his ecclesiastical career and to complete his training at the highest theological school — St. Petersburg Theological Academy. An accident barred the doors of the theological academy to the ambitious youth. So, instead, he enrolled in the IMSA, the country's foremost medical school. After five years of extensive studies he became a physician, switching almost seamlessly from theology to gynecology. Florinskii's teachers noticed his abilities and slated the freshly minted physician to a professorial position at his alma mater. Over the next three years, he successfully passed all the required examinations and defended a dissertation for the Doctor of Medicine degree. With all the formal requirements completed on time and with the highest marks, the IMSA Council sent Florinskii on a two-year, all-expenses-paid tour of European medical schools and clinics for advanced training. Upon his return to St. Petersburg in the early fall of 1863, the young doctor became an adjunct professor at the IMSA Department of Gynecology and Obstetrics and began a successful career as a teacher, a researcher, and a clinician. Less than two years later, he started working on "Human Perfection and Degeneration."

Why did the young gynecologist embark on writing about a subject so remote from his immediate professional duties and scholarly interests? And why did he publish his treatise in *Russian Word*, the most radical "literary-political" journal of the time? In search for answers to these questions, the second chapter, "The Publisher," follows the life and works of Grigorii Blagosvetlov (1824-1880), the journal's editor-in-chief and publisher. It was Blagosvetlov who first serialized Florinskii's treatise in his journal and then released it in book format. And, it was Blagosvetlov, I argue, who probably enticed the young professor to write the treatise in the first place and, to a certain degree, shaped its style and contents. Florinskii's essays were actually part of a broad campaign waged by the journal's editor to popularize science and to promote a scientific worldview that sought to understand and eventually to cure the "social ills" plaguing post-Crimean Russia. The propaganda of Darwin's evolutionary ideas, especially their possible "social applications," became a particular focus of this campaign, with nearly all of the journal's core contributors — Blagosvetlov, Pisarev, Shelgunov, and Zaitsev — publishing articles, essays, and reviews on the subject. Alas, none of them had adequate training in the natural sciences to explore these questions in depth. This occasionally led to embarrassing incidents and bitter polemics that apparently prompted Blagosvetlov's invitation to Florinskii to write for *Russian Word* and the publication of his treatise in four of its 1865 issues.

Unlike Galton's 1865 article based on his original statistical studies of "blood relations" among "British men," Florinskii's treatise was a "thought piece" based on his careful reading and analysis of available literature. The third chapter, "The Book," details the actual contents of Florinskii's essays, tracing their major themes and ideas to a variety of English, French, German, and Russian sources, first and foremost, Darwin's *Origin*. Florinskii applied Darwin's key notions of variability, heredity, and selection to the understanding of the past history, the present state, and the possible future of humanity, examining how Darwin's "laws of selection" might play out in the evolution of the human species. Much like Galton's, Florinskii's major idea was based on Darwin's analysis of different forms of selection (artificial, natural, and sexual) and on implicit analogies between artificial selection (the choice of progenitors in both its "unconscious" and "methodical" varieties) and sexual selection (the choice of mating partners). For Florinskii, these analogies were likely facilitated and amplified by a particular contemporary Russian translation of Darwin's key term "selection" as *podbor rodichei* (matching of kin) and *vybor* (choice), as well as by a popular equation of species evolution with "progress." The essence of Florinskii's idea of "rational" marriage was to replace the "unconscious" choice of marital partners ("masquerading as love," in Florinskii's words) with a conscious, "rational" one. In order to identify the sources and causes of degeneration and to discover the principles of perfection of the human species, Florinskii synthesized available data, ideas, and concepts from an array of scientific fields and medical specialties, including physical anthropology, social hygiene, general biology, and theoretical gynecology. In seeking to understand the "conditions conducive" to perfection and degeneration, he also identified those characteristics — the ideals of human health, beauty, and mind — which together should guide a "rational" selection of spouses. Since he equated beauty with health and mind with the brain, "physical and moral health" emerged as the key criteria of spousal choice in his concept of "hygienic" marriage.

The fourth chapter, "The Hereafter," chronicles the events that followed the appearance of Florinskii's treatise in *Russian Word*. Surprising as it might seem, almost immediately after its publication, Florinskii completely withdrew from any further collaboration with Blagosvetlov. He never returned to the subject of this treatise in any of his later works and did absolutely nothing to promote it among his colleagues or the general public. But his publisher did. Just a few months after the publication of Florinskii's essays, *Russian Word* was shut down by the order of the imperial authorities. In September 1866, however, Blagosvetlov established a new journal, evocatively titled *Deed*, through which he continued his campaign to popularize the role that the natural sciences (especially Darwinism) could play in the understanding and curing of "social ills." And nearly simultaneously, he released Florinskii's essays as a book. He kept the book in print for more than a decade, regularly advertising it on the pages of *Deed*. Furthermore, he continued to publish in his new journal numerous articles and printed several books, including a translation of Darwin's *The Descent of Man*, which explored further various issues raised by and in Florinskii's treatise. Yet despite all these efforts, much as happened to Galton's early eugenic works, Florinskii's book and its main ideas went virtually unnoticed. Similar to Galton's first publications on "hereditary talents," Florinskii's concept of "human perfection and degeneration" held the promise of generating a viable research programme, stirring public opinion, and, perhaps, even initiating policy change. But it proved impotent in arousing the interest of either the Russian scholarly community, or the general public, to say nothing of the imperial bureaucracy. Numerous personal, social, political, and scientific factors have contributed to the sudden end of what at first had seemed a very productive collaboration, to Florinskii's abandonment of a promising line of inquiry, and to its general neglect by its intended audiences.

Part II describes the punctuated life of Florinskii's treatise after the death of its author. The fifth chapter, "Rebirth," documents the active growth during the first two decades of the twentieth century of a British "national" eugenics initiated by Galton and the formation of a transnational eugenics movement that spread rapidly around the world, including to Russia. The infiltration of eugenics (in its Anglo-American "eugenics," French *"eugénnetique,"* and German *"Rassenhygiene"* versions) into professional and popular discourse on human reproduction, heredity, development, and evolution became the major stimulus for the "resurrection" of Florinskii's long-forgotten work in the new, Soviet Russia that emerged out of the firestorms of World War I, the Bolshevik Revolution, and the ensuing civil war, which engulfed the former empire from 1914 through 1921. Mikhail Volotskoi, an anthropologist and a founding member of the Russian

Eugenics Society established in Moscow in 1920, discovered, actively popularized, and in 1926 issued a new edition of Florinskii's treatise. For Volotskoi, this tract was more than a historical curiosity. The young anthropologist found in Florinskii's book a model and justification for what, in contrast to Galton's "bourgeois" eugenics, he envisioned as a "proletarian" eugenics. Building on Florinskii's ideas, he advanced a pointed "Marxist" critique of Galtonian eugenics and its numerous followers in Russia and elsewhere. He elaborated a concept of "bio-social" eugenics and launched several research projects inspired by Florinskii's notion of "conditions conducive" to human perfection or degeneration, investigating the "eugenic" effects of various factors, ranging from occupational hazards to women's fashion.

In contrast to the first publication of Florinskii's book, the new edition did not go unnoticed. The sixth chapter, "Resonance," details the reactions of the Soviet scholarly and medical communities to Volotskoi's "Marxist" critique of Galtonian eugenics, his concept of "bio-social" eugenics, and Florinskii's ideas that underpinned them. This time, actively popularized by Volotskoi, Florinskii's treatise proved influential in shaping discussions on the agendas and directions of eugenics and its "stepsister," genetics, in the Soviet Union, especially the heated debates over the possibility and necessity of creating a distinct "proletarian," "socialist," "bio-social" eugenics. It also informed certain policies on marriage and family adopted by the Soviet authorities, including the promulgation of new laws on "the protection of health of prospective spouses and their progeny" and the establishment of "marriage consultations." If Volotskoi and his like-minded colleagues did not succeed in creating "bio-social" eugenics based on Florinskii's ideas, it was certainly not for the lack of trying. But just four years after the republication of Florinskii's treatise, in the Soviet Union eugenics was condemned as a "bourgeois," "capitalist," "fascist" doctrine. The Russian Eugenics Society dissolved and any references to Florinskii's treatise vanished from the public scene. A peculiar amalgam of ideas, ideals, concerns, and practices embedded in the very notion of "bio-social" eugenics broke apart. In the next few years, the issues of "human perfection" lost their "hereditary component" and were relegated to the purview of social hygiene, physical and general education, psychology, and pedagogy. At the same time, the problems of "human degeneration" were reduced to "hereditary diseases" that became the subject of a new discipline — soon named "medical genetics" — established and actively developed by several former members of the now defunct Russian Eugenics Society. But, during the Great Terror of the late 1930s, the institutional base of medical genetics was destroyed and its main spokesman, Solomon Levit, arrested and executed. A decade later, in 1948, the entire discipline of genetics was banned in the Soviet Union, as the result of a vicious campaign waged by the notorious agronomist Trofim Lysenko and endorsed personally by the "Great Teacher" Joseph Stalin. It seemed that Florinskii's treatise was destined to gather dust in some remote library storage forever.

Yet, Fate is a fickle mistress. Shortly after Stalin's death, during the de-Stalinization campaign launched by his successor Nikita Khrushchev in the mid-1950s and popularly known as the Thaw, medical genetics re-emerged in the Soviet Union. A decade later, both Galton and eugenics were "rehabilitated," and with them Florinskii and his book re-entered Soviet discourse on human reproduction, heredity, development, and evolution. The next chapter, "Afterlife," examines the re-emergence of Florinskii's book in the late-Soviet and post-Soviet eras. In the early 1970s, Ivan Kanaev, a geneticist-turned-historian, published the first Russian-language scholarly biography of Galton and an extensive analysis of Florinskii's treatise. This time, Florinskii's tract was presented as a foundational work not of eugenics, but of medical genetics. References to Florinskii and his book began to appear in popular articles and introductions to the textbooks on the subject. But it took more than twenty years and the dissolution of the Soviet Union, for Florinskii's treatise to be reprinted again in 1995, through the efforts of Valerii Puzyrev, director of the Tomsk Institute of Medical Genetics. In 2012, the book was reissued once more, this time as part of a reader on "Russian eugenics," compiled by the self-styled "racial encyclopedist" and "bio-politician," Vladimir Avdeev. A series of profound social and scientific developments determined the long quiescence of *Human Perfection and Degeneration* and inspired its new revivals in late-Soviet and post-Soviet Russia.

The concluding chapter, "Science of the Future," examines the implications of the peculiar history of eugenics in Russia, as seen through the biography of Florinskii's treatise, for the understanding of the history of eugenics writ large, illuminating its protean nature, its multiple local trajectories, and its global trends, as well as its continuing and contested appeal to very diverse audiences. To date, *Human Perfection and Degeneration* was published five times — in 1865, 1866, 1926, 1995, and 2012. Galton's *Hereditary Genius* (together with his essay "On Men of Science, Their Nature and Their Nurture") was translated into Russian in 1874, nine years after the first appearance of Florinskii's treatise, and to this day, it remains the only work of the "founding father" of eugenics available (in its 1996 facsimile edition) to Russian readers. The contrasting yet intertwined fates of Galton's and Florinskii's concepts in Russia strongly suggest that what we habitually call eugenics is a time- and place-specific *amalgam* of certain ideas, values, concerns, and actions regarding human reproduction, heredity, development, and evolution. Their local proponents created such amalgams through active "domestication" of available native and foreign models to fit their own agendas and interests. What united various "national" versions of eugenics and shaped its subsequent local trajectories and global trends was an explicit preoccupation with *the future*, which linked the problematics of eugenics with the fundamental existential questions of human nature, human origins, and human destiny: who are we, where did we come from, and where are we heading? Fused into all of its local variations, the possibility of "controlling" humanity's future through active intervention in human reproduction, heredity, development, and evolution made eugenics repelling or appealing to very diverse audiences. The dates of repeated reissuing of Florinskii's treatise, as well as the intervals that separate them, reflect not merely the internal dynamics and local imperatives of the development of eugenics in Russia. They also point to the waning and waxing popularity of eugenics worldwide, spurred by certain concurrent "global" scientific and social developments, ranging from industrialization and Darwinian revolution to World War I and the rise of experimental biology, and from World War II and the emergence of molecular biology to the "end" of the Cold War and the inauguration of the Human Genome Project. A heated debate on "genetics and eugenics" at the 1962 London symposium on "Man and his Future" puts into sharp relief the interplay of scientific and social factors that made, and continue to make, eugenics a subject of intense interest and an inexhaustible source of both hopes and fears regarding human nature and humanity's future.

The appended "Apologia" addresses numerous challenges I faced in researching and writing this book and focuses on two major components of the historian's craft: finding necessary sources and translating the past, both literally and figuratively, for the present-day reader.

# I. "HYGIENIC" AND "RATIONAL" MARRIAGE

> "My friend, let us devote to our Fatherland
> All our souls' exalted impulses!"
>
> Alexander Pushkin, 1818

Considering what is generally known about the history of eugenic thought and of Imperial Russia, a resident of mid-nineteenth-century St. Petersburg seems an unlikely candidate for producing a version of that amalgam of ideas, values, concerns, and actions regarding human heredity, diversity, development, and evolution, which just a couple of decades later Francis Galton would name "eugenics." Yet Vasilii Florinskii did produce such a version in his 1865 treatise on *Human Perfection and Degeneration*. This simple fact suggests that, perhaps, we need to re-examine our conceptions of both Russia and eugenics. In order to do so, in the next four chapters I take a close look at the author and the publisher of the treatise and explore its origins, sources, contents, contexts, and reception.

# 1. The Author: Vasilii Florinskii

> "Now I have the only heart-felt desire — to bring to my Fatherland as much benefit as possible, specifically, by acting in that area where I could be most useful."
>
> Vasilii Florinskii, 6 October 1861

The English and Russian progenitors of "eugenics" were born on the same day, twelve years — and a whole world — apart. Francis Galton was born on 16 February 1822 to a family of successful gun-manufacturers and bankers at the family estate "The Larches" in Birmingham, England.[1] By all accounts, the youngest of nine children (seven of whom survived infancy) sired by Samuel Tertius Galton and Violetta Darwin was a prodigy: he started to read at two, by five learned some Greek, Latin, and arithmetic, and by the age of six could write eloquent letters and recite Shakespeare at length. Initially schooled at home by his older sister, Galton attended the famed King Edward's School for boys in Birmingham. At the age of sixteen, at the insistence of his parents, who obviously wanted their son to follow in the footsteps of his illustrious maternal grandfather Erasmus Darwin, Galton began studies of medicine, first by apprenticing for a year at Birmingham General Hospital and then at King's College Medical School in London. As for his cousin Charles Darwin, for Galton, the medical profession proved uninspiring, and in 1840 he took up the study of mathematics at the Trinity College of the University of Cambridge. Four years later, just as he was finishing his degree, the death of his father made him financially independent. From that time on, young Galton led the leisurely life of an educated upper-class English gentleman, enjoying hunting, extensive travels, and occasional pursuits in various fields of science, ranging from geography and anthropology to meteorology, statistics, and psychology. In 1865, at the age of 43, inspired by his cousin's *Origin of Species*, Galton first turned his attention to the issues that "clustered round the central topics of Heredity and the possible improvement of the Human Race"[2] and that some twenty years later he would name eugenics.

Vasilii Florinskii was born on the same date, 16 February,[3] twelve years after Galton, in 1834, in the ancient village Frolovskoe of the Vladimir province in Central Russia, to the family of a low-rank cleric: his father was a deacon in the village church. The boy was meant to continue in his father's footsteps: at the age of nine, he entered a primary theological school and at nineteen graduated from a theological seminary. But young Florinskii clearly aspired to something other than being a village priest and applied to the highest theology school, an academy. Destiny, however, is greater than family tradition. An accident barred the doors of the theological academy to the ambitious youth. So, instead, he enrolled in a medical school and became a physician, switching almost seamlessly from theology to gynecology. His teachers noticed Florinskii's abilities and slated the freshly minted physician to a professorial position at his alma mater. In 1861, he successfully defended his dissertation for the Doctor of Medicine degree and went on a two-year, all expenses-paid tour of European medical schools and clinics for advanced training. Upon his return, he was appointed an adjunct professor and began a successful career as a teacher, researcher, and clinician. Two years later he published "Human Perfection and Degeneration."

Numerous scholars have examined in detail the path that led Galton to eugenics. In what follows I take a close look at the multitude of individuals, events, ideas, places, institutions, and ideals, which all together and each separately shaped the future author of its Russian variant.

## Born to Be a Priest

In mid-nineteenth-century Russia, the Orthodox Church was a nearly 1,000-year-old institution. Its doctrine was the official state religion of the multiethnic and poly-confessional empire. Its numerous clerics constituted a particular social estate (*soslovie*) with its own hierarchy, privileges, and duties.[4] Deacon was the lowest rank of the ordained clergy in the Orthodox Church, and Vasilii's father did not even have a proper family name. He was known only by his given name, Mark, and his patronymic Iakovlev (son of Iakov). Unless they joined monastic orders—the so-called Black Clergy—Russian clergymen were obliged to marry. Indeed, marriage was a requirement for obtaining a parish and the position of a priest. Mark Iakovlev got married right after graduation from Vladimir Theological Seminary in 1820.[5] But he was unable to acquire a parish of his own. In the 1820s and 1830s, Russian theological schools produced many more graduates than there were parishes available. The position of a priest became, in a way, "hereditary": a priest would pass his parish on either to a son, or to a son-in-law. Mark's father was merely a deacon, as was his wife's father. Thus he could not count on "inheriting" a parish and had to be content with serving as a deacon. His wife Maria kept the house and bore the young deacon numerous offspring.[6] Vasilii was the seventh of nine children (though three of his older siblings died in infancy).

The life of low-level clerics differed little from that of their congregations. Deacon's pay was a pittance, barely enough to put bread on the table, and any additional income depended heavily on parishioners' donations and payments for church services. To make matters worse, in 1828, Frolovskoe's old wooden church burned down in a fire ignited by a lightning strike.[7] The construction of a new brick building took ten years and, until 1838, when the new church was consecrated, its clerics could count on very little earnings. To make ends meet Vasilii's parents had to till the land alongside the village's serfs. Mark Iakovlev enjoyed gardening and under his care the church's garden became widely known for the variety and quality of fruit it produced, providing additional income for the deacon's growing family. Vasilii and his older brother Ivan (born in 1832) would have soon joined their older sisters (born in 1821 and 1823) in the family's labors, if it were not for a lucky break in their father's career.

In 1837, Archbishop Arkadii of the Perm province appointed Mark Iakovlev a priest to a yet to be built church in the village of Peski (Sands), located in the south-east corner of the province, some 600 kilometers from his own seat in the province's capital, Perm. Thus, at the age of 37, Vasilii's father finally obtained his own parish. It was a great advancement for the deacon of a rural church, aptly manifested in his acquisition of a proper family name. Mark thought of adopting the name Frolovskii (after his native village), or Zaural'skii (in honor of his new residence beyond the Urals), but in the end, he chose the name Florinskii, to mark — in a Latinized form — his fondness for gardening.[8]

If truth be told, luck had little to do with Mark Iakovlev's promotion. Rather, it was nepotism, pure and simple. Mark's father died when he was still a boy and it was his mother's family who took care of the young widow and her son. Mark's mother Praskoviia Fedorova was the daughter of a deacon at another village church in the same Vladimir province. Her two older brothers, Mikhail and Grigorii, had followed in their father's footsteps. Both had graduated from Vladimir Theological Seminary and became clerics. Mikhail got married and, thanks to his wife's family connections, became a priest at yet another village church in the same province.[9] It was in Mikhail's house that Praskoviia had lived after her husband's death, until her son finished his education and obtained his first position at the Frolovskoe church. Grigorii pursued a different path. He took monastic vows and made an illustrious career within the church hierarchy. The secular name of the Perm Archbishop Arkadii was Grigorii Fedorov.[10] It was Mark's uncle who secured his new appointment.

The Perm province stretched over a huge territory (50 per cent larger than the entire United Kingdom) on both the European and Asian sides of the Ural Mountains. Its subsoils were rich in minerals, ores, gold, and gemstones. Its forests teemed with game and rivers with fish. But it had almost no roads and was very sparsely populated.[11] Since the reign of Catherine the Great in the second half of the eighteenth century, runaway serfs (as well as exiles and prisoners who had served their sentences in the depths of Siberia) were granted freedom and pardon for all their previous crimes if they settled on the "empty" lands beyond the Urals. This policy paid off, especially in the southern parts of the province where fertile soils and a temperate climate created favorable conditions for agriculture. Peski was just such a settlement founded by run-away serfs in the late eighteenth century on the shores of a small lake, Peschannoe (Sandy), that gave the village its name.

In the mid-1830s Peski was home to nearly 800 "souls." The villagers raised cattle and grew rye, oats, and wheat, as well as some vegetables.

They sold their produce at a market in the nearby town of Dalmatov that had grown in the seventeenth century around the eponymous monastery on the shores of the Iset' river. There was plenty of land to cultivate and the village grew and prospered. But it lacked the traditional center of village social life — a church. The nearest one was almost twenty kilometers away. The villagers had to make a long journey every time they needed church services, be it a baptism, communion, marriage, or funeral, not to mention celebrations of such major holidays as Christmas and Easter. In 1836, the village council, which included heads of all households, decided it was time to build their own church. They allotted a plot of land for the church and its priest, and petitioned the province's archbishop to formally establish a parish in Peski and to appoint a priest to serve it. Archbishop Arkadii granted the petition and then promptly gave the new parish to his nephew. Admittedly, Mark Florinskii was well qualified for the position: his experience with overseeing the church construction in Frolovskoe was a great advantage for his new post that first and foremost required building a church.

It took over three months for the Florinskii family to make the 2,000-kilometer journey from Frolovskoe to Peski, most of it along the Great Siberian Tract — the infamous chain-gang route, stretching from the country's historic capital Moscow to the hard-labor prisons of Siberia, and farther on to China.[12] A caravan of horse-driven carts carried the family with all their possessions east, first to the capital of their home province Vladimir. From Vladimir, they crossed the Russian plain, passing through its major cities, Nizhnii Novgorod and Kazan, all the way towards the Urals. They traversed numerous rivers, including the mighty Volga and its largest tributary Kama, upon whose shores stood the Perm province's capital. The caravan stopped for a few days in Perm. Mark Florinskii had to see the Archbishop to be formally ordained as a priest and to get the necessary paperwork and instructions.

From Perm, they trekked across the Urals and turned south to Ekaterinburg, the province's second largest city renowned for its metal works. About forty kilometers before it reached Ekaterinburg, the caravan came to a tall wooden pyramid sitting on the side of the road in the middle of the endless forest. On two of its sides, the pyramid carried one-word inscriptions. The one on the western side read "Europe," the one on the eastern side "Asia." Marking the border between the two continents, the pyramid had been erected just a few months earlier, in the spring of 1837, in preparation for a visit by the imperial heir, Grand Duke Alexander Nikolaevich, who had been travelling through his future empire in the company of his tutor, poet Vasilii Zhukovskii.[13] Perhaps, Florinskii Sr. showed the pyramid to his older children and told them that they were following in the footsteps of their future emperor and that at this point they were entering a new "promised land" — Siberia.[14] After Ekaterinburg, the tract ran along the Iset' river to Dalmatov. And from Dalmatov, it was just forty kilometers more on a smallish side-road to Peski.

At the end of the summer of 1837, Florinskii, accompanied by his mother, wife, and five children (Vasilii's younger brother, also named Ivan, had been born just a few months before the trip) arrived at their destination. The first task of Father Mark, as his parishioners called him, was to oversee the construction of his new church and his own house. The church was to be the village's first stone building and all the villagers took part in its construction: digging trenches for its foundation, making bricks for its walls, and cutting trees for its rafters. A year later, the church's first altar chapel was completed and consecrated, and Father Mark could begin proper liturgical services. The same year, he built a large house for his family and started a new church garden.[15] But it took nearly fifteen years to complete the construction of his church, modelled to a certain degree on the Cathedral of Saints Peter and Paul in St. Petersburg, though of course on a much smaller scale, with a tall bell tower and a large portico adorned with four columns.[16]

As a child, Vasilii very much enjoyed his life in Peski. In the summers, he went exploring nearby forests and meadows, swimming and fishing in the lake, or helping his father in the church garden. In the winters, he loved to listen to the stories of the village's elders about events and times long gone: the conquest of Siberia, the Pugachev rebellion, or the skirmishes with local nomadic tribes, the Bashkirs. He began collecting old coins, buttons, arrowheads, and other objects related to the history of his new homeland, which he could dig out, especially from old burial mounds (*kurgany*), during his excursions in the village environs. As he wrote in his memoirs many years later, "...nearly everything I had achieved in my life, I owe to Peski.... My physical and spiritual development was deeply influenced by its nature and its surroundings" [ZV, 11].

Judging by available materials, Vasilii grew up in a very traditional clerical family, with father, as a kind and benevolent patriarch, mother, as the loving and generous center of the household, and several siblings of various ages who always supported and helped each other. At the age of six, Vasilii's father taught him how to read and the boy fell in love with books. He kept the copy of the *Book of Psalms* that his father had used for his reading lessons for the rest of his life.[17] On Sundays, probably with his father's encouragement and permission, he often read out loud excerpts from some edifying texts, such as the *Lives of Saints*, the *Book of Psalms*, or the *Book of Hours*, to the villagers gathered in the church between the morning prayers and the midday liturgy services. But his personal favorites were Russian epics and folk tales. The cheap (*lubok*) imprints of classic epics, as well as the more expensive editions of Alexander Pushkin's and Petr Ershov's literary renderings of folktales, reached Peski on the backs of traveling salesmen (*korobeiniki*). They were Vasilii's most treasured possessions, even though he soon knew them all by heart. The books fired up his imagination and, on his ventures through the village's environs, he often daydreamed about being an epic knight (*bogatyr'*) fighting off some evil creatures invading his native land and his beloved Peski.

But the worry-free childhood soon came to an end. He was expected to follow in his father's footsteps and eventually became a priest. In the fall of 1843, Vasilii Florinskii tailed his older brother Ivan (in the family he was called "Big" to distinguish him from Vasilii's younger brother also named Ivan, who was called "Little") to a *bursa*, a primary theological school (*dukhovnoe uchilishche*) at the Dalmatov Monastery. He would spend the next five years there, returning to Peski only for summer vacations and brief Christmas and Easter holidays. The Dalmatov Monastery had largely been constructed in the late seventeenth century and its theological school occupied an old, dilapidated, dirty one-story building. The monastery's Father Superior Mefodii served as the school's principal. Students nicknamed him the Monster. He was a mean, cruel man and his main pedagogical principle was the indiscriminate use of corporal punishment. Any misstep — a poorly prepared lesson, talking during class, or shuffling during the prayer — was punished by birching. For repeated misbehavior, students often were put in the stocks or made to wear iron collars. Many years later, after reading a horrifying fictionalized exposé of students' life in Nikolai Pomialovskii's famous

*Bursa Sketches*,[18] Florinskii noted that his school "was even worse" than the one described by Pomialovskii [ZV, 54-55].

Florinskii, fortunately, was spared the worst of the *bursa*'s life. His father could afford to pay for room and board at a private house, so he did not have to sleep in the school cold dorm, nor eat at the monastery nasty kitchen, as did other students. Being a grand-nephew of the Perm Archbishop — the Monster's direct superior — certainly protected him from the habitual cruelty of his teachers, while having an older brother in the same school shielded him somewhat from the customary taunting and hazing by fellow students. Plus, he loved to learn and quickly became the school's star student, thus giving his teachers little cause to punish him. Even so, Florinskii remembered his years at the *bursa* as "the darkest period of my life" [ZV, 53].

The school programme included Russian, Latin, Greek, and Old Church Slavonic languages, arithmetic, geography, the Old and New Testament, catechism, the rules of church service (the *Typikon*), and church singing. Florinskii was a diligent and capable student. Alas, his teachers were incompetent and inept. In his memory, they "were so bad, they could not, and did not really want to, explain anything" [ZV, 57]. Students learned by rote, memorizing assigned pages from their textbooks and paying no attention to their meaning. "The five years at the [Dalmatov] school did little to advance my education," Florinskii later recalled, "only in Latin did we learn something, mostly to translate from Latin to Russian" [ZV, 58].

In the spring of 1848, Vasilii Florinskii graduated at the top of his class and, again following his older brother, he enrolled in Perm Theological Seminary to continue his education.[19] In late August the Florinskii brothers made the 600-kilometer trip to Perm, where for Vasilii "an entirely new life has begun" [ZV, 58]. Founded in the early eighteenth century as the center of mining and metallurgy on the western slopes of the Ural Mountains, Perm became the provincial capital half a century later, during the reign of Catherine the Great. In the 1840s it was a large city with more than 15,000 inhabitants and several metallurgical plants and munition factories. In 1842, a great fire almost completely destroyed the old "wooden" town, occasioning wide-scale reconstruction and rebuilding. As the seat of both the state and the church provincial authorities, the city boasted a number of administrative offices, as well as a hospital, a gymnasium, a theater, and a public library. After a sleepy small-town life in Dalmatov, Perm might have looked to Florinskii like a world capital. Fortunately, he had his older brother to help him navigate the city life. The brothers rented a small apartment in the house of a low-ranking civil servant not far from the seminary. They regularly visited their great-uncle Archbishop Arkadii at his official residence in the center city and often borrowed books from his extensive private library. But, of course, most of their time they spent at school.

Perm Seminary differed drastically from the Dalmatov *bursa*, even in appearance: it occupied a recently constructed three-story brick building in the city center (see fig. 1-1). In addition to a variety of theological courses, its six-year curriculum included such subjects as literature, history, philosophy, psychology, the German language, as well as some basics of the natural sciences, agriculture, and medicine, which were all completely new for Florinskii. More important, the seminary's instructors subscribed to "an entirely different system of education" [ZV, 58]. Unlike his *bursa* teachers, Florinskii's new professors were highly-educated graduates of theological academies: "true teachers and kind

Fig. 1-1. Perm Seminary, a postcard (c.1900). This three-story brick house was constructed across a square from the city's main cathedral in 1843, after the old wooden building had been destroyed in the fire. In the mid-1930s it was completely reconstructed (with addition of two extra floors) to house an air-force school for the Red Army. Courtesy of RNB.

mentors." They "knew their subjects very well and knew how to teach and how to make students interested in learning" [ZV, 59]. Florinskii took to his new school like a fish takes to water, earning top-marks in every subject.

Several teachers won Florinskii's particular admiration and left a profound imprint on his worldview, ambitions, and interests. His junior contemporary, an eminent naturalist and a leader of Russian anarchism, Prince Petr Kropotkin, pointedly noted in his famous *Memoirs of a Revolutionary* that, unlike in western Europe and North America, "there were no prominent individuals in Russia who did not get the first stimulus to their development from a literature teacher."[20] Although Kropotkin's statement might seem exaggerated, it certainly holds true for Florinskii. Since childhood Florinskii loved to read, but the range of books he had had access to had been very circumscribed and limited mostly to religious and instructional literature. The *bursa* library had no secular literature at all. Only at home, during vacations, did he have a chance to read for pleasure, but in Peski books were a rarity and a luxury. All of this changed after Florinskii's move to Perm. His teacher of Russian literature, Alexander Vishniakov, had graduated with honors from St. Petersburg Theological Academy just one year before Florinskii entered the Seminary.[21] His teaching methods were as unorthodox as they were effective (at least for Florinskii). During the classes, instead of boring his students with grammatical rules and rhetorical principles, the young teacher read aloud the masterpieces of Russian poetry and prose. He was a talented lector and enchanted students with his readings of works by the giants of Russian literature: Alexander Pushkin, Alexander Griboedov, Nikolai Gogol, and Mikhail Lermontov. Vishniakov's recitations "had not only showed us all the beauty of belles-lettres," Florinskii recalled, "but, most important, they had ignited in us a love of literature. ... From this time on, for us, reading had become a necessity" [ZV, 62].

Vishniakov also introduced his students to "thick" journals such as *The Muscovite, Reading Library, Annals of the Fatherland,* and *The Contemporary*.[22] These journals not only carried literary fiction by contemporary authors, including the early publications of such future greats as Ivan Turgenev, Fedor Dostoevsky, and Ivan Goncharov. They also published literary criticism and lengthy essays on history, philosophy, politics, economics, and science. As Florinskii recorded in his memoirs:

> I owe the best parts of my spiritual development to *The Contemporary* of the end of the [eighteen] forties and beginning of the [eighteen] fifties. We read every newly-arrived issue from cover to cover. One ought to be a contemporary of *The Contemporary* [and] to remember the harsh times and [social] order, which had existed in the [eighteen] forties, to truly appreciate the influence of the new born literature [ZV, 72].

Vishniakov encouraged in every way his students' desire to read, often lending them books and journals from his personal library. But he did not limit his classes to reading. Every week each student had to hand in an essay in prose or verse about that week's subject. And every week the teacher provided each student with detailed comments on his essay. As a way of motivating students, he sometimes read and analyzed the best essays in front of the class. He often invited students to see him outside the classroom to discuss their essays and always supported their attempts at creative writing. Florinskii was a frequent visitor to Vishniakov's apartment, and under his teacher's patient tutelage, he became quite an accomplished writer of both prose and poetry.

Another influential teacher was Father Makarii — the future Archbishop of the Nizhnii Novgorod and Novocherkask provinces (Nikolai Miroliubov, 1817-1894) — who served as deputy-principal and professor of church history.[23] A graduate of Moscow Theological Academy, at the time of his tenure at Perm Seminary, Father Makarii already was a scholar of considerable repute, a member of the Imperial Archeological Society in St. Petersburg, and the author of several important studies. Along with his teaching and administrative duties, he conducted extensive historical and archeological research on local churches and religious artefacts. Florinskii often visited Father Makarii's private quarters, which were "overflowing with books and manuscripts," for after-class conversations. Father Makarii "was a true scholar who loved his studies and was set on planting the seeds of [historical] scholarship in our young hearts" [ZV, 59]. He had certainly succeeded in planting such seeds in Florinskii's heart: Florinskii's haphazard collecting of various historic artefacts during his childhood grew into a life-long passion for history and archeology.[24]

Florinskii's favorite teacher, however, was Alexander Morigerovskii, a lecturer in logic, philosophy, and psychology.[25] Vishniakov's classmate at St. Petersburg Theological Academy, Morigerovskii had graduated at the top of his class with distinction, which allowed him to pursue an advanced degree. So while teaching at Perm Seminary, he was also working on a dissertation. In 1852, he successfully defended his thesis, which brought him a magister degree.[26] Morigerovskii was an enthusiastic and engaging teacher: he even received a teaching award from the Holy Synod, the highest governing body of the Russian Orthodox Church. He was a devotee of the great German philosopher Georg Wilhelm Friedrich Hegel, and, following his idol, tried to integrate the three subjects he taught (philosophy, logic, and psychology) into a unified conceptual whole. He introduced students to the classics, as well as the recent works, in all of these three fields, using his own research, as well as his notes taken during the lectures he himself had attended just a few years prior at St. Petersburg Theological Academy. He was an eloquent orator and got frequently carried away, deviating from the subject at hand into lengthy discourses on contemporary politics, philosophical doctrines, or his research. Morigerovskii regularly invited students to continue their in-class discussions at his private apartment, where he also regaled them with tales about his studies at St. Petersburg Theological Academy and about the attractions of St. Petersburg "where all of the spiritual interests of Russian life are concentrated" [ZV, 65]. Morigerovskii clearly missed the city and, in late 1852, after he had obtained his magister degree, he left Perm to take a civil servant post in the country's capital.[27]

In the spring of 1852, Vasilii Florinskii's older brother Ivan graduated from Perm Seminary. He soon got married and obtained a parish of his own in a village not too far from Peski. Ivan's departure from Perm undoubtedly prompted Florinskii to think about his own future and career choices. He could follow his older brother, get married, and become a priest. He could give up the priesthood and seek a teaching position at a low level theological or secular school (that was what his younger brother Ivan would do after graduating from the same seminary a few years later). Or, he could go to a theological academy to become a learned scholar, like Father Makarii, or a church hierarch, like his great-uncle Archbishop Arkadii, or even a statesman, like his distant relative Mikhail Speranskii, who had served as an adviser to the Emperors Alexander I and Nicholas I. It is unclear from available materials what exactly Florinskii aspired to at this point. But he was definitely set on continuing his education.

The church educational system was strictly hierarchical and organized on a territorial principle. Thus, Perm Seminary's best graduates were allowed to go to the nearest theological academy in Kazan, which accepted them on the basis of their seminary grades, without entry examinations and with all the expenses paid by the academy. But Florinskii was so taken by Morigerovskii's tales about St. Petersburg that instead of Kazan Theological Academy he decided to go to his teacher's alma mater, which meant forsaking these advantages and overcoming numerous bureaucratic and financial obstacles.

Every other year St. Petersburg Theological Academy accepted about one hundred students, mostly from among the top graduates of St. Petersburg Seminary, as well as qualified monks and priests from the St. Petersburg Diocese, who had to take entry exams. To shield its students from the temptations of the outside world, the academy required them to live on the premises, and hence, admission numbers were also delimited by available dormitory spaces. For starters, since he resided in Perm, and thus did not belong to the St. Petersburg Diocese, Florinskii needed permission from the academy rector to take the entry exams. Morigerovskii, who had by that time moved back to St. Petersburg, helped his star student secure such permission. Although there was no tuition to pay, Florinskii needed money for his trip to St. Petersburg and for his living expenses during the examination period, which in the capital would be substantially higher than in provincial Perm. Fortunately, Father Mark supported his son's decision wholeheartedly. He gave him fifty rubles (a substantial sum for a rural priest) to pay for his trip and expenses during the exams. Morigerovskii also offered Florinskii a room in his own apartment for the duration of the exams.

Yet there was one more problem. Florinskii was to graduate in 1854, but St. Petersburg Theological Academy accepted students only on odd years. So as not to lose a whole year, he decided to take all the required exams in order to graduate a whole year earlier.[28] The seminary's rector allowed such an unheard of breach of the established rules and even granted Vasilii a two-month home leave to prepare for his final examinations scheduled for the summer of 1853. In early July, Florinskii passed his exams with flying colors, thus graduating from the seminary in five, instead of the customary six years. On 17 July, he received his graduation certificate and travel permit and bid farewell to the seminary and its faculty.[29] "So strange," he wrote to his parents, "I have parted with Father Rector and my professors in a very amicable manner. Just two days ago I had been at a very respectable distance from them, but now they accept me almost as their equal and call me by my full name, Vasilii Markovich."[30]

Understandably, Florinskii felt quite anxious. He would have to make a 2,000-kilometer journey on his own, with very limited resources and without his older brother's protective hand that had steered him through both the Dalmatov *bursa* and Perm Seminary. On the eve of his departure from Perm he wrote a long poem (ten four-line stanzas) brooding over his decision:

> What do I seek? What was I missing?
> I could have stayed at home, married and content;
> With relatives around and a full purse,
> With father and brothers always close by [ZV, 88].

He pondered the uncertain future it entailed: "what am I leaving here, what is awaiting there?" The last questions were clearly rhetorical. Obviously, what he was leaving behind was Peski, a small village lost in the vast expanses of the empire. What lay ahead was the imperial capital. Florinskii's decision reflected not only an exalted image of St. Petersburg, where, in Morigerovskii's words, "all of the spiritual interests of Russian life are concentrated." His decision also indicated, quite contrary to the sense of humility a good servant of God and the church should have exhibited, a certain nagging ambition on the part of the young seminarian. Clearly, staying in the Perm province as a priest, with all the outward signs of success and fulfillment — a marriage, a full purse, his family home and relatives nearby — was not enough to make him feel content. He certainly felt that he could do more than that. As his poem makes clear, Florinskii did not quite yet know what it was exactly that he was going to do. But it was certainly something more than being a priest at some rural church.

The entry examinations at St. Petersburg Theological Academy were to begin in mid-August. Even a few years earlier, Florinskii would probably have never made the trip from Perm to the imperial capital in time for his exams. After all, it had taken his family nearly three months to travel a comparable distance (from Frolovskoe to Peski) some fifteen years prior. But in those years much had changed: Florinskii had technical progress on his side. Just three years earlier, in 1850, a steamship line opened between Perm and Nizhnii Novgorod, the seat of Russia's largest wholesale and retail fair on the shores of Volga.[31] The line had three tugboats that pulled barges loaded with goods to and from the fair. It also took passengers. A simple cabin cost twelve rubles and Florinskii booked a passage in "the third class" — on the open deck, which cost him three times less. This, of course, left him open to the elements during the voyage, but he hoped that the summer weather would not be too bad.

On the early morning of 20 July, Florinskii boarded a large barge towed by a steamer tug, enticingly named *Sunrise*. Even though the river route was considerably longer than an overland one and the boat stopped for hours at a time at nearly every town along the way, the 1,400-kilometer journey took only twelve days. Florinskii passed the time watching the scenery, talking to his fellow travelers, reading, and writing long letters to his parents and siblings. On the late afternoon of 31 July, *Sunrise* reached its destination.

Together with several fellow passengers he had befriended on the boat, Florinskii spent the night in a cheap inn near the fair complex. The next morning he went to the fair and got nearly lost in its hustle and bustle, gawking at exotic goods and merchants, stalls and stores. But his foremost concern was to find a way to make the next 450-kilometer leg of his journey — to Moscow. Again, technical progress greatly facilitated his passage. By that time, an old dirt road between Moscow and Nizhnii Novgorod had been converted to a crash-stone highway with regular stagecoach services offered by several state-run and private companies.[32] Alas, a seat on the stagecoach had to be reserved a few weeks in advance and, for Florinskii, the price of a ticket — fifteen to eighteen rubles — was forbidding. He found a cheaper option. With three other *Sunrise* passengers, he hired an independent coachman who promised to deliver them to Moscow on his *troika* in three days for just seven rubles per person. The coachman kept his promise. On the early morning of 4 August, Florinskii was in "Mother-Moscow," the country's historic capital. He spent the entire day walking around the city, visiting its holy shrines and historical monuments, paying homage to the relics of its saints, and praying in the Kremlin cathedrals. In the evening he left for St. Petersburg.

The march of technical progress made the last leg of Florinskii's journey — from Moscow to St. Petersburg — the fastest and the cheapest, but certainly not the most comfortable. Only two years prior, in the fall of 1851, a railroad had connected the country's two capitals, with the twice daily departures of passenger trains in both directions. Ripping through the countryside at an average speed of fifteen kilometers an hour, the train made the 650-kilometer trip in just 48 hours, with regular stops along the way to refuel and to discharge and pick up passengers and cargo.[33] Florinskii, of course, chose the least expensive "open" car, which turned out to be merely a wooden box on wheels, with low walls, no roof, and rows of simple wooden benches for the passengers. The benches had neither back supports, nor armrests. The car was filled to capacity, so there was no way he could lay down. But it cost him only three rubles.

The first day, the weather was sunny and clear, and Florinskii quite enjoyed his "fantastic" voyage. But, in the second night rain came. The temperature dropped. Soaked to the bone, he felt miserable. Luckily, the rain stopped the next morning, the sun and wind dried his clothes, and life looked bright again. Unable to sleep, he struck up a conversation with a young man sitting next to him on the bench. His neighbor, named Vladimir Bogoliubov, turned out to be a student of the Imperial Medical-Surgical Academy (IMSA) returning from summer vacation for his last year of studies. Florinskii told Bogoliubov of his plans to become a student at the theological academy and, of course, had myriad questions about the city, its attractions, student life, and on, and on, and on. Bogoliubov was happy to share his experiences and to satisfy the curiosity of the younger man. Time flew and soon the long journey came to an end — the train arrived in the capital of the Russian Empire. On the evening of 6 August, Florinskii got off the train at the Nikolaevskii Railroad Station that sat at the foot of Nevsky Prospect — the city's most famous street immortalized in the eponymous story by Gogol.

Florinskii was stiff from sitting still for two days and utterly exhausted from lack of sleep. Instead of going straight to Morigerovskii's apartment, as his teacher had instructed him to do, he took a room at a nearby hotel, just across the square from the train station, and collapsed. The next day was Sunday and he decided to rest, clean up, and make himself presentable. Plus, the hotel was much closer to the theological academy, situated on the grounds of the famous Alexander Nevsky Monastery, only one kilometer south-east of the train station, than Morigerovskii's apartment located about five kilometers from the station in the exactly opposite direction.

On Monday morning, Florinskii came to the academy to submit his documents. He learned that entry examinations would begin in less than ten days, on 17 August. In the evening he paid a visit to Morigerovskii. His former teacher was delighted to see him and insisted that Florinskii move in at once. Florinskii was more than happy to oblige: the hotel bill was eating into his much diminished funds. Morigerovskii worked every day from 10am to 5pm at the Ministry of State Properties, so Florinskii had the apartment all to himself to study in peace and quiet all day. He rarely went out, focusing all his energies on preparing for the exams. Only on his way home from the academy after taking an exam did he allow himself to explore some parts of the fabled city.

Compared to provincial Perm and even historical Moscow (whatever he had glimpsed of it during one day), St. Petersburg was imposing and captivating. The former seminarian wandered along the granite embankments and crossed numerous bridges of its main rivers: Neva, Moika, and Fontanka. He marveled at its palaces and monuments built by famous Italian, French, and Russian architects. He strolled on Nevsky Prospect all the way to the Admiralty and the Winter Palace, the imperial residence, admiring recently installed gas street lamps. Along the way, he gawked at richly decorated carriages, opulently dressed dames, and the smart uniforms of the officers of the Imperial Guards, crowding the empire's main street. He walked on the manicured grounds of the Summer Garden and the Field of Mars, eyeing the replicas of antique statues that decorated the imperial playgrounds. Even though the white nights, which had given Dostoevsky the title for his famous novella, were over, the city was every bit as enticing and enchanting as numerous fictional stories and Morigerovskii's tales had promised it to be. Florinskii was eagerly looking forward to having more time to enjoy its wonders after the exams were over.

The exams proved to be not too difficult. Florinskii felt that the assessors were quite pleased with his written and oral responses. He was convinced that he was doing well and would be accepted. On Tuesday, 30 August, he joined all the entrants in the academy's ceremonial hall where the rector announced the names of accepted students. To Florinskii's astonishment his name was not among them. Together with two other applicants, who found themselves in exactly the same situation — Nikolai Dobroliubov from the Nizhnii Novgorod province and Nikolai Markov from the Tambov province — he went to see the rector to find out what had happened. The rector clearly could not even imagine that the academy had just barred its doors to the three most brilliant applicants of that year.[34] He explained that, though all three of them had indeed passed the entry exams successfully, the academy simply had no place for them, because it had to accept three widowed priests from the St. Petersburg Diocese who had decided to take monastic vows and continue their education. According to the rules, these priests had priority in entering the academy.[35]

Florinskii was devastated. Everything he had worked so hard for during the last year came to naught. All his hopes and ambitions were dashed. He was broke, with less than ten rubles left in his pockets. He had no idea what to do.

## Taught to Be a Doctor

After a sleepless night, the next morning Florinskii went for a walk to clear his head or, perhaps, to say goodbye to the city and his dreams. At some point, he found himself in front of a beautiful cathedral dedicated to the Vladimir Icon of Our Lady, one of the most sacred relics of the Russian Church. The icon had reputedly been produced in the province of his birth, which perhaps was what caught Florinskii's attention. There was no service, but the cathedral was open and, on impulse, he went in. He kneeled in the main altar before the icon of the Holy Virgin and prayed for almost an hour, complaining about the unfair fate and asking for guidance. Exiting the cathedral, deep in thought, Florinskii bumped into a passerby. He was surprised to recognize Vladimir Bogoliubov, the IMSA student he had shared a bench with on the train to St. Petersburg. Bogoliubov also recognized his fellow traveler and asked whether Florinskii had passed the exams and been admitted to the theological academy. Florinskii recounted the sad story of his rejection and explained his predicament.

Bogoliubov took Florinskii's troubles to heart. He brought the distraught youth to his dorm at the IMSA campus on the other side of the Neva (see fig. 1-2). There he produced a pen and a sheet of paper and dictated to the bewildered Florinskii a petition to the IMSA president, Ventseslav Pelikan, asking for permission to take entry examinations. Tomorrow afternoon, he explained, the Academy Council (a general meeting of its professors) would meet to discuss the results of the entry exams and decide on admissions. Although the regular exam period had just ended, if Florinskii could persuade the president to allow him to take the exams, he might get a chance to become an IMSA student

Fig. 1-2. A postcard (c.1900) featuring the main building of the Imperial Medical-Surgical Academy. It was built in the course of a decade, 1798-1809, in the then popular classicist style, by the two famous St. Petersburg architects Antonio Porto and Andrei Voronikhin. In 1881, the academy's official name was changed into the Imperial Military-Medical Academy, and the new name placed on the building's frieze, just above its six columns. In front of the building there is the monument to the academy's first president, James Wylie, created by the architect Andrei Shtakenshneider and the sculptor David Iensen and unveiled on 9 December 1859, during the second year of Florinskii's studies at the Institute of Young Doctors. In the late 1940s, the monument was moved into the academy's courtyard, where it stands to this day. Courtesy of RNB.

At Perm Seminary Florinskii had learned some elements of medicine and pharmacy, which were considered useful for rural priests — the destiny of the majority of seminary graduates. But he had never thought about medicine as a profession. All of his dreams had revolved around studying at St. Petersburg Theological Academy. But returning home and waiting for two years to take another shot at entering the academy was clearly not an option he considered seriously. If he were to make it, entering the Imperial Medical-Surgical Academy would certainly be a much better outcome than crawling back to Peski defeated. Not to mention the fact that after graduation he would receive a physician diploma that would automatically entitle him to the ninth civic rank in the Table of Ranks (equal to the military rank of a captain) and open the door to personal nobility.[36] This, of course, would be a huge step up the social ladder from a rural priest — the top job he could only hope to get if he were to return to Peski. There was little time to ponder all of these consequences and possibilities. He made up his mind: if he could not tend to the souls of his parishioners as a cleric, he would tend to their bodies as a physician.

The next morning, with all his papers in hand, Florinskii came to the IMSA and sat in the outer room of the president's office. He was in for a long wait. Finally, at noon, Pelikan arrived, and Florinskii managed to get an audience to present his petition. At first, the president was unsympathetic: "you've failed in the Theological Academy and now want to enter the Medical Academy. ... We don't need damaged goods" [ZV, 98]. But when Florinskii explained his actual situation and presented documents from the theological academy supporting his story, Pelikan changed his mind. He allowed Florinskii to take the exams, "if you could do it in one day, starting right now" [ZV, 98]. On the spot, the president appointed assessors from among the professors present at the academy at the moment and Florinskii began his exam marathon. He was given one hour to write two essays, one in Russian, another in Latin. Even though the time was very short, Florinskii was able to produce acceptable essays in both languages. Then he had to take oral examinations in history, geography, physics, and mathematics. He got top marks in history and geography and a passing grade in physics, but he failed mathematics. Nevertheless, two hours later, the IMSA Council decided to accept him as a "self-supported" student, on the condition that he would retake (and, of course, successfully pass) the mathematics exam before Christmas, at the end of the fall semester.[37]

Florinskii was elated. At that time, education at the IMSA was free: unlike their counterparts elsewhere in Europe, IMSA students did not pay tuition or lecture fees.[38] Florinskii's status as a "self-supported" (*vol'noslushatel'*) student, however, meant that he had to pay all other expenses — room and board, uniforms, textbooks, and so on — out of his own pocket. He could certainly hope that, by successfully passing exams at the end of the academic year, he would become one of the "state-supported" (*kazenokoshtnyi*) students, whose expenses were covered in full by the academy. All he had to do was somehow fund his first year of study. Probably, he expected that his father would be able to help, which the latter did by sending his son 100 rubles (a huge sum for a rural priest).[39] He also counted on earning some income by giving private lessons, as he had done in Perm, or by copying documents for some government office — a common way to earn money for many St. Petersburg students at the time. Together with a classmate, for three rubles a month he rented a tiny room in a house near the IMSA campus and dove into his studies.

Florinskii's first year at the academy proved to be the most challenging, and not only financially. Up to this point, he had spent his entire life among the clergy, the most uniform — ethnically, culturally, and, of course, confessionally — of all social groups in the country. From his very childhood he had imbibed the deep suspicion and mistrust of "alien" nationalities (*inorodtsy*) and "alien" confessions (*inovertsy*), which permeated the Russian Orthodox doctrine.[40] At the academy he entered a Babylon, in both the secular and the biblical sense of the word. Starting with its president Ventseslav (Wacław) Pelikan and vice-president Konstantin Balbiani (both of them Catholics of Polish extraction), the majority of the academy's instructors were "aliens." Ethnic Germans (Lutherans) and Poles (Catholic), who spoke Russian poorly and often taught their subjects in Latin, constituted nearly two thirds of the professoriate. The composition of the student body was not much different. Florinskii's class of 242 included 85 Polish Catholics, 62 German Lutherans, seventeen Jews, and only 78 Orthodox Russians (about half of them, seminary graduates like Florinskii) [VZ, 100]. German and Polish were spoken in the academy corridors — filled with tobacco smoke! — more often than Russian. Florinskii felt as if he were not in the capital of his Fatherland, but in some foreign country.

Florinskii also felt quite disappointed by the IMSA educational system, which he found wanting compared to his studies at the seminary. Historically, the academy's primary function was to prepare medical cadres for the military. It was founded in 1798, during the reign of Paul I, in conscious imitation, and just a few years after the founding of such military-medical schools as Val-de-Grâce in Paris (1789), Josephinum in Vienna (1789), and Pépinière in Berlin (1795). For all intents and purposes, the IMSA was a military school, subordinate to the War Ministry. Military discipline permeated students' daily life, from rollcalls at the beginning of every class to the dress code and daily regimen, monitored and strictly enforced by a specially assigned high-ranking officer (usually a colonel). After the freedom he had enjoyed at the seminary, Florinskii found the academy's regulations oppressive.

Judging by his memoirs, the former seminarian was also quite dissatisfied with his academic studies. The IMSA first year curriculum included zoology, botany, anatomy, pharmacopeia, physics, and inorganic chemistry. The first four subjects required endless rote memorization of innumerable Latin names — of animals, plants, body parts, and medicines — which Florinskii found exceedingly boring and unrewarding. He disliked the professor of physics and mathematics (the very one who had failed him during the entry exams) for his stern demeanor and dry, uninspiring teaching style. Chemistry was the only subject he truly enjoyed, even though it proved to be quite difficult, especially since the academy had no laboratory for hands-on practical learning. But the professor of chemistry Nikolai Zinin (1812-1880) (who also served as the chairman of the Academy Council) enjoyed a European reputation as one of the discipline's founders in Russia. His masterful lectures were always lively and stimulating.[41] As Florinskii put it, he was "an eagle among chickens." Unfortunately for Florinskii, by the time he entered the academy, the two "superstars" on its faculty, Karl von Baer, Russia's foremost embryologist and anthropologist, and Nikolai Pirogov, the country's most eminent anatomist and surgeon, had retired. So, Florinskii's initial impressions of his studies and teachers were far from favorable. As he later noted in his memoirs,

> In the seminary we ... grew accustomed to exercise thinking, not memory. Whether in theological subjects, or in the field of historical studies, not to mention philosophy and psychology, everywhere, we had a cause for personal thoughts and comparisons. Better or worse, but our thought was constantly at work, and in this I saw the main stimulus to my [intellectual] development, all the interest of scholarship. At the Academy I encountered something entirely different [ZV, 107].

Despite all the challenges, Florinskii persevered in his studies and his efforts paid off. He successfully passed end-of-the-year examinations and became one of the state-supported students. This meant that he was now entitled to free room and board at the IMSA dormitory, as well as subsidies for uniforms, textbooks, and other related expenses. His financial security guaranteed, Florinskii allowed himself a luxury: he went to his beloved Peski for the summer vacation, no doubt eager to show off his new uniform and new student status to his parents, siblings, and villagers.

The full course of studies at the academy lasted five years. With each passing year, Florinskii found his studies easier and more interesting. In the second year, along with such subjects as physiology, comparative anatomy, and organic chemistry, he began to learn medicine proper: internal diseases and surgery. Now he attended not just lectures, but practical sessions at the academy clinics, as well as surgical operations and autopsies performed in the academy's anatomical theater. He delved deeper and deeper into his future profession and got more and more inspired.

Florinskii had begun his studies in one epoch and finished them in an entirely different one. Just a few weeks after he had been admitted to the academy, in October 1853, Russia went to war with the Ottoman Empire, ostensibly to protect the rights of Christian minorities under Ottoman rule. Britain, France, and the Kingdom of Sardinia soon joined the Ottoman forces. With Austria and Prussia jockeying for political and territorial advantages by threatening to break their neutrality, the war quickly escalated into a conflict of pan-European, if not global, proportions. Just as Florinskii returned from his summer vacation in Peski, in September 1854, the allied forces landed on the Crimean peninsula and, a few months later, laid siege to Sebastopol, Russia's main naval base on the Black Sea. Although today mostly remembered as the Crimean War, the conflict had actually spread far beyond the Crimea and had been fought in the lower Danube region, the Balkans, the Caucasus, and on the Black, Baltic, and White seas, and even the North Pacific.[42]

The war lasted nearly three years and was a watershed in the country's history. Although appalling incompetence and ubiquitous corruption plagued all the armies involved in the war, the Russian military suffered a devastating and humiliating defeat. Sebastopol fell, the entire Black Sea fleet was destroyed, and Russian forces took nearly half a million casualties, almost twice as many as all of their opponents combined.

The war brought the country to the brink of financial and political collapse, exacerbated by the death of Emperor Nicholas I in the middle of the war, on 18 February 1855. His oldest son Alexander ascended to the throne. In October, after the fall of Sebastopol, the young emperor fired the war minister and opened peace negotiations. The Treaty of Paris, signed on 30 March 1856, ended the war. Faced with the dismal legacy of his father's thirty-year rule, which the Crimean War had so brutally and unambiguously exposed, Alexander II initiated wide-ranging political, military, economic, and social reforms — most importantly, the emancipation of the serfs, which set the country on an entirely new course that would be remembered in history as the Great Reforms.[43] To facilitate the reforms, for the first time in Russian history, the emperor invited his subjects to voice their concerns and ideas regarding the country's future, encouraging limited and carefully controlled *glasnost'* in the matters of state policy.[44]

Many events of Florinskii's life during his studies at the IMSA reflected the momentous events in his country's history. In this respect, the academic year of 1856-1857 proved as fateful for him, as it did for the Russian Empire. In the early fall of 1856, just as he began his fourth year of studies, the IMSA Council selected Florinskii to become one of the two interns appointed annually to the academy clinics, a clear sign of his professors' high regard for his abilities. Normally, this position would have gone to a fifth-year student. But that year the academy did not have any. The war demanded not only soldiers, but also doctors. The supply, however, was sorely inadequate. The Russian government tried its best to fill the demand, going so far as hiring doctors abroad, from the Germanic lands and even the United States. It also speeded up training at the IMSA: the students who entered the academy in 1852, one year before Florinskii, had to graduate after just four years of study in the spring of 1856. If the war had not ended that spring, the same fate would have befallen Florinskii. Instead, he obtained an unparalleled opportunity in his fourth year to become an intern at the academy surgical clinics reformed by the famed Pirogov and now directed by the renowned professor of surgery Ivan Rklitskii (1805-1861). The position came with an apartment attached to the clinics, which afforded Florinskii much more privacy than the overcrowded academy dorm and, hence, peace and quiet for his studies. It also placed him under the mentorship of Petr Platonov (1823-1860), a talented surgeon and author of a monumental, three-volume anatomical manual, who, as Rklitskii's assistant, actually ran the surgical wards.[45]

Florinskii utilized this opportunity to the best of his abilities. His duties included making twice daily rounds of the wards, applying and changing dressings, admitting new patients, and assisting during operations, which gained him invaluable firsthand experience in practical surgery. By the end of the academic year, he himself had performed several complex operations: leg amputations at the knee and at the hip. He had also prepared his first scholarly work: a detailed description of a complicated wound of the knee joint he had observed and treated in the clinic. The next year, this work appeared on the pages of the *Military-Medical Journal*, the IMSA official outlet.[46]

Florinskii also got involved in Platonov's work on the anatomical manual, assisting his mentor with editing and proofreading several parts of the 1,700-page-long manuscript. The manual was one of the first of its kind: it was to be published in Russian, as opposed to the traditional Latin. The good command of the written word that Florinskii had acquired at the seminary under Vishniakov's tutelage came in very handy. Indeed Platonov was so impressed with his student's edits and corrections that he recommended him to Alexander Kiter (1813-1879), the professor of obstetrics and gynecology, who that very year decided to publish a textbook on his own specialty. This recommendation proved decisive in defining Florinskii's further career as a doctor.

An imperial subject, Kiter was an ethnic German, born Justinus Ludwig Alexander von Kieter near Riga in the Baltic provinces of the

Russian Empire. He graduated from the medical school of the German-language University of Dorpat (today Tartu, Estonia), where he had studied under Nikolai Pirogov. Although Kiter had been teaching in Russian medical schools since 1840 (first at Kazan University and from 1848 at the IMSA), he still preferred to speak and write in German and publish his works in German-language journals, such as *Medizinische Zeitung Russlands* and *St. Petersburger medizinische Zeitschrift*.[47] He definitely needed assistance in transcribing his lecture course on obstetrics and rendering it into readable Russian. And this was exactly what he asked Florinskii to do. As Florinskii later recalled, the professor "gave me a free hand — unrestrained by [his] not-quite-Russian spoken phrases — to refine the language of the manuscript according to a literary style [I thought] appropriate for a publishable academic treatise" [ZV, 127]. At the end of the 1856-1857 academic year, the first 300-page volume of Kiter's *Handbook for Studying the Science of Obstetrics* came out, and a manuscript of the second 400-page volume was nearly completed.[48]

Kiter was clearly impressed by Florinskii's contribution to this "collaborative" project, even though it went unacknowledged in its final published products. Florinskii's edits demonstrated not only the fluency of his literary style, but also a keen understanding of the subject. It seems that the collaboration whetted Kiter's appetite. The professor decided to produce another textbook, this one on gynecology, which he could do only with Florinskii's assistance. Perhaps as a way to facilitate the work on the new project, Kiter offered Florinskii an internship at his own obstetrics and gynecology clinic. Florinskii accepted the offer.[49]

During his fifth and final year as a student, Florinskii found himself exceedingly busy. His duties at Kiter's clinic were to be similar to those he had performed at the surgical wards the previous year: making daily rounds, attending to patients, and assisting the professor during clinical lectures and operations. But he ended up actually running the entire clinic. Kiter's assistant Anton Krasovskii, an adjunct professor of obstetrics, that very year went abroad for advanced training, and Kiter himself visited the clinic only to follow some interesting cases and to perform scheduled major operations. So all the routine, as well as emergency, care fell on Florinskii's shoulders. The work on Kiter's textbook also took much of his time. He sat in on Kiter's lectures, during which the professor read aloud — in his broken Russian — some relevant parts from a German book. Florinskii wrote down the lecture, and later, at his apartment, rewrote it in the appropriate style and format. He sent the manuscript for typesetting and then read and corrected the proofs. Towards the end of the academic year, he finished the work and the 500-page *Handbook on Studying Women's Diseases* came out, earning Kiter a special prize from the Academy Council.[50]

The preparation of Kiter's textbooks, supplemented by practical work at the women's clinic, certainly gave Florinskii a much greater knowledge of obstetrics and gynecology than that of any of his fellow students. But, this field was just one of numerous subjects required by the fifth year curriculum (see fig. 1-3). He also had to attend lectures and practical sessions on the pathology and therapy of the internal, skin, and nervous diseases, as well as on legal and forensic medicine. And most important, on top of everything, he had to prepare for his final exams in ten different subjects. The examination session was to last six weeks, from early May to mid-June 1858, and its results would determine Florinskii's future posting. If he passed all the exams with top grades, he could count on being appointed a doctor to a military regiment quartered not in a God-forsaken place somewhere on the borders of the empire, but in a large city — perhaps even St. Petersburg.

Fate, however, had yet another surprise in store for the future doctor.

The seemingly sudden explosion of publishing textbooks and manuals in Russian, which had kept Florinskii so busy during his last two years of study, was actually a sign of important developments both at the IMSA and in the country as a whole. In January 1857, as part of the far-reaching reorganization of the military prompted by the Crimean War fiasco, the IMSA also entered a period of "great reforms."[51] Petr Dubovitskii, formerly a professor of surgery at the academy, was called back from retirement and appointed the academy's president instead of Pelikan.[52] A few months later, Ivan Glebov, renowned professor of physiology at Moscow University's Medical School, replaced Balbiani as the academy's vice-president.[53] With Zinin retaining his position as the head of the IMSA Council, the academy's entire administration was entrusted — for the first time in its history — not to "foreigners" but to Russians. In Florinskii's words, "A new spirit has permeated the academy" [ZV, 146].

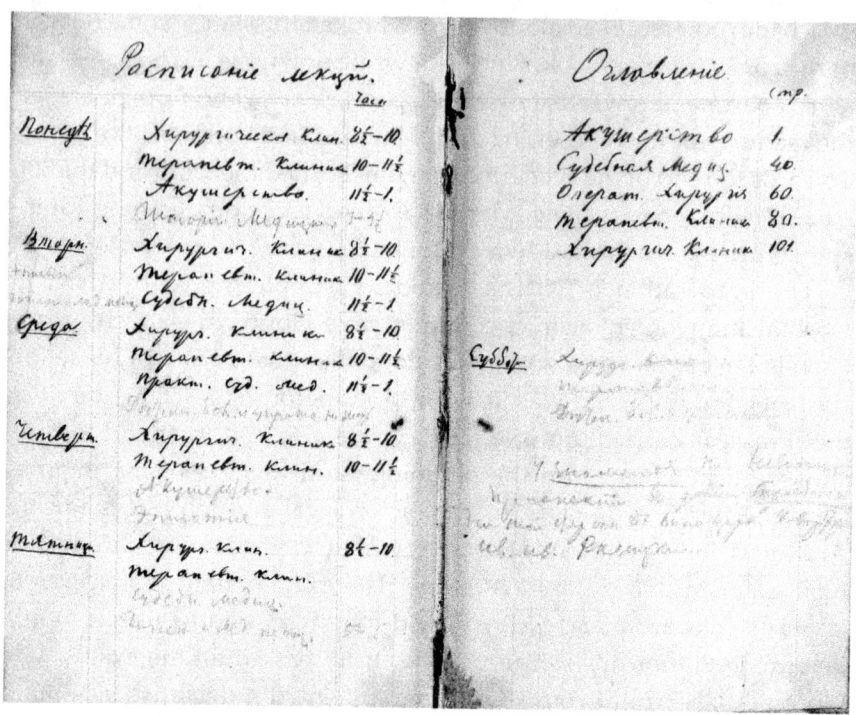

Fig. 1-3. A page from Florinskii's diary with a weekly schedule of classes during his fifth and final year as a student at the IMSA, 1858. From Monday through Saturday, he spent four hours every morning in the clinics (dividing his time between surgery and therapy). Every other day, he had lectures and practical lessons in forensic medicine and obstetrics. He also attended lectures on the history of medicine, as well as on hygiene and medical police. Courtesy of NMRT.

The triumvirate — Dubovitskii, Glebov, and Zinin — immediately launched a series of concerted reforms focused on revising the outdated curriculum, recruiting and preparing new faculty, and creating infrastructural and material support for both teaching and research.[54] The academy turned into a construction site, erecting new buildings for teaching auditoria, clinics, and laboratories, which were supplied with necessary equipment and materials. Teaching in Latin was abolished. Professors were required to teach (and hence produce textbooks) in Russian. The new rules also allowed doctoral dissertations prepared at the academy to be written and published in Russian, instead of the traditional Latin.[55] New departments were organized to teach the new subjects in the updated curriculum and many new professorial positions appeared. In just three years, nearly a dozen old professors were pressed into retirement and replaced with young promising instructors, including such future greats of Russian medicine as Sergei Botkin and Ivan Sechenov. To facilitate the "changing of the guard" among its faculty, in the spring of 1858, the academy also established an "institute of young doctors for preparation to professorial positions" administered by Glebov.[56]

Since his very first days as academy president, Dubovitskii had lobbied the War Ministry to create a new system for preparing professorial cadres. He proposed that each year the ten top graduates remain at the academy for three extra years of advanced training. They were to work on doctoral dissertations on subjects of their choosing in particular medical specialties to prepare them for teaching these specialties to the next cohorts of students. The authors of the two best dissertations would also be awarded a fully-paid two-year-long trip abroad for advanced training in their chosen specialty at European medical schools and research centers. It took Dubovitskii almost a year to push the proposal through the layers of state bureaucracy. Finally, on 25 May 1858, an imperial edict allocated the necessary funds for the "Institute of Young Doctors" and decreed that the ten top graduates of the class of 1858 would become the institute's first members.

The decree came just in time for Florinskii's graduation, but he did not make the cut. He graduated twelfth in his class. However, for various reasons, three of his betters declined to continue their studies at the institute, so Florinskii was offered a place there. He accepted and chose to continue his specialization in obstetrics and gynecology. On 22 June he received his diploma and was appointed a junior attending physician to the women's ward at the Second Army Hospital attached to the academy. Florinskii counted on continuing his work under Kiter's tutelage, but that very fall his professor was appointed head of the surgery department. Kiter's assistant Krasovskii, who had just returned from his year of study in Europe, became the professor of obstetrics and gynecology, and thus, formally, Florinskii's adviser.[57]

From the start of his studies at the institute, Florinskii set his eyes on the top prize — a trip to Europe for advanced training. He did his best to win it. The next three years of Florinskii's life were swallowed in a flurry of activity. The sheer amount of work he accomplished is staggering. He was running his ward at the hospital, assisting Krasovskii at the academy clinics,[58] studying for his doctoral exam, and researching and writing his dissertation. But he was also eager to fill out numerous lacunae in his previous training, learning new subjects and techniques. Florinskii took full advantage of the courses and public lectures offered by recently arrived faculty members, including therapist Botkin, physiologists Sechenov and Glebov, histologist Nikolai Iakubovich, and forensic medicine expert Evgenii Pelikan (the son of the former president of the academy). He was excited by the ideas, methods, and concepts, ranging from Rudolf Virchow's theory of cellular pathology to Claude Bernard's studies of internal secretion and sugar metabolism, introduced by the new professors. Florinskii's recollections of a public course on general biology delivered by Evgenii Pelikan at the Hospital for Menial Workers exemplify his impressions: "His lectures shined with newness and freshness. ... He presented to us science in its true beauty" [ZV, 147]. The young doctor was captivated by this beauty. For the first time since he had begun his study of medicine, he saw his future profession not merely as a set of practical techniques to treat injuries and diseases, but also as a field of inquiry, a science to be advanced. Inspired by the vision of modern experimental medical science, he started to work with a microscope and conduct experiments.

The institute's statute required each member to pass "a full examination for the degree of doctor of medicine." The examination included oral, written, and practical tests in 25 (!) different subjects, which candidates had to complete during their second year. To ensure the fairness of the examination, the required material for each subject was divided into themes, with specific questions written on several dozen "tickets." A candidate taking the exam picked a ticket at random and had to answer all the questions (or perform the specific tasks) listed on the ticket. Florinskii had to pass oral examinations in nine "substantive subjects": physiology, pathology, general therapeutics, special therapeutics, *materia medica*, pharmacology, obstetrics with gynecology and pediatrics, theoretical surgery, and forensic medicine with medical police and hygiene. He also had to complete nine "practical demonstrations." He had to perform an autopsy and produce a forensic protocol; prepare lecture demonstrations on pathological anatomy, pharmacology and pharmacy, and therapeutics; and demonstrate his practical skills in major surgical operation on a cadaver, a few minor surgical operations on patients, and in obstetrics. He also had to deliver two different presentations on physiological anatomy and to take several tests (in Latin) on special pathology, therapeutics, and surgery. It took Florinskii almost eight months, from October 1859 to June 1860, to fulfill all of these requirements.[59] On 12 June 1860 he performed the last task — an autopsy — and was ready to move on to his dissertation. But, utterly exhausted, he decided to take a break. In late June, for the first time in six years, he went on a month-long vacation to his beloved Peski.[60]

Florinskii's workload during his second year at the institute left little time for anything else. Yet, in the fall of 1859, he joined the St. Petersburg Society of Russian Physicians. He diligently attended the society's biweekly meetings and participated in discussions of clinical cases, new techniques, and theoretical concepts presented by the members. He also delivered two reports on his own work at the obstetrics clinic, which soon appeared in the society's journal. In early March 1860, sponsored by Iakov Chistovich, the IMSA professor of forensic medicine, the young doctor was unanimously elected a full member of the society.[61] Florinskii became even more active in the society's gatherings during the following academic year, presenting six reports based on interesting cases he had encountered in his practice at the academy clinics and the women's ward.[62]

Surprisingly, none of these reports dealt with the subject of his dissertation: the tearing of the perineum during childbirth, a very common, but little studied complication that often required intervention by a qualified obstetrician. Florinskii prepared a thorough review of relevant Russian, French, and German literature, which he found in the academy's excellent library. He collected and collated numerous observations in his own clinical practice, describing various types of tears and accompanying complications during the delivery and recovery period. He constructed a special apparatus and conducted experiments on cadavers to investigate the mechanics and mechanisms of tearing the tissues. And on the basis of all these materials, he proposed several original methods of prevention and treatment. In the opinion of a historian of the IMSA obstetrics and gynecology department: "It was the first truly scientific investigation produced at the department, based not only on clinical material, but also on experiments, which showed that its author was a talented scientist with absolutely original and novel ideas."[63]

In addition to his daily duties at the clinics and work on the dissertation, in the fall of 1860, Florinskii also began teaching. A fully staffed IMSA department usually included one ordinary professor (who served as the department's chairman and its representative on the Academy Council); one extra-ordinary professor; and one or two adjunct professors.[64] The teaching load was distributed among all the professors. Large departments (such as surgery and therapy) often also had a cadre of lecturers (Privatdozent) who both taught theoretical courses and oversaw practical clinical sessions. The renewed curriculum required several courses (both theoretical and clinical) on obstetrics, gynecology, and pediatrics to be offered to the fourth- and fifth-year students. But, since Kiter's transfer to the surgery department, Krasovskii remained the only professor at the obstetrics and gynecology department, and he was completely overwhelmed with his teaching duties. He petitioned the Academy Council for permission to transfer some of his teaching to his clinical assistant Florinskii. According to the rules, a candidate for a teaching position had to deliver two "trial" lectures attended and adjudicated by the council members. The subject of one lecture was assigned by the council, the subject of the second lecture was chosen by the candidate himself. Florinskii's "trial" lectures went very well. The council granted Krasovskii's petition and confirmed Florinskii's appointment as a lecturer (Privatdozent). In October 1860, he began teaching a year-long course on theoretical obstetrics and gynecology to the fourth-year students.[65]

Florinskii was the first among his ten classmates at the institute to fulfill all the requirements: he completed his dissertation in the early spring of 1861. On 9 March, just one week before the public announcement of the first and the greatest of the Great Reforms — the emancipation of the serfs — he presented a 300-page manuscript to the Academy Council to obtain permission for its publication, a prerequisite to the dissertation's public defense. Following the established rules, the council appointed two of its members to serve as official reviewers. Not unexpectedly, Kiter and Krasovskii got the job. On 1 April, both reviewers gave their go-ahead, and the council permitted Florinskii to publish his work. A week later, he delivered 225 copies of the printed dissertation to the council.[66]

On 15 April, Florinskii defended his dissertation at a special meeting of the IMSA Council. The council members agreed that this "superbly presented" work was "an original scientific treatise," and its author undoubtedly deserved to be awarded the degree of Doctor of Medicine.[67] Furthermore, after a short discussion, with nineteen members voting for and only one against, the council decided that Florinskii had also earned a trip abroad for advanced training. The same day, the council issued a formal recommendation to the IMSA president to send Florinskii abroad for two years at the academy expense "to study further his chosen subjects: obstetrics, gynecology, and pediatrics."[68] Dubovitskii forwarded the recommendation, along with his own endorsement, to the war minister. On 6 May, the minister presented the petition to the emperor, who gave it his seal of approval.[69]

Florinskii's dream had come true: he would spend the next two years at the best medical schools in Europe. But, although the emperor had authorized his trip, Florinskii could not leave St. Petersburg right away. First, he had to wrap up his teaching and his duties at the academy clinics. During May, he finished his lectures and administered the final exam. He prepared his annual report on the activities of the obstetrics clinic and sent it for publication to the *Military-Medical Journal*.[70] Apparently as a way to familiarize himself with pediatrics — an area he had had little experience with and was supposed to study during his trip — Florinskii worked his way through a voluminous annual report recently published by the St. Petersburg Foundling Home.[71]

Perhaps, initially, Florinskii did not plan to publish the results of this work. But just a few months earlier, in late March 1861, Chistovich founded an independent weekly, titled *Medical Herald*. Modeled to a considerable degree on the *Gazette Médicale de Paris*, this publication was to become a major forum for discussion "on all the areas of scientific and practical medicine and hygiene, as well as on all the problems faced by medical practitioners in their everyday life."[72] It seems likely that Chistovich asked some of his former students to contribute to the new periodical and Florinskii was among the first to do so: his first publication in *Medical Herald* was a thorough review of the medical report for the year of 1857 by the St. Petersburg Foundling Home.[73]

This review was Florinskii's first venture outside of purely "technical" medical writings. For the first time, he exercised in print what he considered "the most essential element of scientific knowledge — original scientific criticism."[74] The volume was the first "annual medical report" the institution had issued in the 75 years of its existence, Florinskii emphasized. Its publication breached "the impenetrable secrecy" that had covered "rich scientific materials on obstetrics, gynecology, and pediatrics" collected by the Foundling Home physicians. The reviewer praised the report's "administrative section" that provided detailed information on the institution's internal organization and administration, and, especially, extensive statistical tables on the mortality and morbidity of its charges. But, he noted, the numbers and tables that dominated the report raised numerous questions that remained unanswered. Florinskii was sorely disappointed by the report's "scientific section," which, contrary to his expectations, contained no materials on "the character of epidemics, the occurrence and development of various diseases," nor a "critical evaluation of methods and medicines" used to treat these diseases. Instead, its "scientific conclusions" were little more than a collection of case histories. He was also dissatisfied with the writing style and lack of clarity, noting a large number of obscure and incomprehensible expressions, especially in autopsy reports. Despite these shortcomings, Florinskii stressed, as the first of its kind, the report certainly deserved the "sincere gratitude" of the medical community that hoped to see it becoming a regular publication.[75]

By early June 1861, Florinskii managed to complete all of his numerous duties and obligations and was finally ready to embark on his "grand tour" of Europe.[76] On the eve of his departure he felt even more anxious than he had when preparing to leave Perm for St. Petersburg eight years earlier. The challenges ahead now seemed even more formidable, the tasks even more daunting. This time he would have to travel through foreign lands, live among foreign peoples, converse in foreign languages, and abide by foreign customs. "The close bonds of [one's] kinship, language, religion, and community are far from empty words," he wrote in his diary, "their meaning is most clearly understood in those cases when a man is removed from his own familiar milieu and feels, as I do, exactly what he is missing" [DZP, 12]. Although he read regularly in French and German — the most frequently used languages of science and medicine of the time — Florinskii was particularly concerned with his limited conversational skills and his ability to follow and understand his foreign instructors. And he had a lot to learn. "Judging by what I had read about foreign universities, instruction at our medical academy is far behind in all its aspects," he recorded in his journal: "In order to reach the heights of modern medical education I would have to learn [almost everything] anew" [DZP, 5]. But at least this time he did not have to worry about money. Compared to the fifty rubles with which he had embarked on his trip to St. Petersburg, the academy provided him with the astronomical sum of 1,700 rubles annually — more than sufficient to cover all of his expenses.[77]

## The Grand Tour

On 5 June 1861, the freshly minted Doctor of Medicine left St. Petersburg for Berlin, his first stop on what would be a 27-month tour through nearly all major medical centers in Europe. Aside from the general instruction to focus on obstetrics, gynecology, and pediatrics, Florinskii had a completely free hand to choose the particular locales and duration of his stays.[78] He visited leading German universities and clinics in Berlin, Halle, Munich, Giessen, Leipzig, Dresden, Wurzburg, and Erlangen. He worked at medical schools and hospitals in Prague, Vienna, Paris, and London. He attended lectures by, and studied with, the leading lights of European medicine, including Claude Bernard, Theodor Bischoff, Eugène Bouchut, Ferdinand Hebra, Carl Rokitansky, Armand Trousseau, and Rudolf Virchow.

Florinskii spent a whole month in Berlin, "largely because of Virchow," even though there was "very little interesting on my own specialties here" [DZP, 23]. The German professor deeply impressed the Russian visitor: "What a force of mind, oratory talent, and pedagogical tact." He was enthralled by Virchow's lectures, which were supplemented by "demonstrations of excellent preparations and practical work in the laboratory." "[Virchow] captivates [the audience] by the newness of his ideas. He is not a copyist, not your ordinary teacher, but a creator of science, its primary mover," raved the young doctor. "I have never [before] seen such individuals," Florinskii admitted, "That's how he is at the lectern, in social affairs, [and] in everything that is touched by his universal mind. But look at his outward appearance — modesty incarnate" [DZP, 27-28]. Florinskii stayed in Berlin until Virchow finished his course in early July and then embarked on a tour of numerous German universities and clinics.

Florinskii stayed almost two months in Prague, but spent most of his time in Vienna and Paris, the world renowned centers of medicine at the time. He found the instructors, as well as clinical and research facilities, in these two cities to be the most advanced and the most suited to his own interests, and stayed more than six months in each city. Indeed in January 1863, from Vienna, he petitioned the Academy Council for permission to extend his trip until September (he was supposed to return to St. Petersburg in early June), so that he could spend extra time in Parisian and Viennese clinics and laboratories. After a nearly five-month delay occasioned by mishandling of his documents by the Ministry of Foreign Affairs, his request was granted at the end of May, allowing him to spend the entire summer in Paris.[79]

Florinskii's daily routine during his stay in a particular locale followed the same pattern he had established for himself in St. Petersburg. Every day he arose at 7am and, after a quick breakfast, went to the clinic to attend a lecture and/or to work with patients. At 2pm he took a dinner break and then returned to the clinic (or went to the laboratory) to pursue independent research projects. He spent several hours a day reading the newest medical journals and monographs. At 8pm he had supper and then caught up on his journal, correspondence with family, friends, and colleagues, and official reports to his superiors at the academy.[80] As he wrote to his parents from Prague, "time flies with a fantastic speed; a week goes by in a blink of an eye. My studies are so interesting and so varied that I am forgetting everything else. I am so happy now that my dreams, about which I had been afraid even to think, are coming true."[81] Indeed, six months into the trip he got so busy that he stopped keeping a journal where he recorded his thoughts and events of his daily life, and the steady stream of letters he had been sending home dwindled to a trickle.

Fig. 1-4. Vasilii Florinskii in Paris, c.1862. This photo was taken by Sergei Levitsky, a patriarch of Russian photography. Between 1859 and 1864 Levitsky operated an atelier at 22 rue de Choiseul in Paris. Almost every Russian visiting Paris in those years (from Levitsky's cousin Alexander Herzen to Emperor Alexander II) went to the studio, which became one of the hubs of "Russian circles" in the city. Courtesy of NMRT.

As in his last two years in St. Petersburg, the sheer amount of work Florinskii had completed during his two years abroad is astonishing. As Glebov noted in his 1863 overview of the accomplishments of the institute graduates, "Florinskii has published much more than anyone else."[82] Indeed, by the end of his sojourn in Europe, Florinskii had published eighteen articles, describing clinics and universities he had visited,[83] surveying the latest German, French, and English literature on his specialties,[84] and reporting results of his own original research.[85] In Vienna, for instance, he conducted an extensive pioneering study of the anatomical and histological changes in the uterus after childbirth.[86]

Florinskii's stay in Paris spurred his interest in ovariotomy — a technique that was just then making inroads in French obstetrics. This operation, which involved a resection of parts of the ovaries to remove ovarian cysts, had been developed and popularized by the British surgeon Isaac Baker Brown and was at the time successfully practiced by a number of his compatriots. Florinskii took a side trip from Paris to London with the express purpose of studying the technique at its source. He prepared a detailed overview of the operation, comparing the British, German, and French indications for, and modifications to, the operation he had personally observed in the three countries.[87] Over the course of his tour, Florinskii frequented booksellers and acquired a substantial library of medical and scientific publications, including multivolume editions of the "collected works" of Georges-Louis Buffon, Alexander von Humboldt, Johannes Muller, Rokitansky, and Virchow.[88] He also bought several sets of obstetrics and surgical instruments, as well as a microscope with the assortment of extra lenses, to continue his research after returning to St. Petersburg.

Despite all the work Florinskii was doing during the trip, he also found time for play. Everywhere he went, he visited museums, historical monuments, art galleries, and concerts. During four weeks in the summer of 1862, he journeyed the entire length of the Apennine peninsula and parts of Switzerland. He marveled at great artworks in the Louvre and the Uffizi Gallery, explored the 1862 London International Exhibition and the British Museum, stood in awe of the ancient monuments of Rome, and admired the breathtaking beauty of the Alps and the Bay of Naples.[89]

Florinskii's initial apprehension of living in foreign lands and among "alien" peoples proved unfounded. He quickly overcame his linguistic handicaps, mastering both German and French, and, in preparation for his trip to London, also learning English. He developed good working relations with his professors and indeed befriended several of his mentors in Vienna and Paris. But most important, in almost every city he visited, he found a sizable group of his compatriots. In fact, he spent most of his free time not among the "aliens" but in the company of fellow physicians who had also come to Europe for advanced training from Russian medical schools.[90]

"Russian circles," which sprung up in every major European center of learning in the late 1850s and early 1860s, enabled Florinskii to maintain those "close bonds of language, religion, and community" he had been so loath to be missing during his tour.[91] With their fluid composition, and members coming and going according to their individual itineraries and interests, these circles became a major "exchange" for professional gossip and the latest news from home. They served as hubs of heated discussions, which often continued long into the night and ventured far beyond the confines of science and medicine into contemporary politics, literature, philosophy, and education. They helped reinforce his identity as a Russian, often defined by deliberate comparisons with, and a conscious opposition to, the surrounding "alien" nations, individuals, institutions, beliefs, attitudes, and mores.

As for many of his compatriots, Florinskii's "grand tour" became much more than merely a means to advance his professional career. It also provided him with a new lens through which to look at his Fatherland and a whetstone to sharpen his ideas, perspectives, and ideals. Already during the first month of his tour, Florinskii was struck by the sharp contrast between his experiences in St. Petersburg and what he encountered in Berlin. Working at Virchow's Anatomical Institute at Charité he could not help but notice that "compared to those of Berlin, our academic clinics are simply stinking barracks ... Compared to Virchow's, our Anatomical Institute is a real cesspool, even though it had been created by Pirogov and is currently run by the no less famous [Ventseslav] Gruber" [DZP, 23-24]. This was not the result of Russian medical scientists' negligence or sloppiness, Florinskii surmised, but rather a direct consequence of "the Russian government's neglect of science in general and medicine in particular." "The previous reign [of Nicholas I] had viewed European science as a bogeyman in the form of an atheist and a revolutionary," he observed, "we had shunned it like a plague, barring to young people any access to foreign lands." Yet, Florinskii felt, the situation had begun to change with the ascendance of Alexander II:

> The doors abroad are now open to each and every one ... The government itself actively assists in the capital reorganizations of the entire life in the country, and, thus, we could hope that our medical institutions will

very soon reach the same heights as those abroad. We have people and funds. All we need is to trust the former and properly administer the latter [DZP, 25].

Florinskii was very enthusiastic about the Great Reforms, which at that very time were reshaping so dramatically the life of his country, as well as his own life. But this enthusiasm did not blind him to numerous problems and issues that plagued his Fatherland. The officially sanctioned *glasnost'* gave him an opportunity to add his own voice to the chorus of suggestions, criticisms, and ideas regarding the country's past, present, and future that had risen in the aftermath of the Crimean War. He seized this opportunity with a vengeance. In October 1861, in a letter to his parents (see fig. 1-5), he recorded what would be his life motto: "Now I have the only heart-felt desire — to bring to my Fatherland as much benefit as possible, specifically by acting in that area where I could be most useful."[92] At the moment, that area, of course, was medical science, practice, and education.

In his "free time" during the first months of his tour, Florinskii continued to exercise "the most essential element of scientific knowledge — original scientific criticism." He wrote an extensive critical analysis of the process, meaning, and purpose of getting a doctoral degree in medicine, based on his recent personal experience, as well as a careful study of available literature on the subject. In August 1861, *Medical Herald* serialized his assessment in four consecutive issues.[93] He began with a brief historical excursion into the institution of doctoral degrees, outlining its Middle Ages origins, development, and expansion in such learned professions as theology, jurisprudence, and medicine. He then described in detail the current system of awarding the degree of Doctor of Medicine in Russia, along with various privileges it accorded to its bearers. According to Florinskii, it was exactly the privileges — the automatic advancement to the eighth rank in the Table of Ranks, with the corresponding increase in salaries, benefits, and pensions — that enticed numerous medical school graduates to seek the degree. Nearly fifteen per cent of all physicians practicing in Russia in 1860 were Doctors of Medicine, he noted. This, Florinskii felt, subverted the very meaning of the degree as the symbol of "highest learning" and "scientific advancement," undermining its prestige and generating mistrust among the public.

### 1. The Author: Vasilii Florinskii

Fig. 1-5. The Florinskiis family, c.1862. An unknown photographer. Upper row: Florinskii's brothers Ivan Markovich (left) and Semen Markovich (right) with their wives, Augusta Petrovna (left) and Aleksandra Alekseevna (right). Lower row: Florinskii's father Mark Iakovlevich and his mother Maria Andreevna, with their grandchildren, Arkadii (left) and Vasilii (right). Upon sending his parents a photo of himself (see fig. 1-4), Florinskii asked them to send him a family photograph. The nearest photographic studio was in Ekaterinburg and apparently the family took a trip there to have this photo taken. The family portrait contrasts sharply with the photo of Vasilii Florinskii, illustrating the profound difference in the lifestyle and status of two social estates: the rural clergy and the country's educated elite. Courtesy of NMRT.

Florinskii critically reviewed the two major requirements for obtaining a degree: a doctoral examination and a doctoral dissertation. He thought that both of them were insufficient, first of all, because they did not include any prerequisites. For Florinskii, the main prerequisite was solid proof — in the form of published scholarly works — of the applicant's commitment to the advancement of medical knowledge and/or practice. Only those who proved their abilities to produce new medical knowledge and to express it in readable publications, he argued, should be allowed to pursue the degree. According to Florinskii, in its present form, a doctoral examination was primarily aimed at testing the applicant's memory. It basically reproduced the format of the medical school final exams, only with many more subjects (25 in the doctoral examination as opposed to ten in the finals). This kind of examination was "a simple formality, boring to the assessors, empty, slightly offensive, and sad to some applicants, and, lately, untrustworthy even to the uneducated public." A doctoral examination, Florinskii suggested, should be specialized and designed in such a way as to allow the applicant to demonstrate his ability to understand and tackle the general problems and issues of his chosen specialty.

He was also dissatisfied with the current requirements, as well as the principles of evaluation, for a doctoral dissertation. A dissertation, he insisted, should be not a compilation of available literature, but an original (clinical and/or laboratory) investigation of an important subject in the applicant's specialty. It should be written and published in Russian to make its results widely accessible to practicing physicians and other researchers. Florinskii was concerned with the style and language of the dissertation: "any truly learned individual must also be a writer." Writing, for Florinskii, was a special skill that required training and practice. In reality, for many applicants, their dissertation was the first and often the last experience in writing a legible text. Too often, he felt, a "Russian Doctor [of Medicine] cannot write in Russian." "Who could believe in my doctoral degree," he wondered, "if I am unable to write two words without a grammatical error [and] to put two sentences in a logical sequence?" Florinskii was convinced that a dissertation must be published well in advance of its public defense to allow its assessment not only by the two "official reviewers" appointed by the medical school, but also by a wider professional community. It is this "public approbation" that, according to Florinskii, must be the ultimate test of any scientific publication and thus a proper evaluation of the significance of facts, ideas, theses, and conclusions presented in a dissertation.

All the criticism Florinskii leveled at the existing system of medical science, practice, and education in Russia, did not mean that he did not believe in the bright future of his beloved country. Quite the contrary, he was a true patriot and was deeply offended by the common perception of his homeland and its people as backward, undeveloped, and retrograde. A telling example of this attitude is a feuilleton that Florinskii published in *Medical Herald* in early 1862 under a revealing title "Notes of a Russian Foreigner about Our Medical Life."[94] The feuilleton responded to a short piece he had read in the Viennese newspaper *Medizinische Halle*, titled "German doctors in Russia and the lawful order of their practice," and written by a German physician residing in St. Petersburg. The report stressed the shortage of Russian physicians and portrayed the country as a mecca for German doctors, a land of great opportunities to make a fortune and achieve a high status, and it described the exalted prestige and huge honoraria accorded to them by their Russian clientele. "Foreigners have always been court physicians in Russia," the author stated, and they have also held the top-level positions in medical institutions, education, and administration. Furthermore, he emphasized, "in case of illness, the highest aristocratic circle, the nobility, and the honorary citizens always prefer to seek the help and advice of foreign doctors." The author described in detail the very accommodating Russian system of certification for foreign physicians, as well as the ranks and privileges accorded to them by the state.

Upon reading the article, Florinskii was outraged, a feeling apparently shared by the editor of *Medical Herald* Chistovich. Just eight days after the publication of this unflattering article in the Viennese newspaper, Florinskii's response graced the title page of its Russian counterpart. Florinskii dismissed the correspondent's claims as outdated and misleading. "Some one hundred-fifty years ago," he sarcastically noted, "there was a time, when, in the simplicity of the Russian soul, we did consider Germans as our teachers, but facts dissuaded us in their pedagogical abilities." "Honest instructors, seeing their charges reaching adulthood, let them act independently," he continued, "but we are still considered [to be] adolescents." He refuted the German author's various statements and declared forcefully:

> God be with you, [our] uninvited and unwelcome helpers; we do not believe anymore in your assistance in the affairs of life and science; take your nomadic swarm somewhere far away, overseas, where life has not yet awakened [and] strengths have not yet developed; offer yourselves to another master, who might, for the first time, as we once had, be fooled by your cunning appearance.

"If our word would ever reach the author of the aforementioned article," Florinskii concluded ironically, "we would ask him to learn more of the true Russia and to advise his compatriots not to rush to pack their bags [and] to tone down their noble intentions of enlightening their kind, but once simple-minded neighbors; he would do it to our pleasure and to his compatriots' forewarning."[95]

As these publications demonstrate, Florinskii did not fully subscribe to the visions of his Fatherland's future advanced by either the Slavophiles or the Westernizers, the two major opposing trends of thought in contemporary Russia.[96] His diaries and letters show very clearly that he did share the views of many Slavophiles on the close similarity and affinity among Slavic peoples. For instance, during his stay in Prague he repeatedly emphasized numerous similarities and mutual sympathy between the Czechs and the Russians, contrasting them with the differences and hostility between the Germans, on the one hand, and the Russians and the Czechs, on the other.[97] Yet, unlike many Slavophiles, he did not believe in an exalted image of a "God-chosen Russian people" or a "special Russian path," independent and separate of that of western Europe. He freely admitted that the empire and its people had a lot to learn from their western counterparts. But, he did not share the viewpoint of many Westernizers who saw the country and its people as backward and completely incapable of innovation and development (in science, technology, medicine, agriculture, etc.) without the guiding hand of their Western neighbors. Rather, Florinskii believed deeply in the talents and abilities of the Russian people. Was not his personal career the best illustration of such talents and abilities? He blamed the country's perceived backwardness on state policies that denied its people access to education and social, political, and economic freedoms. Yet, he was not a rabid nationalist, and believed in a free intercourse among various nations. As he pointedly noted in his diary, the two greatest Russian poets, Pushkin and Lermontov, had both descended from mixed, "inter-national" bloodlines. He would further develop this notion in his treatise "Human Perfection and Degeneration."

As for many others, Florinskii's "grand tour" of Europe became a formative experience. He summed up his appreciation of what it meant to him personally in a letter home:

> I have often dreamed about [a trip] abroad, but could not even imagine that here one could develop so fast and so far. ... Encounters with various nations, a close look at their goals [and] historical events, [and] conversations with the finest Russian circles composed of the best, most advanced individuals (because fools will not come here), could not fail to influence any man.

"If I did not go to the [Medical-Surgical] Academy, I would have become an ordinary priest," he concluded, "and, if I did not go abroad, I would have remained, for the rest of my life, an ordinary physician."[98]

Alas, too soon to Florinskii's liking, his trip came to an end. In late August 1863, he returned to St. Petersburg. Within a few days, he submitted a final report on his work abroad to the Academy Council. And shortly thereafter, he was swept by a whirlwind of teaching, clinical practice, laboratory research, writing, and editing. After an obligatory "trial" lecture, he was appointed an adjunct professor to the obstetrics and gynecology department headed by Krasovskii and began teaching theoretical and practical courses.[99] To continue several research projects he had begun abroad, he created a histological laboratory under the auspices of the academy's gynecology clinic and equipped it with instruments and materials he had bought during his European tour. On 12 September, just two weeks after his return, the St. Petersburg Society of Russian Physicians elected Florinskii as its "scientific secretary" and the editor-in-chief of its journal. He had to arrange the society's meetings, keep its minutes, and edit its *Proceedings*.[100] During the 1863-1864 academic year, he also delivered four reports on his own research to the society's meetings, which soon appeared in its journal.[101] He continued to publish voluminous surveys of the latest foreign literature on his specialties in the *Military-Medical Journal*,[102] as well as reviews

Fig. 1-6. Maria Florinskaia, née Fufaevskaia, c.1863. Photographer unknown. Courtesy of NMRT.

of Russian publications in the *Medical Herald*.[103] In March 1864, in addition to his regular duties at the academy, he was appointed an attending surgeon to the women's ward at the Second Army Hospital, an appointment that increased his workload considerably.

A year after his return to St. Petersburg, in August 1864, Florinskii met Maria Fufaevskaia, the daughter of a retired lieutenant-colonel from the Novgorod province. We do not know exactly when, how, or where they met. Available materials are surprisingly silent on details of this fateful meeting and what followed it. She was seventeen, he — thirty. Was it love at first sight? Was it a calculated decision? We could only guess. But less than half a year later, on 30 January 1865, as was mandated by the rules, Florinskii formally petitioned the IMSA president for permission to marry Maria. He appended to his petition all the required documents, including a copy of his bride's birth certificate and her written statement that she indeed was willing to marry him.[104] Dubovitskii granted the petition. Three days later, on 3 February, Vasilii and Maria were joined in holy matrimony.

The newlyweds moved into a spacious apartment that Florinskii rented in a fashionable neighborhood in the very center of the city. The apartment building was situated on Malaia Ital'ianskaia Street that run parallel to Nevsky Prospect between the Liteinyi and Ligovskii avenues (see fig. 1-7). In choosing the location of their new apartment, Florinskii certainly considered its proximity to the academy. To get to his clinics, all he needed to do was to follow Liteinyi Prospect for about two kilometers to the other side of the Neva.[105] He hired a maid and a cook to help his young wife with running their new household. As Florinskii wrote to his mother-in-law, "What shall I tell you about our married life? We are very happy. ... We are spending most of our time at home, because we do not feel like going anywhere, unless we have to, since it is so nice and cozy here."[106] In the next letter, however, he begged her pardon for not writing more often, "I am now almost constantly busy."[107]

Fig. 1-7. The façade of the Solomko house on Malaia Ital'ianskaia Street in St. Petersburg, where Vasilii and Maria Florinskii lived for fifteen years, from 1863 to 1878. Current address, 22 Zhukovskogo Street. Alas, I was unable to determine exactly which apartment the Florinskiis occupied, but apparently its windows overlooked the inner courtyard. Photo by the author, 2016.

Florinskii was not just making excuses. He was indeed swamped with work. In March, in addition to his regular teaching, clinical, and editorial duties, he was appointed "a junior attending physician" to the Second Army Hospital. A month later, he was promoted to the next, seventh civic rank (equal to a lieutenant-colonel in the army) in the Table of Ranks, which came with a sizable salary increase and an "Order of St. Anne."[108] At the same time, he was finishing and writing up two substantial pieces of research: one, on the histological changes produced in the uterus by chronic inflammation,[109] and another, on the methods of revival of apparently stillborn babies.[110] He was also editing and annotating a Russian translation (prepared by his students) of the *Obstetrics Compendium* published a year earlier by Gustav August Braun, one of the professors he had befriended in Vienna.[111] He, of course, continued writing reviews of current literature on his specialties[112]

And, most likely, it was during this very busy, but very happy time in the late spring and early summer of 1865 that Florinskii began his collaboration with *Russian Word* and started working on his treatise "Human Perfection and Degeneration."

# 2. The Publisher: Grigorii Blagosvetlov

> "The Crimean War has convinced us that Russian society lacks a lot [of things] to stand on the same level with Europe. [The war's] failures and losses have forced us to look closely at what exactly we are lacking, and it turned out that we lack knowledge."
>
> Nikolai Shelgunov, 1865

Florinskii's treatise on "human perfection and degeneration" represented a clear departure from everything he had written before. It was not a review or a translation of some published work. Nor was it a report on a piece of research he himself had conducted, or a textbook summary of a medical specialty he himself had practiced. Rather, it was a "thought piece" that sought to synthesize the newest ideas, concepts, and data from a variety of knowledge fields, ranging from gynecology and hygiene to biology and anthropology. Furthermore, Florinskii had published all of his previous works in specialized periodicals addressed to his fellow physicians, such as *Medical Herald*, *Protocols of the St. Petersburg Society of Russian Physicians*, and *Military-Medical Journal*. In contrast, his discourse on "human perfection and degeneration" first appeared in *Russian Word*, a leading "thick" journal of the time and, thus, targeted a very different audience. What induced the young professor to take up this new subject and to present it to the new audience? What did he try to accomplish? Why did he publish it in *Russian Word*? On the other hand, why did the "literary-political" monthly lend its pages (and quite a few of those pages!) to Florinskii's essays that dealt with such unusual, seemingly obscure scientific issues?

The venue and the timing of the publication of Florinskii's treatise offer some clues to answers to at least some of the numerous questions it raises. It seems likely that the initial inspiration for and the overall theme of Florinskii's work came from Grigorii Blagosvetlov (1824-1880), the editor-in-chief and publisher of *Russian Word*. It was Blagosvetlov who serialized it in the August, October, November, and December 1865 issues of his journal. Florinskii's treatise was actually part of a broad campaign waged by the journal's editor to popularize science and to promote a "scientific worldview" that sought to explain, and thus, eventually to cure the "social ills" that plagued post-Crimean Russia. A key focus of this campaign was the popularization of Darwin's concept of the origin of species that promised to shed "light ... on the origin of man and his history."[1] Building upon Darwin's concept, Florinskii's essays attempted to illuminate not just the past but also the future of mankind.

## Grigorii Blagosvetlov's *Russian Word*, 1860-1866

Initially, nothing distinguished *Russian Word* from the nearly 100 new periodicals that appeared in Russia during the five years between the end of the Crimean War in March 1856 and the March 1861 announcement of the emancipation of the serfs — the first of the Great Reforms.[2] Nikolai Shelgunov, one of the journal's core contributors, described this period in his memoirs:

> It was an amazing time, a time when everyone aspired to think, read, and learn and when everyone who had anything on his mind wanted to say it out loud. [Russian] thought that until this time had been asleep has awakened, shaken up, and gone to work. Its impulse was great and its tasks immense. What [everyone] thought about and debated were not some fleeting affairs, but the future fate of the entire country.[3]

What was even more important, for the first time in Russian history, the emperor and his ministers (the "enlightened bureaucrats," in historian W. Bruce Lincoln's apt characterisation)[4] seemed willing to listen to what the Russian educated public had to say about the country's "future fate."

The reign of Alexander II had opened a new era of communications between the rulers and the ruled, manifested in the rapid growth of the Russian popular press that provided a wide new venue for the educated public to express its opinions.[5]

It was right in the middle of this "amazing time" — when the newly sanctioned *glasnost'* seemed to offer the Russian educated public at least a say, if not yet an actual role, in defining the country's future — that the first issue of a new "learned-literary" (*ucheno-literaturnyi*) monthly journal, titled *Russian Word*, was launched in St. Petersburg in January 1859 (see fig. 2-1).[6] Its founder, publisher, and editor-in-chief Count Grigorii Kushelev-Bezborodko (1832-1870), the last scion of two rich noble families — the Kushelevs and the Bezborodkos — and himself an aspiring litterateur, was clearly responding to the call of the times.[7] In a letter to Fedor Dostoevsky, inviting the writer to contribute to his journal, the count explained his ambition: "All persons to whom God has granted strength and talent should now join into one close-knit family, not divide into separate parties of Slavophiles and Westernizers, and work together for the common good."[8] A few months later, on the pages of St. Petersburg's major newspaper, he enthusiastically announced his intention to unite in the new journal all "thinking people," be they conservatives, liberals, or radicals: "The very name 'Russian Word' allows no singlemindedness in our views and obligates us to take such a vantage point, from which all [things] accessible to Russian thought and the Russian heart, no matter their various shades, will be [seen as] a complete picture, filled with thought and meaning."[9]

Alas, despite his enthusiasm and ambition, Kushelev-Bezborodko proved an inept editor and even worse businessman. Although he did manage to secure the participation of several well-known poets and writers, including Dostoevsky, Afanasii Fet, Apollon Maikov, and Iakov Polonskii (who became the journal's co-editor), the young count failed to give his journal a memorable "face." Even his closest co-workers considered his venture merely "a whim of the golden boy."[10] Reviewers in contemporary periodicals were much less charitable.[11] The journal was losing subscribers and only the count's immense fortune kept it afloat. All this began to change when the count invited Blagosvetlov to take over the journal in June 1860.

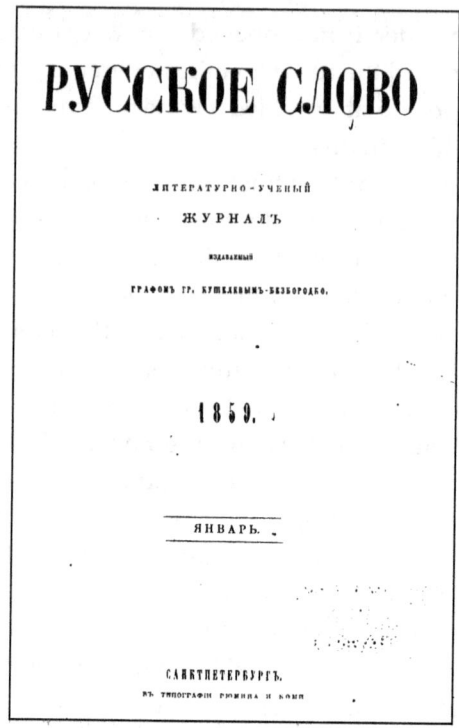

Fig. 2-1. The title page of the first issue of *Russian Word*, identifying it as a "literary-learned journal" and carrying the names of its publisher "Count Gr. Kushelev-Bezborodko" and its typographer, "Riumin and Co." Courtesy of RNB.

Blagosvetlov was ten years older than Florinskii, but the early life trajectories of the two men look in many ways similar.[12] Like Florinskii, Blagosvetlov came from the low-level clergy (his father was a priest in a military regiment stationed in the Caucasus) and graduated first from a *bursa* and then from a theological seminary.[13] He too came to St. Petersburg from the provinces with no money, no friends, no patron, and no one who could have given him a helping hand. Like Florinskii, he enrolled in the Imperial Medical-Surgical Academy. But unlike his future author, after some seven months at the academy, Blagosvetlov transferred to St. Petersburg University. As he himself later explained, "I had particularly enjoyed studying the natural sciences, [but] unfortunately, could not overcome my revulsion towards surgical operations."[14] Yet at the university he chose to study not the natural sciences, but rather literature, history, and languages, enrolling in its "historical-philological" school. After graduating with first-class honors and a "candidate" degree, Blagosvetlov became a literature teacher in elite military schools in St. Petersburg. But his teaching career soon came to an abrupt end. In late 1855, a student at the School of Pages — the country's most exclusive military school — denounced Blagosvetlov's "anti-government views" to his uncle, who happened to be the head of the all-powerful Third Department of the Imperial Chancery — the secret police. Blagosvetlov was fired and a few months later barred from occupying any teaching position in the empire.[15] Denied what he considered his true vocation, in the spring of 1857, he went abroad.

Like Florinskii's trip a few years later, Blagosvetlov's three-year stay in western Europe proved a formative experience. Supporting himself by writing for various Russian periodicals and occasional private teaching, he attended lectures in universities and actively participated in the "Russian circles" he encountered in European capitals. It was in just such a circle in Paris, in 1858, that he met Polonskii, who invited him to write for *Russian Word*. Blagosvetlov eagerly agreed and began contributing articles to the journal from its very inception.[16] A year later, in London, he joined a different circle. He became a disciple and a good friend of Alexander Herzen, "the father of Russian socialism" and the most famous Russian political writer of the time, whose weekly newspaper *The Bell* (*Kolokol*), printed in London and clandestinely smuggled to Russia, played a critical role in awakening and shaping public opinion in the aftermath of the Crimean War.[17] For nearly a year, Blagosvetlov worked as Herzen's secretary, assisted with smuggling his newspaper to Russia,[18] and tutored Herzen's daughters in the Russian language and literature. But, in early June 1860, apparently on Polonskii's suggestion, Kushelev-Bezborodko summoned Blagosvetlov to his estate in the south of France and offered the former literature teacher the position of "managing editor" at *Russian Word*. Blagosvetlov accepted. Within a few weeks, he was back in St. Petersburg.[19]

Despite his modest official title — managing editor (*upravliaiushchii redaktsiei*) — Blagosvetlov soon became the de facto editor-in-chief, and two years later the publisher, of *Russian Word*.[20] As he later confessed, "I had never felt a particular calling to becoming a publisher and entered this trade by pure accident."[21] But he certainly had a talent for it. He quickly turned the journal from a financial sinkhole into a profitable enterprise: in less than a year, the number of subscribers had doubled, from 1,200 to 2,400.[22] Blagosvetlov refashioned the "faceless" *Russian Word* published by Kushelev-Bezborodko into a formidable competitor of *The Contemporary* — at the time arguably the country's most influential "thick" journal, published by Nikolai Nekrasov and headed by Nikolai Chernyshevskii and Nikolai Dobroliubov, the apostles of what Soviet historians later affectionately termed "revolutionary-democratic literature."[23] Blagosvetlov stood firmly at the helm of *Russian Word* through external storms and internal tempests for nearly six years, until the minister of internal affairs suspended the journal's publication for five months in February 1866; and three months later, the imperial edict finally closed it for good.

Petr Tkachev, a young jurist-turned-journalist, who joined the *Russian Word* editorial team in the last year of its existence, later described the "extraordinary, fabulous success that *Russian Word* has achieved in just a few months under Blagosvetlov's editorship." "It was always sold out," Tkachev marveled, "it was read through and through; the appearance of every new issue was eagerly awaited and greeted as a literary event. How much noise, heated disputes, debates, polemics, sometimes thunderous applause, and sometimes venomous cursing did it entice in both literature and society!"[24] One could dismiss Tkachev's accolades as understandable exaggerations forgivable in a tribute to his late friend and editor. But there are numerous very similar accounts written by individuals who could hardly be counted among Blagosvetlov's friends. Perhaps the most "impartial" valuation of the social impact of Blagosvetlov's journal came from the censorship agencies, which in June 1862 suspended the publication of both *Russian Word* and *The Contemporary* for eight months.

*Russian Word*'s "extraordinary success" stemmed not so much from Blagosvetlov's own writings, though it was his articles that first garnered the close attention of the censors as early as September 1860, just three months after he had taken over the journal. It was primarily a result of his determined efforts to gather around the journal a group of like-minded contributors. Blagosvetlov firmly believed that "there must exist consent and accord between an editor and his authors; without them there is no idea, no results."[25] Like any successful editor, he had a good nose and was on constant prowl for fresh talent. Over the years, he recruited to his journal poet Dmitrii Minaev (1835-1889), critics Dmitrii Pisarev (1840-1868) and Varfolomei Zaitsev (1842-1882), historian Afanasii Shchapov (1830-1876), journalists Élie Reclus (1827-1904), Nikolai Shelgunov (1824-1891), Nikolai Sokolov (1835-1889), and Petr Tkachev (1844-1886), and writers Vsevolod Krestovskii (1840-1895), Nikolai Pomialovskii (1835-1863), Fedor Reshetnikov (1841-1871), and Gleb Uspenskii (1843-1902), to list just a few of the better-known names.[26]

Most of the journal's regular authors were young men, just 20-25 years of age, who grew and matured under Blagosvetlov's tutelage.[27] An 1865 statement by Pisarev, *Russian Word*'s most celebrated contributor, is very revealing in this respect: "If ... I now understand to a certain degree the duties of an honest litterateur, I must admit that this understanding was awakened and nurtured in me by Mr. Blagosvetlov ... [my] friend, teacher, and mentor to whom I owe my advancement and whose guidance I still need to this very day."[28] Blagosvetlov helped his younger co-workers find their voice, often recommending how best to present their materials and arguments. "The first and foremost condition of success," he advised one author, "is to enliven each article, to paint it in bright colors, so that it would catch the eye from afar. [Your article] doesn't have such bright colors, try finding them. ... The article will become livelier and will shine. ... Pepper, add more pepper." He also required them to be critical and bold. "Expose, attack any falsehood. Reveal abuses and smash the abusers with your words," he told the same author.[29] "Strike. Burn this rotten society," he urged another author, "hit them harder, harder."[30]

Each author brought to the journal his own particular talent, interests, style, and expertise. But it was Blagosvetlov who created a well-tuned orchestra out of this diverse group of soloists and carefully "conducted" each monthly "performance." He often suggested specific themes and assigned certain tasks to his "staff" authors, at the same time, granting them a considerable degree of responsibility and creative freedom. He defined the general direction of his journal and carefully planned the composition of each issue.[31] He shaped the journal's contents in terms of both the subject matter it addressed and the particular ways each subject was analyzed and organized.[32] In short, he was the real engine behind the journal's successes — and its failures (see fig. 2-2).

Fig. 2-2. Grigorii Blagosvetlov, c.1860s. A lithograph from the posthumous edition of his collected works. From G. E. Blagosvetlov, *Sochineniia* (SPb.: E. A. Blagosvetlova, 1882). Courtesy of BAN.

Nobody understood Blagovesvetlov's critical role in *Russian Word* better than the censorship agencies. Just three months after he had taken over the editorial office, a censor duly recorded, "the September [1860] issue of *Russian Word* clearly exhibits the anti-government direction that the journal is taking under Blagosvetlov's administration."[33] Thereafter, censors kept a watchful eye on the journal and its new editor. Blagosvetlov's surviving letters of 1860-61 are full of complaints about the interference of censors. 7 September 1860: "In the September issue, six articles have been forbidden, thanks to our executioners"; 16 October 1860: "We have suffered a pogrom at the hand of censors"; 4 March 1861: "*Russian Word* has nearly suffocated under the censorship pressure"; 12 December 1861: "The censors are eating us alive. Again [I've received] an admonishment. Again, [there are] threats to close the journal, once more, repressions against our contributors."³⁴ Extensive files devoted to *Russian Word* in the archives of various censorship agencies bear witness to the fact that Blagosvetlov's complaints had not been exaggerated, but fully justified.³⁵ Yet Blagosvetlov did not give up. A secret police report noted astutely: "He is a man of strong will and tough character. No matter how much the Main Directorate for the Affairs of the Press [the top censorship agency at the time] squeezes and pressures him, he nonetheless stands his ground and does not deviate from his goals and aspirations."³⁶

The following year the pressure increased even further. Annoyed by the rising tide of public criticism — the unintended consequence of the officially sanctioned *glasnost'* aptly manifested on the pages of *Russian Word* — Alexander II approved new "Temporary rules of censorship" issued on 12 May 1862.³⁷ The new rules allowed the minister of internal affairs and the minister of people's enlightenment (who at that time shared responsibility for censorship), by mutual agreement, to suspend the publication of any periodical suspected of "antigovernment direction" for a period of up to eight months. Five days later, the minister of people's enlightenment commanded the ministry's censorship bureau to intensify its control to prevent periodicals from "systematically denouncing everything done by the government and enticing public dissatisfaction with its actions."³⁸ A month later, together with *The Contemporary*, *Russian Word* was suspended for the maximum eight-month period allowable by the new rules.

To make matters even worse, several core contributors to the journal were arrested, imprisoned, and/or exiled. In April 1862, just a few weeks before the new censorship rules came into effect, Shelgunov was arrested and, after eight-month imprisonment in the infamous Peter-Paul Fortress, was exiled to a small town in the desolate north of the empire in the Vologda province (where he would spend the next thirteen years). In July, *Russian Word*'s star writer Pisarev also was arrested, for a pamphlet he had written in Herzen's support, and placed in solitary confinement in the Peter-Paul Fortress (where he would spend more than five years).³⁹

The same month, apparently panicked over the suspension of *Russian Word* and the arrests of its key authors, Kushelev-Bezborodko, whose name still graced the journal's title page as its publisher, decided to completely withdraw from the journal. He attempted to transfer the de jure rights to publish and edit *Russian Word* to Blagosvetlov (obviously with the latter's consent) and sent an appropriate petition to the censorship office.[40] The attempt stirred quite a storm in the government agencies that oversaw the press. Urgent consultations among high-level officials in the St. Petersburg censorship office, the Ministry of People's Enlightenment, the Ministry of Internal Affairs, and the secret police resulted in the explicit and unconditional prohibition for Blagosvetlov to assume the journal's editorship. But it turned out that under the existing regulations government agencies could not stop a purely "commercial" deal of transferring the ownership and publishing rights. Blagosvetlov immediately found a proxy editor, who would sign the necessary paperwork[41] and proceeded with the publication of the journal, officially only as its publisher.

Blagosvetlov's trick of hiding behind proxies to keep the journal running did not fool the authorities. On 9 January 1863, in response to a report by the secret police, the censorship bureau of the Ministry of Internal Affairs opened a special file, "On the *editor* of 'Russian Word' Blagosvetlov."[42] But, it seemed, at this point nothing could stop him: on 30 January 1863, after an eight-month hiatus, the journal's subscribers received a new issue. Furthermore, in June 1863, Pisarev, still in solitary confinement in the Peter-Paul Fortress, was permitted to resume writing for *Russian Word*, thanks to Blagosvetlov's herculean labors done largely behind the scenes, with Pisarev's mother acting as his proxy. Of course, before it could be published, every piece Pisarev had written had to go through a multi-layered screening: first, by the commandant of the Peter-Paul Fortress, then, by the St. Petersburg governor, then, by the Ruling Senate (which conducted the investigation of Pisarev's "crimes"), and finally, by the censor. Notwithstanding these hurdles, the July and August issues of *Russian Word* defiantly opened with lengthy essays by Pisarev on "Our university science," even though — due to censorship delays — both issues only came out (nearly simultaneously) in September.[43]

Fully justifying his aforementioned characterization by the secret police, Blagosvetlov stood his ground no matter what ammunition the government apparatus threw at him. In an announcement printed in the first 1863 issue of his journal he stated bluntly: "Beginning the fifth year of its publication, *Russian Word* is not changing its previous programme, its volume, or its moral character."⁴⁴ Yet certain things did change. The "resurrected" *Russian Word* acquired a new subtitle: the "learned-literary" (*ucheno-literaturnyi*) turned into a "literary-political" (*literaturno-politicheskii*) journal, an unsubtle statement of Blagosvetlov's intentions and goals.

Blagosvetlov kept the previous basic format that included three separate parts, each with its own pagination. The opening section was devoted to "Belles-lettres." But despite its name, it carried not only poetry and literary fiction, but also extensive analytical articles on history, politics, education, jurisprudence, economics, science, and philosophy. The second section, titled "Literary Review," dealt in its entirety with criticism and bibliography (the latter was collected in a special subsection, named "Bibliographical Leaf"). Again, despite its title, the actual contents of this part went far beyond literary criticism and presented pointed commentaries on recent publications in every field of scholarship, from politics, history, and economics to science and philosophy. The third part, named "Contemporary Review," consisted of three well-defined subsections: "Politics" discussed foreign affairs; "Domestic Chronicle," as its name made clear, surveyed the Russian scene; and the "Diary of an Ignoramus" contained satirical feuilletons on the "hot" subjects of the day.

Blagosvetlov himself wrote for nearly every section. But he divided the responsibility for particular sections among his core group of writers who contributed to the journal on a regular basis. Minaev kept "the diaries of an ignoramus"; Pisarev and Zaitsev supplied the bulk of the "literary review"; Reclus surveyed foreign politics; while Shchapov, Shelgunov, Sokolov, and Tkachev commented on historical and contemporary issues in economics, politics, law, science, administration, and philosophy both in Russia and abroad.

In running his journal, Blagosvetlov shared Herzen's conviction that "any successful polemical journal definitely must have flair for contemporaneity, [it] must have that delicate ticklishness of the nerves that is immediately irritated by everything which irritates society."⁴⁵ Critical, feisty, polemical assessments of every significant event in contemporary literary and social life became the trademark of *Russian Word* under Blagosvetlov's direction. The journal covered extensively the preparation, content, and implementation of the Great Reforms: the 1861 emancipation of the serfs, the 1863 new university statutes, the 1864 reforms of secondary education, and the complete reorganization of both the local government and the judiciary system launched the same year. It responded to the appearance of such controversial novels as Turgenev's *Fathers and Children* (1862) and Chernyshevskii's *What is to be Done?: From the Stories about New People* (1863). It did not miss the publication in Russian of major works by such eminent western intellectuals as Henry T. Buckle, Charles Darwin, John Stuart Mill, Carl Ritter, and Rudolf Virchow.

Yet *Russian Word* went way beyond mere reactions to that which "irritated" contemporary Russian society. As Tkachev recalled, "*Russian Word* served as a commonly accessible source of new ideas for social consciousness. Every new idea, every new word, every new, original answer to old questions found a place on its pages."[46] The journal identified and defined what society must consider "irritating" and, indeed, itself became a major "irritant" — and not only to the powers that be. *Russian Word*, according to Pisarev,

> has so many writing and reading enemies, it inspires such fiery and univocal hatred from the newspaper and journal doctrinaires who read it so attentively and intently that, in just a week after the publication of each new issue, all thoughts and even all separate expressions it contained have already been counted, measured, weighted, sniffed at, felt over, and taken under consideration.[47]

Indeed, the journal was constantly engaged in fierce polemics on almost every issue with not only Katkov's "conservative" *Russian Herald* (*Russkii vestnik*), Andrei Kraevskii's "liberal" *Annals of the Fatherland* (*Otechestvennye zapiski*), and Dostoevsky's "patriotic" *Time* (*Vremia*) and *Epoch* (*Epokha*), but also Nekrasov's "democratic" *The Contemporary* (see fig. 2-3).[48]

Much of the polemic centered on sociological and aesthetic aspects of Russian literature, which served as a convenient substitute for open discussions of Russia's pressing political, economic, and social issues. Continuing the tradition set by Vissarion Belinskii and further developed by Chernyshevskii and Dobroliubov, the journal's contributors argued militantly against "art for the sake of art" and for a utilitarian approach to music, literature, and the arts, as well as to the notion of beauty more generally. In an 1865 article, provocatively titled "The Destruction of Aesthetics," Pisarev forcefully declared: "aesthetics, or the science of beauty, has the right to exist only if *beauty* has any meaning independent of the limitless variability of personal tastes."[49] Following in the footsteps of John Stuart Mill, one of the contemporary intellectuals most cited and discussed on the journal's pages, *Russian Word*'s authors found such meaning in the "utility" of artistic works. They considered only "socially useful" literature to be truly beautiful — and thus true art. As Pisarev famously quipped in one of his manifestos, "*art cannot be its own goal, ... life is superior to art.*"[50] They even claimed that the greats of Russian literature — Pushkin and Lermontov — were completely useless to the modern reader: "To bring up young people on [reading] Pushkin is to make them into drone bees or sybarites."[51] In accord with these aesthetic principles, *Russian Word* published almost exclusively "socially responsible" fiction, while its leading critics hailed those literary works, which, like Chernyshevskii's *What is to be Done?*, could, in Pisarev's words, "facilitate the intellectual or moral perfection (*sovershenstvovanie*) of humanity."[52]

Not only was the journal largely filled with the writings of the young, it addressed and appealed to Russia's young generation — the "children" and the "new people" immortalized in Turgenev's and Chernyshevskii's novels, respectively. The portrayal of Bazarov and Rakhmetov, the main protagonists of the two novels, became the sparring ground for the literary critics of the day, while their names became a battle cry of the youth. In the essays of its leading literary critics, Pisarev and Zaitsev, *Russian Word* consistently presented Bazarov and Rakhmetov as the rolemodels young Russians should emulate. The journal published a number of short stories and novels, which in various ways "recreated" the major protagonists (new people) and the major conflicts (between fathers and children) first depicted by Turgenev and Chernyshevskii.[53] *Russian Word* also mercilessly criticized those contemporary authors who in their own fictional works "attacked" and "mocked" the ideas and ideals expressed in the images of Bazarov and Rakhmetov.[54] In a special report on the journal to the Main Censorship Directorate, censor Alexander Nikitenko noted, "it will not be an exaggeration to say that the present young generation to a considerable degree is getting first educated on the ideas of *The Bell* [and] *The Contemporary* and is finishing its education on the ideas of *Russian Word*."[55]

Fig. 2-3. A large-format cartoon of the Russian journalistic scene that appeared as a supplement to the satirical journal *Splinter* (*Zanoza*) on 3 March 1863, under the title "A concerto in C-dur(nom) tone." The title was a play on words: "C-dur," a major musical scale, and a Russian expression, *"durnoi ton,"* a loan translation of the French *mauvais ton*. Drawn by Petr Borel', a well-known artist, it depicts an orchestra made up from the editors of major journals and newspapers, whose titles are written either on the musical instruments they play or on their clothing. In the upper left corner is a ringing bell – a clear allusion to Alexander Herzen's eponymous newspaper. A Muse sitting in the chair in the upper right and closing her ears in disgust apparently signifies Russia. Petr Valuev, the minister of internal affairs (second row on the right) holds a tuning-fork, while Vasilii Tsee, the head of the St. Petersburg Censorship Committee (on Valuev's left), conducts the concerto. Three censors (in the upper row, right to left) Alexander Petrov, unidentified, and Vladimir Beketov moderate the performance.

Periodicals and their editors. First row, left to right: 1. Grigorii Eliseev, on top of him Amplii Ochkin, *Essays*, a daily newspaper; 2. Vasilii Kurochkin, *Spark*, a satirical journal; 3. Petr Men'kov, *Military Collection*, an official journal of the War Ministry; 4. Ivan Balabin, *People's Wealth*, a newspaper; 5. Unidentified; 6. Viktor Askochenskii, *Home Conversation*, a journal; 7. Nikolai Strakhov, *Time*, a journal published by the Dostoevsky brothers, the cartoon depicts not the editor, but the journal's leading contributor, whose publications led to shutting the journal down in just a few months after the appearance of the cartoon; 8. Mikhail Rozengeim, *Splinter*, a satirical journal; 9. Alexander Rotchev, *Herald of the City Police*, a newspaper; 10. Ivan Goncharov, *Northern Post*, a daily newspaper of the Ministry of Internal Affairs; 11. Il'ia Arsen'ev, the editor of the political section of *Northern Post*, who in 1865 would found the first Russian tabloid, *Petersburg Leaf*; 12. Dmitrii Romanovskii, *Russian Invalid*, a newspaper of the War Ministry.

Second row, left to right: 1. Valentin Korsh, *St. Petersburg News*, a daily newspaper; 2. Konstantin Trubnikov, *Burse News*, a daily newspaper; 3. Nikolai Pavlov, *Our Time*, a newspaper; 4. Viktor Kapel'mans, *Journal de St.-Pétersbourg*, a French-language newspaper; 5. Nikolai Pisarevskii, *Modern Word*, a daily newspaper; 6. Mikhail Katkov, *Russian Herald*, a journal, and *Moscow News*, a newspaper; 7. Ivan Aksakov, *Day*, a weekly newspaper; 8. Aleksei Pisemskii, *Reading Library*, a journal; 9. Grigorii Blagosvetlov, *Russian Word*, a journal; 10. Nikolai Nekrasov, *The Contemporary*, a journal; 11. Andrei Kraevskii, *Voice*, a daily newspaper; 12. Pavel Usov, *Northern Bee*, a daily newspaper; 13. Al'bert Starchevskii, *Fatherland's Son*, a daily newspaper; 14. Vladimir Zotov, *Illustration*, a journal. Courtesy of RO IRLI.

The journal was indeed read by the students of gymnasiums, seminaries, universities, and other schools of higher learning, as well as their recent graduates who had become civil servants, military officers, teachers, and doctors — in short, by the fledgling Russian intelligentsia. It was received by the public libraries recently established throughout the empire. Its subscribers resided all over Russia's vast territories, from Warsaw in the west to Nerchinsk in the depths of Siberia in the east and from the recently "pacified" Caucasus in the south to Vologda in the north. Russia's future Nobelist Ivan Pavlov, who in 1864 entered a theological seminary in Riazan' (a small town some 150 kilometers south of Moscow), fondly remembered many years later: "Can one forget the passion with which you captured a long-desired book? I can now see clearly the scene as several of us seminarians and gymnasium students stand for hours on a dirty, cold autumn day before the locked door of the public library in order to be the first to capture an issue of *Russian Word* with an article by Pisarev."[56] As a professor of Kharkov University described the journal on the eve of its demise in January 1866, "It is a 'New Testament' for the majority [of our students]."[57]

Yet the popular characterization of the journal's contributors (and readers!) as "nihilists," "insolent boys," "school drop-outs," and "whistlers,"[58] who rejected things for the sake of it and who were set on critiquing and destroying that which their "fathers" held dear, was little more than a polemical tool employed by the journal's numerous enemies.[59] Indeed, the "nihilists" called themselves "realists."[60] As Pisarev explained in a lengthy article innocently titled "A Stroll through the Orchards of Russian Literature," neither he, nor his fellow journal authors were engaged in "criticism for criticism's sake." *Russian Word*'s critical publications aimed, first and foremost, at "the establishment and consistent deployment in the assessment of all current events in life, scholarship, and literature of a particular worldview."[61] A major component, indeed, the very foundation of that "particular worldview," was science.

## Science in *Russian Word*

Students of Russian history have noted regularly that during the late 1850s and the 1860s science moved from the periphery of Russian thought to its very center.[62] As many other changes in the country's life that were embodied in the Great Reforms, this shift was largely spurred by Russia's crushing defeat in the Crimean War, which had challenged the empire's major, if not its only, claim to the status of a great world power — its military might. Searching for the causes of the Crimean War debacle, many contemporary commentators pointed to the technical superiority of the French and, especially, the British armed forces: a steam-powered fleet, advanced weaponry (guns, cannons, and ammunitions), military engineering, and communications, to name but a few of the most often cited examples. Indeed, the emperor in St. Petersburg not infrequently learned about the latest developments on the war's frontlines from British newspapers. Thanks to the telegraph lines built during the war, by April 1855 *The Times* had been printing the latest news from the Crimea nearly instantaneously, and, within just ten days or so after the publication, commercial steamers would deliver the London newspaper to the Russian capital. By contrast, special carriers dispatched by Russian commanders from Sebastopol took nearly a month or even longer (especially, in the fall and spring months, when what in Russia counted as roads became virtually impassable) to make the 2,200-kilometer journey to the headquarters on horseback. Needless to say, similar timetables applied to the delivery of necessary supplies, reserves, and orders dispatched from the headquarters to the opposing armies in the Crimea.

Many contemporary commentators attributed this indisputable technical superiority to the extensive development of science in France and Britain over the course of the prior six decades. A few months after the death of Nicholas I, while the war was still being fought, the minister of people's enlightenment Avraam Norov reportedly told his subordinates: "Science, gentlemen, has always been one among our most important needs, but now *it is the first*. If our enemies have an advantage over us, it is only due to the power of [their] knowledge."[63] Numerous contemporary publications, as well as later memoirs, expressed the same sentiment. As Shelgunov stressed on the pages of *Russian Word*: "The Crimean War has convinced us that Russian society lacks a lot [of things] to stand on the same level with Europe. [The war's] failures and losses have forced us to look closely at what exactly we are lacking, and it turned out that we lack knowledge."[64]

Shelgunov's conviction was widely shared by many diverse groups that formed the Russian educated public. Certainly, different groups and individuals had their own reasons for this new-found interest in, and the exalted appraisal of, science. For some, especially among Russia's budding capitalists and "enlightened bureaucrats" (like Norov), it was the perceived importance of scientific knowledge in the modernization of the empire's dysfunctional administration, antiquated military, and obsolete industry, agriculture, transport, communications, and trade.[65] For others, it was the perception of science as the newest embodiment of the Enlightenment belief in *reason* as the motive force of human history, which, through the spread of knowledge and learning, would guide the people to "happiness and perfection."[66] Still for others, it was science's role as the source of knowledge that could liberate the people from archaic traditions, ignorant superstitions, and religious dogmas — from "metaphysical platitudes," as one *Russian Word* author euphemistically named them.[67] This and other euphemisms employed by contemporary writers, however, did not fool the censor. As the 1864 censor report cited above emphasized, the Russian public turned to reading scientific literature "not with the goal of learning the positive conclusions of science per se, but in search of the repudiation of everything that constitutes another area that stands above science, the area of [religious] Faith," which, according to the same report, "cannot but lead to shaky political views."[68]

All of these reasons certainly played a part in the expansion of science in post-Crimean Russia.[69] From the late 1850s through the 1860s, this expansion was most visibly manifested in the proliferation of new scientific societies, such as the Moscow Society of Russian Physicians (1858), the Russian Entomological Society (1859), the Kharkov Medical Society (1861), the Society of Enthusiasts for the Natural Sciences (1863), the Russian Society for the Acclimatization of Plants and Animals (1864), the Moscow Mathematical Society (1864), the Russian Technical Society (1866), and the Russian Chemical Society (1868).[70] In the same period, the number of books and periodicals devoted to the natural sciences increased rapidly, as did the number of students who graduated annually from natural science departments at the empire's universities (and a few other schools of higher learning, such as the IMSA, the Mining Institute, and the Technological Institute in St. Petersburg). According to some estimates, that number had more than tripled from 1856 to 1865. And it was this very faith in science that guided the profound reforms of the IMSA during Florinskii's final years as a student.

More important for our story, however, was the concurrent move of science from the rarefied circles of professional academics, inhabiting the small number of scientific institutions the enormous Russian Empire could boast at the time, to a variety of new venues, which made "science," "knowledge," and "learning" the buzzwords of the Russian educated public.[71] In the late 1850s, lectures on scientific subjects became available to the public and spilled from university auditoria to theaters, bookstores, private study circles, Sunday schools, libraries, museums, and hospitals (recall, for instance, Evgenii Pelikan's public lectures on biology, which Florinskii attended at the Hospital for Menial Workers in 1860).[72] Such lectures were delivered not only by well-known professors, but also by students, school teachers, military officers, civil servants, and practicing doctors. Numerous visitors flocked to scientific museums and exhibits. As a result of the 1864 educational reforms, the natural sciences entered secondary school curricula, previously thoroughly dominated by the "classics" — languages, history, and literature.[73] Publications on science subjects began to appear not only in the form of specialized monographs, textbooks, and articles in professional journals issued by government institutions and learned societies. New commercial presses and privately-owned periodicals, which mushroomed during the same decade, started to play a major role in publishing scientific texts.

A significant portion of these texts was written, compiled, translated from foreign languages, edited, and commented upon not by professional academics, but by journalists, litterateurs, teachers, physicians, civil servants, and students. The rapid growth of this "scientific-popular literature" (*ucheno-populiarnaia literatura*), as it was called at the time,[74] clearly reflected mounting public demand. Ever sensitive to market trends Mavrikii Vol'f, founder and owner of one of the most successful private publishing houses, began to issue popular magazines, such as *Around the World* in 1861 and *Nature and Earth Sciences* in 1862.[75] Edited by Florinskii's classmate Pavel Ol'khin both magazines were devoted specifically to "earth studies, natural sciences, latest discoveries, inventions, and observations."[76] An 1863 advertisement for the latter magazine appended to *Russian Word* emphasized: "The names of the authors whose works have previously been published in this journal, such as [Carl] Vogt, [Charles] Darwin, [Jacob] Moleschott, [George H.] Lewes, and others, provide the very best indication of its character and direction."[77] An 1864 censorship report, "On the direction of periodicals," duly noted: "Lately, our society took to literature on the natural sciences with particular eagerness."[78]

Although from time to time practically all "thick" journals carried "scientific-popular" texts, *Russian Word* far outpaced all others, especially after its revival in 1863. "It could be said without a hint of exaggeration that the popularization of science constitutes the most important, world task of our century," declared Pisarev, emphasizing that "a good popularizer, especially in Russia, could bring more benefits to society than a talented researcher."[79] Nearly all of the journal's core contributors, including Blagosvetlov, Shelgunov, Shchapov, and Zaitsev, shared Pisarev's conviction and actively participated in this endeavour. Their scientific-popular texts took two distinct forms: an extensive essay (often serialized in several consecutive issues) on a particular subject and a shorter review of a specific publication (mostly in Russian, but occasionally in foreign languages). In 1863-1865, *Russian Word* published no fewer than 39 lengthy essays and 63 book reviews on various scientific subjects.

The sheer number and volume of scientific-popular texts clearly attest to their importance to the journal's editor and contributors. As a rule, the essays were published in the first part of the journal, the "Belles-Lettres," and, occasionally, as a stand-alone piece in the "Literary Review." Sometimes they figured prominently on the journal's opening pages, thus emphasizing and testifying to their particular significance. The shorter reviews were collected in the "Bibliographical Leaf," which during this period was largely written by Zaitsev.[80]

A student of the Moscow University Medical School, Zaitsev moved to St. Petersburg for family reasons in December 1862.[81] He continued his education by attending classes at the IMSA and soon joined the

*Russian Word* editorial team. He quickly became one of the journal's leading literary critics, and, thanks to his medical studies, a "staff" reviewer of various scientific publications. In one of his reviews, Zaitsev candidly explained his approach to this task: "journalistic reviews [of a publication] could acquire their own separate significance, if, relegating purely bibliographical goals to a secondary plane, they expound certain ideas, which — though only tangentially touching upon the reviewed publication — have their own independent importance."[82] What were those ideas that Zaitsev and his fellow writers expounded upon in their scientific-popular texts and what "independent importance" did they have?

The subject matter of these texts points to possible answers to these questions, for not *all* of the natural sciences found equal place on the pages of *Russian Word*. Unlike, other "thick" journals (*Annals of the Fatherland*, for instance), *Russian Word* carried no essays and very few reviews dealing with mathematics, physics, astronomy, and chemistry, i.e. those disciplines that studied the "physical" universe. By contrast, fields that focused primarily on the "human" universe, such as biology, geography, physiology, psychology, medicine, hygiene, and anthropology, commanded the close attention of the journal's authors. This particular focus suggests that, unlike for Vol'f's magazines, for *Russian Word*, the popularization of science was not an end in itself. It was primarily a means to tackle a variety of other issues.

As historians of the journal have noted, discussions of science in *Russian Word* often served as a substitute for discussions of contemporary politics that became virtually impossible under the censorship pressure.[83] They promoted the utilitarian view of science as a tool of modernisation (of industry, agriculture, trade, state administration, transport, and armed forces). They also served as a vehicle for advancing Enlightenment ideals of education, learning, and knowledge and advocating for a "materialistic worldview" to undermine religion, one of the three cornerstones of the Russian Empire, clearly expressed in its official slogan — "Orthodoxy, Autocracy, Nationality (*Pravoslavie, Samoderzhavie, Narodnost'*)."

Yet, first and foremost, such discussions aimed at substantiating the role of the natural sciences as *the* foundation for understanding contemporary (especially, Russian) society and for discovering "natural" laws that govern its history, economics, and politics and that must ultimately define its future. In an 1863 article titled "The Foundations of Social Life," Shelgunov clearly articulated this aim:

> Ultimately, there is only one [kind of] knowledge — the study of nature, i.e. [the study] of its constituent matter and its properties. Humans are only a part of nature, hence, without studying nature, [one] cannot learn the laws that govern both separate individuals and entire societies. This is why the natural sciences have lately acquired such significance and have attracted the attention of all thinking people.[84]

The very titles (to say nothing of the contents) of numerous other publications appearing on the pages of *Russian Word*, such as Zaitsev's "The Natural Sciences and Jurisprudence" (1863), Shelgunov's "The Conditions of Progress" (1863), Petr Bibikov's "The Boundaries of Positive Knowledge" (1864), Nikolai Serno-Solov'evich's "Does the Contemporary State of Knowledge Require a New [Social] Science?" (1865), and Shchapov's "The Natural Sciences and the People's Economy" (1865-1866), clearly demonstrate that for the "literary-political" *Russian Word*, science had become "political" in more than one sense. In the views of its authors, science held the keys to answering all the questions and solving all the problems in Russian life.[85]

A historical coincidence made it possible for *Russian Word* contributors to utilize science for "political" purposes. In the middle decades of the nineteenth century, just as Russia was preparing and then implementing its Great Reforms, several "scientific revolutions" unfolded in western Europe.[86] These revolutions redefined science as an intellectual and social activity aimed at acquiring knowledge about nature in *all* of its manifestations. They erected rigid epistemic and social boundaries — embedded in the explicit practices of observation, measurement, and experiment, undergirding "objective," "exact," "real," "scientific" method — that separated science from other, "subjective," "ambiguous," "fictional," and "speculative" forms of inquiry, such as religion, philosophy, and literature. They involved the emergence of the *scientist* as the sole purveyor of "true" knowledge that need not be simply taken on faith, but could be tested, verified, and proved (or disproved) by other scientists. A series of startling discoveries made during this period, ranging from the laws of thermodynamics to the synthesis of organic compounds from inorganic ingredients, convincingly demonstrated the material — physical and chemical

— unity of nature. More important, however, new discoveries firmly placed humans *into* nature — not as its God's anointed kings, but as its constituent part, subject to the same laws that govern all of nature's other parts, be they stars, rocks, plants, or animals.

From the 1840s through the 1860s, a new field of "experimental physiology" demonstrated the physical and chemical unity of humankind with the rest of the universe, presenting the human body as a physical and chemical machine dependent on the flows of energy and built from the same chemical elements and structural units (the cells) that comprised all other organisms. The budding field of "experimental psychology" began to bridge the gap between human and animal "souls," indicating that the faculties of reason, will, and speech were not exclusive, God-given human qualities, but mere extensions of similar capacities in other animals. The last and possibly the most impactful field to develop during this period was "evolutionary biology," which strongly suggested that humans were not God's favorite creation, but merely a branch of the animal kingdom that had evolved over countless millennia in the process of natural selection and the struggle for existence, as did all other now-living organisms.

It was exactly these three areas — human physiology, psychology, and evolution — that became a special focus of *Russian Word*'s authors. They promoted and debated the works of biologist Darwin, geologist Charles Lyell, anthropologist Armand de Quatrefages, geographer Ritter, cytologist Matthias J. Schleiden, physician Virchow, physiologist Carl Vogt,[87] and psychologist Wilhelm Wundt. They regularly reviewed and referred to the writings by such well-known authors as Alfred Brehm, Ludwig Büchner, Thomas Henry Huxley, George H. Lewes, Jacob Moleschott, and Carl G. W. Vollmer (under the pseudonym "Dr. W. F. A. Zimmerman"), who actively popularized the latest developments in these three fields. These works became their sacred texts, while the authors of these texts became the apostles of their new-found faith in the ability of the natural sciences to solve all social problems and cure all societal ills.

A decade later, referring to his work in *Russian Word*, Zaitsev declared passionately, "every one of us would have gladly gone to the scaffold and given his head for Moleschott and Darwin."[88] This fanatical zeal clearly did not stem from specific scientific contributions by Darwin and Moleschott, no matter how important they might have been in and of themselves. Rather, the names of these and other eminent scientists and science popularizers became potent symbols of an entirely new worldview generated by the scientific revolutions of the mid-nineteenth century. As a commentator in *The Contemporary* perceptively observed in 1864: "the revolution (*perevorot*) effected by the natural sciences in the worldview of [our] society — in its opinions, notions, mores, in a word, in its inner world — is likely even greater and more fertile than all the material goods and comforts [they had brought about]."[89] This scientific worldview discarded the traditional answers offered by scholastic religion, speculative philosophy, and fictional literature to the key existential questions that had boggled humanity's best minds from time immemorial: Who are we? Where did we come from? Where are we going? It provided clear-cut naturalistic answers: humans are a particular kind of animal, which had evolved from some other, "lower" animals and which, in the course of time, will either go extinct, or "progress" further.

This new worldview offered a possibility of answering numerous questions about the life and death of not only individual human beings, but also societies, states, civilizations, and humanity as a whole. Its logic seemed incontrovertible. Humans are a part of nature and, hence, knowledge generated by the natural sciences must be applicable to *all* phenomena of human life. Not only individual — biological, physiological, psychological — life, but also social — economic, political, moral — life must obey the inexorable "laws of nature," which the natural sciences had discovered and were still discovering almost daily.

Many Russian intellectuals of the time began to advocate for the transfer of methods, insights, and knowledge from the natural sciences to the fledgling "science of society."[90] As early as 1848, Nikolai Kashkin, a member of the "Petrashevskii circle," a study and debate society devoted to the spread of socialist ideas whose membership included St. Petersburg young intellectuals (such as Dostoevsky), delivered a speech to his "co-conspirators" gathered at his apartment for their regular Friday dinner. He declared that "owing to the natural sciences, man has learned the laws of nature; what remains is to apply them to the science of society. When man accomplishes this, humanity will have the law of happiness and perfection."[91] This speech figured as the main proof of Kashkin's "crimes" at the Petrashevskii trial and cost him four years of hard labor in Siberia.[92] A decade later, Nikolai Chernyshevskii expounded similar ideas in print, in his 1860 article "Anthropological Principle in Philosophy" published, though anonymously, in *The Contemporary*. Chernyshevskii noted pointedly that, unlike the English word "science" that referred exclusively to the natural sciences, its Russian analogue "*nauka*" was applied to all fields of scholarship. But, he stated, "In the not too distant past, the moral sciences (*nravstvennye nauki*) could not actually possess the content that would justify the title of science they have been given." "Now the situation has changed considerably," he continued, "the natural sciences have developed to such a degree that [they] provide a mass of material for the exact solution of moral questions."[93] Five years later, in an article published in the first 1865 issue of *Russian Word*, Serno-Solov'evich argued for the urgent need to create a "social science" (*obshchestvennaia nauka*) modeled after and based on the natural sciences. Surveying the current state of research and theoretical work in economics, jurisprudence, history, criminology, and demography, he came to the conclusion that "between a social science and the natural sciences must occur — the sooner, the better — an exchange of their main strengths, i.e. the former should borrow from the latter their perfect method and, in turn, give the latter its [own] perfect aspirations." "No economist, no historian could possibly continue their research," he insisted, "without, at least, some basic knowledge of the natural sciences."[94]

A leading knight in *Russian Word*'s crusade to promote and popularize the natural sciences was Pisarev. Blagosvetlov met the twenty-year-old student of his own alma mater, the "historical-philological" school of St. Petersburg University, sometime in the fall of 1860. On Polonskii's recommendation, Pisarev brought for the editor's consideration his translation of Heinrich Heine's poem "Atta Troll." After a lengthy conversation, Blagosvetlov bought the poem, but urged the young man to write literary criticism instead of poetry. Pisarev enthusiastically agreed.[95] The December 1860 issue of *Russian Word* carried both the poem and Pisarev's first review of a collection of poetry translations.[96] Obviously, Blagosvetlov saw Pisarev's potential and liked his writing style: beginning with the February 1861 issue, *Russian Word* carried

Pisarev's reviews of various publications,[97] as well as his articles on historical and philosophical subjects, in practically every issue.

However, in July, along with a more than 100-page-long essay on the Apollonius of Tyana,[98] Pisarev published a lengthy overview of Jacob Moleschott's *Physiologische Skizzenbuch* (*Physiological Sketches*) that had come out in German only a few months prior.[99] It is unclear from available materials what prompted Pisarev, who had had no training in the natural sciences whatsoever, to write on a subject so far removed from his personal interests and expertise.[100] But, from this time on, he would divide his attention almost equally between reviewing belle-lettres and science publications. Two months later, under the title "The Process of Life," the journal carried Pisarev's digest of the latest edition of Vogt's monumental *Physiologische Briefe* (*Physiological Letters*).[101] And in February 1862, Pisarev published another lengthy article, titled "Physiological Pictures," based on the first volume of Büchner's *Physiologische Bilder*, which had appeared just a few months earlier.[102]

Taken together, Pisarev's overviews of the three books presented the latest achievements in the understanding of the physical and chemical foundations of "vegetative life," — "*das vegetative Leben*," Pisarev put in brackets, lest his readers misunderstood the term. He stressed that in their analyses of basic physiological processes — breathing, digestion, and blood circulation — the three authors viewed the human body as both a complex mechanism and a chemical laboratory, which functions according to known physical and chemical laws. These laws, Pisarev emphasized, leave no place for mysterious "vital forces" and similarly outdated "folk traditions and superstitions." He astutely summarized Moleschott, Vogt, and Büchner's views by defining "vegetative life" as "nothing more than a constant change of matter, while preserving a certain form" [9: 8]. In each essay, Pisarev emphasized that many specific questions regarding the basic mechanisms of life so far remain unanswered, but in time the science of physiology would be able to find the answers. He was absolutely certain that similar physical and chemical processes (i.e. the exchange of matter and energy between the organism and its environment) would soon explain not only "vegetative life" but also "animal life, i.e. perceptions and processing of impressions, the work of the nervous system" [9: 6]. He sympathetically assessed, for example, Vogt's "attempts to bring together the fields of psychology and physiology" by explaining the role of the nervous system in blood circulation. "We can assume and hope," he concluded, "that in due time such notions as psychical life and psychological phenomena will [also] be disaggregated into their constituent parts. Their fate has been decided: they will go the way of the philosopher's stone, the vital elixir, and the vital force... Words and illusions die out, facts remain" [9: 15].

Pisarev's essays effectively popularized the newest physiological knowledge, stressing its importance for medicine and the correct "realist" understanding of the human body. But it was Shelgunov who first attempted to apply this knowledge of "vegetative life" to "social life."[103] Two months after the last of Pisarev's essays had appeared, Shelgunov published in two consecutive issues of the journal a 100-page article on "The Unprofitability of Ignorance."[104] The bulk of his treatise dealt with the role that lack of knowledge had presumably played in defining various phases in human history. But his approach to human societies was decidedly "physiological." He was convinced that:

> As physiology defines the conditions necessary for the development of every separate organism, defines what is harmful and what is useful for it and what conditions enable its separate, continuous, and healthy existence, so too history — the physiology and pathology of the collective man — defines those inescapable laws that regulate the healthy development of a social organism [4: 5].

Physiology tells us, Shelgunov emphasized, that "human life is completely dependent on and completely defined by the necessary conditions of the organism itself and its mutual relations with the external environment." "Society is an assemblage of separate organisms, it is a collective human-being," he postulated, "the more advanced is every individual, the more strength and the larger capital of intellect, talents, and knowledge the assemblage has, while the weaker and poorer are the individuals, the weaker is the collective organism" [5: 24]. But "an isolated physiological fact" about the life of an individual, such as the fact that humans need food to survive, he explained, "says nothing about the impact of food shortage on the entire society, or its separate parts. Here to the rescue comes statistics — this physiology of the collective human-being" [5: 25].

Developing further his analogy between a "separate physiological human being" and a "social organism," in the next article that appeared in the June issue under the title "The Conditions of Progress," Shelgunov attempted to describe those "physiological laws" which could account for the differences in the "progress" of such social organisms as nations.[105] As the difference between "a genius and an idiot" is obviously defined by "different qualities of their brains," he proclaimed, so too the difference between developed and undeveloped nations must be defined by the differences in their "collective brains." He rejected the customary explanation of national differences as the result of differences in food and/or climate, arguing that a "national type" remains unchanged, despite the people's movement to different climates or consumption of different foods. For Shelgunov, a "national type" was defined by "intellectual and moral strengths," which derive directly from the development of "brains and nerves."

Since the most important conditions for such development in an individual, according to Shelgunov, are "personal freedom" and "material wellbeing," the same conditions would define the development of collective "brains and nerves" in a nation. However, since only a certain number of individuals in the course of their life fully develop their brains and put them to good use in the "advancement of civilization," he surmised, it is the relative number of such individuals in a nation that defines the nation's "collective brain." To illustrate this point, Shelgunov actually calculated the total mass of brains in the populations of Britain, France, Austria, and Prussia. Then he estimated the number of individuals who, in his opinion, contribute to the national development (including in this category what we today would call a "middle class") and the total mass of their brains. By dividing the first number by the second he created an index of "national intellectual development" and discussed the means by which this index could be increased, thus ensuring the progress of not only separate nations, but also humanity as a whole.

Yet the application of physiological knowledge to the understanding of human societies could take the *Russian Word* authors only so far. The newest addition to the scientific revolutions — Darwin's concept of the origin of species — offered seemingly a much better opportunity for the transfer of knowledge generated by the natural sciences to the understanding of human nature and human societies.[106]

## Socializing Darwin

Published in London in late 1859, under the title *On the Origin of Species by Means of Natural Selection, Or the Preservation of Favoured Races in the Struggle for Life*, Darwin's opus magnum first became popularly known in Russia through its 1860 German and 1862 French translations prepared by Heinrich Georg Bronn[107] and Clémence Royer,[108] respectively.[109] To be exact, the very first extensive introduction of Darwin's theory to the Russian educated public came from the Russian translation of a French-language review of his book by eminent Swiss anatomist and zoologist René-Édouard Claparède. This voluminous review appeared (without identifying either its author, or its Russian translator) in the last two 1861 issues of *Reading Library*, a popular literary journal.[110] But the first Russian translation of *The Origin* by Sergei A. Rachinskii, professor of botany at Moscow University, appeared only two years later, in early 1864.[111]

Presenting his "abridged" concept to the world, Darwin focused exclusively on "the origin of species *in the plant and animal kingdoms*," as both the German and the Russian translations of his book emphasized in their very titles. He merely noted in passing at the very end of the book that his concept could also throw "light ... on the origin of man and his history."[112] Long before Darwin himself would shed that light in his 1871 *The Descent of Man*, almost immediately after the publication of *Origin*, Lyell, Huxley, Vogt, and many others took on the task. In just a few years, they demonstrated the "antiquity of man," striking anatomical similarities between humans and apes, and the "unity of the human type."[113] New paleontological, embryological, and anatomical evidence uncovered and compiled by these scientists strongly suggested that humans had evolved from ape-like ancestors over the course of hundreds, if not thousands, of millennia as a result of natural, sexual, and, perhaps, even artificial selection arising from the continuous struggle for existence. This evidence also implied that the processes of divergence and extinction, which, according to Darwin, accompany the origin of new species in plants and animals, had also occurred in the course of human evolution. It indicated that all of the processes, conditions, factors, and mechanisms that Darwin had discovered in his study of plant and animal species — the "laws" of evolution

— had likely been at work in the past history of the human species and would probably continue to play a role in the future development of humankind.

Darwin's theory, as well as works by his followers, became a special focus of attention for the *Russian Word* authors. Rachinskii's translation of *Origin* came out in the first week of January 1864, a fact enthusiastically greeted barely a week later on the pages of the journal's first issue of that year.[114] In a six-page announcement, Zaitsev excitedly proclaimed that "Darwin's work is destined to create a new epoch in science, and, by demolishing the crumbling old foundations of the natural sciences, to open for them the unlimited [new] horizon ahead." He foresaw "numerous new discoveries, which till now have been beyond the limits of science and human efforts, … made possible by Darwin's theory." Zaitsev illustrated "the exactness of observations and the tireless checking of experiments," which, in his opinion, "constitute the supreme qualities of the naturalist," by excerpting from Darwin's book a section on the slave instinct in ants. He lamented the "poor quality" of the Russian translation, noting that Rachinskii's text "sometimes requires its own translation into understandable Russian." He promised that "in the subsequent issues of *Russian Word*, we would try to present the essence of this new theory and to highlight its application that had already been made by the London professor [T. H.] Huxley in his wonderful lectures *On the Place of Man in Nature*."[115]

*Russian Word* made good on this promise. Already in the March "Bibliographical Leaf" Zaitsev alerted his readers to the appearance of two Russian editions of Vogt's *Lectures on Man, His Place in Creation and in the History of the Earth*, issued nearly simultaneously by two different publishers.[116] More important, however, starting with the April issue, the journal published five (!) lengthy essays by Pisarev under the common title "Progress in the World of Animals and Plants."[117] The essays totaled 214 pages — more than half the length of Darwin's entire book in its Russian translation — and offered the readers not just the "essence of Darwin's theory," but in fact much of the content of his book.

Pisarev carefully explained Darwin's notions of natural, artificial, and sexual selection (*vybor*)[118] and the "struggle for life." He detailed all of the extensive evidence (anatomical, geographical, geological, embryological, etc.) that Darwin had compiled in support of his concept.

More to the point, Pisarev did "translate" Rachinskii's text into "readable Russian." His easy-going and lively style certainly made Darwin's ideas and arguments much more accessible and comprehensible to a general reader than Rachinskii's academic translation (which Pisarev did not neglect to criticize bitterly in his essays). Given the journal's circulation, it would not be too far off the mark to suggest that many more people in Russia first learned about Darwin's theory from Pisarev's essays than from Rachinskii's translation.[119]

Although in his treatise Pisarev stayed mostly within the contents of Darwin's book and did not venture into its possible "social applications," one can argue that he effectively "socialized" Darwin's entire concept already in the very title chosen for his essays: "*progress* in the world of animals and plants." As Shelgunov's 1863 article on "The Conditions of Progress" discussed on the preceding pages makes abundantly clear, at the time the word "progress" was first and foremost applied to human history to describe the perceived stages in its unfolding from "low," "primitive," and "barbaric" to "higher," "more advanced," and "more civilized" human societies.[120] Pisarev's title implied that the progression from "lower" to "higher" was indeed the essence of Darwin's *Origin*. As he put it in the text: "progress is a direct consequence of the struggle and competition [for existence]" [7: 14]. It seems quite likely that Pisarev's focus on "progress" was inspired by the French version of Darwin's *Origin* produced by Royer, who actually "translated" the book's original subtitle "The Preservation of Favoured Races in the Struggle for Life" as "*des lois du progrès chez les êtres organisés*" (the laws of progress in organized beings).[121]

Darwin himself, however, used the word "progress" less than twenty times in the 500-plus pages of his book, and in most cases it denoted nothing more than a forward movement of time or human thought, as in such expressions as "the progress of geology" and "the progress of events."[122] Only twice in the entire volume, first in its "summary" and second in its "conclusions," did he use the word to mean development from a "lower" to a "higher" form.[123]

Pisarev's title then unobtrusively equated the origin of species in animals and plants, as described and analyzed by Darwin (and interpreted by Rachinskii and Royer), with the "progress" of human history, as described and analyzed by contemporary historians such as, for instance, Pisarev's favorites Henry T. Buckle and August Comte, or Blagosvetlov's favorite Thomas B. Macaulay. "Progress undoubtedly exists in the organic world," Pisarev stated bluntly, "this fact is indisputable" [7: 35]. He, however, did not apply Darwin's theory to explain human history. Instead he used human history to reinterpret Darwin's descriptions and explanations of certain facts and processes observed in nature. Not surprisingly, such anthropomorphic reinterpretations appeared particularly "fruitful" in Pisarev's retelling of Darwin's analysis of animal instincts, especially the instincts of such "social" insects as ants, to which Pisarev devoted the entire fourth essay (i.e. more than twenty per cent of his treatise: 46 out of 214 pages of its text).[124]

Pisarev managed to ascribe "personal intelligence, individual inventiveness, the variety of characters and preferences, rational upbringing, the succession of generations that leads to the change of habits, a developed social life with mistakes and deviations, the ability to exploit various situations, [and] the capacity to participate with the conscious efforts of reason in the progress of its own kind" to ants [7: 46]. Furthermore, he applied to ants the same categories he had used in his 1863 "Essays on the History of Labor":[125]

> The principle of the division of labor and the unification of efforts manifests itself everywhere where a society is formed and where collective labor appears. Who forms the society and who labors — peoples or animals — is completely irrelevant. The laws of labor and the characters of association remain the same under all conditions [7: 23].

Admitting, diffidently, that he did not know why Darwin himself had not done so, he simply equated the "civilizations" of humans and animals: "When we look at certain phenomena of ants' social life without our prejudices, then we discover the remarkable meaning of these phenomena and then we understand that conscious progress and purely historical development constitute the unalienable possession of all higher kinds (*porod*) of the animal kingdom" [7: 38].

Pisarev concluded his treatise with a highly praising overview of the latest additions to "Darwinist"[126] literature, such as German geologist Friedrich Rolle's popular interpretation of Darwin's concept and German philologist August Schleicher's extension of Darwin's evolutionary principles to linguistics, both of which had just appeared in Russian translation.[127] He was especially impressed with Schleicher's use of Darwin's concept as a fruitful way to understand the emergence and development of different human languages. Reiterating the major theme of his own "scientific popular" texts, Pisarev particularly stressed the philologist's "deep respect for the natural sciences" and his suggestion that their methods "be applied more and more to the study of languages," thus, emphasizing once again the utility of Darwin's theory in the understanding of "man's history."

*Russian Word* supplemented Pisarev's essays by printing in its "Bibliographical Leaf" reviews of books aimed at popularizing and/or developing further Darwin's ideas. In the same April issue that carried Pisarev's first essay, Blagosvetlov reviewed the Russian translations of two books by Quatrefages, on "the unity of men" and on "metamorphoses in men and animals," noting pointedly that "after the publication of Darwin's book, we should consider [these] works to be completely superfluous and, taking into account [their] exceedingly bad translation, useless in our literature."[128] In July, he enthusiastically greeted the appearance in Russian of Huxley's lectures "On our Knowledge of the Causes of the Phenomena of Organic Nature."[129] In August, Zaitsev praised "the clarity of expression, the abundance and excellent selection of facts" in Rolle's interpretation of Darwin's theory, though noting numerous "inaccuracies and absurdities" in its recent Russian translation.[130] In October, he alerted the readers to the publication in Russian of Huxley's *Evidence as to Man's Place in Nature*.[131] Furthermore, in March 1865, the journal complemented Pisarev's essays with a lengthy article by Shelgunov, which summarized the materials presented by Lyell, Huxley, and Vogt in support of Darwin's assertion regarding "the origin of man and his history."[132] Published under the innocuous title "The Development of the Human Type in Relation to Geology," the article discussed the newest paleontological and archeological evidence that extended Darwin's conclusions regarding the *Origin of Species* to the evolution of the human species.

Yet the attempts to apply Darwin's evolutionary views to the understanding of social issues — what later would be termed "social Darwinism" — had begun even before all this evidence was collected and interpreted,[133] and the Russian reading public was well aware of this fact. One of the first to articulate the implications of Darwin's concept for "man's history" was its French translator Clémence Royer.[134] In a sixty-page "Preface" to her 1862 translation of *Origin*, she praised "Mr. Darwin's theory" as "especially fruitful" in its "humanitarian and moral consequences." Indeed, she hailed it as "the natural social science par excellence," "the codex for living beings of every race and every age" that "contains in itself a philosophy of nature and a philosophy of humanity." Royer claimed that when we apply "the law of natural selection (*l'élection naturelle*) to humanity," it "becomes surprisingly, painfully obvious just how flawed our political and civic laws, as well as our religious morals, are."[135] She pointed to the "unintelligent protection" that contemporary societies accorded to "the weak, the sick, the incurable, and the wicked," as the most obvious example. In her opinion, such social protection runs contrary to the "law of natural selection" by thwarting the struggle for existence (*concurrence vitale*) and allowing the least favorable individuals of the species to survive and multiply. Extending to humans Darwin's discussion of variations in plants and animals as the major source of evolution, she was quick to equate variability (*variabilité*) with inequality (*inégalité*), asserting that "nothing is more obvious than the inequality of different [human] races." "The races are not separate species," she admitted, "but they are well-marked and very unequal varieties." Therefore, she insisted, "we should think twice before proclaiming political and civic equality in a nation composed of a minority of Indo-Germans and a majority of Mongols or Negros." Royer urged the "legislators" to consider seriously the implications of Darwin's theory and strongly advised against the "mixing of different races," which could result in the superior races being "absorbed" by the lower ones and, thus, in the lowering of "the average level of the species."[136]

As historians of biology and anthropology have noted, Darwin's *Origin* profoundly reshaped the old debate over such issues as the monogenic or polygenic origins of humankind and the interrelations among human races by reformulating its key questions.[137] Were human races different species or merely varieties, subspecies of the same species? Did they originate in a particular locale from a single common ancestor and then spread over the world? Or did the human races emerge and evolve in different "centers of origin" from several different progenitors? Were present-day races the result of divergence of some "primordial" human species accompanied by the extinction of intermediate forms? These and many other questions arising from Darwin's concept now guided new research into comparative anatomy, embryology, paleontology, geography, and the taxonomy of the human species.

Royer's "Preface" showed just how easily these scientific questions could be translated into a *political* question about the "natural" superiority of the "white," "European," "Caucasian" race over all other races.[138] At the time of Royer's writing, this political question acquired a particular significance as the liberation of African slaves and the abolition of slavery became a rallying cry of the north in the American Civil War. But it attained special import in post-Crimean Russia. Even before the start of the civil war, discussions of slavery in the United States on the pages of Russian "thick" journals had served as a thinly veiled surrogate for the open debates about Russia's serfdom and the emancipation of the serfs, which had been impossible under the watchful eyes of the censors.

North American slavery and abolitionism commanded the close attention of the Russian educated public.[139] To give but one example, two different translations of the most famous abolitionist novel — Harriet Beecher Stowe's *Uncle Tom's Cabin* (1852) — appeared in late 1857, simultaneously, as supplements to both *Russian Herald* and *The Contemporary*.[140] The same year, Vol'f published an abridged children's version of the novel in book format.[141] In the horrors of slavery vividly depicted in the novel, Russian readers readily recognized the sufferings of the Russian serfs. Quite tellingly, the same March 1861 issue of *The Contemporary*, which carried a nearly 100-page excerpt from the just announced imperial edict on the emancipation of the serfs, also included a complete translation of Henry W. Longfellow's *Poems of Slavery* (1842) under the title "The Songs of Negroes" and an extensive digest of John S. C. Abbott's travelogue *South and North* (1860) under the title "Slavery in North America."[142] As the author of the digest pointedly remarked, "Under the circumstances in which Russia finds herself now, of course, no other kind of literature could be more interesting to her than the books dealing with the relative importance of forced and free labor." Expectedly, *Russian Word* also responded to this interest. In its April 1861 issue the journal carried a lengthy article on the history of slavery and abolitionism in the United States, which perceptively concluded that "the most enduring foundation of slavery is white Americans' scorn for the black race."[143]

Royer's interpretation of Darwin's ideas as the "scientific" foundation and justification for such scorn caught the attention of Russian commentators almost immediately. The French translation of *Origin* came out in summer 1862 and, just a few months later, in November, Dostoevsky's *Time* published a scathing critique of Royer's views by the well-known journalist Nikolai Strakhov (1828-1896).[144] Ostentatiously titled "Bad Signs" and ostensibly written as a review of Darwin's original publication, as well as its German and French translations, the article paid much more attention to Royer's "Preface" than to Darwin's book (the German translation was mentioned only once in the entire article). Indeed, it seems that the review had been inspired not so much by Darwin's concept as by Royer's interpretations of its "social implications."

Strakhov praised Darwin's book as "a great revolution" (*perevorot*) in the "understanding of organisms, i.e. plants and animals" that has profoundly changed "the most fundamental, most essential notions, which till this time had been prevalent regarding living organisms." Pointing as a proof to several successive English editions and the speedy appearance of the German and French translations, he greeted it as "great progress, a huge step forward in the development of the natural sciences." Its significance, according to Strakhov, stemmed from Darwin's discovery of "one of those laws that govern the origin of species" — the "law of natural selection and struggle for existence."[145] The reviewer also noted that, "expectedly," the book generated strong opposition, especially from the British clergy. But he devoted most of his review to Royer's "strange," "monstrous," "improbable" opinions, providing lengthy excerpts from her "Preface."

Strakhov's critique aimed first and foremost at the very idea of using the natural sciences to address social issues. He found the claim of the natural sciences to "domination, to directive importance" in the understanding of human life highly objectionable:

> At the present time the study of nature is draped in a bright halo of hopes and beliefs. Many [people] expect much of it and believe in it as a means to solve all problems, as the spring of wisdom. There are constant

references to the natural sciences as the ultimate authority; their method, their techniques are transferred to other sciences, [and] become rules for those areas of knowledge, which obviously are distanced the farthest from them in their subject, for example, for history, for philosophy.

The main deficiency of Royer's views, according to Strakhov, was that "she has hastened to draw the most far-flung and most general consequences from the great revolution in the natural sciences," and "has assigned to Darwin's theory much more significance and knowledge than it actually possesses." "The main center of gravity in the historical development [of humanity] does not coincide with the field of the natural sciences," he insisted, for "human life is governed and directed by other, deeper fundamentals." Humanity, he continued, "has established for itself a different law, a different norm, a different ideal than those laws and ideals which nature is following." One thing Strakhov saw as absolutely self-evident: "The study of nature is not everything we need [to understand humans]." "As soon as we do not separate human beings from nature," he maintained, "put them on the same level with nature's other creatures, and begin to consider them from the same viewpoint as animals and plants, we cease to understand human life, we lose its meaning."

Strakhov passionately objected to Royer's conclusions regarding the inequality of human races. He contended that if we look at humans as animals, we could easily see that they do differ among themselves in numerous characteristics (in height, weight, color of skin, level of intelligence, sharpness of senses, and so on). Yet these variations do not negate the idea of "equality among humans as humans, not animals." "Human dignity," something clearly "imperceptible, unmeasurable, and undefinable by any distinct trait," he stated forcefully, "overshadows all those obvious differences which separate the most illiterate of Negroes from the most enlightened of Europeans."

Strakhov's sentiment resonated strongly with the attitude of many educated Russians towards slavery and abolitionism. On the eve of the emancipation of the serfs, in February 1861, commenting on the reported failure of the "Liberia Colony" despite its inhabitants being freed from slavery, Blagosvetlov asked rhetorically: "What rational basis could there be for thinking that a man with black skin and stiff hair is less capable to govern himself than some fair-haired Anglo-Saxon?"[146]

In his account of the Caribbean and Central American "future republic" inhabited by the descendants of numerous indigenous tribes, Spanish conquistadors, English and French colonists, and African slaves, Élie Reclus, *Russian Word*'s political observer, pointed out: "The question about the origin of [human] tribes (*plemen*) has long been and for a long time yet will be debatable; but an answer provided [to this question] by science could neither strengthen, nor weaken the ties that unite us with our brothers of all colors and all climates." "Whether humans had originated from one family or sprung from many different roots," he concluded, "they are nevertheless all connected to one another, and their common unity lays in one shared understanding of justice and freedom! Even though their past were full of hatred, it would not matter if eventually they would join together and form one happy and free humanity!"[147]

Neither Strakhov's sharp critique of Royer's assertions, nor humanitarian sentiments of many Russian supporters of abolitionism did much to cool the enthusiasm of some *Russian Word* authors regarding the utility of the natural sciences, and of Darwin's theory in particular, for the "scientific" understanding of human societies. In August 1864, Zaitsev picked up the glove thrown into the ring of Russian polemics by Strakhov's article. Following his principle of "only tangentially touching upon the reviewed publication," Zaitsev used as a pretext the Russian translation of Quatrefages's *Unité de l'espèce humaine*, even though just a few months earlier, in April, Blagosvetlov had already dismissed this book as "completely superfluous."[148] In the eight-page "review" Zaitsev presented his own opinion about such "immense phenomena as slavery and colonization," ostensibly in light of Quatrefages's and Darwin's ideas as he had understood them.

In applying Darwin's "laws" to humanity Zaitsev went even further than had Royer, asserting categorically: "it is impossible to deny that at the present time human races represent species as distinct as, for instance, horse and donkey." "By acting upon human beings over the course of many millennia, heredity, natural selection (*podbor*), and environment had erected between the races a boundary so complete," he alleged, "that, at the present, it is impossible even to show intermediary types, and the crosses between the races are infertile" [97]. He claimed emphatically that "there is not a single European scientist who does not consider the colored tribes to be lower than the white one in the very structure of their organisms" [94]. According to Zaitsev,

> anatomy and the observations of the psychological abilities of aboriginal races in Africa and America demonstrate such a great, fundamental difference between the red-skinned, the Eskimos, the Polynesians, the Negros, the Caffres, the Hottentots, on the one hand, and the white man, on the other, that only sentimental ladies like Ms. Beecher Stowe could insist on the fraternity of all these races [95].

He elaborated this point: "The difference existing between the white race, on the one hand, and the Negros, the [native] Americans, and the Polynesians, on the other, is far too evident to talk seriously about a possibility of existing between them relations even remotely similar to those existing among people of the same race" [98]. Following this argument to its logical end, he proclaimed: "Undoubtedly, and this is recognized by everyone, slavery is the best outcome that colored men could wish for when encountering the white race ... for the majority of them could not exist alongside the Caucasian tribe at all and soon go completely extinct" [94].

Zaitsev delivered a long diatribe against "those people, who, like a philanthropic lady in one of [Charles] Dickens's novels, are preoccupied with the enlightenment and liberation of their black brothers, while their own children are falling from staircases and burn themselves with boiling water" [96]. Citing Quarterfages's claim that centuries of oppression by their English masters had reduced the Irish poor in their physical and moral features to looking like "the most degraded Australian tribes," Zaitsev stated vehemently:

> Instead of championing the equality of the black tribe with the white [one], where millennia, and perhaps the very origins, had carved an ineradicable, organic boundary in both physical and moral characteristics, it would be better to turn philanthropic attention onto those who are in fact our brothers, but whom our political and social conditions degrade to the point that they lose the features and qualities of their [own] tribe and approach the lower races [96].

"A few more centuries of existence in such conditions," he continued, "and Europe will have a new race that would have forever lost those higher abilities that distinguish the Caucasian [race]. But the philanthropists find it more suitable to advance the emancipation of the

Hottentots and the Bechuans and the protection of animals than to take care of their real brothers" [96-97].

Zaitsev certainly was not trying to entice the journal's readers to buy and read Quaterfages's book "devoted to the question that since Darwin has lost any significance," as he stated bluntly in the last paragraph of his "review." He noted that, although the book "is not without interest and presents many remarkable facts," it was priced "too high," especially since "the translation is astonishingly bad" [100]. It seems that Zaitsev's real goal was to redirect the attention of his readers from abolitionism to the miserable conditions of "their real brothers," Russia's newly "emancipated" serfs. If that was what he had set to accomplish, he failed miserably.

Zaitsev's attempt to use Darwin's name and ideas to substantiate his rampant anti-abolitionism did not sit well with the Russian educated public: he received several letters from perplexed or outraged readers regarding his review. *The Contemporary*'s leading critic Maksim Antonovich did not miss the opportunity to pick at his constant opponent: "*Russian Word*, which considers itself progressive and humane, defends the slavery and discrimination of Negroes." "Truly humane people," he continued sarcastically, "especially, the realists, as *Russian Word* calls itself, must be concerned with reducing even the slavery of animals [and must] defend even animal rights, to say nothing of Negroes, who are after all humans."[149] Other observers pointed out an obvious contradiction between Zaitsev's statements and the abolitionist pronouncements of other contributors to the journal, such as the unyielding support for the equality of all human races by Reclus cited above. *Spark*, a popular satirical magazine, even published a mocking cartoon on the subject (see fig. 2-4).

Yet Zaitsev remained unrepentant. In the December 1864 "Bibliographical Leaf" he answered his critics, stating defiantly that there was no contradiction between his views and those of his fellow contributors to *Russian Word*: "One could be an anti-abolitionist without being an obscurantist."[150] He claimed that in his review he "had discussed not the political question about [the emancipation of the] Negroes, ... but had pointed out a conclusion reached by the natural sciences regarding slavery." "Since there is a huge difference between a scientific conclusion and its application to the political life of various

Fig. 2-4. A cartoon published in *Spark* on 23 December 1864 (p. 640) under the title: "A good-natured editor whose left hand does not know what the right hand does." It depicts a two-faced man of Blagosvetlov's considerable likeness patting a black man on the head with his left hand, while pulling the hair of another black man with his right hand. The left sleeve of the man's jacket is labeled "Politics" and the right "Bibliography." A caption underneath the cartoon reads: "All these sections are united into a cohesive whole, by the unity of thought represented by the journal, and by the convictions of its regular contributors ([from] Announcement. 'Russian Word')." Courtesy of RNB.

people," he continued, "it would be fairly strange to mix one with the other." Scientific facts are not the same as political aspirations, Zaitsev asserted, repeating that "the slavery of the black race represents an absolutely natural and normal phenomenon, because it is defined not by some accidental, but by natural-historical causes."

"The point of the matter," he continued, "is that scientists had debated endlessly whether human races are different species or different subspecies." Darwin's work should end this debate, for his concept had erased the strict distinction between species and subspecies. "The dissimilarities between human races are quite substantial and constant," Zaitsev explained, "they differ from the distinctions between the human and other animal species not in the quality of any discrete characteristics, but in their quantity, to be exact, in their degree." According to Zaitsev, "all scientists, including those whom nobody could accuse of obscurantism, such as Huxley and Vogt, adhere to this view." These scientists consider the black race to be "lower in its organization than the white man" and indeed "representing an intermediate stage between the latter and other mammals." On every page, according to Zaitsev, "we encounter in the writings of these scientists the following progression: European, Negro, and so on." But, he surmised, if we accept this view, we must accept the logical conclusions it leads to, namely, that "when two races, one of which is superior to another in its organization, coexist, [then] any equality between them is impossible — the inferior race will inevitably be a slave to the superior one." It was this "logical conclusion" that laid the foundation for his statement: "undoubtedly, and this is recognized by everyone, slavery is the best outcome that colored men could wish for *when encountering the white race,*" Zaitsev insisted, emphasizing the last part of the sentence. Of course, he continued, some people could object to his expression "everyone," for there are certainly many people "who sympathize more with Madame Beecher Stowe's laments than with Vogt's opinion." But he dismissed such objections by saying that "there are still many people who believe that the Earth is sitting on the backs of three whales." "Like E. Reclus, I wish the Negroes all the best and resent those awful happenings that accompany slavery in North America," he reiterated, "but I also point to the opinions of respected progressive scientists and to the conclusions from these opinions regarding the issue of slavery." "Is it necessary, in order not to be an obscurantist," he asked rhetorically, "to shut your eyes and harp on something, while science says something different?"

Zaitsev's insolence did not remain unnoticed. A month later, in February 1865, *Spark* published a scathing anonymous "review" of Zaitsev's articles.[151] The review was actually written by Nikolai Nozhin

(1841-1866), a talented young biologist, who had just returned to Russia from a long sojourn in western Europe, where he had studied zoology at the University of Heidelberg and conducted research on the embryology of the invertebrates in the Mediterranean Sea.[152] Nozhin was well acquainted with Darwin's works. Reportedly, he even had translated German-Brazilian biologist Fritz Müller's book *Für Darwin* (1864), which provided numerous new facts in support of Darwin's evolutionary concept.[153] He also knew quite well the writings of Vogt, the main source of Zaitsev's claims and assertions. Nozhin's review mercilessly mocked Zaitsev's style of argumentation, his "scientific conclusions," and his (mis)use of the authority of "great scientists."

Nozhin derided the propensity of Russian journalists for "cultivating the flowers of oratory," "burning with the flames of noble indignation," and invoking the names of eminent scientists to support "their poetic thoughts," while discussing subjects about which they knew very little, if anything at all. By taking apart several statements Zaitsev had made regarding slavery, he demonstrated that much of Zaitsev's text was little more than "phrase-mongering." "A phrase," he explained, "is an innocent combination of words, mostly very good ones, that has the external appearance of a thought and delights the ears of its writer and readers with its sound play," but is utterly devoid of any real meaning [115]. Nozhin picked as an example Zaitsev's assertion that the enslavement of the black race by the white one "is defined not by some accidental, but by natural-historical causes." He pointed out that this assertion makes no sense whatsoever, since "there are no accidental causes," all causes could be interpreted as "natural-historical."

In a similar fashion, he ridiculed Zaitsev's "scientific conclusions." After stating a certain "fact," such as, for instance, the existence of differences between the white and black races, Nozhin posited, "Mr. Zaitsev thinks he has the right to attach to this fact the most improbable *therefore*, a *therefore* that unconditionally justifies slavery." But, he continued, Zaitsev had misunderstood the essence of Darwin's concept, "which denies the *constancy* of species and subspecies, and hence [the constancy] of human races, and, to the contrary, accepts the possibility of their development into a higher type." "It directly follows from Darwin's theory," he asserted, that "one *should not consider* the marked distinctions between the races as something constant."

Nozhin saved his most venomous comments for Zaitsev's attempts to support anti-abolitionism by references to Vogt and Darwin. "Our thinkers," he quipped, "use an authority to substantiate exactly those opinions which correspond the least to the actual views of the authority." He provided an excerpt from Vogt's book "on harmful and useful animals"[154] to demonstrate that, contrary to Zaitsev's assertions, the scientist himself had vehemently objected to both slavery as a social institution and the use of "zoological facts" for its justification. Nozhin also supplied two very long passages from Darwin's *Journal of Researches into the Natural History and Geology of the Countries Visited During the Voyage of H. M. S. Beagle Round the World*, which showed, in Darwin's own words, the scientist's indignation at and highly negative attitude towards slavery.[155] Nozhin concluded his review on a high note:

> ... [T]he source of inhumane articles and appalling actions against people and even animals (no matter how much Mr. Zaitsev laughs at humane attitudes towards the latter), always and everywhere, is one and the same: the insensitivity to the suffering of others that reaches its uppermost ugliness under the cover of meaningless phrases and [that] has nothing in common, either with the authorities of thought, or with any, let alone the latest, conclusions of science [117].

Nozhin's article delivered a heavy blow not only to Zaitsev's pride, but also to *Russian Word*'s crusade to promote the utility of the natural sciences for scientific understanding of social issues. Zaitsev had apparently swallowed his pride, for he did not respond to Nozhin's review, as he had to Antonovich's remarks a few months prior, even though he had ample opportunity to do so. For instance, in the May 1865 "Bibliographical Leaf," he reviewed the just published Russian translation of the first volume of Darwin's *Journal*, from which Nozhin had extracted Darwin's views on slavery. But Zaitsev ignored this fact, noting that "it is difficult to find another book that contains such a wonderful wealth of interesting facts in all fields of the natural sciences." This time, he abstained from commenting on the "social implications" that could have been drawn from those facts.[156]

Blagosvetlov, however, would not, and could ill afford to, let the subject that constituted one of his journal's signature themes drop. Although nearly all of *Russian Word*'s "staff" authors, in one way or another, addressed this theme, none of them had adequate training in the natural sciences, let alone the necessary expertise in such areas as anthropology, physiology, or evolutionary biology. Thanks to his previous medical studies, Zaitsev was Blagosvetlov's most qualified author to confront the issues raised by Darwin's "revolution." But Nozhin's review had undermined Zaitsev's reputation in this respect. Blagosvetlov desperately needed someone else capable of demonstrating that Darwin's theory, and the natural sciences more generally, did have a role to play in not only understanding, but also curing the "social ills" that plagued the Russian Empire. Vasilii Florinskii and his "Human Perfection and Degeneration" certainly fit the bill.

## Blagosvetlov and Florinskii

Florinskii's extensive treatise appeared in four installments in the August, October, November, and December 1865 issues of *Russian Word*. It actually opened the first three issues, and was the second (immediately following Shchapov's article on "Natural Sciences and the People's Economy") in the December one, clearly attesting to the particular importance assigned to it by the journal's editor. But how did it get there? Had Florinskii brought a manuscript of his treatise to Blagosvetlov requesting him to consider it for publication? Had Blagosvetlov commissioned Florinskii to write it? Why would the busy professor take up the burden of writing an extensive text on something that dealt with a subject quite remote from his immediate interests and required much of his time and effort?

Available materials are completely silent on the circumstances of Florinskii's contacts with *Russian Word* and its editor. I was unable to find any information on exactly when, where, how, and on whose initiative the two men had met, and who and how, individually or collectively, had come up with the idea of "Human Perfection and Degeneration." In the entire collection of Florinskii's personal documents there is not a single sheet of paper with Blagosvetlov's name on it. Florinskii never mentioned Blagosvetlov, *Russian Word*, or this treatise in any of his diaries and memoirs. Nor are there any plans, drafts, notes on sources and references, or indeed anything that could illuminate the process of writing and editing it. Similarly, Blagosvetlov's surviving materials contain no trace of Florinskii or his book. Nor does Florinskii's name appear in Blagosvetlov's surviving correspondence and various reminiscences about Blagosvetlov and his journal, such as, for instance, Shelgunov's lengthy memoirs.[157] Nevertheless, what little materials exist allow us to make certain suppositions.

It is possible that it was Zaitsev who made the initial introductions. Zaitsev had attended lectures at the Imperial Medical-Surgical Academy and had many friends among its students. He had moved to St. Petersburg in December 1862 and had transferred to the IMSA after several years of study at the medical school of Moscow University. He might have taken Florinskii's lecture course on gynecology offered to fourth and fifth year students in the 1863-1864 academic year, or attended the professor's clinical demonstrations, and thus might have known Florinskii personally. Or perhaps, one of Zaitsev's friends among the IMSA students had mentioned the young professor's reputation as someone with a definite talent for writing clearly and succinctly on a variety of complicated subjects.

It is also possible that Florinskii's former teacher Alexander Morigerovskii had a hand in the matter. While Florinskii had been making his way to becoming a professor, in 1858-59, in parallel with his civil service job at the Ministry of State Properties, Morigerovskii had begun teaching literature at several secondary schools.[158] A year later, Morigerovskii decided to return to fulltime teaching. He quit his job at the ministry and took a position as lecturer on Russian literature at the Technological Institute. But, like Blagosvetlov, he was unable to keep his new job for long. In early 1862, he was fired for "inciting student unrest," and the secret police placed him under surveillance for suspected "revolutionary activities." He supported himself by teaching in a gymnasium and by taking on translating and proofreading jobs for various periodicals, including *The Contemporary* and *Russian Word*. He built an extensive network of contacts in St. Petersburg's literary community, among publishers, writers, and editors. He was on good terms with Nikolai Nekrasov and developed a close friendship with Nikolai Chernyshevskii: it was Morigerovskii who accompanied Chernyshevskii's wife to the civic execution ceremony that preceded the writer's exile in May 1864. Morigerovskii also knew Blagosvetlov quite well.[159] Indeed, he became something of a confidant and even served as one of the editor's proxies in running the *Russian Word* printing shop.[160]

Over the years, Florinskii apparently kept in touch with his former favorite teacher, who had so profoundly influenced him at the seminary and had done so much to bring him to St. Petersburg. It seems quite likely that it was Morigerovskii who introduced his former star student to Blagosvetlov.

Whoever introduced the two men to each other, Blagosvetlov and Florinskii must have met (probably more than once) sometime in the late spring or the early summer of 1865. Given the similarity of their backgrounds, they must have hit it off pretty well. Perhaps Blagosvetlov had read a few of Florinskii's critical articles in *Medical Herald* and liked his writing style. It seems certain that the editor would have explained his predicament to the prospective author in some detail. He probably suggested that Florinskii use his personal expertise in writing a series of essays that in some way would address the role of the natural sciences in the understanding of social life and would discuss possible "social implications" of Darwin's theory and its "laws."

Judging by his previous publications, at that time, Florinskii had no particular interest in (and perhaps not even much acquaintance with) Darwin's evolutionary concept or its "social implications." The topic was far removed from his immediate duties, interests, and preoccupations. But he agreed to take on the task, perhaps because, like Blagosvetlov, he certainly believed in the power of knowledge as a tool in improving the life of his country and his compatriots. After all, did he not use that power on a daily basis in his own work at the IMSA clinics to improve the health of his patients? Indeed, he would pointedly end his first essay written for Blagosvetlov with a paraphrase of Francis Bacon's famous statement: "correctly organized knowledge becomes power!"

Florinskii's was likely flattered by the invitation to write for the most influential journal of his time. Blagosvetlov offered him a chance to contribute to the current heated debates on the future of his Fatherland and its people, to make his personal views on important social concerns known in the farthest corners of the country, and to reach a much broader audience than the one he had addressed in his previous writings (which was limited to his fellow physicians and medical students). It seems that the young physician accepted the invitation, at least partially, because he felt that he could do it. As he remarked in one of his articles, "anyone, who feels his inner strengths, who recognizes his literary abilities, cannot remain silent. The need to write, as the need to speak up, is irresistible."[161]

Florinskii was undoubtedly aware of the considerable public resonance that the essay on "Reflexes of the Brain" written by his fellow IMSA professor Ivan Sechenov had generated less than two years earlier in 1863. In this essay, initially slated to appear in *The Contemporary*, Sechenov used his expertise in physiology to discuss much broader issues of the human psyche, such as free will, desires, and consciousness.[162] Even though the censorship had prohibited its publication in *The Contemporary*, the essay did appear on the pages of Chistovich's *Medical Herald*.[163] Sechenov suggested that all of the phenomena of "psychic" life could be explained by the simple reflexes he had observed in his experiments on the inhibition of nervous impulses in a frog's brain, which provoked a prolonged debate among the Russian educated public. Perhaps, Florinskii hoped that he too could stir the public opinion by using his professional expertise to illuminate another vitally important biological aspect of human life — reproduction, and its social embodiment in the institution of marriage.

Florinskii's intellectual style well prepared him to take on the new task. Even though his main scientific interests focused on the seemingly narrow medical specializations — gynecology, obstetrics, and pediatrics — he was far from a narrow specialist. He read widely. He researched and published on a variety of very diverse subjects, from physiology to therapy and clinical diagnostics, from histology to the ethics of medical practice, and from the principles of medical education to surgery. Most important, unlike Blagosvetlov's "staff" writers, Florinskii was well aware that one particular branch of his own profession, named variously "public hygiene," "social medicine," or "social hygiene," had already begun to address the health issues of, in Shelgunov's terminology, not only "separate physiological human beings," but also such "social organisms" as families, professions, occupations, classes, and nations. In contrast to the focus of clinical medicine on health and disease *treatments* of an *individual*, the nascent field of social hygiene focused on the issues of health and disease *prevention* in particular *groups* of people through various legislative measures. Florinskii was also well aware that Virchow, his idol whom he had admired since attending the

German professor's lectures in Berlin in 1861, had already identified "social conditions" as one of the most important determinants of health and disease.

From this awareness it was but a short step to connecting Darwin's "laws of selection" with social issues through the discussion of "human perfection and degeneration" and "marriage hygiene," as the subjects of Florinskii's essays would be defined in their title and introduction. And it was perhaps an even shorter step to connect the contemporary legislative initiatives to prevent the spread of communicable disease and to promote the health of a nation to the proposal of legislative interventions in marriage aimed at preventing degeneration and promoting the perfection of humankind.

Moreover, Florinskii definitely had a personal interest in examining these subjects in some depth. As we saw in the previous chapter, just a few months prior to his meeting(s) with Blagosvetlov, in early February, Florinskii got married. In late April or early May, his wife got pregnant, and, most certainly, Florinskii monitored her pregnancy very carefully.[164] He must have thought about their future child and pondered his responsibilities as a husband, a father, and a physician. His specialization in gynecology, obstetrics, and pediatrics provided him with the necessary knowledge to see these subjects not merely as a personal matter but as an important social issue, which he, as a physician, could address in detail. And this is exactly what he set out to do in his essays.

The author and the editor must have discussed Florinskii's ideas and the possible ways of presenting them. Since no drafts or proofs of, nor any correspondence regarding "Human Perfection and Degeneration" have survived, all we have to go on in our analysis is the published text. We thus can only guess at just how much or how little Blagosvetlov influenced its author and the actual writing. Based on what we have learned about Blagosvetlov, the editor, on the preceding pages, it is safe to assume that he allowed his newest "find" of an author considerable freedom in choosing the specific topics and composition of the essays. But he must have approved the general shape and goal of the entire treatise as it was outlined on the opening pages of the first essay: to discuss the key components of Darwin's evolutionary concept — variability, heredity, selection, extinction/

degeneration, and progress/perfection — and the ways that they could be applied to humans.

The writer and the editor must have also agreed on the total length and number of essays, for this would determine the journal's size, contents, and composition for several months running, as well as the amount of Florinskii's honorarium. At the time Russian publishers paid their writers per each "typographical" list, which, in the then most popular "octavo" format, meant sixteen printed pages.[165] Since the total length of the published essays is 142 journal pages, which translates into almost exactly nine lists, it is quite likely that Blagosvetlov had asked Florinskii to write three, not four, essays, each about three lists in length. The first and the second essays are roughly three lists in length each. And it certainly looks like the third essay was split into two parts and published in the two consecutive (November and December) issues for some editorial reasons that had nothing to do with its contents. The two parts actually have continuing pagination, which indicates that they might have even been typeset as a single piece. Combined, they match exactly the length of the second essay. Blagosvetlov paid his lead authors, like Pisarev and Shelgunov, fifty to sixty rubles per list.[166] To Florinskii, as a novice, he might have offered a bit less, but considering the total length of the published essays, this still amounted to a considerable sum (about a third of Florinskii's annual salary), which must have been discussed and agreed upon.

It is also quite likely that the editor and the author had agreed on the specific timing of publication and, hence, the delivery schedule. The contents of Florinskii's work strongly suggest that he had written it piece by piece during the summer and early fall of 1865. Given the time necessary for typesetting, proofreading, and getting a censor's permission, in order to appear in the August issue the first essay should have been delivered to the editor no later than mid-July. The next essay had perhaps been delivered at the end of the summer, before Florinskii had to resume his teaching at the academy, and the last — sometime in October. It is likely that all these technical details had been discussed and settled in May or early June.

But there was one more issue that Blagosvetlov and Florinskii must have discussed — censorship. Since all of his previous publications had appeared in professional journals or as academic monographs and textbooks, which had summarily been exempted from the general censorship, Florinskii had had no experience in dealing with censors.[167] Blagosvetlov, on the other hand, dealt with them on an almost daily basis. For the editor, avoiding the censor's merciless pen — that could not merely cripple a piece of writing but actually forbid its publication altogether — was a matter of utmost importance. At the time of his discussion(s) with Florinskii, Blagosvetlov definitely knew that new censorship rules would come into effect on 1 September 1865, which might have influenced his choice of the exact timing of publication for each essay.

The new laws promulgated in early April changed the Russian censorship system profoundly.[168] The old rules of "preventive" censorship had required that every piece of writing be screened before publication and had placed responsibility for permitting something "unallowable" to appear in print on the shoulders of the censor first, and the publisher second. Indeed in 1862, one of the censors responsible for monitoring *Russian Word* had actually been fired for permitting the publication of several articles that his superiors found totally unacceptable and which led to the journal's eight-month suspension. The new laws instituted a system of "punitive" censorship, whereby a publisher could print virtually anything, but, if the censor found a certain published piece "objectionable," the publication would be arrested and destroyed, while its publisher — in addition to losing initial investment — along with its author, would incur stiff penalties imposed by the court, ranging from large fines to imprisonment. In the case of a periodical, the situation could also result in an official "warning," and, after the periodical received three such "warnings," it could be suspended or even shut down completely. Blagosvetlov certainly wanted to avoid such an eventuality, and most likely advised the novice author regarding various ways to evade the censorship's clutches.

The first essay appeared in the August issue, and thus Blagosvetlov had to abide by the old rules and to get the censorship approval prior to its publication. Perhaps this is why it contained only a synopsis of the latest scientific views on "the variability and heredity of the human type." Based on his previous experience in publishing "scientific-popular" texts, Blagosvetlov might have hoped that in this form the first essay would unlikely attract much attention from any censor. As far as we know, it did not. But the subsequent essays came out *after* the new rules had come into effect, and their contents put the editor on much shakier ground. The second and third essays discussed various factors, from "material well-being" to "rational marriage," which, according to Florinskii, could produce the perfection or the degeneration of the human kind, but which, from the viewpoint of the authorities, also presented certain potentially subversive issues. To begin with, the whole matter of marriage in Russia was the exclusive domain of the church, and thus constituted a particularly sensitive subject, especially considering Florinskii's advocacy of inter-confessional marriages, which were explicitly forbidden by Orthodox rules. Moreover, "materialism," "rationalism," "realism," and other similar "isms" had long been telltale watchwords that the censors saw as undermining the authority of the church and its doctrine.[169] The contents of the second and third essays were bound to invite the censorship attention and, perhaps, intervention.

Florinskii and/or Blagosvetlov took certain steps to deflect the unwanted attention, which proved quite effective.[170] Although it was the October, November, and December 1865 issues that elicited the three "warnings" to the journal under the new censorship laws and resulted in its suspension for five months in February 1866, Florinskii's essays published in these issues were not to blame. Pisarev's and Zaitsev's articles printed in the October issue provided the foundation for the censor's wrath that led to the first "warning." Similarly, the main reasons for the second "warning" issued after the appearance of the November issue were Pisarev's and Shelgunov's articles. The third "warning" announced after the publication of the December issue was, in turn, provoked by Shelgunov's and Tkachev's essays.[171] In the censorship reports that justified these warnings and the ensuing suspension of *Russian Word*, Florinskii's treatise was not even mentioned.[172]

In the end, however, no matter how much input Blagosvetlov had in defining the overall theme of "Human Perfection and Degeneration," or how much guidance he gave to its author on its particular parts and issues, the treatise was the result of the extensive effort and careful thought of Vasilii Florinskii.

# 3. The Book: Darwinism and Social Hygiene

> "She married (I forget the pedigree)
> With an hidalgo, who transmitted down
> His blood less noble than such blood should be;
> At such alliances his sires would frown,
> In that point so precise in each degree
> That they bred in and in, as might be shown,
> Marrying their cousins, nay, their aunts, and nieces,
> Which always spoils the breed, if it increases.
> This heathenish cross restored the breed again,
> Ruined its blood, but much improved its flesh;
> For from a root the ugliest in Old Spain
> Sprung up a branch as beautiful and fresh.
> The sons no more were short, the daughters plain.
> But there's a rumour which I fain would hush;
> 'Tis said that Donna Julia's grandmamma
> Produced her Don more heirs at love than law.
> However this might be, the race went on
> Improving still through every generation..."
>
> Lord Byron, *Don Juan*, 1819

The excerpt from Lord Byron's famous poem used as the epigraph to this chapter exemplifies the popular notions of human reproduction, heredity, degeneration, and improvement, as well as the role of marriage in these processes, current among the British educated public in the first half of the nineteenth century. One could surmise that the Russian educated public subscribed to the same notions. A complete Russian translation of Byron's poem by Dmitrii Minaev, one of *Russian Word's* regular contributors, had begun to appear in *The Contemporary* just a few months before Florinskii started to work on his treatise.[1] The "First Canto," from which this epigraph is taken, was published in Russian in January 1865, and its translator did not think it was necessary to explain its references to the role of "blood" in passing certain features, such as beauty or stature, from parents to offspring. Nor did he explain the role of "breeding in and in" — marrying cousins, aunts, and nieces — in "spoiling the breed" and of bringing "fresh blood" through a "heathenish cross" in "restoring" and "improving" the breed. Minaev apparently counted on his readers' understanding of the poem's "biological" references.

Florinskii, however, did not. He set out to provide his readers with a detailed overview of the latest scientific views on human reproduction, heredity, variability, and development. This overview laid a foundation for his thorough analysis of the role Charles Darwin's evolutionary ideas — combined with the precepts of fledgling social hygiene — could play in preventing the degeneration, and advancing the perfection, of the human species.

## The Variability and Heredity of the Human Species

The first, nearly sixty-page-long essay appeared in the August 1865 issue of *Russian Word*, which came out only on 4 October due to censor's delays.[2] Florinskii opens his treatise with a simple statement that defines the thrust of his entire work:

> Hardly anyone will doubt that the wellbeing of the population increases with the development of hygiene and the growing application of its rules to social life. Every educated nation understands this very well and strives to apply the results of this science to public health (*narodnoe zdravie*). The people and the government endeavor to eliminate miasmas, better the quality of food supply, improve the salubrity of housing, and so on, but, surprisingly, they pay very little attention to the root of public health — *marriage hygiene* [1].[3]

He laments that in this respect domesticated animals fare much better than humans. Farmers and animal breeders pay special attention to, and develop "whole doctrines" about, perfecting "the stock of cattle, sheep, horses, dogs, [and] even chickens and pigeons." As a result, they indeed produce "wonderful perfections." But humans, in their successive generations, breed diseases and physical weakness rather than perfection. "Historical evidence shows us," he claims, "how the stocks of privileged estates (*sosloviia*) and even of entire nations degenerate and diminish physically and morally." Even when certain perfections do occur, according to Florinskii, they happen by accident, and not as a result of deliberate efforts. "However," he states, "science has already obtained so much data that, by applying it to life, [we] could count on the conscious perfection of the [human] breed (*porody*)." Since "the goal of a marriage from the physiological and civic points of view is the production of the progeny," the main instrument of such perfection should be "arranging marriages not according to the unconscious attraction of the sexes, [and] even less according to mercantile considerations, but more or less according to the goal of producing better progeny" [2]. Florinskii emphasizes that "physical beauty, health, and, partially, moral qualities depend not so much on the upbringing and nurture of a new individual as on heredity" [4]. He notes that "poorly-matched parents often complain that God did not grant them good children, without realizing that their reproductive failure stems not from a [bad] fate but from a badly thought-through marriage" [4].

On the first four pages of his treatise, Florinskii sketches his main lines of inquiry and his two-pronged approach. The first half is to focus on "human variability and perfection in general," with special sections on "heredity as the main cause of human variability and perfection" and on "conditions conducive to stock perfection," including "taste and demand for certain qualities, influence of external conditions of life, rational marriage, and sex life." The second half is to discuss "the degeneration of the human stock in general" and "conditions facilitating such degeneration," among which Florinskii lists "incest and lack of stock renewal, inequality [of partners] in marriage, [and] the influence of drunkenness, debauchery, diseases, poverty, and slavery" [4].

## Variability

The first section, titled the "variability of humankind,"[4] begins with a clear allusion to Darwin's *Origin*:

> Before we speak about the betterment of the human stock by means of the natural, or rational, matching of kin, it is necessary to demonstrate that human anatomical and physiological characteristics are to a certain degree variable, volatile; hence, under the influence of heredity and the physical conditions of external environment, they *could* either be perfected, or degenerate [4-5].

Florinskii warns his readers that he is not going to engage in "anthropological debates about human origins." Nor is he interested in "the question of the origin of human races and tribes." His main concern is human variability. On the next twenty pages, he discusses the forms, causes, and limits of such variability. He illustrates his arguments with both paleontological data and the anthropological/ethnographic descriptions of various groupings (races, tribes, nations, families, types, etc.) among "the peoples inhabiting the globe," and, particularly, among the population of the Russian Empire.

Florinskii notes that humans vary in all of their features: anatomical, physiological, mental, and moral. But at first he focuses almost exclusively on the variability of anatomical characteristics, for the paleontological evidence he brings to his discussion is based on anatomy. He uses the notion of "type" in such expressions as the "Slavic-Russian type" and the "Italian type" as shorthand to describe certain patterns of anatomical organization that differentiate various human groups.[5] He discerns two main causes of variability: heredity and environment. In his opinion, the "intermixing" (*pomesi*) of individuals, tribes, breeds, nations, and races, each one of which is characterized by a particular combination of hereditary features, is the most important cause of human variability. Certain conditions could either increase, or diminish such "intermixing," especially the isolation and migration of particular groups, thus contributing to considerable differences in their heredity.

Florinskii begins by demonstrating that some anatomical characteristics are more variable than others. Thus, in his opinion, the characteristics of the "soft parts," such as the color of eyes, hair, and skin, the size and form of mouth and ears, the form of women's breasts, and the quantity of skin fat, are all easily changeable. Other, skeletal characteristics, such as the size and form of the skull, the relative (to the skull) size of the face, the form, direction, and degree of the development of facial bones, and so on — "those features, which give a particular shape and expression to the face and actually characterize every race and nation, are more stable and, if they change, this is only as a result of blood mixing and, for the most part, not suddenly, but gradually, over the course of several generations" [5]. He explains in a footnote that the expression "blood mixing" is actually incorrect, for "the blood, in the exact meaning of the word, does not play any role in the transmission of anatomical characteristics from parents to children." But since the expression is commonly used and understood, he will occasionally use it "in the sense of mixing of races and tribes," and, in the same sense, he will use such expressions as "pure-blooded, half-blooded (*polukrovnyi*), and such" [5, fn. 1]. He asserts that although the limits of skeletal variability are unknown, for his purposes it is sufficient to demonstrate that "tribes and races are able to transform one into another and that the skeletal form can change [and] acquire a different type, becoming [in the process] perfected or degenerated."

Florinskii proceeds to show that such transformations had indeed occurred in the course of human history. He takes as a starting point available paleontological/archeological evidence: "those primary forms [of the human skeleton] which we know through the fossilized remnants of human bones from the most remote epoch." Following closely Charles Lyell's and Carl Vogt's works on the subject, he lists numerous recent discoveries of human fossils in France, Sweden, Denmark, and Switzerland.[6] He compares them to "the forms of the contemporary inhabitants of the same countries" and correlates "the forms of present-day wild and semi-wild peoples with the forms of civilized peoples." These comparisons convince him "that humans had made a huge step towards progressive physical development." It is a historical fact, he concludes, that "the form of the human body improved gradually and is still being improved, even though up to this time people took no efforts, no conscious actions, which could speed up and strengthen such perfection" [6]. This, however, does not mean that "human physical perfection reached its final limit." To the contrary, he argues, such development will continue into the future: "We can expect

[that] humans [will attain] a new, better form ... by means of changes in races and tribes, their transformation one into another, [and] the development of new types from combinations of existing [types]" [6-7]. The result of these processes could be "progressive or regressive," he concedes, but "the general, imperceptible movement tends, in the end, towards the perfection of the human type."

After the analysis of human "historical" variability, Florinskii turns to the present. He claims that this "movement towards perfection" in human physical development to a certain degree "also exists in our epoch, [unfolding] before our eyes," illustrating its existence through anthropological/ethnographic data. He surveys the general principles and specific measurements that anthropologists use to classify different human groupings. Referring to an extensive treatise on "Man in Natural-Historical Relations" by Karl von Baer, Russia's foremost anthropologist of the time,[7] Florinskii notes that anthropologists disagree on how exactly and into how many major groups humanity might be divided.[8] Some recognize six main groups, others four, and still others eight or nine. He follows the simplest classification, derived in part from the biblical tradition,[9] and identifies four "main human tribes": "white or Caucasian," "yellow or Mongolian," "red or American," and "black or African." He also describes various "branches" into which these main tribes are subdivided by ethnologists, referring to the studies of the "national particularities of many European tribes," based largely on cranial measurements of Gypsies, Hungarians, Italians, Bohemians, Slovenians, Ruthenians, Poles, Croats, and others.[10]

Florinskii claims that though the four major human "tribes" are easily identified, they are changeable and the boundaries between them are quite fluid. He refers to the observations of Pierre Trémaux, a prolific French traveler, photographer, and amateur ethnographer, that "under the influence of hot climate, the white type could transform into the black one, while in moderate and cold climates, the black type softens and even completely disappears in several generations" [7].[11] Florinskii states that the influence of climate on certain human characteristics, such as the color of skin and hair, or height, is indisputable. But variations of a human type due to "mixing" (*pomesi*) are much more significant. He cites the "father" of French ethnology William Frédéric Edwards: "All tribes, whose history is known, have more or less experienced such mixing. This cause [of the variability of the type] is all the more important since it affects internal organization. If this cause were acting without limitation, it might have eliminated all tribal differences" [10].[12] "Any tribe that is left to itself, whose blood is not refreshed and renewed, usually diminishes physically and morally," Florinskii states, "sooner or later, a pure type degenerates" [13].

It is common knowledge, Florinskii claims (directly refuting Zaitsev's statements to the contrary), that "all races, as well as nations and families, could interbreed and produce the progeny of an intermediary type." He illustrates this statement with the gradual "whitening" of the descendants of mixing between black and white races, identified in the then current classification as mulattos, quadroons, and octoroons. The same process occurs with other "racial mixes," such as between Mongolians and Caucasians. Florinskii demonstrates this phenomenon on the population of Siberia, referring to his own observations, as well as reports of other travelers: "Crossing the Urals, every traveler could see how the Slavic-Russian type of inhabitants begins to change and vary, acquiring Finnish and Tatar-Kalmyk features, in south-eastern Siberia — Mongol-Buryat ones, and in Yakutia — Yakut ones" [10].[13] He takes a historical excursion into the times of the Mongol yoke and describes its effects on the "Slavic-Russian type," which, he emphasizes, could easily be seen in all of the social estates of the empire. In the same vein, he discusses the historical origins of the Cossacks, and depicts their current "type" as the direct result of intermixing with various tribes they had encountered in their settlements on the borders of the empire.

Furthermore, in Florinskii's opinion, "in the present time, due to her geographical and ethnological situation, Russia, perhaps more than any other country, provides conditions for the variability of her inhabitants" [14]. He offers numerous examples of "mixing" among the various "tribes" inhabiting the empire, including, "Poles, Jews, Gypsies, and tribes of the Caucasus and the Baltic," as well as the intermixing of imperial subjects with foreign nationals, such as Germans, English, French, Italians, and so on. The results of such intermixing are particularly visible in the capitals, large trading centers, and seaports, as well as in the border regions, he observes. "Encountering beautiful Russian brunets and brunettes with an elongated face, a wide forehead, and a straight, narrow nose, reminding one of the Italian type," he writes, clearly portraying his wife and her family (see fig. 1-6, 4-3, and 4-7), "we, if possible, sought to analyze their genealogy." Such analyses demonstrated that "for the most part, ... in the line of their ancestors, sometimes several generations prior, there had been such invading elements as Jewish, Armenian, Georgian or Italian" [20]. Florinskii continues his depiction of intermixing among various human groups with the examples of the mixed populations of Vienna, Budapest, and Prague, as well as the inhabitants of northern and southern Italy, which he had himself observed during his European tour. "These examples," he concludes, "prove sufficiently that tribal and national types are variable. ... The variation of a type as a result of intermixing among nations could fluctuate greatly, depending on the elements participating in the mixing"[19].

Florinskii notes that a "lack of intermixing" helps maintain stability of certain "types." He points out that such "isolation" could result from different causes, geographical and social. He demonstrates the existence of geographical barriers to intermixing by referring to the visible geographic variability in the populations inhabiting different regions of the Russian Empire. Thus, he claims, if one were to compare the peasants of the Moscow province with the inhabitants of other provinces, "with sufficient experience, one could even determine quite accurately from which particular province a peasant comes" [21]. To illustrate the influence of social barriers, he refers to the enclaves of the "old believers" (for instance, in the St. Petersburg province) who married only within their faith, and thus did not mix with their neighbors, preserving their own "type."

If geographical and social isolation prevents the intermixing of different types, Florinskii observes, the fracture of such isolation could facilitate it. Therefore, regular or occasional migrations of certain groups, in his opinion, could influence the variation of "the type." As examples of such influences, he discusses seasonal relocations of peasants to the cities and billeting of military regimens (especially the Guardsmen, whose members have been specially selected for their physical features, such as height and strength) in various garrisons all over the empire. In both cases, he claims, one can see clear marks of the "invading blood" in the physique of local inhabitants.

Florinskii concludes his treatment of human variability with a forceful statement: "The aforementioned facts and discussions, I hope, prove clearly that human types are variable and changeable [and] that they are affected to a very considerable degree by intermixing and partially by external environment. These ideas must serve as the cornerstone of our further arguments" [24-25].

## Heredity

Florinskii's "further arguments" focus on heredity (*nasledstvennost'*). "We have to establish the causes and conditions of the variability of the type, therefore, [we have] to begin with heredity," he declares, "because without the knowledge of its laws, it is impossible to understand how exactly the human type improves or degenerates" [24-25].

This was a bold statement, for the notion of heredity at the time was quite ambiguous. As in other European languages, in Russian the very noun "heredity" was relatively new and not yet widely used.[14] Indeed, it is absent from contemporary dictionaries. Vladimir Dal's massive *Explanatory Dictionary of the Living Russian Language* (1863-66) has an entry for the verb *nasledovat'* (to inherit), which also explained the meanings of its derivatives, the adjective *nasledstvennyi* (hereditary) and the noun *nasledie* (inheritance), but not *nasledstvennost'* (heredity).[15] Similarly, the monumental, three-volume (each more than 1,000 pages long) *Table Dictionary for Inquiries in All Fields of Knowledge* compiled and published in 1863-66 by one of *Russian Word*'s authors, Feliks G. Toll', has an entry only for the adjective "hereditary."[16]

As both Dal's and Toll's dictionaries make clear, all of these words were used almost exclusively in two areas: in jurisprudence they denoted inheriting property or title; in medicine — inheriting a particular ailment (such as gout, insanity, or hemophilia). Florinskii was obviously familiar with all of these Russian words and their popular meanings. But as a trained physician, he also certainly knew the word *hérédité* that in the prior few decades had acquired terminological status in French medicine and had begun to make inroads in specialized medical vocabularies elsewhere, as the contemporary usage of *Heredität* in German, *heredity* in English, *eredità* in Italian, and *nasledstvennost'* in

Russian readily indicates.[17] Thanks to his specialization in gynecology, obstetrics, and pediatrics, Florinskii was well versed in ongoing debates regarding *nasledsvennost'* (heredity) and set out to explain its meanings to his readers in considerable detail.

Florinskii uses the word "heredity" and its various derivatives in two related but separate meanings, both of which had been articulated in the contemporary literature. The first refers to the *process* of inheritance. The second denotes *what* is being inherited. Yet he does not simply follow the common views, as expressed, for instance, in Prosper Lucas's popular 1847 *Traité philosophique et physiologique de l'Hérédité naturelle*, which Darwin had used in his *Origin*.[18] For Florinskii, heredity as a process means exclusively the *transmission* of the "hidden potentials (*zachatki*)" of *all traits* — anatomical, physiological, mental, and moral — from parents to children. And it is the sum of such "hidden potentials" that, according to Florinskii, constitutes the heredity of an organism. But, he argues, even though these "hidden potentials" are transmitted from one generation to the next, their actual realization in offspring depends on a variety of environmental conditions.

Unlike his numerous predecessors, including Darwin, Florinskii provides a detailed analysis of "heredity as *the main cause* of human variability and perfection." A common understanding of heredity, according to Florinskii, is limited to the well-known phenomena of familial resemblance between parents and children, expressed in a popular belief that "like produces like." However, he insists, we need to understand the causes, mechanisms, and conditions, which produce not only similarities, but also dissimilarities between parents and offspring. He provides a succinct thirty-page overview of the contemporary understanding of three interconnected but different sets of phenomena: *reproduction, development,* and *heredity,* clearly delineating their different roles in generating both familial resemblance and divergence.

Obviously building upon certain parts of his course on theoretical gynecology, Florinskii begins with a general outline of human reproduction, providing thorough descriptions of the microscopic, anatomical, and physiological features of the ovum and the sperm, the main stages of ovo- and spermatogenesis, the menstrual cycle, fertilization, and the first stages of embryonic development in the mother's womb. He is clearly expanding on the recent developments in the newborn science of cytology, especially Rudolf Virchow's

*Cellular Pathology* (1858) that had popularized the notion of *omnis cellula e cellula* (every cell comes from another cell).[19] Florinskii stresses that without such detailed knowledge "the reader will not be able to follow subsequent discussions."

Both the ovum and the sperm carry "the hidden potentials (*zadatki*) of all the individual particularities of a person they belong to," he emphasizes, "this is what the phenomena of heredity and the resemblance between parents and children are based upon" [26]. On the other hand, he asserts, the process of individual development, and hence the realization of these "hidden potentials" is affected by the environment — the conditions of nourishment, growth, and organ use — in such a way that "inherited features" could "acquire a different direction, and, as a result, the resemblance between children and parents might get either less or more pronounced, especially in the course of several generations" [26-27].

Florinskii briefly describes the well-known facts of familial resemblance, pointing out that a new individuum could carry the features of only one parent or a mix of features from both parents. In turn, this new individuum could transfer to his/her children the features inherited from both parents or from only one of them, which accounts for the resemblance between grandparents and grandchildren. We still do not know exactly, he states, why in certain cases the progeny resembles more the father than the mother and, in other cases, the other way around. He surveys several current hypotheses that explained such cases as resulting from the "age, mobility, and strength" of sperms and ova, or from the time of fertilization in relation to the menstrual cycle, but emphasizes that "this regularity is still far from being fully investigated and substantiated in either humans or animals, which creates significant difficulties for the rational perfection of stocks" [27].

Unlike many of his contemporaries, Florinskii emphasizes a principal difference between heredity and development. He limits the phenomena of heredity exclusively to the *transmission* of the "hidden potentials" of parents' characteristics to the offspring. He goes to considerable length to dispel a popular belief that, since an embryo develops inside the mother's organism, "maternal influence" must play a larger role in heredity than paternal one. He explains that in this respect, the maternal organism provides only the necessary environment ("as the soil does for a plant seed") for the development of forms and features, which lay hidden in the reproductive elements (i.e. sperms and ova). Thus the maternal organism "plays no role in the replication (*vosproizvodstve*) of the features in the new individuum," and affects not the *heredity*, which is determined at the moment of fertilization, but the *development* of a new organism.

Florinskii next discusses the oft-observed phenomena of the "*stability and constancy of the breed* (porody)" [30]. He postulates that the degree of a "breed's" stability is in inverse relation to "intermixing": "the longer over the generations a breed has not mixed with other tribes and races, the more stable it is and the truer it transmits its qualities to the offspring of mixed marriages." He recounts well-known facts from animal husbandry about the constancy of certain breeds of cattle and sheep and about their "degeneration" that results from mixing with other breeds or wild forms of the same animals. Similarly, he notes, human stocks could also be quite stable. He states that different nations exhibit different degrees of stability and even provides a hierarchical list of nations according to their decreasing ability to preserve their features in mixed marriages: Jews, Armenians, Georgians, French, English, Germans, Slavic-Russians, Tatars, and Finns.

Florinskii then turns to an analysis of the causes of dissimilarities between parents and children. He derides a popular belief in the so-called "*zagliad*," according to which children could resemble not the husband but a relative or an "*ami de famille*," whom the mother saw frequently during the pregnancy. Noting that the roots of this belief could be found in the biblical story of Jacob and Laban,[20] Florinskii dismisses it as "absurd fables, created and perpetuated either by ignorance, or by the trickery aimed at hiding marital sins and digressions" [32]. He reiterates that "the type is transmitted to the offspring *exclusively* through the sperm and the ovum," and "if we see in a child a feature different from the ones that both of the parents possess, such an exemplar would force us to doubt the actuality of a lawful parent rather than the general laws of heredity" [32]. He treats a popular theory, known at the time as "telegony," with similar disdain: it was believed that a child born in a second marriage could still resemble in certain features the mother's first husband. "A child can never carry features of two fathers" [33], he asserts, since only one sperm and one ovum take part in the fertilization process.

Florinskii discusses another popular belief — "a return to the ancestral type" — that was often used to explain dissimilarity between parents and children. He admits that "the transmission of characteristics not directly from parents to children, but by jumps over one or two generations" does not contradict available scientific evidence and constitutes "a very interesting but poorly studied physiological fact." He assesses available data on familial "diseases and developmental defects," which have been observed to skip a generation or two, but calls for caution in interpreting such data as evidence of hereditary transmission, "since such diseases and defects could reappear in different generations due to accidental causes, independent of heredity." He offers his own interpretation of such cases. There are no "jumps in the exact meaning of the word." What we perceive as a jump is "actually the reappearance in the second or the third generation of those characteristics of the grandparents, which in the first generation were expressed unclearly, [were] concealed, but nevertheless existed." Florinskii refers to several genealogical studies he had conducted, which supported "the possibility of the transmission of certain characteristics over generations, with the complete absence of such characteristics in the intermediary generations." He insists, however, that such transmission is only possible through a direct line of inheritance, and not, as popularly believed, in "a sideline or a spiritual relation," such as uncles and aunts or Godfathers and Godmothers.

As a part of his discussion of heredity, Florinskii also touches upon the question of the determination of sexes. Using as his major source an 1863 report delivered to the French Academy of Sciences by Jean-Marc Boudin, eminent physician and the president of the Anthropological Society of Paris, Florinskii presents several existing hypotheses and focuses in particular on the alleged influence of the age of a parent on the sex of his or her child.[21] In his opinion, "the sex of an embryo, as well as its hereditary features and characteristics, depends on the predominant influence of one parent, or to be exact, on the predominant influence and strength of one of the elements of reproduction [the sperm or the ovum]" [40].

Along with the majority of his contemporaries, Florinskii to a certain degree subscribes to the notion of "blended heredity," according to which many characteristics of the progeny represent a blend of parental features. Such blending is most clearly manifested in a "mixed" skin color in the offspring of parents with different colors of the skin. He admits that the exact mechanisms of such blending are unknown, noting: "The mixing of types, as the mixing of colors, has its own laws and limits, which [one] might expect, will be defined precisely only in future investigations." But he is thoroughly convinced that "the foundation of such laws, that is, of the mixing of colors and forms, must be purely physical and mechanical" [33].[22] Following the then widely held view that acquired characteristics could be inherited, he contends that "not only natural but also artificially developed features could be transmitted" from parents to offspring, and provides examples of such transmission in domesticated animals and humans.

After outlining his general views on heredity, Florinskii moves to discuss the hereditary transmission of particular traits. He states that "parents could transmit to their offspring all [of their own] anatomical, physiological, and even partially, mental qualities" [40]. He draws heavily from Darwin's *Origin* in illustrating this statement with numerous instances of breeding various domesticated animals "by means of a rational selection, according to a preconceived plan with known-in-advance stripes and color spots" [41]. He then provides examples of the hereditary transmission of particular anatomical features in humans, including obesity, the size and shape of women's breasts, and the density and color of men's beards. Referring to anecdotal evidence he found in the literature[23] and personal observations from his own clinical practice, Florinskii demonstrates that "physiological" characteristics, such as the speed and easiness of child delivery, exceptional fecundity, and longevity, could also be transmitted to progeny: "physiological qualities and particularities are transmitted to the offspring in exactly the same way as anatomical ones" [45].

Florinskii next discusses the inheritance of "mental" (what we would call behavioral) characteristics. He again follows Darwin's views on the inheritance of such "mental qualities" as habits and instincts in animals, stating that "the majority of animal instincts are nothing but habits, which had been developed by exercise and strengthened by heredity." He borrows from Darwin various examples of such "inherited habits" in animals and provides a number of his own observations. He notes that human habits could also be transmitted by heredity: "who haven't seen that sometimes children reflect not only the basic features of their parents, but also [their] particular gaits, habitual postures, manners of speech, and many other small particularities." In many cases, "this is a result of imitation during upbringing," he cautions, "but it is very likely that heredity exerts here some influence as well."

The inheritance of specifically human *"moral qualities and mental abilities"* holds Florinskii's particular attention. He admits that "the issue has not been resolved yet. Some say that these qualities develop only under the influence of upbringing and regular exercise of organs, [and] that by means of upbringing one can make anything of a human being." But, according to Florinskii, "facts convince us otherwise" [46]. He rejects the usual objections against the heritability of "special talents," namely that "smart families give their children better upbringing, thus, no wonder, that [in such families] the children get smarter, while particular inclinations to, and abilities for, this or that science or art occur as a result of accidents, preferential training of such abilities, and hence, their more prominent development" [47]. According to Florinskii,

> Since moral and mental abilities must be considered exclusively as the result of a particular structure and development of the brain, it is clear that the material particularities of the brain's anatomy, and hence, the potentials for certain mental qualities, could, in fact, must be heritable, in the same way as all other human anatomical and physiological characteristics are heritable [47].

One cannot deny, he claims, the inherited abilities to mathematics, painting, poetry, and so on. If one accepts that the sperm or the ovum could transmit to the child certain physical features that appear many years after the birth, "there is no basis not to allow that the potentials of mental growth are also transmitted to the offspring" [47]. He clarifies his arguments: "It is self-evident that what is transmitted hereditarily is not mental qualities in their full development, but only the *potentials for such qualities*, i.e. the ability to attain a better development [of such qualities] with determined training" [48]. He gives an example, "a colt of a trotter (*rysak*) does not have the particular trot [of its breed], but it is very easy to develop it by exercise, while a colt of a draft horse cannot be trained to trot, no matter the efforts."

Florinskii takes special issue with the elitist interpretation of the inheritance of mental qualities. Many of his contemporaries, including Galton in his 1865 article, claimed that only elites were the bearers of "hereditary talents." But for Florinskii, "a genius could be born to a peasant, and a fool to a nobleman." In fact, he declares, "we see very often that the best members of our society came from an undeveloped or underdeveloped [social] milieu (*sreda*), for instance, from the peasantry, the town-folk, or the provincial clergy." He insists that this fact "in no way contradicts our above statements about the hereditability of mental qualities in general. It is self-evident, that in this [underdeveloped] milieu too, if not in larger numbers, one could find persons talented by nature, with a healthy and strong, if little developed, mind." He calls for distinguishing "an inherited, so to say, natural mind" from "a mind shaped by upbringing, in the same way as we distinguish a natural beauty of the body from an artificial beauty achieved by [physical] culture and manners and sustained by cosmetic potions and various accessories of refined care" [49].

Florinskii does not deny the role of upbringing in the *development* of "natural" talents. Quite the contrary. But he rejects the idea that the upper strata of society — "the blue bloods" — are more capable of such development than people from the lower strata. "It is enough to compare the level of learning at different educational institutions where people from the upper, middle, and lower strata of society are educated and to notice a relative percentage of capable, talented, and inadequate students," he states (perhaps referring to his own experience at the Medical-Surgical Academy), "to discover on whose side the advantage is," implying, of course, that it is on the side of the lower strata [51]. "We can only regret," he observes, "that not all societal groups have the same opportunity for the development of their natural mind, [and] that many excellent, talented individuals remain hidden in the mass of the people as wasted, unproductive capital that has neither purpose, nor use" [49].

From the inheritance and development of *individual* mental and moral qualities, Florinskii moves to the discussion of such qualities in "social organisms," — families, nations, and tribes. "Considering the general mental development of an entire nation, we must reach the same conclusion we have reached considering national anatomical types," he declares, "namely, that the mental form of a nation is developing, improving, and strengthening gradually, through its own training and favorable intermixing" [49-50]. Florinskii discusses how "mental types" develop in various tribes, nations, and civilizations. He again emphasizes the anatomical basis of mental capacities, noting that individual changes in the brain (whether attained through exercise or inherited from parents) are, in turn, transmitted to the next generation, thus facilitating a further increase in the "level of mental development" and, as a result, "the moral and mental level of a [national] type increases little by little" [49-50].

Florinskii provides three examples of this process by analyzing the historical development of the North American, Slavic-Russian, and Jewish "mental types." The "rapid progress of civilization" in the northern United States, in his opinion, had resulted from the transatlantic transfer of not only European books and learning, but also of the "European brain" that has been "multiplied and dispersed throughout the country by the best breeding stock (*proizvoditeli*) — the talented, smart, and energetic people of European ancestry." Similarly, he explains the stagnation of "Russian civilization" in the pre-Petrine time as a result of unfavorable intermixing with "less civilized nations," meaning the Tatar-Mongol tribes that had conquered and ruled the Russian lands for nearly 300 years from the thirteenth through the fifteenth centuries. After Peter the Great had re-established contacts with Europe, according to Florinskii, "the Slavic-Russian brain began not only to receive more fodder for its development and began to develop faster, but, on top of that, by means of physiological mixing with Europeans, it began to directly receive from them the already developed fruits of modern civilization."

The development of the Jewish "mental type" offers a different example. According to Florinskii, the brains of Jews had developed and strengthened in the course of millennia and the "race" became extremely stable in the transmission of its hereditary qualities. "Despite all the vicissitudes of life, all the repression and disadvantages of their social standing," he asserts, "the Jews not only preserve their national type, but also maintain its relatively high intellectual level." Even though they are dispersed all over the globe, unlike other "physiologically less stable" nationalities, they remain unabsorbed by surrounding tribes, and "have their own enviable representatives among scientists, diplomats, artists, capitalists, and so on" in their adoptive countries. "If one could calculate the percentage of exceptional personalities among the Jews relative to their general population and compare it with the percentage of similar personalities among other, younger nations," he suggests, "in all likelihood, a huge difference would be found in favor of the Jews." He concludes that "the mind of every nation is developing over time and every step towards development taken by the brain is imprinted [on it] as an indelible footprint that enters into the totality of national hereditary characteristics" [51].

Florinskii, once again, rejects the notion that elites constitute the best of a nation and thus could be taken as the embodiment and representatives of national development: "Not for nothing, the public opined that the moral and mental strengths of a nation could be expected not from privileged groups, but from the ordinary people (*narod*)." This is not a matter of simple numbers, he claims, the people are not just numerically larger than the elites, they "are mentally and morally stronger" [52]. He promises to discuss in detail the reasons for such a disparity in the subsequent sections dealing with the degeneration of races and families. Here he just notes that the main reason resides in the different ways of life. As any other organ, he declares, the brain atrophies if it does not engage in a regular work. Since the upper classes "live by exploiting the mental and physical labors of others and turn into parasites," their way of life involves very little, if any, mental work, which leads to diminished mental capacity among aristocratic families and even entire nations. The people, on the other hand, due to their oppressed existence, face numerous challenges and, thus, have to exercise their mental abilities on a daily basis. He acknowledges that "poverty sometimes represses human beings, kills their intellectual labor, but sometimes it also stimulates such labor."

The last issue Florinskii discusses in relation to the inheritance of moral and mental qualities is the question of "*the mental development of women*" [53]. In his opinion, there is no need to reiterate the fact that, with the exception of reproductive organs, a woman has exactly the same anatomical organization as a man, and hence she is "capable of exactly the same physiological functions" [54]. The question he wants to examine is "at what level of progressive mental development our woman stands, or better, to what degree her brain has been affected by intellectual exercise and strengthened by heredity to make her capable of original and fruitful intellectual labor." To answer this question Florinskii applies the same principles he has used in his discussion of national and class differences in mental and moral qualities. He states that historically, "in the course of centuries and millennia," the situation of women "as regards their intellectual development had been most unfavorable," because men's domination denied them participation in any important intellectual work. "Under the influence of such historical circumstances, women's brains, due to lack of exercise, developed differently from men's brains," he posits, and as a result, "women are less receptive to and less productive in intellectual labor" [55].

Florinskii admits that this line of reasoning might seem faulty: since every child receives his/her hereditary features from both parents, the female offspring should be endowed with the same capabilities as the male offspring. But, he states, we know that certain traits are transmitted only along the female, and others only along the male, lines of descent (what today we call sex-linked heredity). It is clear, he claims, that many differences — including mental ones — between males and females in both animals and humans are hereditary. This is further supported by the facts of "pathological heredity," namely that certain parental diseases are transmitted exclusively to female and others exclusively to male children. The exact causes of these phenomena are unknown, he admits, but they do suggest that in the distribution of parental — especially moral and mental — hereditary features between sons and daughters, such features tend to cluster in the progeny according to their sex, which allows him to suppose that a "women's brain type is also transmitted along the female line"[56].

According to Florinskii, this is the only possible explanation of "such minuscule percentage of women's mental capital that we have so far seen manifested in literature, sciences, and the arts." He refuses to accept that this is a result of exclusively social conditions. He reiterates that "habits for this or that way of life and for this or that kind of work are strengthened by heredity if they are repeated continuously in the course of many generations." But, he states, this is not to say that women are "inherently incapable" of intellectual work: "under different circumstances, women's brains will make a rapid move ahead and will in time become as receptive and productive as men's brains" [57]. "We can already notice a step in this direction," he observes optimistically, "a modern generation of women has among its ranks many representatives of this gratifying movement; we can only wish that the conditions of life did not hinder it [and] did not obstruct women's striving for this noble goal that makes humans the kings of existence." "Knowledge, says Bacon — only correctly organized [knowledge], we should add — becomes power," he concludes demurely.

## Conditions Conducive to the Changing of the Human Breed

The second essay appeared in the October issue of *Russian Word*, with the subtitle "Conditions Conducive to the Changing of the Human Breed."[24] Florinskii opens with a simple statement that in "seeking to perfect the human breed" one must consider three essential necessities: 1) the need to improve *health*; 2) the need to improve *beauty*; and 3) the need to improve the *mind* ("mental and moral qualities," in Florinskii's exact words) [1]. This troika of needs is so fundamental, he claims, that "every parent, consciously or unconsciously, strives for their fulfillment in the offspring." But, he observes, not every parent wants to, and is capable of, using "the rules of science, which could make such a fulfilment possible." He proceeds to outline those "rules of science," which, in his opinion, should define the *ideals* of health, beauty, and mind and, thus, guide human perfection.

Even though he listed *health* as the first component of his troika on the previous page, Florinskii begins with a discussion of *beauty*, emphasizing that every nation and even every social group has their own ideals of beauty, which not only differ widely, but also change over time. "There is no arguing about taste, the saying goes," yet he is convinced that there must be some common ground to these varying ideals. Answering Pisarev's call for the meaning of beauty "independent of the limitless variability of personal tastes,"[25] he finds such common ground in close association between the notions of health (both physical and moral) and the perceptions of beauty: "true beauty is inseparable from health." He refers to the statues of Greek and Roman antiquity as prominent examples of such true beauty. "Everyone," he claims, sees in the Apollo Belvedere or the Venus de Milo "a full harmony of physiological life," not only a complete harmony of the physique, but also the perfect development of each separate organ.

Florinskii clearly follows the utilitarian aesthetics of Dobroliubov and Chernyshevskii, much admired and popularized by *Russian Word*'s contributors including Blagosvetlov, Pisarev, and Zaitsev in their discussions of Russian literature.[26] "Beauty that goes against the physiological functions of an organism or an organ should not be considered a true beauty," he asserts, "as the notion of the beauty of a dress should be related to its utility" [2]. He castigates certain popular notions of beauty, which are based on admiring the pathological, as opposed to the healthy: "A consumptive woman, despite all the loveliness of her luminous eyes and her flushed face [well-known symptoms of tuberculosis], cannot be considered representative of women's beauty, as a rotting grouse cannot serve as the example of this bird's flavor, even though there are numerous admirers of both." Similarly, he claims, "too small and narrow a head, despite all the graciousness that some want to see in it, cannot be beautiful, since it contains too little of the brain."

All the divergence of individual tastes notwithstanding, Florinskii insists, our attractions to and repulsions of particular persons are not mere whims: "they only seem such, because the process of appraisal based on various features of a person unfolds too fast, imperceptibly." As regards beauty, the words "like" and "dislike," which we often utter without thinking, always have a foundation. This foundation, according to Florinskii, is sexual instinct, "a very powerful and very important instinct" that ensures the propagation of every kind of animal, as well as humans. It is sexual instinct, whether "camouflaged by pink colors as heart relations," or "adorned by the halo of love and morality," that underlies the notion of human beauty. In its essence, then, the notion of beauty is based on "the ability or inability of an individual to reproduce" [4], he declares, which cannot possibly exist without health, "without the normal organization and functioning of all bodily organs" [5]. Florinskii does not dwell much on the last member of his "troika," merely asserting that "the need to improve mental and moral qualities is as characteristic of humans as the need to improve beauty and health."

Florinskii next defines three sets of conditions conducive to changing of the human breed, which thus could affect perfections in health, beauty, and mind: a) "taste and demand for certain qualities"; b) "external circumstances of life"; and c) "rational marriage."

## Taste and Demand

"Everyone," declares Florinskii, "consciously or unconsciously, strives for seeing in oneself and in one's offspring better qualities, be it external features or moral qualities." And this is exactly what guides the choice of marital partners, leading, as a result, to the perfection of humans, in the same way as "methodical and unconscious selection" described by Darwin had led to the perfection of domesticated animals. "*Love*, in a physiological sense, is the manifestation of not merely sexual attraction," he reiterates, "but also of aesthetic attraction, that is, not merely the need to reproduce, but a barely conscious or completely unconscious need to produce better progeny" [5-6]. He cites Darwin's description of sexual selection (*polovoi podbor*) among animals to illustrate the universality of the process by which "brides and grooms" choose each other, according to the preference for the better over the worse. "This is natural selection (*estestvennyi podbor*) by means of which better breeds reproduce," he asserts, "but worse and weak ones became rare and disappear." The awakening of sexual instincts, he claims, leads to the development of the feeling of love, "that is the preference for a better (according to personal tastes) individual to a worse one and the drive to enter into marital sexual relations with this [better] individual."

If we accept that "love is a sexual aesthetic choice (*vybor*) with the goal of producing more perfect offspring," we must recognize, Florinskii states, that "the principles, upon which this choice is made, differ among different individuals, different estates, and different nations" [7]. These principles, according to Florinskii, are largely defined by *"demand and personal taste for various qualities."* He illustrates this statement by examining preferences for a particular physique in different social estates in Russia. The peasant, for whom a wife is first of all a laborer, "in choosing a bride looks most of all for physical strength, health, and a well-built [body]." "In the olden days," this was an all-Russian ideal, shared by all social groups, Florinskii claims, citing as evidence a lengthy description of an "ideal woman" taken from the eighteenth-century *Pis'movnik*, arguably the most popular and widely read Russian book.[27] This old ideal, Florinskii notes, now is preserved almost exclusively among the "lower class" and it is among this class that a corresponding type of women could still be found. He observes that a similar ideal, emphasizing physical strength, height, and a thickset body, was applied to men. As evidence, he refers to the image of a mighty Russian knight (*bogatyr'*) perpetuated in numerous folktales.

If we look at the current ideals, Florinskii asserts, we could see that "among the middle and the upper classes notions of beauty, and hence taste [for particular features] have changed completely." According to Florinskii, this change occurred to a large degree under the influence of "romantic" belles-lettres, which propagated the image of pale, thin, unearthly men and women [8]. "What kind of offspring this type of men and women can produce," he exclaims: "The constitution of a child is the result of heredity; upbringing could only maintain and preserve it, from a weak seed cannot spring strong progeny, no matter how much effort we put into its nurture" [9]. Florinskii does not claim to know which ideal of beauty is better or more natural, "the readers could decide for themselves." But, he warns his readers, "taste and demand are reflected in human generations in exactly the same way they are reflected in the perfection of domesticated animals according to a certain set goal (fine-fleece sheep, trotter horses, and so on) by means of artificial selection (*iskusstvennyi podbor*)" [10].

Florinskii continues along the same line of reasoning in assessing the influence of demand on the changing of human minds: "Along with the demand for physical beauty, naturally, there must exist a demand for moral and mental strength." He claims that until the early eighteenth century such demand was largely applied to men: "all that was required of women was not mind, but kindness and beauty, slavish submissiveness and modesty" [10]. "Only in very recent times, when women came to be seen as equal to men in their mental prowess," he observes, "did [we] begin to expect and demand [from women] more serious education and more serious mental activity." Consequently, he states, "there began to emerge among women persons who more or less satisfy such demand" [11]. According to Florinskii, "There is no need to dwell on the influence of demand for mental development in men — it is self-evident." Nevertheless, he observes, we can notice a differential application of such demand to various national and social groups within a particular society, and the effect of such differential application is profound.

"There was a time when in Russia the largest portion of mental labor, at least in certain specialties, was given to foreigners," Florinskii notes, "since native Russians were considered incapable of this [kind of] labor, in the same way women are still considered incapable." As a consequence, he claims, the country has fallen behind in the development of such specialties: "until quite recently, it was thought that only a German [man] could be a mechanic, an apothecary, a physician, a professor, and so on, but that a Russian [man] lacks both patience and the mind for this kind of work." However, he insists, this attitude has changed and "now the public is convinced that a Russian [man] too could be a good doctor or a mechanic," which, according to Florinskii, "guarantees that these specialties will be developed on our native soil." But, he continues, "the public is still unconvinced that women could work in certain professions as successfully as men do and that not only a German [man] could be an apothecary." And as a result, "we still do not have either women professionals, or Russian apothecaries, since there is no demand for either" [12-13]. He believes that demand for "mental labor" will necessarily drive "everyone capable of such labor" to "aiming their life activity in this direction" [13], thus ensuring the further mental development of his compatriots.

## Conditions of Life

Florinskii devotes the next section to "the external circumstances of life" which could influence "the changing of the human breed." He discusses very different factors such as climate, food, "life comforts," and "material wellbeing." In many ways, this section runs contrary to his previous arguments, since heredity rarely, if at all, enters his discussion. In assessing the influence of the "external circumstances of life" Florinskii merely lists numerous instances of human variation observed when such circumstances differ for different groups. He does not address the question of how, or even whether, such variations are hereditary and could be transmitted to the next generations. One could perhaps construe this omission as a reflection of his implicit belief in the inheritance of the characteristics acquired under the influence of all these factors, yet he never addresses this issue directly in his text.

"The influence of climate on the structure and external features of humans and animals," Florinskii reiterates, "is indubitable." A walk through a zoology museum, he says, perhaps remembering his own visits to the Natural History Museums in Paris and London, affords an easy comparison of animals of the same kind (bears, for instance) inhabiting polar and tropical regions. Such comparison clearly shows the influence of climate on fur color and height. The same observations hold true for humans: "travelling through Russia from north to south, one cannot fail to notice that a blond and short population gradually changes into a taller and more pigmented one" [14]. However, according to Florinskii, "the influence of climate, though reflected in the height [and] the color of skin and hair, does not extend to the bones of the skeleton." Climate does not change typical tribal characteristics, he asserts, for they "depend only on blood mixing."

As an illustration, Florinskii refers to the preservation of the "European type" in various settlements around the world. In several centuries since the beginning of colonial expansion, he claims, Europeans "did not undergo any substantial changes" whether they settled in Asia, Australia, Africa, or the Americas. If we look at English, French or Spanish settlers, we could easily see that they all "preserve the characteristic features of their ancestors" [15]. Similarly, he asserts, the Jews "constantly preserve the same characteristic forms and proportions, which constitute their national type," no matter whether they live in northern or southern Europe, even though such secondary traits as the color of eyes and hair do vary. If structural features are little influenced by climate, Florinskii states, the same cannot be said about physiological characteristics, for instance, longevity and fecundity. Apparently drawing from the Belgian statistician Adolphe Quetelet's monumental, two-volume "treatise on social physics," he notes that in hot climates life expectancy is considerably shorter than in the cold ones, while fecundity is noticeably higher.[28]

In a similar fashion, Florinskii describes numerous variations, which could be observed in both animals and humans, under different conditions of feeding (frequency as well as the quality and quantity of food), a sedentary or an active way of life, the quality of housing and clothing, use and disuse of certain muscles, and so on. In discussing such variations in humans, Florinskii particularly emphasizes the role of "material wellbeing," by which he means a combination of political, social, and economic factors. "There is no doubt," he asserts, "that material wellbeing exerts enormous influence on the perfection of the physical and moral qualities of an entire nation, as well as separate individuals, and that such perfection goes hand in hand with the success of a true civilization." Perhaps relying on his own observations in Siberia and Central Russia, he notes that it is enough to compare a village of free peasants (in Siberia) with a village of serfs (in Central Russia) to notice marked differences between their inhabitants: "An oppressed, degraded, impoverished population strikes one at the very first glance by its physical and moral underdevelopment." But, he contends, "place the same population under more favorable conditions of material life, and they will quickly catch up with the development of their more fortunate neighbors." "Impoverishment and slavery are the first steps to the degeneration of a nation," he insists, "therefore, those countries and those institutions, which grant the population more freedom and wealth, could expect to succeed better in the progressive development of the human breed" [19-20]. He promises to return to "the influence of poverty and slavery" on human perfection and degeneration in the later sections of his treatise.

## Rational Marriage

More than half of the second essay addresses "the most important condition for the perfection of the human breed": "the selection of spouses." Florinskii states categorically: "the qualities of offspring depend directly on the qualities of spouses" [20]. People have long recognized this truth, he claims, which is reflected in a number of proverbs and folk sayings, such as "an apple doesn't fall far from the tree" and "you reap what you sow."

But before he goes any further in his examination of what he terms a "rational marriage," Florinskii writes a long explanatory "aside" addressed to his "female readers." In fact, this aside looks more like a ploy designed (probably with Blagosvetlov's assistance) to divert the attention of the censor from the subversive nature of the subsequent discussions. Florinskii slyly presents "a torrent of admonitions and objections" that "the beautiful sex" might unleash upon him:

> How could [he], they would say, look so materially at marriage — the most sacred of life mysteries; so harshly and coldly view love, subordinating it — this whimsical and capricious child — to the laws of science and

reason. Finally, how is it possible to see marriage in the same way as the breeding of domesticated animals with the goal of improving the stock! This is beyond even the vagaries of materialism, this is cynicism! [20]

He claims that he understands this sort of indignation, but sees "its source in a series of misapprehensions," which he wants to clarify. "There is neither cynicism, nor materialism" in his views, he insists: "I respect the sanctity of marriage and love and do not deny the latter its enchantments." But, Florinskii warns his "female reader," "an unnaturally developed sexual love could have very bad consequences for the progeny." "Love, as any other need," he asserts, "must submit to limitations, to the control of reason, because this feeling could develop abnormally and irrationally" [21]. He does not in the least deny the necessity of love in marriage, but insists that "not every love is natural, rational" and that sometimes love could lead to a "positively harmful" marriage, not to mention "those marriages that go against science and reason." He maintains that a correct choice of spouses depends on "knowledge of those conditions under which it could be accomplished more easily (what's cynical about that?)." He admits that "some could say that the issue of marriage cannot be considered from a purely hygienic viewpoint, that moral and societal conditions are equally important." But focusing on the physiological and hygienic sides of marriage, he does not reject other sides, "as the physiologist, examining functions of the human organism and ignoring human social and other relations, in no way denies the existence of such relations and undermines their importance" [22].

Florinskii reminds his "female readers" that "we need to remember that marriage must be seen not as a matter of personal gratification, but as a very important act of civic life, as a mystery of reproducing the human breed." Therefore, the issues of marriage "must interest not only the two people [entering the marriage], but the entire society, science, and the law." He is convinced that if people would "understand that a marriage between an old man and a young girl, or between a TB sufferer and an epileptic, is not merely a folly, but a crime," they would avoid such irrational marriages, while "understanding the [necessary] conditions of a normal marriage" would help "develop a true, rational taste" and thus lead to "rational marriage."

The professor calls on his "female readers" not to "take offence" at his regular comparisons of "human sexual reproduction with the similar phenomena in animals." Such comparisons, he states, "would seem improper" only to those who are "accustomed to thinking of themselves as the kings of nature, [who] imbibed with mother's milk a conviction that there is nothing in common between them and animals." But "a thoughtful reader" must recognize, Florinskii insists, that "the organic life of humans and animals is so close to one another that the majority of their physiological processes are nearly identical." Therefore, while investigating these processes in humans, it is not only necessary, but very important "to use facts taken from animals, because they are much more accessible to experimental investigations and because in animals we could not only observe physiological processes, but also deliberately expose and direct them according to a particular purpose" [23]. The results of such experiments, he continues, could with full confidence be transferred to humans, thus we should not take offence, but instead value them. After this detailed exposition, he expects that his "female readers" "will not be horrified by the thought of the perfection of the human breed, of the rational matching of spouses, and such."

Florinskii proposes to look at three sets of conditions that should be taken into account in "considering the issue of spousal choice (*vybor suprugov*)": 1) the age of people entering marriage; 2) their moral and physical qualities, and 3) blood mixing. He states that "the question of at what age people should marry is as important as it is difficult to answer" [23]. Nevertheless he is ready to offer some guidelines. Generally speaking, his advice is that people should enter marriage only after the organism's development is completed. He suggests a median age for completing this process as 25 years for men, and twenty for women. He claims that these figures do not mean that neither men, nor women are incapable of sexual life before that age. He provides numerous examples, many from his own clinical practice, of much earlier sexual maturity in both men and women. But is it sexual capacity (manifested by the onset of menstruation in females and nighttime ejaculations in males), he asks rhetorically, that is the real sign of their readiness to reproduce?

Florinskii presents evidence in support of two opposing viewpoints, first by those advocating for early marriage, and then by those arguing against it. Many observers have noted, he states, that "children of very young parents (not completely formed and strengthened) are distinguished by slow growth, weakness, and strong predisposition to disease." Citing a voluminous article on "Animal husbandry as an argument for Darwin's theory," Florinskii notes that animal breeders have reported similar observations for a variety of domesticated animals, from dogs to chickens, and have also described various harmful consequences for the progenitors bred at too young an age.[29] His personal observations of pregnancy and birthing in young women of around 16-17 years of age, however, do not fully support these views; by all clinical indications, pregnancy, fetal development, and delivery were completely normal in this age group. Furthermore, he claims, "all obstetricians know that the younger the woman who gives birth for the first time, the easier the delivery," while "a woman who gives birth for the first time after the age of twenty-five rarely escapes needing some surgical obstetrics assistance" [25]. He suggests that, if we look at marriage only from a physiological point of view, "that is as sanctified by the law means of producing offspring," and consider only the ability or the inability to reproduce, "then nothing would preclude marriage at the age of 17-18 for men, and 15-16 for women." But, he cautions, we must also take into account "the moral side of marriage and the social conditions of family and society." "Aside from sexual capability," he states, people entering marriage "must have abilities and strengths for civic life, a certain amount of education, [and] a certain maturity of reason" [26]. Russian law allows marriage for women at sixteen and for men at eighteen, which, he contends, is not contrary to the physiological requirements, but not entirely corresponding to the social ones.

Many observers, Florinskii notes, argue that late marriages are often harmful to offspring, "despite the stable material conditions" that usually accompany such marriages: "Children born to the parents of advanced age ... almost always have inborn (*vrozhdennye*) deficiencies of constitution — the result of the weakness of the parental organisms."[30] He then discusses the influence of "*the age difference between husband and wife*" on the quality of their offspring. He notes that "in the lower classes this difference usually ranges from two to three years, and in the educated class — from five to ten." He considers these ranges to represent a physiological norm, since women age faster than men. "The ability to have children in women lasts till 40-45 years," he states, "in men the exact limit is unknown, but in any case it lasts much longer than in women." Therefore, he concludes, a marriage between spouses of the same age is not ideal, while a marriage between an older woman and a younger man is positively unfavorable.

But he saves most of his indignation for marriages between old men and very young women: "Everywhere in our society we see the sad facts of men of advanced age, even greybeards, marrying youngish girls," which, he states, "in equal measure defile the sanctity of matrimony and undermine public health"[31]. Florinskii's indignation resonated with and perhaps was further amplified by the societal response to a painting by Vasilii Pukirev, titled *The Unequal Marriage*, that created a furor in the fall of 1863 in St. Petersburg (see fig. 3-1).

The next set of conditions to be taken into account in a "rational marriage" is the morbidity (both physical and moral) of potential spouses. Since moral and physical characteristics are inherited, Florinskii posits, marriage should be arranged in such a way that relative deficiencies of spouses do not strengthen but weaken each other, "thus neutralizing the morbid (*boleznennaia*) heredity of both spouses" [32]. He declares forcefully that "from a medical viewpoint, marriages between two sickly, cachectic individuals must be forbidden. Otherwise, the weakening and degeneration of the stock will be unavoidable" [32]. At least one of the spouses, according to Florinskii, should have "a healthy constitution," thus "suppressing the morbid heredity of the weaker element." He is convinced that the growing number of people afflicted with "syphilis, tuberculosis, and scrofula, which weaken our population, speak volumes as a vivid reminder of our neglect of the rules of hygiene in marriage."

Florinskii gives a long list of practical advice to prospective brides and grooms aimed at balancing out the weak and strong elements in their constitutions, temperaments, functioning of particular organs, heights, talents and moral characters, and hereditary predispositions to certain diseases. This lengthy list basically comes down to the suggestion that everyone should "strive to marry a person with contrary qualities" and that "spouses' physical deficiencies should not be identical with one another." To support his views he cites the recommendation of the eminent French gynecologist and social hygienist Louis Alfred Becquerel: "a strong, stout man, with dark skin and well-developed musculature should marry a blond [woman] with blue eyes, white and fine skin, and a lymphatic temperament."[31] Florinskii particularly emphasizes the importance of anatomical and physiological capability to producing

3. *The Book: Darwinism and Social Hygiene* 155

Fig. 3-1. Vasilii Pukirev's painting, *The Unequal Marriage*, 1862, oil on canvas, 173 x 136.5 cm. The painting was first presented at the 1863 annual exhibition by the St. Petersburg Imperial Academy of the Fine Arts, which opened shortly after Florinskii's return from his European tour. It earned Pukirev both professional (he was granted the title of the academy's full member) and public acclaim. From https://commons.wikimedia.org/wiki/File:Pukirev_ner_brak.jpg. The original is in the Tretyakov Gallery, Moscow, Russia.

progeny, stating sternly that women with "a wrong form of pelvis that prevents normal birth should not get married" [33]. He is equally stern on the issue of hereditary disease: "Individuals suffering from serious hereditary diseases, such as, for instance, hereditary insanity, epilepsy, and so on, should abstain from the pleasures of married life" [33].[32] All in all, he advises prospective spouses to choose "strong over weak, beautiful over ugly, smart over stupid, not sacrificing these significant qualities to petty benefits and calculations" [34].

"The rules of hygiene exist to distinguish the organism's actual need from false and perverse ones," Florinskii states categorically: "Following these hygienic rules could be very beneficial to individuals, families, and society at large." Alas, he laments, "physicians are not invited to participate in discussions of marriage, or in creating laws pertaining to this subject." Except for forbidding marriage among close relatives, he observes, "our laws do not prescribe any measures for the perfection of the physical status of humanity." He calls on the public to take notice of this important issue: "society should proscribe, as it does in the cases of serious crimes, those anomalies of marriage which bring about deadly consequences for the next generation." Public opinion, in Florinskii's views, should be unequivocal: "why can't we call a murderer a fifty-year-old man marrying a sixteen-year-old girl, or a parent suffering from TB or syphilis, as we call a person who takes somebody's life or health for his personal gain." He admonishes parents who are not concerned with passing on to the next generations their own afflictions and diseases, and promises to discuss in more detail "the consequences of irrational marriages in the essay on the degeneration of human stocks" [35].

Another set of considerations Florinskii offers as important for "rational marriage" concerns "*blood mixing*" (*pomes' krovi*). Generally speaking, he contends, "the inbred stocks of both animals and humans diminish." He cites Darwin's opinion that breeding close relatives decreases both the strength and the fecundity of a breed. He offers an example of the European aurochs, whose current population in Russia were descendants of just 500 animals that had remained alive in 1824. In the last forty years, due to interbreeding, the weight and height of the animals significantly decreased, attesting to their visible degeneration.[33] Similar facts could be observed in human stocks, he maintains, when the lack of blood mixing leads to the degeneration and extinction of families and nations. "There is no doubt that in general a blood mixing has very favorable effects," he states. "Isolate any estate," and, as a result, "in its midst soon there will be accumulated so many physical and moral deficiencies that it will diminish, degenerate, burn out, and go extinct, as every organic body burns out" [36].

We do not know exactly how the blood mixing works, Florinskii concedes, but we can discern certain conditions that could facilitate or hinder it. The main among them is isolation. When isolation is absent, a "continuous exchange of physical and moral features" between different groups of people leads to the renewal of national strengths and to the leveling off of their particular deficiencies. Revisiting his earlier discussion of human variability, he identifies two main kinds of barriers that could affect blood mixing, which could work separately or in combination: geographical and social. He observes that the variety of climates, landscapes, and other environmental conditions in the Russian empire were conducive to the variable development of inhabitants in different locales. Until recently, he states, "the lack of roads and the close attachment of our people to a concrete locale" — he is obliquely referring to the serfdom here — worked against blood mixing among them, preserving these local varieties. "The way of life of our people (agriculture), insufficient entrepreneurship, and the lack of [investment] capital and freedom (as a consequence of the serfdom)," he states, "were the causes of a far rarer mixing than there could have been" [38]. He hopes that "with the improvement of the transportation system and the increase in wealth, development, and freedom, the interactions among the inhabitants [of different locales] will improve, and then blood mixing and the renewal of provincial types will be much more noticeable." He is convinced that the Russian people are in no danger of degeneration, for "the diminution and degeneration of the type is possible in the future only under a very unfortunate political organization, namely, if the people's initiative were suppressed completely, the interactions among separate provinces stopped, the existing disjunction of the masses increased even more, and separate locales became even more isolated," which, he states optimistically, "of course could not happen" [39].

Florinskii next considers various social barriers to blood mixing, stating that "social (*soslovnye*) prejudices" of particular groups could hinder blood mixing among them. He observes that, motivated by "true or false notions of their own worth," these groups "think it improper to mix their own noble blood with a less noble, plebeian blood." According to Florinskii, nothing could be farther from the truth, for "in the mixing of different estates (*mezhsoslovnye pomesi*), the better qualities of combining elements would dominate over worse ones, and therefore the more mixing occurs the more an estate's deficiencies will be purified." Under opposite conditions, naturally, the consequences will too be opposite: "the mass of physical and moral defects will accumulate more and more, and in the end will lead to the diminution and degeneration of the estate." This is why, he maintains, many aristocratic stocks have disappeared and continue to disappear. But, he states, "the fewer the prejudices against the plebeian origins, the better the chances for the moral and physical perfection of the stock" [41].

Political rights and freedoms, according to Florinskii, profoundly influence the success of mixing between different estates: "During serfdom one could not have expected the people to mix with the nobles." He is convinced that "the spread of literacy, the granting of the right to enter secondary and higher educational institutions, and the bestowing of a certain level of wealth and equal rights will serve as the Archimedes lever" in both promoting intercourse between different estates and stimulating the growth of the nation's productive forces. As an example, he considers the fate of the social estate he himself had come from, the rural clergy. He notes that the majority of the clergy had come from the people (*narod*) and their way of life had been (and in many instances still is) not much different from that of the people. But since the clergy had access to education and the right to enter universities, they became much more important to the life of the country: "They occupy a prominent place in the ranks of scientists, litterateurs, artists, civil servants, and so on." "But numerically the clergy is just a fraction of the people!" — he exclaims: "If the latter had the same access to education and civic activity, it goes without saying that a hundred-fold quantity of fresh and energetic forces would have entered our educated society," which, he implies, would tremendously benefit the entire nation.

Florinskii notes that religious prejudices are also very powerful in preventing blood mixing. As a rule, each confession allows marriages only within its own faith. Thus, the Orthodox marry only the Orthodox, not the Catholics, the Lutherans, the Muslims, or the Jews. As a result, he observes, despite the fact that Russian peasants and "alien nations" (*inorodtsy*) often live side by side in the same village, each remains totally isolated. For this same reason, he states, the German colonists who settled in various parts of the Russian Empire do not "disappear" within the surrounding populations. "Religion serves as a shield against absorption by a stronger tribe," he states, but at the same time, it prevents "borrowing from 'aliens' the particularities of their physical stature and ways of life by means of mixing." When religious barriers are removed, he observes, "aliens" disappear in the general population. To illustrate this point, Florinskii refers to the story of a military (male) choir, which Emperor Alexander I had given as a present to his Prussian counterpart king Frederick Wilhelm III after the end of the Napoleonic wars. The members of the choir had been settled in a village (named after the emperor, Aleksandrovka), on land near Potsdam, and had been allowed to marry local (Lutheran) women. Florinskii had visited the village during his stay in Berlin in 1861 and observed that "in the second generation [of the descendants of the original settlers], a Russian element remains visible only in appearance, but even there just barely" [43]. "The same could happen to an entire nation," he concludes.

## The Degeneration of the Human Breed

For some reason, the third essay was split between the November and December issues and, unlike the second essay, it did not have a subtitle.[34] The two parts, however, have continuous pagination, and their contents make clear that, as Florinskii indicated in the introduction to his treatise, the essay should have been titled "the degeneration of the human breed."[35] Echoing Shelgunov's analogy between a "physiological individuum" and a "social organism," it opens with a simple statement:

> Through all times, the history of mankind has shown endless examples of progressive as well as regressive movement. Whether we observe the human life *en masse* — in entire nations, or in separate groups — estates and families, everywhere we notice the same fact of the [initial] gradual increase, then the gradual or quick decrease, and, finally, the complete disappearance of life. The life of a nation, an estate, and a family has its own periods and limits, as does the life of an individual. Everywhere there are periods of youth, maturity, and old age. The duration of these periods for different nations and estates differs, as it differs for every individual human being, [and] depends on the supply of organic forces, on the strength of the physical and moral constitution of a nation, and on accidental disorders of state and social life [1].

But, Florinskii continues, "the degeneration of peoples (*narodov*), to which history bears witness, occurs in the same way, though on a smaller scale, during our own epoch, before our own eyes" [2]. He describes two major ways in which such degeneration takes place: first, "*by means of the fusion with, or the transformation into, another nation,*" and second, "*by means of diminution, weakening, and extinction (vymiranie)*" [2]. In the first case, "a tribe or a nation does not disappear in the direct sense of the word, but only changes into a different, usually better, form." Florinskii sees this not as regressive, but to the contrary, as progressive movement, which allows a less developed tribe "to catch up with" the more developed ones. In direct opposition to both Royer's and Zaitsev's views, he considers this a much more humane and just outcome of the contact between tribes of different degrees of development than the alternative: "slavery and inescapable extinction." The second path of the degeneration of the human breed, according to Florinskii, has its main causes in "deplorable tyranny, enslavement, poverty, and moral failings." He backs this statement with mortality statistics among slaves of African descent. According to numerous observations, he states, whether in New York, or in French, English, or other colonies, slaves have twice the mortality rate of free people, while "the emancipation of slaves in Saint-Domingue led to the doubling of their numbers" in a very short time [4].

But it is not only colonialism and slavery that drive extinction. According to Florinskii, poverty in general is a leading cause of degeneration. A comparison of mortality among different classes and social groups in European countries shows vast disparities: "In England the [median] life expectancy of the upper and educated classes is fifty-eight years, while among the working and poor classes it is only thirty." "Poverty and misery act most deadly on children," he asserts, providing statistical evidence for his claim. In London, child mortality before the age of ten constitutes among the gentry only 2% of all mortality, among merchants and retailers — 6%, and among the poor — 28%. In Dublin, among the children of the working class up to 33-36% die before the age of two, and in bad housing this number rises to more than 50%.[36] These examples are sufficient, he argues, to show that "the health and longevity of the people closely depend on all conditions of life, and especially their material wellbeing."[37] "In an oppressed race," he postulates, "the number of deaths will be higher than the number of births" [5].

Florinskii likens the extinction of human tribes to the displacement and extinction of "weak animals and plants" in competition with animals and plants that are "more advanced and endowed with better means for the self-preservation and dispersion of its kind." So, it is the competition with the strong (in the case of humans, exploitation by the strong) that is the real cause of extinction. "The more equal is the distribution of wealth and estate's rights and privileges, and the fewer exploiting parasites there are in a society," he reasons, "the greater the harmony and success in the development of people's forces and the more protected the [poorer] estates will be against weakening and degeneration" [6].

Florinskii is not attempting to solve "the issue whether such harmony is possible in complex human societies." For him, it is enough to point out the link between degeneration and poverty — this "social sore that threatens the life of entire estates and even races (colored)." "How to prevent or diminish this threat is not a question for natural history," he admits, "but a social question." Nevertheless, he insists, a rational society must recognize that the oppressed classes need "not [our] laments regarding their helpless situation, but such social institutions that would support them and prevent their impending degeneration." "If we merely state, even on the basis of historical facts, that a lower race is destined by nature to die out and just rest the case," he remarks in a clear allusion to Zaitsev's statements about the "destiny" of the black race, "we would have sinned both against the truth and against humanity." If all we have seen so far is that the white race had enslaved and displaced the colored ones everywhere and that the serfs had suffered under their masters, he maintains, "this is not the fault of Nature and Fate, but the bloodthirsty inclinations of humans themselves." "It is not completely impossible to reign in such inclinations, to protect the weak against the strong, and to equalize the rights and means of existence," he declares forcefully, "to the contrary, trying to do so is the sacred duty of everyone who can help by word or deed."

The process of degeneration, according to Florinskii, affects not only races and tribes, but smaller groups, as well: "The alternate rise and fall of families, the flourishing and declining of privileged estates represents a common and commonly known phenomenon" [7]. To support this statement, he cites *Don Quixote* (in French!):

> Il y a dans le monde deux sortes de races; l'une tire son origine des rois et des princes, mais peu à peu le temps et la mauvaise fortune l'ont fait déchoir, et elle finit en pointes, comme les pyramides; l'autre, partie de bas, a toujours été en montant, jusqu'a faire naitre de très grands seigneurs de manière que la différence qui

existe entre elles, c'est que l'une a été ce qu'elle n'est plus, et que l'autre est ce qu'elle n'était pas.[38]

Already Aristotle had noticed the degeneration of old aristocratic families, he claims. History is full of similar stories about the Egyptian and Syrian kings, the Venetian patricians, and the French aristocracy. "The examples of such degeneration of aristocratic families and privileged classes more generally could be seen wherever these classes formed closed circles" [8], he explains echoing Byron's poem, since in making spousal choices they value the "noble origins" over the intellectual and physical qualities of a spouse. The inbreeding amplifies progressively all possible defects and gives "to a whole estate a particular stamp of moral impotence and emptiness." "This atrophy, this racial suicide," he observes, "happens gradually and therefore the individuals involved do not notice their [own] self-destruction" [8].

Extending his previous discussion of the ideals of health, beauty, and mind, Florinskii rejects the popular notion that the aristocracy represents the best examples of beauty in particular nations. The aristocratic beauty, he states, "only concerns the external features and largely does not harmonize with the [natural] physiological functioning of organs." "In the same way as English racing horses, merino sheep, Berkshire pigs, miniscule dogs, and so on, cannot be considered the [natural] zoological examples of these animals," he elaborates, "so too, the aristocratic type, perfected exclusively in terms of external beauty and gentility, cannot be considered the example of human physical perfection." Various breeds of animals were produced for special purposes, he observes, but to the detriment of "the physiological functioning and harmonious development of more important organs," and thus represent "degeneration, not perfection."

## Consanguineous or Kin Marriages

The next section of this essay deals with "conditions conducive to the degeneration of the human stock." In the treatise's introduction, Florinskii promised to discuss under this heading such phenomena as "incest and lack of stock renewal, inequality [of partners] in marriage, [and] the influence of drunkenness, debauchery, diseases, poverty, and slavery" [4]. But, almost the entire essay dealt with only one, according to its author, the most important, condition: "kin marriages (*rodstvennye braki*)" [10].[39] Florinskii states that people had long recognized the danger of marriages between close relatives, supporting this statement with a long quote from the Bible that prohibits sexual relations among kin (Leviticus 18: 6-18). "Humanity has a kind of instinct against incest in close degrees of kinship," he claims: "Whether we consider this instinct as natural or as a consequence of centuries-old traditions, it nevertheless exists" [11]. He notes that in recent years, "particularly from 1859" (that is from the publication of Darwin's *Origin*), the question of consanguineous marriages (*krovnye braki*) has attracted scientists' serious attention: "science has already collected so many facts that this issue could now be seen not from a religious or a philosophical, but from a purely scientific point of view." Yet, scientists strongly disagree on the subject. Some argue heatedly against such marriages. But some advocate for them. Florinskii details the debate on the subject, providing lengthy description of the pro and contra arguments.

What he did not tell his readers, however, was that the debate had largely unfolded in the Anthropological Society of Paris, between its President Jean-Marc Boudin (1806-1867) and its Secretary Eugène Dally (1833-1887), exactly at the time he was in Paris in 1862 and 1863, and that this debate was reflected in various Russian periodicals.[40] Furthermore, Konstantin Tolstoi (1842-1913), one of Florinskii's students who had graduated from the Imperial Medical-Surgical Academy the previous year, published in *Medical Herald* a lengthy article about the debate just before Florinskii's first essay appeared in *Russian Word*. Indeed, from 26 June to 31 July, under the general title "On Consanguineous Marriages," the newspaper carried six weekly installments of Tolstoi's article, which probably prompted Florinskii to devote such a disproportionate attention to the subject in his last essay.[41]

On the next fifteen pages Florinskii recounts the arguments of the *opponents* of consanguineous marriages. Both individual observations and statistical data show, he states, that the offspring of such marriages often exhibit the increased appearance of various deficiencies and diseases, including "deaf-mutism, idiocy, albinism, anatomical defects, weakness of development, decrease in fecundity, predisposition to miscarriages, and so on." He begins his discussion with deaf-mutism.

Since the condition was easy to detect, it had commanded attention of the medical profession since ancient times, and became a popular subject of research in the nineteenth century.[42] Florinskii summarizes numerous studies of deaf-mutism conducted by eminent French physicians, including Boudin,[43] Francis Devay,[44] and Prosper Ménière,[45] which appeared in such influential periodicals as *Annales d'hygiène publique et de médecine légale*, *Gazette médicale de Paris*, and *Journal de la Société de Statistique*.[46] He also cites similar findings by British, German, Swiss, and North American doctors. All of these studies, he states, have found a much higher proportion of deaf-mutism among the progeny of consanguineous marriages than among the offspring of "normal" marriages.

Many authors, Florinskii maintains, have reported various psychiatric disorders, such as hereditary insanity, low intelligence, and epilepsy, to be directly associated with kin marriages: "It has been noted for a long time, that people, who marry only within one closed circle and carefully protect their blood from any external impurities, engender in this way the degradation of mental abilities in the subsequent generations. Such degeneration is not at once manifested in idiotism or insanity, but gradually develops into these [afflictions]" [18-19].[47] Florinskii further recounts the discovery of the association between consanguineous marriages and certain diseases of the eyes (*retinitis pigmentosa*) by the German ophthalmologist Richard Liebreich.[48] He also provides numerous examples of albinism observed as a result of inbreeding in both humans and animals.[49]

Among other morbid conditions "indicating the degeneration of the stock," Florinskii lists decreased fecundity, anatomical deformities, and weakened physique. He recounts reports on various anatomical deformities in the offspring of kin marriages, which had been observed in fetuses and newborns, including extra or missing digits and late teething, citing Devay's finding that in 121 consanguineous marriages, seventeen had offspring with extra digits.[50] Florinskii provides numerous examples demonstrating that "consanguineous marriages not only weaken and disfigure the progeny, but, by diminishing their numbers, lead to the complete disappearance of entire families" [22]. He relates Devay's observations that out of 39 kin marriages, eight were completely sterile and among the rest pregnancies were often accompanied by miscarriage.[51] He sums up the findings of numerous authors: "The decrease in fecundity is a constant phenomenon in consanguineous marriages" [24]. "The fatal, weakening influence of consanguineous marriages on the offspring," he concludes, "is proven by many positive facts" [25].

## Consanguineous Marriages

The final part of the third essay, and thus, the conclusion of the entire treatise, appeared in the December issue, but unlike the previous installments, it did not open the issue.[52] Subtitled "The Influence of Consanguineous Marriages," it continued the discussion of the subject began in the previous part. Here Florinskii surveyed the arguments of the *advocates* of consanguineous marriages and presented his own views on the subject. He concluded the treatise with a discussion of the role of the law in regulating marriages in order to forestall the degeneration and to promote the perfection of the human type.

Florinskii recapitulates the conclusions of the previous installment, that those scientists who argue against consanguineous marriages present numerous facts demonstrating the "harmful influences of incest on the progeny." Yet, many equally reputed scientists maintain that consanguinity in and of itself is not the cause of those various diseases and deformities people ascribe to it. These scientists claim that this issue is much more complex than it seems at first glance and that the harmfulness of consanguineous marriages should be considered a result of "morbid heredity."

The proponents of consanguineous marriages offer three types of evidence in support of their views: observations of inbreeding in animals, observations of individual families that for many generations inbred, and observations on the populations of certain very isolated geographical locales. According to Florinskii, evidence gathered from animal breeding provides the main foundation for the support of consanguineous marriages. Many veterinarians and agriculturalists insist that crossing (*skreshchivanie*) of pure-blooded animals, despite their close kinship, not only produces good results but actually improves the breed. By crossing very close relatives — fathers and daughters, sons and mothers, brothers and sisters — what English breeders named "breeding in and in,"[53] they have created a number of very good breeds of horses, cattle, sheep, dogs, and other domesticated animals, developing them to near "perfection" [28]. This led many breeders to adopt a rule of breeding a stock *"in itself,"* following the idea that "the best qualities of progeny will be not only preserved but even improved in the course of generations."

If one accepts these facts for animal stocks, some scientists argued, there was no reason not to apply them by analogy to human stocks. Based on these facts, the advocates of consanguineous marriages formulated a special law: *"consanguinity strengthens heredity to its highest power."* Thus consanguinity is favorable, if the procreators-relatives are healthy, but harmful if they have defects and deficiencies that could be transmitted and thus "strengthened" by heredity. Florinskii relates the opinion of André Sanson, an eminent French veterinarian and member of the Anthropological Society of Paris: "The better every deficiency, as well as every good quality, is transmitted to the progeny, the closer the relation among the progenitors (*proizvoditeli*)."[54]

Florinskii states that this view contradicts many observations made by agriculturalists, naturalists, and physicians, including Boudin, Devay, Darwin, and many others.[55] He relates their shared opinion that the methods of producing purebred stocks by means of crossing close relatives pioneered by British breeder Robert Bakewell[56] are useful only when the goal is to preserve a certain feature of the stock that is valuable to the owner, "but not for the perfection of a race in a zoological sense" [28]. Florinskii presents the view that, for the most part, what is considered the perfection of stock in many domesticated animals is at the same time a physiological defect: "these animals are nothing but physiological monsters" [29]. He sums up the objections against evidence from animal breeding: "If the breeding of relatives will continue for many generations, then the race will necessarily weaken and degenerate. This is a general law for both animals and humans." "Every pure breed, without renewal of blood, in the course of time, not only loses its qualities, but becomes sterile," he continues. "A temporary usefulness of consanguinity (in a few generations), which is manifested in the increased development of certain qualities to the detriment of the organism's general perfection, could only be applied to domesticated animals and only for special purposes," he asserts forcefully, "but not to humans" [30].

Florinskii then proceeds to consider certain observations on humans made by the numerous proponents of consanguineous marriages,[57] which were summarized in a special report by the Secretary of the Anthropological Society of Paris Eugène Dally.[58] All of these authors claimed that such marriages were *not always* accompanied by harmful consequences. Each of them reported cases in which marriages between close relatives, even continuing in the same family for several generations, not only were fertile, but did not cause any of the diseases and deformities described by their opponents. Florinskii finds Dally's examples not entirely convincing, since "to clarify the issue it is necessary to investigate the comparative frequency of appearing diseases and deformities in kin and non-kin marriages. Only such comparative statistics will allow a sound conclusion" [30]. He recounts Boudin's objections to Dally's arguments,[59] and finds them much more convincing, since Boudin used such "comparative statistics." According to Florinskii, all the examples collected by the proponents of consanguineous marriages "in a scientific sense, prove neither usefulness nor harmfulness" [32] of such marriages, since they do not provide enough comparisons and enough statistics.

Florinskii then examines the third type of evidence offered by the advocates of consanguineous marriages, namely the absence of harmful consequences in the interbreeding of inhabitants of isolated locales, summarized in Dally's report [33]. Florinskii notes that the opponents of human inbreeding such as Devay used the materials from the same locales to prove exactly the opposite conclusion, namely that consanguineous marriages are harmful. Both opponents and proponents use the same materials, mostly related to the so-called accursed races (*le races maudites*) in France and Spain, such as "the Pyrenees Cagots," "the Vaqueros of Asturias," and "the Alpine cretins."[60] Florinskii states that "all these facts … do not offer anything concrete and conclusive. So many different conditions influence the development of the inhabitants of isolated locales that, even with the most precise analysis, it is impossible to say what is the result of consanguineous marriages and what depends on climatic conditions, hereditary diseases, material and moral ways of life, and so on" [33-34]. He comes to the same conclusions regarding the facts related to the consanguineous marriages among another two groups — the aristocracy and the Jews — often cited by both sides of the debate: "to isolate the influence of consanguinity from numerous physical and moral conditions not only difficult, but completely impossible."

Despite all the uncertainties, Florinskii feels that the advocates of consanguineous marriages did not present statistical proof of their views, while their opponents did. He offers a thorough discussion of statistics as a method of investigation. The method is not perfect, he admits, in the process of collecting statistical data, certain facts are sometimes grouped incorrectly, making it difficult to discern all the external influences, and thus sometimes statistics generate "imprecise and flimsy" results [36]. But, he insists, statistics are absolutely necessary and very convenient in addressing many issues that could not be solved by any other method. He agrees with many observers that collected statistical evidence demonstrates convincingly that kin marriages produce unfavorable consequences with much greater frequency than non-kin marriages. Yet, according to Florinskii, this evidence does not answer the most important question of why this is so: "Is this unfavorable influence a result of consanguinity, ipso facto, even if the spouses-relatives do not have any physiological and pathological potentials (*zadatki*) for the weakening of the stock? Or is it a result of the doubling of their morbid heredity?" [36]. He notes that the advocates of kin marriages accentuate a difference between "healthy and morbid consanguineous marriages," arguing that, if the procreators are healthy, such marriages would be beneficial, and, if they are not, "hereditary diseases [would] grow and perpetuate in the descending generations." In their opinion, the bad consequences are not the results of consanguineous marriages per se, but of hereditary diseases.

Florinskii deduces that this line of reasoning does not explain all the facts. Whether morbid or normal, for him, heredity is a replication of parental characteristics in the progeny. But, the defects of development and diseases, which are observed under the conditions of consanguinity, often appear *only* in progeny and, moreover, many of these deficiencies are not hereditary. For instance, deaf-mutism is "almost never hereditary." Florinskii refers to the expert on deaf-mutism

French physician Prosper Menière, who demonstrated that "in the overwhelming majority of cases, a marriage of two deaf-mute parents gives birth to children who speak and hear."[61] He approvingly cites the opinion of Boudin, that to invoke heredity to explain the unfortunate consequences of consanguineous marriages is to completely distort the notion of heredity: "Consanguineous individuals, who are full of strength and health [and] do not have any deformities and deficiencies, transmit to their children not something they have, but something they themselves do not have, and this is called heredity!" [37].[62]

According to Florinskii, this rebuttal forced the proponents of consanguineous marriages to consider questions about heredity in general and, particularly, "about the transformation of hereditary diseases during their transmission from one generation to the next." They came up with the idea that a particular disease in one generation could produce "a completely different, separate disease in the next." The foundation of this view on "hereditary diseases," Florinskii asserts, is a "theory of the gradual degeneration of the human type and the formation of morbid races" recently proposed by French physician Bénédict Augustin Morel.[63]

Florinskii gives a brief summary of Morel's theory.[64] Morel allows the possibility of the transformation of "nervous diseases" into "mental disorders" and of the latter into "physical disorders," and so on, up to "the complete extinction of a race" [37]. "Hereditary morbidity," according to Morel, not only causes infertility in parents and early mortality in their progeny, but also "arrests the physical and mental development of apparently healthy children, leading to the appearance of brain diseases, which, in turn, lead to idiocy or epilepsy." Morel postulates that various "harmful influences, having generated in ancestors a nervous disease, in the progeny that is still exposed to such influences produce, sequentially, hysteria, epilepsy, hypochondria, idiocy, or insanity." In such a way "any disease, strengthening and changing with every generation, could produce the diminution of height, scrofula or English disease, an arrest in the development of certain organs, the inborn deformities of the scull, nearsightedness, strabismus, Saint Vitus Dance, and so on" [38].

Florinskii finds Morel's theory "ambiguous and jumbled" and not very useful in explaining the "unfortunate consequences of consanguineous marriages," especially the statistically proven more frequent appearance of such consequences in kin marriages as compared to non-kin ones. Even if we accept Morel's scheme of "hereditary morbidity," he contends, it should work the same way in all marriages, and the number of affected progeny should be "roughly equal" in both kinds of marriages. "As we saw, however, the facts show something completely different," he states. In his opinion, no theory could overturn statistical evidence: "the fact that consanguineous marriages very often result in infertility, various serious diseases, and deformities remains indisputable" [39].

In the end, Florinskii admits, whether one explains this fact exclusively by heredity or exclusively by consanguinity, all physicians would advise a society *"to avoid kin breeding and to renew the stock* with a new, alien blood in order to prevent the weakening and degeneration of the stock." "Since there is hardly a family in our society," he continues, "that does not have some hereditary diseases, because similar diseases and deficiencies predominantly cluster in the same social class, then hygienic advice to mix different social strata and different families by means of marital ties in order to renew the stock with stronger and healthier elements or to paralyze and equalize morbid heredity with the heredity of an opposite quality [i.e. healthy heredity] is in any case sound."

This long discussion of consanguineous marriages serves as a basis for Florinskii's "glance at our civic laws on marriage from a hygienic point of view."[65] He explains that, in Russian law, the question of whether a marriage between related individuals is allowable or forbidden is decided according to the rules of the individual's religion.

He goes on to explain that the Orthodox Christians cannot marry a relative in "the fourth degree of relationship inclusive,"[66] while Jewish law is much more permissive and allows not only a marriage between an uncle and a niece, but also between an aunt and a nephew, forbidding only marriages between brothers and sisters (see fig. 3-2). Although Christian doctrine generally forbids marriages between relatives, in "Lutheran and Catholic countries, civic laws allow marriages between the first cousins, and sometimes even between uncles and nieces or between aunts and nephews." Thus, what is allowable for Jews, Lutherans, and Catholics, is forbidden to the Orthodox. These differences, he declares,

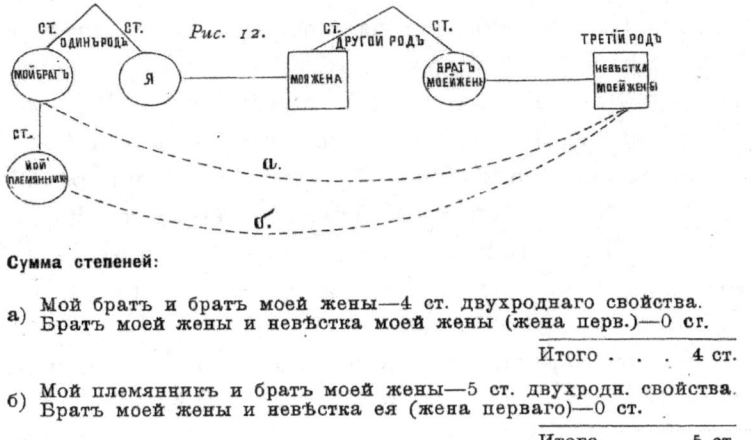

Fig. 3-2. A table illustrating the degrees of kinship according to the Russian Orthodox Church regulations. Circles represent male and squares female members of the families. First row (left to right): my brother, myself, my wife, brother of my wife, his wife. Second row: my nephew (son of my brother). The kinship "distance" between my brother and the wife of my wife's brother is four degrees (therefore, a marriage between them would be prohibited). The kinship "distance" between my nephew and the wife of my wife's brother is five degrees (therefore, a marriage between them would be permitted). From S. P. Grigorovskii, *O rodstve i svoistve* (SPb.: Trud, 1903), 6th ed., p. 25. Courtesy of RNB.

derive from historical or purely theological views of various faiths. "But everyone knows that physiological laws are absolutely the same for all religious confessions," he insists, "hence everything that is harmless for a Lutheran is harmless for a Catholic, and vice versa" [40]. On the other hand, he asserts, everyone should agree that every law, whatever issue it addresses, "could remain strong and unshakeable only when it is based on an actual need and aims at the positive guarantee of the moral and material wellbeing of the people." Therefore, Florinskii surmises, "a law on marriage must justly be based on physiological and hygienic data and must limit only what is, in whatever way, harmful." All other restrictions that are "based not on scientific, positive principles, but on conscience and conviction, and whose obeying or disobeying is not injurious to the people's wellbeing, should be left to the good will of citizens themselves." He is convinced that "beliefs should be free and voluntary" [40].

While defending the freedom of religious beliefs and civic convictions, Florinskii does not condone absolute freedom of actions. Every civic law, he argues, is created with the purpose of balancing personal benefits and life comforts of all citizens, and hence must, to a degree, limit their freedom. To be effective, a law must derive from a strong and rational foundation and its restrictions should be rationally recognized by all members of a society as a necessary condition of their personal wellbeing. Therefore, laws on marriage must put forward only those restrictions that are necessary to preserve public health. This position makes clear, he states, that a law prohibiting marriage between consanguineous individuals is well founded. But it should regulate only direct bloodlines, i.e. marriages between brothers and sisters, first cousins, uncles and nieces, aunts and nephews. Marriages among side-line relatives — "relations of the wife of my brother or of the husband of my aunt" — he asserts, should be excluded from the law, to say nothing of "spiritual relations" such as Godmother and Godfather, which were actually forbidden by the current Russian law.[67] "Free action of every person, based on freely-held convictions," would be "the best formula for a happy marriage of this kind," he states. Florinskii is adamant that civic laws should be the same for all citizens, hence, the Russian law on marriage for Lutherans and Catholics should be extended to forbid consanguineous marriages in the direct lines of relations.

If we recognize the right of the law to regulate consanguineous marriages, Florinskii continues, we should allow the legislation "to intervene into other hygienic aspects of marriage." For instance, he declares, we know that "many hereditary diseases (TB, epilepsy, hereditary insanity, and so on) are transmitted from parents to children," therefore, "legislation could take certain measures against the weakening and degeneration of the stock in this way." He foresees, however, that the implementation and enforcement of such laws would be quite difficult. Indeed, it is nearly impossible, he admits, "to establish the exact limits and concrete regulations regarding hereditary diseases: in which cases morbid heredity is detrimental to marriage and in which it is not." But this difficulty cannot be used as "an argument against the legislative regulation of consanguineous marriages." The blood relations of a bride and a groom, as well as the age of prospective spouses, can be determined before marriage; therefore, Florinskii argues, in these two instances, certain rules could be established, which would not infringe on individual freedom. But "regarding the questions of health and hereditary predisposition, those questions to which at the present time it is difficult to give a judicially positive answer," he insists, "the legislation so far cannot intervene without the burdensome limitations of the free choice of families and individuals" [41]. He suggests that "in such doubtful cases" prospective spouses should seek a doctor's opinion, which "without limiting personal freedom could do more good than any law."

Florinskii concludes his expose with a discussion of the prohibition of marriages of Orthodox Russians with Jews and Muslims as per both church rules and state laws. He finds such prohibitions "to be too strict." "We have already demonstrated," he declares, "that nothing brings peoples closer together than marital relations [and] nothing facilitates the perfection of the stock so noticeably as various mixes." He is convinced "that the freedom [for orthodox men] to marry Muslim, Jewish, and heathen women could bring substantial benefits, even from political viewpoints."[68] Such freedom, he asserts, would facilitate to a substantial degree "the fusion of Russians with peoples of 'alien' confessions" and "the absorption of 'alien' peoples by the mass of the Russian tribe." This would be beneficial to both Russians and "aliens," for, according to Florinskii, "marital ties are the best path to the initiation and diffusion of civilization." "Everywhere and everyone is set upon this path," he concludes, "except perhaps for the Chinese who fear that the freedom of marriage with the Europeans could deliver a fatal blow to their immobility" [43].

Thus ended Florinskii's treatise.

## A "Russian" Eugenics?

The contents of Florinskii's essays demonstrate that he forged a particular amalgam combining the contemporary ideas about human reproduction, heredity, variability, development, and evolution with his personal beliefs, values, and ideals, along with a specific set of measures that addressed certain perceived social concerns and which, he thought, could avert possible degeneration and lead to perfection of humankind. In short, he created his own version of what Galton would perhaps have little trouble recognizing as "eugenics."

Florinskii's treatise was a "thought piece." Although he claimed that to support his ideas he had traced a "few genealogies" (most likely the ones of his wife and her family), unlike in Galton's 1865 article, these genealogies did not provide a foundation for his arguments. He used them merely to illustrate certain points of his thinking. Unlike Galton, he did not describe these studies in any detail, presenting only his conclusions, and thus did not offer a template other researchers could follow. His treatise focused on the "political" significance of the natural sciences and, more specifically, on the possible implications of Darwin's concept for "social issues," a subject of intense interest to the editor and writers of *Russian Word*. Florinskii did not emulate Royer's or Zaitsev's simplistic, "mechanical" transfer of Darwin's ideas about varieties, species, and speciation to the current political discussions of racial and/or class inequality. Rather, he sought to apply Darwin's key notions of *variability*, *heredity*, and *selection* as the fundamental principles of species evolution to the understanding of the past history, the present state, and the possible future of humanity.

Darwin's book presented a monumental synthesis of numerous facts in the anatomy, embryology, reproduction, behavior, geography, paleontology, heredity, physiology, breeding, and taxonomy of animals and plants. Florinskii extended this synthesis by applying Darwin's analytical scheme to similar facts about the *human* species. Furthermore, he supplemented these facts with the new knowledge regarding differential morbidity, mortality, and fecundity of particular well-defined (by age, sex, class, ethnicity, location, and occupation) human groups furnished by rapidly developing social hygiene. Darwin identified three basic phenomena that could occur in the course of the existence of any species: *divergence, extinction,* and *preservation* ("survival with improvement," in Darwin's own understanding, and "progress," in the understanding of his numerous followers). He considered them to be the *inevitable outcomes* of various forms of selection (natural, artificial, and sexual) driven by the struggle for existence or by the deliberate goals of plant and animal breeders. In his own synthesis Florinskii set out to examine how Darwin's "laws of selection" would play out when applied to humans.

There was clearly no doubt in Florinskii's mind that humans had evolved and were continuing to evolve, which meant that they had to face the possibility of outcomes "predicted" by Darwin's theory. The possibility of divergence in the human species was undeniably manifested in its existing "racial" variations, which travelers, ethnographers, and anthropologists had described in numerous accounts. The possibility of its extinction was also clearly visible in various signs of "moral and physical degeneration" in individuals, families, tribes, nations, and civilizations, which had been "documented" by historians and were currently "observed" by contemporary commentators, from physicians to litterateurs and from anthropologists to theologians. Following Darwin's logic, both of these phenomena must have resulted from some form of *selection* that had been and was still acting upon human beings; left unchecked, this selection would inevitably lead to either the divergence, or the extinction of the human species.

Yet, according to the numerous publications of *Russian Word*'s authors, unlike other living creatures, humans were no longer helpless subjects to "nature's inescapable laws." They had developed a powerful instrument that could allow them not only to escape but also to subjugate these laws to their own will — science. To forestall the extinction/degeneration and to assure the survival/perfection/progress of the human species, then, scientists only needed to identify and halt those particular forms of selection that had produced, and were still producing, divergence and degeneration. They needed somehow to counteract this "negative" selection with some "positive" measures that would "improve" the species and thus assure its preservation and progress. And this is exactly what Florinskii sought to accomplish in his work: to identify the sources and causes of degeneration/extinction and to discover the principles of perfection/progress of humankind. Florinskii personally did not consider the divergence of the human species to be a problem. Quite the contrary, he saw human variability as an advantage, for the "mixing" of different families, tribes, races, nationalities, nations, etc., in his opinion, actually advanced the perfection of the human type as a whole.

In preparation for writing his treatise, Florinskii likely read everything published in *Russian Word* and other "thick" journals on the interrelations of the natural and social sciences, as well as on the more specific topic of Darwin's theory and its possible "social" applications. His essays contain numerous direct and veiled references to, hidden quotations from, and open dialogues with these publications. Judging by the content of his essays, he cast his net wide and read nearly everything then available on the topic: the amount of literature he had consulted and cited in his essays is massive.[69] Seen through the lens of *Russian Word*'s publications and discussions, Florinskii's treatise offered an entirely new approach to this topic and its key issues. It indicated that, combined with the principles of social hygiene, Darwin's theory could not only shape the understanding of certain social issues, but also uncover effective tools for solving them. Among such social issues, he addressed the ubiquitous "women's question," the empire's perceived backwardness, the exploitation of one social group by another, and the impact of the law on "human perfection and degeneration."

Much like Galton's, Florinskii's major idea was based on obvious analogies/association among the three forms of selection identified by Darwin: natural, sexual, and artificial (in both its "unconscious" and "methodical" varieties). This association was likely facilitated and amplified by Rachinskii's translation of Darwin's term "selection" as "matching of kin" (*podbor rodichei*), as well as by Pisarev's consistent use of the word "choice" (*vybor*) as a synonym of "selection" and his equation of the origin of species with "progress."[70] Furthermore, Florinskii's own choice of the word "*usovershenstvovanie*" (perfection) for the treatise's title was perhaps prompted by Rachinskii's use of the same word (in adjective form) in his translation of the phrase "favoured races" (*usovershenstvovannykh porod*) in the subtitle of Darwin's *Origin*. In this "collective" translation, Darwin's theory in its Russian version amounted to "matching" — by various forms of selection — those animals and plants that possessed any favorable characteristics, thus ensuring their survival, avoiding their extinction, and producing new ("modified and improved," in Darwin's favorite expression)[71] individuals, varieties, breeds, subspecies, species, and ultimately, in Pisarev's interpretation, "progress."

Florinskii proposed to "combine" artificial and sexual selection by matching spouses who possessed certain favorable characteristics to advance perfection and stave off degeneration of humankind. He suggested replacing "unconscious selection," which, according to Darwin, had led to the initial improvement of domesticated animals and, which, according to Florinskii, had so far directed — in the guise of love

— marital choices among humans, with "rational marriage," paralleling what Darwin had named the "methodical selection" of animal and plant breeders. Florinskii identified those "favorable" characteristics — the ideals of human health, beauty, and mind — which together should guide a "rational" selection of spouses. Since he equated beauty with health and mind with the brain, "physical and moral health" emerged as the key criteria of spousal choice in his posited "hygienic marriage." Furthermore, he identified certain "wrong choices," such as, for instance, consanguineous marriages, which were based not on the ideals he described, but on social (financial, national, racial, religious, class, etc.) biases, and which, therefore, led to degeneration instead of perfection. The prevention of such wrong choices and the propagation of correct, rational choices constituted the essence of his concept of "marriage hygiene," while changing existing laws and social mores around marriage served as the main instrument for its actual implementation in his Fatherland.

Given the differences between their life trajectories and personal experiences, coupled with the profound dissimilarities in nearly every feature of their respective homelands (from political organization to laws and from economy to social structures), it comes as no surprise that Florinskii's version of "eugenics" differs substantially from Galton's. Although both had common roots in Darwin's *Origin* and its detailed analysis of artificial, sexual, and natural selection, domestication and speciation, heredity and variability, Florinskii's medical background profoundly shaped his vision. To give but one example, if Galton's "extraordinary gifted race" is defined exclusively by its "hereditary talents" and "hereditary genius," Florinskii's "perfection of the human type" hinges on the trinity of "beauty, health, and mind" and ultimately translates into the perfection of "physical and moral health."

Furthermore, Florinskii's specialization in gynecology, obstetrics, and pediatrics gave him a much deeper and much more elaborate understanding of human reproduction, heredity, and development than that expressed in either Darwin's *Origin* or Galton's early writings. His expertise and experiences allowed the young professor to disentangle the customary notion of heredity as "like begets like" into three interconnected, but separate processes: the *transmission* of hereditary potentials (*zadatki*) for particular physical, mental, and moral traits from parents to offspring through the fusion of such parental potentials contained in ova and sperms during *fertilization,* and the subsequent *realization* of these potentials in the course of individual development. This pioneering view of heredity — as both a process of transmission and a set of certain potentials being thus transmitted, but realized under and according to the particular circumstances of the organism's development — directed the professor's attention to various factors that could influence one, or another, or all three of these fundamental processes. This in turn allowed him to discern possible individual and collective actions that could direct such influences to his desired outcome: the elimination of degeneration and the perfection of humankind.

At the same time, Florinskii's roots in the clergy, combined with his early theological education, made him quite sensitive to the traditional attitudes towards marriage embodied in the Orthodox Church's regulations and rules. In fact, his concept sought to explain and modify these attitudes in light of the contemporary knowledge about human reproduction, heredity, variability, and evolution. Thus, to a degree, Florinskii endorsed the church's prohibition of "kin" marriages, presenting them as an important source of degeneration. But he vehemently opposed the prohibition of "mixed," especially inter-confessional, marriages that he saw as a major instrument of perfection.

In certain respects, Galton's and Florinskii's programmes of action look quite similar. Both, for instance, are attentive to issues of individual liberty and choice and emphasize the need for educating the public about "correct views" on human reproduction, heredity, variability, and development. Both stress the necessity of state intervention in certain aspects of human reproductive decisions and propose legislative regulation of marriage. But, as evidenced by their respective goals, the norms, values, and ideals underpinning the two programmes differ substantially, reflecting the social origins and positions of their authors.

Aimed at the creation of "an extraordinary gifted race" and "men of a high type," Galton's programme embodies his bourgeois, atheist, imperialist, individualist, elitist, racist, and sexist views, which were well-entrenched among the British upper-middle class and were largely supportive and protective of established social stratifications, hierarchies, and roles under Queen Victoria. In contrast, directed more generally at "the human breed," Florinskii's programme is imbued with norms, ideals, and values highly critical and subversive of the social arrangements of the Romanov Empire under Nicholas I. This set of beliefs and values was advanced by a relatively new, small, well-educated, and very vocal social group, to which Florinskii himself belonged, the *raznochintsy* — literally, persons of various ranks.[72]

As the very name of this group makes clear, the *raznochintsy* came from a variety of low-level social backgrounds (the clergy, the peasantry, petty civil servants and military officers, impoverished gentry, merchants, town folks, etc.). They filled a social space between the empire's two major social estates — the landed gentry (*dvorianstvo*) and the people (*narod*) — and most of them identified with the latter. A defining feature of their value system was vocal opposition to existing social stratifications, hierarchies, and roles, first of all, serfdom, absolutist monarchy, hereditary nobility, and the tight grip of the Orthodox Church on many aspects of social life, from education to marriage. *The Contemporary* and *Russian Word* served as the major venues for the articulation and propagation of the *raznochintsy's* beliefs, norms, ideals, and values,[73] while the journals' major contributors — Chernyshevskii and Dobroliubov in *The Contemporary* and Pisarev in *Russian Word* — became the apostles of this new faith.[74]

Grounded in the famous triad of "liberté, fraternité, et égalité" of the French revolution and the deliberate opposition to the Russian Empire's official motto — Orthodoxy, Autocracy, and Nationality (*Pravoslavie, Samoderzhavie, Narodnost'*)[75] — this value system included the *raznochintsy's* strong commitment to advancing the interests of the people. It embodied their particular notion of patriotism that equated the Fatherland not with the monarchy, and even less with the person of the emperor, and least of all with the Orthodox Church, but with its people. A key element of this value system was the *raznochintsy's* firm belief in the tremendous potential that could be unlocked by granting civil liberties, economic independence, and, especially, access to education traditionally enjoyed by the nobility to the people. Belief in education, and its pinnacle — *nauka* — as the motive force of human progress and a means to make real the ideals of *liberté, fraternité, et égalité* in Russia became firmly embedded in this value system. It is easy to see all of these values expressed in one form or another in Florinskii's treatise.

A comparison of the perceived social concerns and the measures they advocated for addressing these concerns brings into sharp relief the contrast between the value systems underpinning Galton's and Florinskii's concepts. For Galton, the upper stratum of British educated society — exemplified by the "hundred illustrious men" (but no women) of science, the law, the state, the arts, the military, and the cloth he studied in *Hereditary Genius* — represents the "hereditary wealth" of the nation.[76] The "lower classes," on the other hand, represent both the source and the locus of "hereditary degeneration" manifested in crime, pauperism, insanity, and so on. Accordingly, Galton focuses on a set of social measures that would increase the reproduction of the upper classes and decrease that of the lower ones, and thus assure both the arrest of degeneration in the British nation and the creation of "men of a high type."

In Florinskii's concept, the situation is almost in complete reverse. It is the upper stratum of society, the aristocracy and the nobility, that is both the major source and the major locus of "hereditary degeneration." According to his views, such degeneration derives in large part from breaking the principles of "rational" or "hygienic" marriage in obeisance to social mores, economic interests, confessional biases, and perverted tastes "inherent" to the upper classes. It is the people that constitute the "hereditary wealth" of the nation by safeguarding the "hygienic" ideals, tastes, and customs guiding their marital choices. The fact that this "capital," as he calls it, "remains hidden" and "wasted" is not a sign or result of degeneration, but the direct consequence of the social, political, and economic order that denies the people the possibility for expressing and developing their "hereditary talents," especially by blocking their access to education. Unlike Galton who treats the "hereditary degeneration" of the lower classes as a leading *cause* of poverty, Florinskii instead considers poverty an important *condition* conducive to such degeneration and, thus, for him, eliminating poverty offers a way to forestall and prevent degeneration. The same logic and the same values undergird Florinskii's discussions of "women's mental capital" and the "hereditary mental types" of various nations. Not surprisingly, for him, the political, economic, and social equality of all people is the necessary foundation for human perfection, while "mixed" marriages between nobles and commoners, and between individuals of different confessions and ethnicities provide a powerful instrument to avoid degeneration and promote perfection.

The subsequent fate of Florinskii's treatise would be determined not only by the ideas articulated in his essays and the programme of actions he offered to address the perceived social concerns and anxieties, but also by the particular value system providing the scaffolding for his views.

# 4. The Hereafter: Words and Deeds

> "...Every new endeavor has to wait for its time, only then will it come into its own and fulfill its highest purpose."
>
> Veniamin Portugalov, 1870

After the appearance of "Human Perfection and Degeneration" on the pages of *Russian Word*, their author and their publisher parted ways as suddenly as they had come together just a few months earlier. In February 1866, the imperial authorities suspended publication of *Russian Word* and, in early June, shut down the journal altogether. Nevertheless, in late August, Grigorii Blagosvetlov released Vasilii Florinskii's essays in book format. He kept the book in print for more than a decade, regularly advertising it on the pages of *Deed*, a new journal he created after the prohibition of *Russian Word*. Florinskii, however, completely withdrew from further collaboration with Blagosvetlov and never published anything in his new journal, even though it continued to carry numerous articles that explored various issues raised by Florinskii's treatise.

Much as with Francis Galton's early works, at the time of their publication, Florinskii's essays went virtually unnoticed. Similar to Galton's first studies on "hereditary talents," Florinskii's concept of "human perfection and degeneration" held a promise of generating a viable research programme, stirring public opinion, and even initiating policy change. But it proved impotent in raising the interest of either the scholarly community, or the general public, to say nothing of the imperial bureaucracy. Furthermore, in contrast to Galton — who spent several decades and a large portion of his personal fortune gathering

© 2018 Nikolai Krementsov  https://doi.org/10.11647/OBP.0144.04

support for his eugenic ideas, methods, and concepts in the course of his long and distinguished career — Florinskii never returned to the subject of his eugamic treatise in any of his later works and did absolutely nothing to promote it among his colleagues or the general public. He found other avenues for applying his talents and "bringing benefits" to his Fatherland.

Numerous personal, social, scientific, and political factors contributed to the sudden end of what at first had seemed a very productive collaboration, as well as to Florinskii's abandonment of a promising line of inquiry and to its general neglect by its intended Russian audience.

## Blagosvetlov's *Deed*

Blagosvetlov was more than pleased with "Human Perfection and Degeneration" and had certainly counted on Florinskii's continuing collaboration with *Russian Word*. In the December 1865 announcement of the journal's plans for the next year, the publisher stated that Florinskii would become a "regular contributor,"[1] which would have been hardly possible without the professor's explicit permission and commitment. The first 1866 issue came out on 15 February, but it did not contain anything written by Florinskii. The very next day, by order of the minister of internal affairs, publication of *Russian Word* was suspended for five months.[2] Inventive as ever, Blagosvetlov decided to circumvent the order and to fulfill, as best he could under the circumstances, his obligations to journal subscribers by publishing at least some of the materials he had promised would appear in *Russian Word* as a two-volume "learned-literary" collection, under the general title *The Ray*. Each volume was about 800 pages in length, so that together the two volumes roughly equaled the length of the five issues of *Russian Word*, which subscribers would not be receiving due to the journal's suspension. An announcement of the new collection published in early March again listed Florinskii, along with Dmitrii Pisarev and Petr Tkachev, as a contributor.[3] Yet when the first volume came out a few weeks later, Florinskii was not among its authors.[4] Nor did he contribute to the second volume that was scheduled to appear in early May, but was seized by the censorship office and eventually destroyed.[5]

Perhaps, Blagosvetlov was hoping that Florinskii would resume writing for *Russian Word* after the journal's suspension was to end in July. But on the morning of 4 April, Dmitrii Karakozov, a 25-year-old member of a small "revolutionary circle" of Moscow University's students, tried to assassinate Alexander II.[6] His shot went wide and he was arrested on the spot. Karakozov's attempt on his life profoundly shook the emperor and his closest advisors. It opened the gates to a country-wide search for "revolutionaries," especially among the students and recent graduates of the empire's schools of higher learning. Not unexpectedly, *Russian Word* — the oracle of the young generation — became one of the first victims of this witch-hunt. Ten days after Karakozov's assassination attempt, on 14 April, the secret police arrested Blagosvetlov and, two weeks later, Zaitsev. Although both were soon released (Blagosvetlov in early June and Zaitsev in early July) because the secret police could not find any direct links between the assassin and *Russian Word*, on 3 June, the journal, together with *The Contemporary*, was shut down for good.[7]

We can only speculate on why Florinskii stopped writing for Blagosvetlov and his journal before this happened. Once again, available documents shed no light on the subject. Various considerations might have influenced this decision. To begin with, during the academic year of 1865-1866, Florinskii was busier than ever with his work at the Imperial Medical-Surgical Academy. He was teaching his regular courses and running the women's ward. On 6 November 1865, on top of these duties, the IMSA Council put him in charge of lecturing on pediatrics and establishing a pediatric clinic.[8] He also continued writing his regular 100-plus-page surveys of current foreign literature on his specialties for the *Military-Medical Journal*,[9] as well as other scholarly works.[10] Indeed, he got so busy that he declined to continue as the scientific secretary of the St. Petersburg Society of Russian Physicians, even though the society's members had unanimously re-elected him to the post.[11]

Given Florinskii's incredible productivity, if he really wanted to, he probably would have found time to write for *Russian Word*. But, at the beginning of January 1866, just as the November 1865 issue of the journal that carried the opening part of his third essay came out, Florinskii found himself at the center of a public scandal. On 11 January, *Petersburg Leaf*, Russia's first tabloid,[12] published a letter by a certain

Neonilla Kondrat'eva.[13] The letter claimed that a month earlier, on 13 December 1865, Florinskii had rudely refused to visit the terribly sick three-year-old child of the Andreevs family that occupied an apartment two stories up from Florinskii's own. Kondrat'eva alleged that Florinskii had demanded a payment of 25 rubles for the visit, which was way beyond the means of low-level civil servant Andreev, and even the tears and pleadings of the child's mother could not move him to change his mind.

Florinskii was outraged. The same evening he knocked on the Andreevs' door and demanded an explanation. Mr. Andreev, the child's father, declared that he did not know anything about the matter and even gave Florinskii a written deposition to that effect. Florinskii sent an indignant letter to the newspaper's editor, with Mr. Andreev's deposition attached. Florinskii explained that the first time he had ever heard about the sick child was from the newspaper and that the entire story was nothing but a fabrication. He demanded that *Petersburg Leaf* publish a retraction and provide him with Kondrat'eva's address, so that he could ask her personally where she had gotten her story. Four days later, the editor did publish Florinskii's letter along with Mr. Andreev's deposition, but he refused to give Florinskii the address of his accuser.[14] Furthermore, in the same issue, the tabloid carried a long letter signed by Mrs. Andreeva, the child's mother, who repudiated her husband's deposition. She claimed that Kondrat'eva had told the truth, while Florinskii had lied about the whole affair and had actually threatened her husband, forcing him to write the deposition.[15]

Apparently, Florinskii again demanded that the tabloid provide him with the address of Kondrat'eva and was again refused. He clearly began to suspect that "Neonilla" was fictional because, two days later, he wrote to the St. Petersburg Censorship Committee, asking for help in finding out the real name and address of "Neonilla Kondrat'eva."[16] He obviously thought that the censorship officials would know who was hiding under the pseudonym. But the censors replied that they did not and that he could only find out the real name of this person through a court decision, if he were to press charges against the newspaper and the author of the slanderous publication. Florinskii decided to do exactly that.

In the middle of this sordid affair, on 23 January, Florinskii's wife gave birth to their first child, a daughter christened Olga. This joyful event perhaps helped calm Florinskii's outrage. But he did not let the matter slide. He hired a lawyer and filed charges against *Petersburg Leaf* with the St. Petersburg Criminal Court, which in due time began an investigation.[17] Nevertheless, the tabloid continued to harass the doctor by publishing "is he guilty, or not" commentaries and letters "from our readers."[18] The court investigation dragged on for nearly three months, during which Florinskii himself, his students, his colleagues, and even his superiors at the IMSA gave their testimonies. The testimonies proved that the stories of both Neonilla Kondrat'eva and Mrs. Andreeva were indeed a fabrication. On the morning of 13 December 1865, when, as Mrs. Andreeva claimed, she had pleaded with Florinskii to see her sick child, the professor had actually been lecturing and spent the rest of the day, until the late afternoon, at the IMSA clinics.[19] The court held hearings on Florinskii's case and on 20 May delivered its verdict: the editor of *Petersburg Leaf* was sentenced to four months and Mrs. Andreeva to two months in jail, while publication of the newspaper was suspended for one month.[20] But the author of the first slanderous letter, a certain Mr. Balabolkin, who had hidden under the pseudonym of "Neonilla Kondrat'eva," was acquitted.[21]

The court investigation and decision cleared Florinskii's name but did not answer the key question: who had initiated the whole affair and why. In his testimony to the court, Balabolkin claimed that he did not write the letter, but only "corrected its literary style." Yet he refused to explain where and how he had obtained the letter in the first place. The newspaper's editor declared that he did not know either Florinskii or Mrs. Andreeva and hence had no personal stake in the matter. Balabolkin was a regular contributor to his newspaper, he explained, and he had relied on the reporter's judgement.[22] At the trial, which was held *after* Blagosvetlov's arrest and imprisonment, the editor pointedly noted that Florinskii was a "former contributor to the liberal journal *Russian Word*." He suggested that in accusing the professor, Mrs. Andreeva had "either been insane or deliberately served as a tool of someone's revenge." But he did not venture a guess at who exactly that "someone" might have been and why that "someone" would seek revenge against Florinskii.

The city's major newspapers, *Voice* and *St. Petersburg News,* published reports on the case, along with indignant editorials against the tabloid and its editor.[23] *Medical Herald* also published the court verdict to let the medical community know that all the accusations against one of its members were nothing but slander.[24] Florinskii was vindicated, but he probably could not help wondering why he had been targeted by the newspaper. He might have thought that the major reason for the whole affair was that publishing his essays in *Russian Word* made him a public figure, and thus "fair game" for public scrutiny, even if it took the form of slander. Perhaps this encounter with the "free press" made him wish to stay away from the public eye altogether and to retreat to the relative obscurity of his profession.

A month later Florinskii received a further and much more compelling incentive to do just that. As a result of the Karakozov Affair, on 13 May, Alexander II issued a special edict on "measures to counteract the spread of false doctrines that undermine the very foundations of Faith."[25] The edict in part stated: "[Divine] Providence mercifully opened the eyes of Russia to what consequences could be expected from strivings and thinking that dare to encroach on everything sacred, on religious beliefs, on the foundations of family life, on property rights, on obedience to the law, and on respect towards the powers that be." The edict commanded top-level state functionaries to pay special attention to "the upbringing of youth" and to make sure that such upbringing was directed by "the spirit of religion, respect of property rights, and maintenance of the basic principles of social order." The emperor demanded that "the higher schools of all agencies allow neither open nor secret propaganda of those destructive doctrines that are alien to all the conditions of moral and material wellbeing of the people."

Less than a month later, on 7 June, war minister Dmitrii Miliutin sent a special letter, accompanied by a copy of the imperial edict, to the IMSA president.[26] The minister commanded Dubovitskii "to take every measure necessary [to ensure] that the entire IMSA staff without fault obeys the edict to the letter." He emphasized that professors must teach their subjects "on the strict basis of religion, morality, and the precise fulfillment of all the duties of a true servant to His Majesty." Miliutin stressed that they must uproot "recently developed pernicious doctrines." The instructors must, he elaborated, "serve as examples for their students in such a way that, along with learning science, the students imbibe the spirit of discipline and love for the emperor and the Fatherland. Only under this condition could they become fully useful practitioners of their chosen vocation." The minister demanded that the president "explain to all your subordinates that anyone deviating from the principles of the Imperial edict" would be severely punished. "If any individuals do not feel that they are able with clear conscience to carry out their required duties," he concluded, "they better leave on their own [volition] the institution that by its very name is placed under Imperial patronage." The same day, Dubovitskii distributed copies of the imperial edict and the minister's letter among all members of the faculty and staff (requesting each one to sign a receipt), along with his own injunction: "do take these documents as guides to all your actions."

Along with every other member of the faculty, Florinskii signed the receipt. We can only imagine how he read and reacted to these documents. Given the contents and especially the venue of its publication, his "Human Perfection and Degeneration" could easily be construed as one of those "recently developed pernicious doctrines," to which the war minister had referred in his letter to Dubovitskii and which, according to the imperial edict, "dare[d] to encroach on everything sacred, on religious beliefs, [and] on the foundations of family life." Even if he personally believed with all his heart in the ideas and ideals of his treatise, Florinskii certainly could not afford to resign his post at the academy for the sake of those ideals. All he could do was to obey his superiors. He clearly decided to stay away from any public engagements. He stopped contributing even to *Medical Herald*,[27] confining his publications to the highly technical *Protocols of the St. Petersburg Society of Russian Physicians* that had very limited circulation.[28] Only a decade later, after his resignation from the IMSA, would he resume publishing in non-professional periodicals.

It seems likely that sometime in the late spring of 1866, Florinskii notified his publisher about his decision to withdraw from any further collaboration with *Russian Word*. Blagosvetlov might well have been disappointed, but he was not willing to give up the fight. As early as December 1865, after his journal had received the first "warning," anticipating the possibility that it would not survive the censorship onslaught, the editor began thinking about possible ways to deal with the situation. In January, after receiving the second "warning," he wrote to Shelgunov: "Here is what we shall do: pick up a new title for a journal just like *Russian Word* and continue publishing with the same contributors and subscribers."[29] On 17 February, the next day after the suspension of *Russian Word*, a certain "staff-captain N. Shul'gin" applied for permission to publish "a new learned-literary journal, titled 'Deed'."[30] As subsequent events suggest, the "staff-captain" was just another proxy whom Blagosvetlov used to continue his crusade. Blagosvetlov's arrest in April slowed down the organization of the new journal. But the moment he was released from the Peter-Paul Fortress, he redoubled his efforts. On 7 June, just one day following his release, Blagosvetlov wrote to Shelgunov:

> As you no doubt already know from the newspapers, *Russian Word* is definitively forbidden. In a week, I will write to you in detail on how our general situation will be resolved. One has to work, because one has to live, but [they] do not leave us any opportunity to live and work. Yet when pressed hard, a person becomes inventive and therefore I think that *Russian Word* will be resurrected in another form.[31]

Indeed, within a week, Blagosvetlov had signed an agreement with Shul'gin to publish *Deed* as a continuation of *Russian Word*, "with the same contributors and subscribers." In late September of 1866 the first issue of the new journal came out.[32]

Busy with organizing a new journal, Blagosvetlov also searched for other means to spread the message of the now forbidden *Russian Word*. In late August, clearly undeterred by the continuing witch-hunt for "revolutionaries" and the bacchanalia of "love for the emperor and the Fatherland" unleashed by the press, Blagosvetlov released Florinskii's essays in book format.[33] Although Blagosvetlov had previously reprinted under a separate cover several novels first published in *Russian Word*, this was the first time that a piece of non-fiction writing that had appeared on the pages of the journal was reissued as a book.

The slim volume had no table of contents, but contained the entire text of Florinskii's three essays with only minor technical edits. The subtitles "The second essay" and "The third essay," which had followed the general title in the October and the November issues of the journal, were, expectedly, removed, but all the subheadings that had appeared in the individual essays were kept in place. The text was typeset anew, word for word and paragraph for paragraph, but in a different format and type size, which accounts for the book being longer (206 pages) than the three essays published in the journal combined (144 pages).

Although no name of the publisher appeared anywhere in the book, its title page bore a logo of the *Russian Word* printing shop owned by Blagosvetlov through yet another proxy.[34]

Most likely, the publisher issued the book without any involvement of its author. Several features of the publication support this supposition (see fig. 4-1). The title page announced that the author of the book was "Professor F. Florinskii," thus giving Vasilii Florinskii the wrong initial, which probably would not have happened if the author had seen the proofs.[35] Furthermore, the title page, for the first time, identified Florinskii as a professor, but did not provide his institutional affiliation, which was then customary. Moreover, the book had no "preface,"

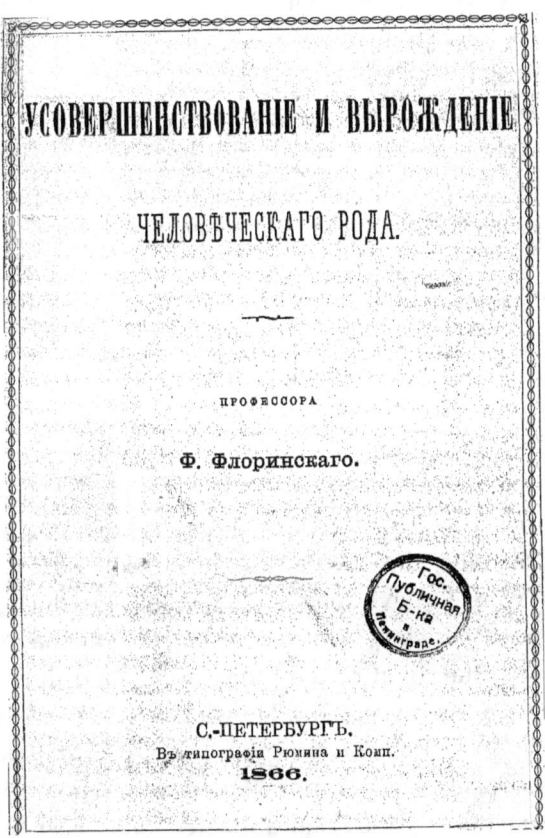

Fig. 4-1. The title page of the first book edition of Florinskii's essays, 1866, with a typo in its author's name. This is apparently a copy from the first printing, for it carries the name of the printer "Riumin and Co." that produced *Russian Word*. Courtesy of RNB.

"foreword," or "note from the author" that would have indicated to its readers that the text had previously been published in *Russian Word*, along with the reasons for its appearance as a separate volume, which usually accompanied such publications.[36]

We can only guess at why Blagosvetlov decided to issue the book. Perhaps, he simply hoped to earn some money. The prohibition of *Russian Word* and the arrest of the second volume of *The Ray* put Blagosvetlov's publishing enterprise in deep financial trouble. He had to fulfill his obligations to all the individuals (and institutions) who had prepaid their 1866 subscriptions to *Russian Word*. Fourteen rubles (including shipping costs) per annual subscription multiplied by approximately 4,000 subscribers made for a very large sum. The publisher had to either return the money, or compensate subscribers with other publications, which was exactly what Blagosvetlov had tried to do by publishing *The Ray*. He had certainly paid honoraria to all contributors, as well as the costs of production of 4,000-4,500 copies of each of *The Ray*'s two volumes. Given that each volume was about 800 pages (fifty typographical lists) in length, Blagosvetlov must have invested quite a sum in their production. The seizure of the second volume "froze" (and eventually forfeited) at least half of Blagosvetlov's investment and certainly disrupted his expected cash flow.

Perhaps, since as its publisher, he "owned" the manuscript,[37] Blagosvetlov counted on earning some money in reprinting Florinskii's essays with minimal investment (only the costs of production). However, the pricing of the book — 75 kopeks at a book store, or one ruble via mail order — seems too low to suggest that he hoped for a substantial profit. Plus, the book's title did not promise a quick sale. If Blagosvetlov were to have renamed the book, giving it a new, less obscure and more enticing title, for instance, "Marriage Hygiene," that would have certainly attracted more buyers and thus guaranteed a quicker return on his investment. He was likely aware that Auguste Debay's book, *Hygiene and Physiology of Marriage* (first published in 1848), had at that time sustained more than thirty editions in its original French, and from 1861 to 1865 had appeared in at least two different Russian translations and numerous printings.[38] Yet, Blagosvetlov did not follow this simple marketing strategy, which indicates that his motivations for republishing Florinskii's treatise were not monetary but rested elsewhere.

There is little doubt that Blagosvetlov was genuinely interested in the main themes of Florinskii's essays: the applications of Darwin's evolutionary concept to humans and, more generally, the role that the natural sciences could play in understanding social issues and curing "social ills." As we saw, *Russian Word* had carried numerous essays that in various ways examined these themes on a regular basis, and this was why Blagosvetlov had invited Florinskii to contribute to his journal in the first place. The prohibition of *Russian Word* did not change Blagosvetlov's interests — he was determined to continue publishing articles (and books) that addressed these themes. Indeed, the announcement of *Deed*'s publication that appeared in the fall of 1866 in *Voice* stated unambiguously:

> "Knowledge is power" — the modern generation is called to work on this simple and great task [of converting knowledge into power]; its better present and its very future depend on the success of this work. Undoubtedly, the [complete] fulfilment of this task lies far ahead, even for the peoples leading the intellectual development of the nineteenth century; but its results are already so great that to reject this task means to understand neither the demands of [our] time, nor the needs of [our] life.
> Knowledge becomes actual power only when it is aimed directly to the benefit of humanity, when there is a fundamental and solid connection between the thoughts and the deeds of society. This is what is demanded by societal conscience, the logic of events, and progress of human societies; we, Russians, especially need this, [for] there is so very little in common between [available] knowledge and actual needs of our life that, for a great majority, a journal and a book is not as much a necessity as a shot of vodka before dinner. Therefore, we do not yet know the *utilitarian* side of knowledge, even though this is one of the most beneficial sides of intellectual development, because an idea without utilization is nothing but a buried treasure. The journal *Deed* chooses this side as its direction and, in it, will seek its own strength and meaning.[39]

But first, as had happened after Zaitsev's debacle with writing about the applications of Darwin's evolutionary concept to "human races," Blagosvetlov needed to find authors willing and capable of taking on this task of converting knowledge into power.

Apparently, when Blagosvetlov learned that Florinskii was not going to continue writing for his journal, he took certain steps to find a suitable replacement. Initially, he probably hoped that Pavel Iakobii

(1841-1913), Zaitsev's brother-in-law, who at the time was enrolled in Zurich University's Medical School, could take on this role.[40] Most likely on Zaitsev's recommendation, Iakobii had begun writing for *Russian Word* nearly simultaneously with Florinskii in the summer of 1865. His first publication was a lengthy article on "the development of slavery in America," which had come out in the July issue. Although the article promised "to be continued" in the next, August issue, its conclusion appeared only in December, immediately following the last installment of Florinskii's treatise.[41] In between, in the October issue, Iakobii published a highly critical, nearly thirty-page-long review of the Russian translation of Heidelberg psychologist Wilhelm Wundt's *Lectures on the Soul of Men and Animals*.[42] Unlike Florinskii's second essay that opened the issue, Iakobii's review caught the censor's eye and became one of the pretexts for the first "warning" issued to the journal.[43] The censor pointedly commented that Iakobii "laughs at the beliefs in the existence of the soul and ironically portrays the individuals who do not sympathize with a materialistic worldview."[44]

Blagosvetlov clearly liked Iakobii's writing style and invited him to continue his collaboration with the journal. Building on his current studies of psychiatry at Zurich University, Iakobii wrote a lengthy essay modeled to a certain degree on, and carrying almost exactly the same title as, Jacob Moleschott's *Psychological Sketches*. Blagosvetlov likely planned to publish the essay in one of the spring 1866 issues of his journal, for he included it in the second volume of *The Ray*.[45] After the arrest of *The Ray* and the final demise of *Russian Word*, Blagosvetlov asked Iakobii to write for *Deed*. The very first issue of the new journal carried the future psychiatrist's lengthy article on the "physical conditions of the primeval human civilization."[46] On 14 November 1866, just before the second issue of *Deed* was to come out, Blagosvetlov sent Iakobii a long letter in response to the latter's question: "what and how should I write for the journal?" "Until pressure from the government is eased," the editor advised his author, "write serious articles on the natural sciences. However, do not touch religion. This is, for now, a strictly forbidden fruit."[47]

Iakobii followed this advice: the fifth and the sixth issues of the journal that came out in April and May 1867, respectively, carried his lengthy "Chronicle of Natural Science Discoveries."[48] But he apparently felt more comfortable writing about his own specialty: the ninth issue contained Iakobii's extensive overview of French psychiatrist Louis-Francisque Lélut's *Physiologie de la pensée*.[49] In the next issue, he reviewed at length recent Russian translations of several books on physiology, while in the eleventh issue he published a new version of his own "Psychological Sketches."[50] The following year, however, Iakobii stopped writing for *Deed*, probably because he was too busy working on his doctoral dissertation.[51] As a result, the entire 1868 run of the journal contained not a single article on either science or medicine.

Blagosvetlov was certainly not happy about the absence of his trademark subject on the pages of his journal and searched for another author who could fill the niche. In early 1869, he recruited Veniamin Portugalov (1835-1896) to become a science/medicine commentator for *Deed*.[52] A graduate of Kazan University's Medical School, Portugalov since his student days had actively participated in various "revolutionary circles" and, during the late 1850s and early 1860s, he had been arrested and imprisoned several times. In 1863, after spending several months in solitary confinement in the Peter-Paul Fortress, he was exiled to a small town in the Perm province (not far from Florinskii's beloved Peski). Prohibited by the conditions of his exile from practicing medicine, Portugalov began publishing articles on medicine and social hygiene topics in various journals and newspapers.[53] And these articles were probably what sparked Blagosvetlov's interest.

It seems likely that the initial stimulus for Blagosvetlov's invitation was a series of articles on "the causes of diseases," which Portugalov published during 1868 in the recently established *Archive of Legal Medicine and Social Hygiene* and also issued in book format at the beginning of the next year.[54] Blagosvetlov apparently asked Portugalov to write a synopsis of his book for *Deed* and published it, under the title "The Sources of Disease," in the March 1869 issue.[55] What might well have prompted Blagosvetlov's invitation was that Portugalov's book addressed several major topics presented in Florinskii's treatise.[56] Portugalov too attempted to "synthesize," in his own way, Darwin's evolutionary concept, physical anthropology, and social hygiene and to discuss certain issues related to "human perfection and degeneration."

Unlike Florinskii, however, at this point, Portugalov was concerned not with "marriage hygiene," but with disease causality. He built much of his reasoning on the recently published (under Ivan Sechenov's editorship) Russian translation of Darwin's latest work, *The Variation of Animals and Plants under Domestication*.[57] Portugalov claimed that "human perfection and degeneration" depend first and foremost on the development of "human culture." As he put it, "culture, in the most general sense of the word, leads to the perfection of the breed in animals as well as in humans, [while] lack of culture leads to the degeneration of entire breeds."[58] For Portugalov, the most obvious sign and a leading cause of degeneration were numerous diseases that plagued humanity. Following closely the works of Rudolf Virchow and Austrian physician Eduard Reich,[59] two leading proponents of social hygiene, he thought that the most important causes of disease, and, hence, of human degeneration, were "the two main deficiencies of human culture: famine and poverty." As he noted in the conclusion of his article: "only by the force of culture could [we] eliminate humanity's scourges — epidemics, infections, miasmas, [and] parasites."

Portugalov further elaborated these ideas in a lengthy essay, enticingly titled "The Limitlessness of Hygiene," which appeared in the August issue of *Deed*.[60] The essay was based on a voluminous treatise on *Popular Hygiene* published by Karl H. Reclam, Leipzig University professor of legal medicine and the founder of *Deutsche Vierteljahrsschrift für öffentliche Gesundheitspflege*, the first German journal for "social hygiene."[61] Unlike Florinskii, Portugalov was not well versed in the biology and physiology of human reproduction, including the concepts of the cell, embryonic development, and, especially, heredity. Apparently, at this time, he was also unfamiliar with Florinskii's treatise. He subscribed fully to the notion of the inheritance of acquired characteristics, which Darwin had elaborated in the second volume of his latest work in the form of a "provisional hypothesis of pangenesis."[62] As a result, in his arguments about "perfection and degeneration," Portugalov paid little attention to reproduction and heredity, focusing instead on "environment," and especially "social environment."

Blagosvetlov definitely found Portugalov's writings worthy of publication in *Deed* — in private correspondence, he even likened Portugalov to his former star-writer Pisarev — and invited him to become a regular contributor to the journal.[63] Furthermore, it seems that Portugalov prompted the editor to produce a Russian translation of

Reclam's treatise on *Popular Hygiene*, which stayed in print for more than a decade with Portugalov's essay "The Limitlessness of Hygiene" as its introduction.[64] But, even though he was very happy with Portugalov's articles on social hygiene, Blagosvetlov clearly wanted more from his newfound science/medicine expert. The editor asked him to explore in much greater details the possible implications of Darwin's evolutionary concept for social hygiene and for human affairs more generally. He even sent Portugalov copies of several publications necessary for examining the topic, which were unavailable in the small town where Portugalov was serving his sentence.[65] Portugalov enthusiastically took on the job. From November 1869 to July 1870, *Deed* carried a series of six lengthy essays (more than 200 pages in total) under the general title "The Latest Word in Science." The essays presented Portugalov's analysis of Darwin's works (both *Origin* and *Variation*) and their import for understanding human health, society, and evolution.[66] None of these essays, however, even mentioned Florinskii's pioneering study.

It seems likely that sometime in 1870, his publisher alerted Portugalov to the existence of Florinskii's book and perhaps even sent him a copy. In November and December 1870, *Deed* published Portugalov's extensive essay on "Development and Deterioration,"[67] and, in the first three issues of 1871, serialized Portugalov's lengthy tract "On Degeneration."[68] As their titles indicate, both publications dealt with a major theme of Florinskii's treatise — the degeneration of humankind. The first presented an overview of more than twenty different works that had addressed the issue. Florinskii's book did appear, as number twelve in the list of these works, but Portugalov did not cite it in his text. As he explained in a footnote, the presence of a particular work in his list of sources "does not mean that we fully share the views and beliefs of its author. We take from him facts and facts only, [but] whenever we actually share the views and beliefs of an author, we cite him directly in the text."[69]

Judging by his in-text citations, just three books served as Portugalov's main sources in addressing the subject. The most frequently cited was Eduard Reich's nearly 500-page volume, *On the Degeneration of Men, its Sources and Prevention*,[70] (which Portugalov had already used in his 1869 book on the causes of disease). The second was the first volume of Reich's *System of Hygiene* that had recently come out and was devoted to "Moral and Social Hygiene."⁷¹ And the third was Bénédict Morel's 1857 treatise.⁷² Although Portugalov did not even mention Florinskii in his text, and, hence, supposedly did not share Florinskii's views, he devoted the second series of his essays "On Degeneration" specifically to one issue that Florinskii had promised to (but for some reason did not) address in his "Human Perfection and Degeneration" — drunkenness and alcoholism. Blagosvetlov was very pleased with the job Portugalov had done in his articles. In May 1873 he reprinted all of them as a sizeable tome of more than 600 pages, titled *Issues in Social Hygiene*.⁷³

Blagosvetlov's continuing interest in Darwin's evolutionary concept and especially its applications to humans was manifested not only in his publication of Portugalov's essays. In 1871-1872, Blagosvetlov also edited and published a Russian translation of Darwin's own long-awaited take on human evolution — *The Descent of Man*. Darwin's two-volume work came out in London at the end of February 1871.⁷⁴ Almost immediately, an abridged Russian translation began to appear in installments under the auspices of *Knowledge*, "a scientific and critical-bibliographic journal," established just a few months earlier by Florinskii's IMSA colleagues, professor of physics Petr Khlebnikov and professor of chemistry (and future famous composer) Alexander Borodin.⁷⁵ From April to September 1871, each issue of *Knowledge* carried a portion of the translation as a supplement. In September, the supplements were bound together and released as a 440-page single tome.⁷⁶

Blagosvetlov was deeply disappointed by *Knowledge*'s translation.⁷⁷ He described it as yet another example of the typical situation when "a remarkable foreign author, at the hands of an ignorant or unscrupulous translator, appears before our public in such a clownish costume as to make it impossible not only to read his work, but even to recognize that this is indeed the very author whose name is printed on the cover of the Russian edition."⁷⁸ He saw "such treatment of Darwin" as "not merely a literary impropriety, but a crime, a forgery in the direct sense of the word." "Instead of Darwin," he continued indignantly, "our public is given a collection of grammatical errors, deliberate distortions, [and] omissions, [compounded by] complete ignorance of [both] the language from which [the book] is being translated and the language to which it is being translated. This is akin to selling to an inexperienced buyer a tin spoon instead of a silver one."⁷⁹

Blagosvetlov surmised that the only reason the editorial board of *Knowledge* had committed such a "crime" was to scoop the readership (and hence, profits) by putting out a translation of Darwin's book before any other publisher. He was particularly infuriated by the arbitrary and, as he saw it, totally unjustifiable, abridgement of Darwin's work, which prompted him to publish a complete translation.[80] Blagosvetlov commissioned and personally edited a new translation, with Florinskii's former literature teacher Morigerovskii acting as its nominal publisher. The first volume of what eventually became a three-volume set came out in July 1871, with the second following shortly on 1 September. The two volumes contained a complete translation of the first volume of the English original.[81] The third volume of Blagosvetlov's translation (containing the entire second volume of the English edition) was promised to be published by December, then by March of the next 1872 year, but, in the end, appeared only in May.[82]

Blagosvetlov's reviews make clear that he saw *Knowledge* as an unscrupulous but formidable competitor who exploited his trademark themes: Darwin's theory's applicability to humans and the utility of the natural sciences for curing Russia's "social ills."[83] This attitude found further expression in Blagosvetlov's reaction to *Knowledge*'s release of a Russian translation of Galton's *Hereditary Genius* (see fig. 4-2).

*Hereditary Genius* had come out in London in November 1869. Compared to the furor generated by Darwin's *Origin*, Galton's book was accorded a lukewarm reception in both England and abroad.[84] As far as I was able to ascertain, no Russian periodicals even remarked on its appearance. But it did impress Darwin who cited and referenced it extensively in his own *Descent of Man*. Most likely, it was Darwin's references that directed the attention of *Knowledge*'s editors to Galton's work and prompted them to commission a translation of the book five years after its original publication.

While the translation of *Hereditary Genius* was in progress, in the late spring of 1874, in its fifth (May) issue, *Knowledge* published a translation of Galton's report delivered to the Royal Society just a few months earlier "On Men of Science, Their Nature and Their Nurture."[85] This translation seems to have been the first appearance of Galton's name and work in a Russian publication. The translation of *Hereditary Genius* came out as a supplement to *Knowledge*'s last double (November-December) issue of the same year.[86] The supplement had no title, name of the author or

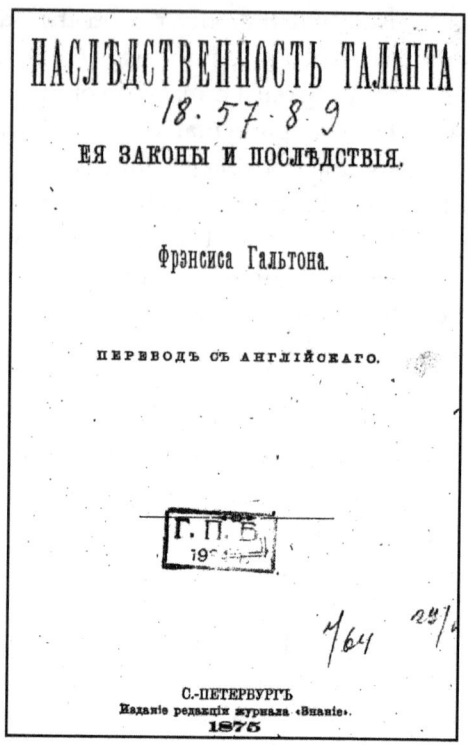

Fig. 4-2. The title page of the Russian translation of Francis Galton's *Hereditary Genius*, issued in book format by the publishers of *Knowledge* in 1875, under the title *The Heredity of Genius, Its Laws and Consequences*. Courtesy of RNB.

the translator, table of contents, nor any editorial comments, just the abridged text of Galton's book. Its author was identified only in the journal's table of contents. In March 1875, the supplement was released as a separate volume, under the title "Heredity of Talent, Its Laws and Consequences," and appended with the text of Galton's report "On Men of Science."[87]

In spite of being advertised continuously in *Knowledge*, the translation attracted very little attention. Indeed it seems that the only periodical that even noticed its publication was *Deed*: Blagosvetlov apparently asked his natural science expert Pavel Iakobii to write a review of Galton's book. In its May 1875 issue, *Deed* carried Iakobii's 25-page essay, provocatively titled "Modern Lack of Talent."[88] The bulk of the review recounted the main conclusions of, and emphasized the pioneering use of statistics in, Galton's study of the heritability of special talents. Its last section assessed the implications of Galton's findings for "the failings of modern civilization," and his proposals to remedy the situation by changing contemporary social mores regarding such issues as marriage and children, support for talented young individuals, and immigration, which Iakobii perceived altogether as "far too remote an ideal." "But, anyhow, it is pleasant to dream about [such an ideal]," he concluded his review. A few months later, in its regular section "Foreign literature," *Deed* published an extensive critical review of Galton's newest book, *English Men of Science*. The reviewer's general attitude was unambiguously expressed in pairing the assessment of Galton's statistical study with a review of John Timbs's purely anecdotal *English Eccentrics and Eccentricities*.[89]

To the end of his days on 7 November 1880, Blagosvetlov continued to examine the role that the natural sciences, and, especially, their latest achievements — Darwinian evolutionary theory and social hygiene — could and should play in understanding and curing Russia's "social ills." In 1878, for instance, he translated into Russian and published the second edition of *Contributions to the Theory of Natural Selection* — a collection of essays on biological evolution and its applications to humankind by Alfred Russel Wallace, Darwin's co-discoverer of the principle of natural selection.[90] For nearly fifteen years, he kept Florinskii's treatise in print and continued to advertise it on the pages of *Deed*. Tellingly, it was only after Blagosvetlov's death that *Human Perfection and Degeneration* disappeared from the list of publications available for purchase at the journal's bookshop.

Blagosvetlov seemed to be the only contemporary who fully grasped the import of Florinskii's essays as a synthesis of Darwinism and social hygiene, which offered a scientific solution — "rational" or "hygienic" marriage — to a number of perceived social problems in post-Crimean Russia and outlined its possible effects on the country's future. This seemed to be the reason why he reprinted Florinskii's essays as a book and kept it in print until the end of his life. Indeed, Florinskii's treatise had launched a whole series of books produced by Blagosvetlov, which in one way or another explored further various issues raised in Florinskii's tract: these included translations of Reclam's *Popular Hygiene* (1869), Darwin's *The Descent of Man* (1871-1872), and Wallace's *Contributions* (1878), as well as Portugalov's *Issues in Social Hygiene* (1873).

Florinskii, however, did not take any part in the varied activities his former publisher undertook to promote and explore further the themes and subjects of *Human Perfection and Degeneration*. He never partook in the polemics between *Deed* and *Knowledge* over Darwin's and Galton's works. He kept silent even when, as if to mark the anniversary, exactly ten years after the publication of his book, the September 1876 issue of *Deed* opened with a voluminous essay by Portugalov on "Hygienic Conditions of Marriage."[91] Although, without once mentioning Florinskii's name, the essay was nothing more than a — sometimes nearly verbatim — recapitulation of his own discussions of "kin" and "mixed" marriages, the role of the law in preventing degeneration and promoting perfection, and many other basic ideas of his treatise, "updated" with materials drawn from the latest works by Darwin, Galton, John Lubbock[92] and Edward B. Taylor,[93] Florinskii simply ignored it.

Apparently, he found other ways "to bring benefits to his Fatherland."

## Florinskii's Deeds

Neither the public scandal over *Petersburg Leaf*'s slanderous articles, nor Blagosvetlov's "unauthorized" release of Florinskii's treatise as a book seem to have had any ill effects on the professor's life and career. During the late 1860s and early 1870s, he continued teaching his assigned courses on gynecology, obstetrics, and pediatrics, working on a voluminous textbook on gynecology and obstetrics,[94] running the IMSA women's ward and the children's clinic, and carrying on various research projects, for instance, examining the use of chloroform during childbirth and obstetrics operations.[95] He built the reputation of a very knowledgeable physician and a successful private practice. His family continued to bring much joy to his life: on 6 October 1867, his wife gave birth to a son (see fig. 4-3). The happy parents christened him Sergei, probably in honor of Sergii Radonezhskii, one of the Russian church's most venerated saints. In November 1867, Florinskii was nominated and, a few months later, duly promoted, to the position of an extraordinary professor at the IMSA Department of Gynecology, Obstetrics, and Pediatrics. The promotion brought with it advancement to the next, sixth rank (equivalent to a colonel in the army) and an increase in salary and benefits. It also made Florinskii a full voting member of the Academy Council.

Fig. 4-3. Maria Florinskaia with her children, Olga (left) and Sergei (right), 1870. The photo was taken at the St. Petersburg studio of Ludvik Cluver, a well-known photographer. Courtesy of NMRT.

But then, Florinskii's career stalled. Unlike nearly every one of his classmates at the Institute of Young Doctors, he never got the position he had been groomed for — that of an ordinary professor and a department chairman at the IMSA. This frustrating situation apparently had nothing to do with Florinskii's contributions to *Russian Word* or the contents of his treatise. Rather it was a manifestation of the old animosities between the so-called "Russian" and "German" factions within the academy.

Like modern science, modern medicine in Imperial Russia was largely a product of western imports.[96] Since its creation, the IMSA was thoroughly dominated by "the Germans," a label that came to encompass all "aliens" (both ethnically and confessionally) among its administration, faculty, and student body. Indeed in its first years, the newborn academy even had two separate sections for Russian- and German-speaking students. In 1805, the Russian government invited Johann Peter Frank (1745-1821), prominent German physician and the leading ideologue of state medicine and medical police, to become the academy's first rector.[97] Although his tenure lasted only three years, it was Frank who laid down main principles that guided the academy's operations and curricula for nearly half a century.[98] Frank was succeeded by James Wylie (1768-1854), a graduate of Edinburgh University's Medical School who had come to Russia in 1790 and made meteoric career, rising to the position of personal surgeon to the imperial family.[99] In 1808 Wylie was appointed the academy's first president and occupied the post for thirty years. But he left most of the actual administration to his "deputies," such as Johann Georg Wallerian, Elias Gustav Eneholm, Johan Orlay, and Friedrich August Wilhelm Heuroth.[100] After Wylie's retirement, the presidency went to Johann Gottlieb Schlegel, and after the latter's death to Ventseslav Pelikan.

As this (far from exhaustive) list of names clearly shows, for half a century the academy's top administrative positions were occupied by individuals of Austrian, British, Dutch, German, Polish, and Swedish origin (and often training), even though during the same period a number of ethnic Russians joined the faculty, which could not but give rise to mutual resentment and considerable tensions between the two groups. As Florinskii bitterly remarked in one of his essays, "until quite recently, it was thought that only a German [man] could be a mechanic, an apothecary, a physician, a professor, and so on, but that a Russian [man] lacks both patience and mind for this kind of work."[101]

Towards the end of Florinskii's time as a student, the appointment of Dubovitskii and Glebov to the presidency and vice-presidency, respectively, placed the academy's governance — for the first time since its creation — in the hands of ethnic Russians. Joined by Zinin who headed the IMSA Council, the triumvirate actively fostered the "Russification" of the academy and increased considerably the Russian contingent among both faculty members and students. In early 1867, however, partially in response to the newfound emphasis on "the upbringing of youth" incited by the Karakozov Affair, the positions of both president and vice-president were eliminated and replaced with a "commander" appointed by the War Ministry. Dubovitskii and Glebov were "promoted out" of the academy. The former came to head the Military Medical Department of the War Ministry, the latter became a permanent member of the department's Scientific Council.[102] With the new command structure in place, the German faction began to reassert its influence, especially in the matters of new appointments to the faculty positions.[103]

The dividing line between the two factions certainly did not fall purely along lines of ethnicity and/or religion.[104] In fact, several ethnic Germans (for instance, Alexander Kiter and Ventseslav Gruber) supported the Russian faction, while certain ethnic Russians (such as Academy Commander Nikolai Kozlov) sided with the German one. Rather, the major source of contention was a conflict between "patriotism" and "science for science's sake." The Russian faction considered service to the Fatherland and its people the foremost duty of the medical profession[105] and advocated the advancement of "true" Russians — those individuals who upheld this ideal — to leadership positions in medical administration and education. They accused their opponents of pursuing exclusively financial and career interests to the detriment of their obligations to the country and its population and of promoting individuals according to "ethnic" and "confessional" rather than patriotic and professional criteria and objectives.[106] The German faction, in its turn, loudly advocated the notion of "science for the sake of science" and valued scholarly engagement over public service.

Florinskii's 1862 article "Notes of a Russian Foreigner About Our Medical Life"[107] clearly attests that he wholeheartedly and very vocally subscribed to the ideals and ideas of the Russian faction. The same ideals also informed to a considerable degree his concept of "human perfection and degeneration." This, of course, did not make him very popular among the "opposition." Plus, Florinskii's scholarly reviews, such as his polemics with the "German" author of the St. Petersburg Foundling Home medical report[108] — with their highly critical and uncompromising style, and total disregard for positions and connections held by the authors of reviewed books, dissertations, and presentations — did not win him many friends among his "German" colleagues.

The first sign of trouble appeared already at Florinskii's promotion to the position of extraordinary professor. Usually, the chairman of the department that had a vacancy advanced the nomination for a professorial position. In Florinskii's case, however, the nomination was proposed not by Anton Krasovskii, chairman of his department and former supervisor of his doctoral dissertation, but by Nikolai Iakubovich, chairman of the histology and embryology department. Since Kiter's transfer to the surgery department and Krasovskii's promotion to the chairmanship in 1858, the gynecology department had for years had a vacancy for an extraordinary professor. Yet, despite Krasovskii's constant complaints about being overburdened with his duties at the department, the chairman did not hasten to grant Florinskii the well-earned promotion. Perhaps he saw the younger colleague as a serious threat to his own authority within the academy and even his own extensive private practice. Reportedly, Florinskii's decidedly critical review of Krasovskii's own textbook on obstetrics[109] and a similarly scathing review of a doctoral dissertation by Krasovskii's protégé Roman Bredov[110] played an important role in Krasovskii's attitude.

So, it was not Krasovskii, but a leading member of the Russian faction, Iakubovich, who in November 1867 presented Florinskii's nomination to the Academy Council. On 13 January 1868, in its regular session, the council considered Iakubovich's highly praising recommendation and voted in its favor. A few days later, in his capacity as the head of the Military Medical Department, Dubovitskii approved Florinskii's appointment as extraordinary professor.[111]

Two years later, however, when the Russian faction initiated Florinskii's promotion to the position of ordinary professor, it was effectively blocked by "German" opponents.[112] On 3 January 1870, the leader of the Russian faction Sergei Botkin[113] presented Florinskii's nomination at a regular session of the Academy Council. This time, the nomination met serious opposition, mobilized and led by Florinskii's immediate superior — Krasovskii. As a result, the vote was postponed. The next session on 24 January turned into a heated dispute that lasted for several hours, and voting was again postponed. The debate continued during the next session, held on 16 February, and in the end the council took a vote that came very close, with twelve members casting their votes "for" and eleven "against" Florinskii's promotion.[114]

According to the rules that required a simple majority for a vote to pass, Florinskii was elected to the post. The council's decision, however, had to be approved by the Military Medical Department of the War Ministry, and the "opposition" took further steps to stop the appointment. Nine out of the eleven council members who had voted against it prepared three "special opinion" letters with three signatories each (which certainly looked much more impressive than one letter signed by the nine individuals would have), protesting against the council's decision. They sent the letters to the War Ministry, along with the protocols of the council's deliberations and vote. The maneuver proved effective. The protest made it all the way up the bureaucratic ladder and landed on the desk of war minister Miliutin himself. In early May, after consulting with head of the Military Medical Department Nikolai Kozlov (the former commander of the academy and a staunch supporter of the German faction), the minister overturned the council's decision. As a historian of the academy has noted, "with sorrow," some forty years after the fact, "having obtained a department at the Academy, an 'alien' (*inorodets*) strove as much as he could to promote another 'alien'. ... Kiter promoted Krasovskii, Krasovskii obstructed the advancement of Florinskii, but promoted Karl [sic] Bredov."[115]

Florinskii remained an extraordinary professor. This was a heavy blow to whatever career ambitions he had entertained, but it was certainly not the end of the world. A year later, fate delivered Florinskii a far more terrible blow. His son Sergei passed away, just a few days before his fourth birthday. We know nothing about the circumstances and causes of this tragedy.[116] But whatever they were, Florinskii, who taught pediatrics and ran the children's clinic, must have been devastated by his inability to save his own child. The tragic loss, along with the failure of his promotion, dampened Florinskii's enthusiasm for both teaching and research. He completely withdrew from pediatrics, by first assigning his lecture course on the subject to one of his assistants, and, a few months later ceding to the same assistant the direction of the pediatric clinic he had created.[117]

The number of Florinskii's publications dropped precipitously: over five years, from 1870 through 1874, he published fewer than ten articles, nearly all of them based on clinical observations, not research. Of all his previous duties, Florinskii retained only lecture courses on theoretical obstetrics and gynecology, which he had been teaching since 1860, and supervision of the IMSA women's ward. But he stopped publishing his annual surveys of foreign literature and abandoned work on his textbook: its promised second volume on obstetrics remained unfinished.[118] Perhaps as a consolation prize, on 12 September 1871, the St. Petersburg Society of Russian Physicians elected Florinskii as its vice-president,[119] and on 18 January 1872, the IMSA administration advanced him to the next, fifth civil rank. Despite these accolades, he apparently gave up his ambition of becoming an ordinary professor and department chairman at his alma mater.

In the spring of 1873, however, Florinskii's career took an unexpected turn. While retaining his post at the IMSA, he became a "permanent member" of the Scientific Committee of the Ministry of People's Enlightenment.[120] Formally (re)established in 1863 as part of the sweeping reorganization of the state apparatus in the course of the Great Reforms, the Scientific Committee advised the minister on various facets of the agency's operations, ranging from secondary school curricula to research and teaching programmes at universities to the contents of textbooks and books "for the people."[121] Florinskii was invited to join the committee to oversee all issues related to medicine. He mostly reviewed popular "self-help" medical publications and various educational and research projects submitted by the medical schools subordinate to the ministry. His duties were not too onerous, and he could carry out most of them in the comfort of his own home, coming to the ministry once in a while to attend a general session of the committee. The new job paid quite well (1,000 rubles per annum) and nearly doubled his regular income. For almost two years, Florinskii managed to serve the two masters — the academy and the ministry — without any problems. But, willingly or unwittingly, his new position drew him into ministerial intrigues.

According to the results of the secret police's hunt for "revolutionaries" in the aftermath of the Karakozov Affair, nearly fifty per cent of all the individuals suspected of "anti-government activities" were students of various secondary and higher schools, thus making the issues of education and upbringing one of the government apparatus's top priorities. Indeed, the day after Karakozov's attempt on his life, Alexander II fired the head of the Ministry of People's Enlightenment and appointed the Chief-Procurator of the Holy Synod Count Dmitrii Tolstoi to the post.[122] Students' protests and unrest that swept through the Russian Empire's higher schools in 1869-1870 added considerably to the government's anxieties over the "university question," as it was referred to in official parlance. In this situation, the importance of the Ministry of People's Enlightenment that oversaw most primary, secondary, and higher schools in the country rose significantly.[123] In the best bureaucratic tradition, the ministry's officials attempted to capitalize on the situation and extend their control to the schools that were administered by other state agencies.

One such school was the Imperial Medical-Surgical Academy. Sometime in the fall of 1874, the officials proposed to subordinate the academy (once again)[124] to the Ministry of People's Enlightenment. Since St. Petersburg University had no medical school, the officials argued, the ministry had very little insight and even less input into the matters of medical education in the capital and the country as a whole. Expectedly, the War Ministry was loath to cede its control over the empire's premier medical school, which led to a serious conflict between the two agencies and their respective heads, Dmitrii Tolstoi and Dmitrii Miliutin.[125] It so happened that Florinskii got caught right in the middle of it.

Florinskii thought that moving his alma mater out from under the War Ministry would be very beneficial to both its faculty and its students. As his diaries make clear, he had thoroughly disliked the military spirit (discipline, uniforms, inspections, and so on) that permeated the school during his student years. It is doubtful that this attitude changed after he became a faculty member. Perhaps, it was even strengthened by the further militarization of the academy manifested in the appointment of a "commander" instead of a president. Tellingly, there is not a single photograph of Florinskii's dressed in his official military uniform, either from his student days or professorial tenure. Furthermore, Florinskii might have hoped that the IMSA's transfer to civic authority would improve the provision of medical services to the people. The academy was the country's largest medical school, but the majority of its graduates served first of all the military, not the populace. The IMSA's "demilitarization" would increase dramatically the supply of doctors to the new *zemstvo* system of health services, which was taking shape and was Russian physicians' major concern at this very time.[126] As Florinskii's various publications, including *Human Perfection and Degeneration*, demonstrate convincingly, he fully shared this concern.

But why would the personal opinion of a relatively low-ranking, part-time official matter in the power struggle between the two ministries? As the only expert on medical issues employed by the Ministry of People's Enlightenment, Florinskii certainly had the ear of minister Tolstoi, with whom he developed good working relations. More important, however, sometime over the summer of 1874, by sheer accident, Florinskii made a personal acquaintance with, and became a close confidant of, the second most powerful man in the Russian Empire — the Grand Duke Konstantin Nikolaevich Romanov (1827-1892), the younger brother of Alexander II, the head of the State Council (the highest government agency), and one of the main architects of the Great Reforms.[127] Florinskii met the grand duke in his professional capacity as a physician, but under very unofficial and sensitive circumstances: he was called in to attend to the duke's "second family."[128]

As all dynastical marriages, the grand duke's was prearranged.[129] In 1846, the nineteen-year-old duke met his bride-to-be, Princess Alexandra of Saxe-Altenburg (1830-1911). By all accounts, the sixteen-year-old princess was incredibly beautiful, and Konstantin fell in love at the first sight.[130] Two years later, they were married. The first years of their married life were happy and filled with joy: from 1850 to 1862, Grand Duchess Alexandra bore her husband six children. But, towards the end of the decade, the couple drifted apart and the marriage began to crumble: the grand duke had a series of affairs, while his wife sought solace in mysticism.[131] Sometime in the early 1870s, Konstantin found his true love — Anna Kuznetsova (1847-1922), a young ballerina of the Imperial Mariinskii Theater. Divorce was, of course, out of the question, and Anna became the grand duke's wife in all but name. Reportedly, the duke himself referred to Grand Duchess Alexandra as his "government-issue wife" and to Anna as his "true wife."[132] Konstantin bought for Anna a large, comfortable house near his personal estate in Pavlovsk, the former imperial palace built for Emperor Paul I at the end of the previous century. Situated just a stone's throw from his official residence, Anna's house became the duke's real family home where he spent every moment he was free of official duties. On 17 July 1874, Anna gave birth to their first child, a boy.[133] It was during the summer of 1874 that the duke met Vasilii Florinskii.

As did many well-off St. Petersburg families, in the summer the Florinskiis rented a dacha so that they could spend a few months away from the "miasma-laden," treeless capital, in the much healthier atmosphere of a rural setting. As it happened, the Florinskiis had for years rented their dacha in Pavlovsk, perhaps attracted by the easy access to the town and its renowned imperial parks opened to the public. One day in the summer of 1874, apparently in some emergency,

Florinskii, as the nearest reputable physician, was called on to attend to a patient among the grand duke's "second family." It is unclear from available documents whether Florinskii was called in as a gynecologist/obstetrician to see Anna herself or as a pediatrician to take care of her sickly firstborn, or both. Whatever the case, Florinskii clearly made an impression on Anna and the duke, who doted on his "true wife" and their firstborn. The emergency call led to regular visits. Although, in the end, Florinskii was unable to save the infant (six months later, the boy died), the duke obviously trusted the doctor and, reportedly, consulted him not only on health issues of his "second family," but on many other medical matters.

According to the diaries of war minister Miliutin, it was Florinskii who advised the State Council chairman Konstantin Nikolaevich in the spring of 1875 during the "territorial" conflict between the War Ministry and the Ministry of People's Enlightenment over which would have control of the IMSA.[134] Indeed, probably on the grand duke's recommendation, Florinskii was appointed a member to a special commission struck by the State Council to look into the matter. Florinskii managed to swing the commission's decision in favor of the transfer, and the State Council decreed that the academy be transferred to the authority of the Ministry of People's Enlightenment. The council's decision, however, had to be approved by the emperor, and, by threatening to resign his post, the war minister prevailed. Alexander II overturned the State Council decree, and the academy remained under Miliutin's control. A few days later, Florinskii resigned from the IMSA and moved to a full-time position at the Ministry of People's Enlightenment.

The official reason for this move was that Florinskii's duties at the ministry got vastly expanded. In early 1875, the ministry initiated a "revision" of the 1863 University Statute in the spirit of "love for the emperor and the Fatherland" engendered by the Karakozov Affair. Adopted at the height of the Great Reforms, the 1863 Statute had granted considerable autonomy to the faculty in governing universities, but now it was seen as "too liberal" and thus responsible for students' "anti-government" opinions and actions.[135] In April 1875, Alexander II approved the membership of a special six-member commission charged with preparing the draft of a new university statute. The emperor appointed Ivan Delianov, a career bureaucrat, who had served in the

Ministry of People's Enlightenment in various capacities since 1857, the commission chairman.[136]

Florinskii was appointed a member. As a preliminary step, the commission was to conduct a thorough review of all universities and Florinskii's responsibility was to inspect their medical schools. The inspection tour was to begin in the early fall of 1875 and to last several months, since, in addition to St. Petersburg University, the commission had to visit universities in Dorpat, Kazan, Kharkov, Kiev, Moscow, Odessa, and Warsaw. During the fall, however, Florinskii had to teach his regular courses at the academy. Delianov asked the IMSA administration to release Florinskii from his teaching duties, but, apparently after consulting with war minister Miliutin, the Academy Commander refused.

The refusal meant that Florinskii could no longer serve the two masters. He faced a choice: to resign his post at the academy or at the ministry. Perhaps he decided to follow his motto: "to bring to my Fatherland as much benefit as possible, specifically by acting in that area where I could be most useful." At this point, he probably felt that the area where he "could be most useful" was medical education writ large, not his research, clinical work, or lectures attended by just 100 or so students. Career considerations might well have played a role in his decision too. After the 1870 debacle with his promotion and the war minister's personal hostility spurred by his role in the "territorial dispute" between the War Ministry and the Ministry of People's Enlightenment, Florinskii's further prospects at the IMSA looked bleak indeed. On 17 June 1875, he submitted his letter of resignation to the Academy Commander. The letter was promptly accepted.

On 11 July, Florinskii's transfer to the Ministry of People's Enlightenment was made official. He became the ministry's expert on all matters related to medicine and its representative in the Medical Department of the Ministry of Internal Affairs, the country's highest medical agency. Two weeks later, he was promoted to the next, fourth civil rank (equal in military terms to a major-general), which required the emperor's personal approval and brought him not only a substantial increase in salary and benefits, but also *hereditary* nobility. Alas, Florinskii had no male heir to pass on this highest mark of his achievements — after the death of his son Sergei, he and his wife had no more children.

In September, the commission on the new university statute began its inspection tour. Since at the time only five Russian universities — in Dorpat, Kazan, Kharkov, Kiev, and Moscow — had medical schools, Florinskii's duties were less time-consuming than those of other commission members. But it did not mean that he did not give them his full attention. During the tour, he compiled a huge amount of data on every aspect of education at each school, ranging from the composition of the faculty to the necessary equipment for research and teaching to the physical conditions of buildings and auditoria. He spent several months working through the information he had collected. In the fall of 1876, he published the results of his inspection as a 300-page volume titled *Materials on the Conditions and Needs of Russian Medical Schools*, which presented a detailed plan of improving medical education throughout the empire.[137]

The same year, Florinskii resumed writing for the general public. His resignation from the IMSA apparently re-awakened the "irresistible" need to speak up he had felt at the height of the Great Reforms, some fifteen years earlier. He became a regular contributor to *New Time*, a popular newspaper run by Aleksei Suvorin, a successful journalist, theater critic, and publisher.[138] In the course of 1876, Florinskii published more than two dozen articles, reviews, and opinion pieces not only on various subjects in science and medicine, but also on history, geography, and politics. He continued his collaboration with Suvorin's newspaper in the next year.

Florinskii's work at the Ministry of People's Enlightenment also involved him in another large project — the establishment of a university in Siberia.[139] The idea of creating such a university had been around since the first expansion of the country's university system during the reign of Alexander I and had been promoted by Florinskii's distant relative Mikhail Speranskii. But for nearly seventy years it did not get beyond the stage of being merely "a good idea."[140] In the early 1870s, however, just as Florinskii began working for the ministry, the idea gained support from various local groups and leaders (including from the Governor of Western Siberia), as well as from St. Petersburg officials (most importantly, Minister Tolstoi).

Since Florinskii had spent his early years beyond the Urals, his colleagues and superiors saw him as something of an expert on

"Siberian matters." In the spring of 1875 Tolstoi put him in charge of preparing the ministry's preliminary plans, budgets, memoranda, and all other paperwork to get the "Siberian University" off the ground. Florinskii enthusiastically dove into the project. He developed a network of informants throughout Siberia, gathered all sorts of necessary information, and mobilized several local groups of merchants and civil servants in support of the project. He prepared an extensive position paper for the State Council, arguing for the establishment of a university in Tomsk, a large trading center in Western Siberia.

But navigating the seas of imperial bureaucracy proved much slower and more difficult than Florinskii had anticipated. The project got bogged down in the contrary interests of numerous central and local officials. The exact location of the projected university turned to be a major point of contention. Some wanted the university established in Omsk, the administrative center of Western Siberia, others in Tobol'sk, the former capital of Western Siberia, still others in Irkutsk, the administrative center of eastern Siberia, yet others in Tomsk. Florinskii was convinced that Tomsk presented the only viable option. In the course of the next year he campaigned in favor of the city, even publishing several articles in Suvorin's *New Time* defending his position.[141] But the debate dragged on and eventually led to the appointment by the emperor of a special commission to decide on the matter.[142] Expectedly, Florinskii became a member. Tolstoi fully supported Florinskii and clearly groomed him for the post of rector at the projected university. But Florinskii apparently got fed up with endless meetings, countless papers, and bureaucratic infighting. He began looking for an opportunity to leave the ministry.

Such an opportunity soon presented itself. In 1877 the chairman of the gynecology and obstetrics department at Kazan University's Medical School retired. One of Florinskii's former students, who worked at the school as a professor of pathology, suggested that the University Council offer the position to Florinskii. Florinskii immediately agreed, and on 17 October 1877, the council elected him to the vacant post. The appointment, however, required ministerial approval. Tolstoi was disappointed by Florinskii's decision to leave the ministry and return to teaching. But Florinskii promised the minister that he would not discontinue his involvement with the Siberian University, and, if and when a final decision on its establishment would come to pass, he would gladly accept responsibility for its implementation. Tolstoi approved Florinskii's appointment to Kazan University, but Florinskii stayed in St. Petersburg for another six months to see the decision on the Siberian University through. He continued to lobby for Tomsk as the site of the new university and reportedly even used his personal connections to the Grand Duke Konstantin Nikolaevich to swing the decision in the city's favor.

Only in mid-February 1878 had Florinskii wrapped up all his remaining work at the ministry and come to Kazan. As he wrote in his memoirs: "After twenty-five years of life in the capital, the quiet province enveloped me with a life-giving atmosphere of a bright sunlight and a piece of mind. ... Here I have disengaged from countless commissions and dry bureaucratic work, belonging only to myself and my department, and can do what I wanted."[143]

But what did he want? He was 44 years old and in 25 years had traveled a long way from a poor seminarian contemplating his uncertain future in Perm to a civil-service general occupying a high office and rubbing shoulders with the rich and powerful in St. Petersburg. From a career perspective, leaving the capital made little sense. But, at this point in his life, Florinskii was apparently not very interested in advancing further his already quite remarkable career. Following his motto, he was looking for a place where he "could bring the most benefit" to Russia and its people. Shuffling papers and fighting bureaucratic wars in the capital's corridors of power apparently was not what he had in mind. In Kazan, he began working on his *Thoughts and Notes on My Childhood and Education* — a clear sign of his desire to reflect upon his own origin, choices, and life trajectory. What, for a time, remained unspoken and unwritten, but likely also occupied his thoughts, was his legacy. He had lost his son, the natural heir to his accomplishments, who could carry his name and further his legacy. What would he leave to posterity? What would he be remembered for by his Fatherland and his people? Florinskii's varied activities undertaken after he had left St. Petersburg provide some clues to what he considered to be the answers to these questions.

Immediately upon arrival in Kazan, Florinskii took up his official duties as the department chairman. He began to teach lecture courses on gynecology and obstetrics, reorganize the department's clinics, and set up a laboratory for his research. He joined the university's Society of Physicians and in May delivered his first presentation to the society's meeting. He became involved in extensive studies of various local health issues, especially the plague.[144] He resumed work on the textbook that he had abandoned after the debacle with his promotion, and a few years later, its second volume devoted to obstetrics finally came out.[145]

Florinskii's new job gave him plenty of free time — the entire summer, from mid-May till mid-September while university students were on vacation — to take up several new large scholarly projects. One of them was writing a "domestic medicine" manual. His work in the Scientific Committee of the Ministry of People's Enlightenment had alerted him to the poor quality of available "self-help" medical literature. He decided to remedy this situation by producing a book that would provide any literate person with all the necessary knowledge to diagnose and treat the most common medical conditions and ailments. In 1880, the 900-page volume came out to much public acclaim — it won Florinskii a prestigious prize named after Peter the Great from the Ministry of the People's Enlightenment.[146]

The second project was also connected to the "self-help" genre, but was more historical in nature. Those "seeds of [historical] scholarship," which his professor of church history Father Makarii had planted in Florinskii's heart at Perm Seminary, germinated and sprouted numerous shoots. He joined Kazan University's Society of Archeology, History, and Ethnography and became very active in its meetings, publications, and expeditions. One of the first ideas he offered to the society was to create a "historical-ethnographic museum" open to the public.[147] The idea did not find much support at the time, but Florinskii found another way to pursue his renewed interest in history.

Since childhood, Florinskii had been a passionate collector. His profession gave this passion a particular direction: he began collecting old medical books and manuscripts, combing antiquarian shops, booksellers, and libraries wherever he went. Now he decided to share some of his finds with the public. He edited and annotated nine rare manuscripts of sixteenth- and seventeenth-century Russian herbal and medical manuals from his personal collection. In 1879, he published each manuscript as a supplement to *Scientific Memoirs* — Kazan University's official outlet — and, at the beginning of the following year, he issued them together in book format.[148]

Alas, the delightful freedom to pursue whatever scholarly project he fancied did not last long. After more than two years of bureaucratic delays, the decision to build the Siberian University in Tomsk was finally made official. In May 1880, keeping his promise to Tolstoi, Florinskii travelled to Tomsk to take up the duties of the ministry's representative on the Building Committee of Tomsk University.[149] Although the Tomsk Governor was the Building Committee's nominal head, Florinskii became its main engine. He came to oversee every facet of the herculean task of creating from scratch a modern university in Siberia, from its budget (which, of course, had to be revised several times) and architectural plans to the hiring of a workforce and the procuring of building materials. In August, as part of a publicity and fund-raising campaign, Florinskii arranged a ceremony to mark the symbolic placing of the first cornerstone in the foundation of the future university's main building.[150] For the next five years, while continuing his professorial duties at Kazan University, every summer Florinskii made the journey of nearly 3,000 kilometers (one way) to Tomsk to oversee personally the construction, fund-raising, and furnishing of the new university.

Fig. 4-4. The main building of Tomsk University at the time of its official opening in 1888. From *Pervyi universitet v Sibiri* (Tomsk: Sibirskii vestnik, 1889), insert. Courtesy of RNB.

In the late summer of 1885, the construction was completed and Tomsk University was ready to open its doors to the first cohort of students (see fig. 4-4). But its opening was put on hold. The five years that Florinskii spent building the university saw drastic changes in the life of his Fatherland and, especially, its educational system. The assassination of Alexander II and the ascendance of his son Alexander III in the spring of 1881 signaled the final end to the policies of the Great Reforms and the return to "reaction" reminiscent of the reign of Nicholas I. Over the same five years, the Ministry of People's Enlightenment witnessed a succession of three different heads. Florinskii managed to maintain good working relations with each of them, which certainly was a decisive factor in his success in building the university. The new university statute, whose contours Florinskii had helped define, was finally promulgated in 1884, after nearly a decade of numerous revisions and edits, and amply manifested the oppressive tenor of the new times. It radically curtailed the autonomy of the faculty and the student body and instituted strict administrative control of bureaucracy in St. Petersburg over all university affairs (from hiring and firing of personnel to admission policies).[151] Students' protests and unrest that shook the Russian Empire after the adoption of the 1884 statute made the central bureaucracy wary of creating in Siberia a "center of revolutionary propaganda," as all universities came to be seen in the aftermath of the 1881 regicide.[152]

The fate of Tomsk University hung in the air, and with it, the fate of Florinskii. He had fulfilled his promise to Tolstoi, yet the post of the university's rector — that Tolstoi had promised him — remained unavailable. However, the new minister Ivan Delianov (who had known Florinskii since the two worked together on the 1875 imperial commission on the university statute) found an even better use for Florinskii's administrative talents. In the fall of 1885, Delianov appointed him "supervisor of the Western Siberia educational district,"[153] a de facto deputy-minister in charge of all matters related to education in a territory six times the size of France. Florinskii resigned his professorial position at Kazan University and moved to Tomsk to oversee not only the still-unopened university, but also nearly fifty other educational institutions (primary and secondary schools) scattered throughout the region (see fig. 4-5).[154]

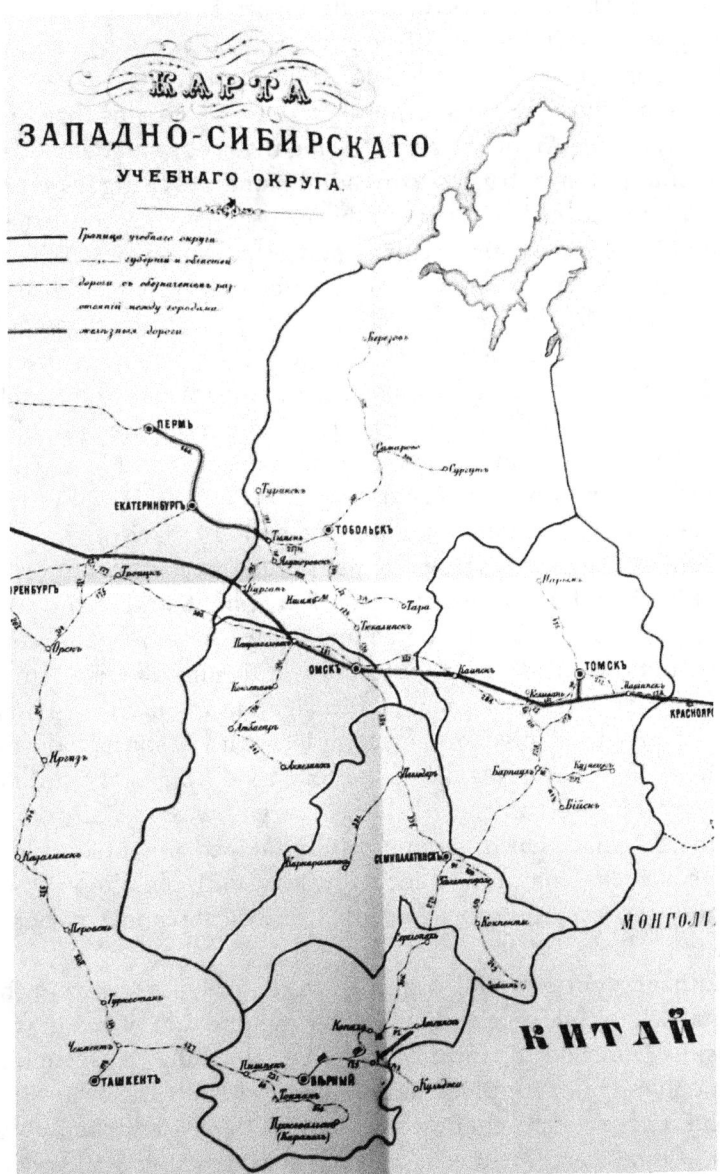

Fig. 4-5. Map of the Western-Siberian educational district supervised by Florinskii that stretched from China to the Arctic Ocean. From *Pamiatnaia knizhka Zapadno-Sibirskogo uchebnogo okruga*, 5th edn. (Tomsk, 1897), insert. Courtesy of RNB.

Florinskii attended to his new duties with his customary efficiency and dedication.[155] He certainly saw expanding access to education — which he had presented in his treatise as a major condition for the realisation of the Russian people's "hidden hereditary potentials" — as a worthy cause. He made regular inspection tours of his domains that stretched from China to the Arctic Ocean, assessing the needs and demands of the educational system entrusted to his administration. Using his experience with building Tomsk University, he focused on the construction of new buildings for existing schools and establishing new schools. He founded a new gymnasium in Vernyi (today's Almaty, Kazakhstan), erected new buildings for gymnasia in Tobol'sk and Tomsk, and created new primary and low-level technical schools throughout the region. In a dozen years during Florinskii's administration, the number of schools (and their students) in Western Siberia more than doubled. His efforts were certainly appreciated by his superiors and in 1892 earned Florinskii a promotion to the next, third civil rank in the Table of Ranks (equivalent to a lieutenant-general in the military service).

Opening Tomsk University remained his overriding priority and Florinskii lobbied the Ministry of People's Enlightenment and the State Council for permission. It took him nearly three years to convince government officials that he could prevent turning the new university into a "nursery" of revolutionary propaganda in Siberia.[156] Finally, in July 1888, Tomsk University — complete with a student dormitory, apartments for the professoriate, an extensive Botanical Garden with several greenhouses, and an excellent library — was officially opened.[157] Alas, the original plan of having a fully-fledged university with four separate schools (natural sciences, history and philology, law, and medicine) got severely truncated: only its medical school was permitted to open.[158]

Another crown jewel of Florinskii's work in Siberia was a new higher technical school in Tomsk. Shortly after the opening of the university, Florinskii launched a campaign for establishing a Technological Institute that would supply Siberia — and especially the Trans-Siberian Railroad whose construction was to begin in 1891 — with badly-needed cadres of engineers, architects, and other technical specialists. It took Florinskii almost a decade to push the project through the layers of state bureaucracy. Only after the death of Alexander III in the fall of 1894 and the ascendance of his son Nicholas II to the throne, did Florinskii manage to obtain imperial permission for its implementation. In 1896, the construction of the first higher technical school in Siberia began.[159]

Fig. 4-6. Vasilii Florinskii, the supervisor of the Western-Siberian educational district, c.1888. From *Pervyi universitet v Sibiri* (Tomsk: Sibirskii vestnik, 1889), insert. Courtesy of RNB.

Each of these projects required visits to the chosen construction sites, lengthy negotiations with the local administration, continuous fundraising campaigns, extensive correspondence with ministry officials and other state bureaucrats, and regular trips to the capital. Florinskii had a small support staff to assist him with technical work, but every year he spent four to six months on the road. Yet, busy as he was with his numerous administrative duties, the former professor did not abandon his scholarly pursuits.

His new job profoundly influenced his scholarly interests. Immediately upon opening the university, Florinskii spearheaded the establishment of the Tomsk Society of Naturalists and Physicians under the university's aegis and was unanimously elected its first chairman.[160] He presided over the society's regular meetings and delivered several reports on the endemic and epidemic diseases affecting the region, such as influenza and the plague.[161] But the main focus of his scholarly work shifted from medicine to history, archeology, and ethnography.

Even before construction of Tomsk University was completed, Florinskii had started to implement his idea of creating a public museum of history and ethnography, which he had first voiced at the meetings of the Society of Archeology, History, and Ethnography in Kazan.[162] Already in 1882, he began collecting various objects that would form the future museum's exhibits. He persuaded several wealthy local patrons and private collectors to support this endeavor, and by the time the university finally opened in 1888, its Archeological Museum had acquired a large collection of Siberian archeological and ethnographic materials. For the opening of the university, Florinskii published an extensive, 500-page catalogue of this collection, complete with detailed descriptions, annotations, and photographs of its various holdings.[163] In the subsequent years, this collection became the major foundation of his personal scholarly work.

Since childhood Florinskii had been fascinated by the old burial mounds (*kurgany*) scattered all over the Southern Urals and Western Siberia. Now he made the artefacts dug out from these mounds into a subject of intensive study. On his numerous travels through the region he mapped out the location of various mounds and recruited local enthusiasts to conduct their excavations and send all the materials they dug out to Tomsk. In just two years, 1889 and 1890, he published nearly a dozen articles on burial mounds found in various locales.[164] These studies culminated in a monumental, nearly 1,000-page-long treatise on "Slavic archeology." From 1894 to 1898, the treatise appeared in installments as a supplement to the *Herald of Tomsk University*, and simultaneously as a multi-volume monograph under the general title *The Primordial Slavs according to the Monuments of their Prehistoric Life*.[165] Based on careful comparison of various objects (weapons, jewelry, ceramics, sculls, etc.) from the Stone, Bronze, and Iron Ages found in different mounds, Florinskii advanced a hypothesis of the early migration of the prehistoric "Slavic tribes" from India to Siberia and farther on to Europe.[166]

In June 1898, just as the last installment of his "Slavic Archeology" came out, Florinskii's career reached a point — forty years of service — that allowed him to retire with a full pension. He was 64 and, under the stress of constant travel, perpetual administrative conflicts, and non-stop scholarly work, his health had begun to deteriorate. The new minister of people's enlightenment, Nikolai Bogolepov, offered him a post at the ministry, but he declined. He and his wife wanted

Fig. 4-7. Maria Florinskaia, c.1877. The photo was taken at the St. Petersburg studio of Iulii Shteinberg, a well-known photographer, and the tentative date is established by the style of the photographer's logo. Note the black weeper she still wears on her neck, likely marking her mourning for her lost child. Courtesy of NMRT.

to move back to Kazan. Their only child, Olga, resided in the city with her husband and a ten-year-old daughter, christened in honor of her grandmother, Maria (see fig. 4-7). Upon learning of his retirement, the Tomsk City Duma bestowed on Florinskii the title of "honorary citizen" to mark his contributions to education in the city and Siberia writ large.

Over the summer, the Florinskiis packed up their belongings, including Vasilii Florinskii's extensive collections of paintings, china, manuscripts, and personal papers, and shipped them to Kazan. On 30 August, taking advantage of a new railroad that connected Tomsk to the Trans-Siberian Railway, they boarded a train to St. Petersburg.[167] Florinskii had to formally present his final reports, along with resignation and pension paperwork, to the ministry. On arrival at the capital, they took up residence in a fancy hotel in the center of the city. But shortly thereafter, Vasilii Florinskii fell ill. On 3 January 1899, a few weeks short of his 65th birthday, he died of a heart failure in his hotel room.

## A Superfluous Synthesis?

As the "superfluous man" of classic nineteenth-century Russian literature, whose views and ideas were misunderstood, unwanted, and ultimately marginalized by his contemporary society,[168] so too Florinskii's treatise did not find a response or a place in any of the scholarly fields it attempted to unify in his concept of "human perfection and degeneration." As happened to Galton's 1865 article, at the time of publication Florinskii's essays went virtually unnoticed. Much like a "superfluous man," Florinskii's synthesis had a definite potential that went unrealized. Similar to Galton's first publications on "hereditary talents," Florinskii's concept of "rational" or "hygienic" marriage had the potential to generate a viable research programme and even initiate policy changes, but it failed to raise the interest of either the scholarly community or the general public, to say nothing of the government bureaucracy. In contrast to Galton, who personally carried out his research programme and spent several decades gathering support for his ideas, methods, and concepts, Florinskii never returned to the subject of his treatise in his subsequent works and did absolutely nothing to promote it among fellow scientists and physicians or the general public.

Perhaps, an important factor in Florinskii's abandonment of any further work on "Human Perfection and Degeneration" was the near total silence that greeted the first appearance of his treatise. Contrary to Pisarev's assertion that "in just a week after the publication of each new issue [of *Russian Word*], all thoughts and even all separate expressions it contained have already been counted, measured, weighted, sniffed at, felt over, and taken under consideration,"[169] the publication of Florinskii's essays attracted no attention either from the censor, or from reviewers and commentators in contemporary periodicals. In late October 1865, in its regular review column, St. Petersburg's leading newspaper *Voice* carried a lengthy overview of the August issues of the three major "thick" journals: *The Contemporary*, *Russian Herald*, and *Russian Word*.

But it did not even mention Florinskii's essay that opened the issue of *Russian Word* under review.[170] The same situation was repeated in December when *Voice* reviewed the contents of *Russian Word*'s October issue.[171] Only on 17 February 1866, in a review of the last two 1865 issues of the journal did *Voice*'s columnist briefly mention Florinskii's treatise: "In the December issue, the conclusion of a voluminous investigation by V. M. Florinskii on human perfection finally came out. The essay is very remarkable in its goal [and] clearly imbued with love for humanity whose wellbeing is undoubtedly very close to its author's heart."[172]

The release of Florinskii's essays as a book did nothing to break the silence that had greeted their publication in the journal. On 6 August, a member of the St. Petersburg Censorship Committee reported to his superiors on the book's contents. He had read the book very closely, identifying its subject as "the hygiene of marriage, i.e. those physiological and hygienic conditions, which, if ignored at the time of marriage's arrangement, [could] carry fatal consequences for the next generation."[173] His report is the only explicit evidence that Florinskii had succeeded in making his very complicated subject accessible to any educated reader. The censor did note "a certain democratic (philanthropic) tint that expresses, in some places, its author's sympathy for the simple, uncorrupted masses of the people, and, in other places, his disapproval of the aristocracy and his contempt for degenerating aristocratic families." He also noticed Florinskii's comment about "the perfection of the peasant type" by billeting elite military regiments in the provinces and by the migration of peasant laborers to the cities. Nor did he miss Florinskii's critique of church regulations of "kin" and "mixed" marriages as "too strict." He clearly bought the double-talk devised by Florinskii and/or Blagosvetlov to avoid the censorship's wrath. He stated that the book's "direction" cannot be defined as "purely materialistic," pointing out that its author himself had specifically refuted such reading of his text and offering as proof a long quote from Florinskii's "aside" to his "female readers." According to the censor's assessment, the book as a whole "did not contain any infringements on censorship laws, which could provide a reason for its arrest or a court injunction."

On 31 August 1866, *Voice* ran an advertisement announcing that Florinskii's treatise "has been published as a separate volume and is available for purchase ... at a new bookstore, established at the former editorial office of *Russian Word*."[174] The book's publication was also duly announced on the pages of several other newspapers and journals.[175] Starting with its very first issue, Blagosvetlov's *Deed* carried an advertisement of the book in its back matter on a regular basis (see fig. 4-8).[176] Yet I did not find a single review of the book in any of the numerous contemporary literary-scholarly, medical, and scientific periodicals I examined.

Various factors might have been responsible for this silence. The reaction to the Karakozov Affair in the highest echelons of the imperial government led to the shutting down of not only *The Contemporary* and *Russian Word*, but also several other periodicals, which otherwise might have reviewed Florinskii's work. Thus, *Book-lover* (*Knizhnik*), a major bibliographic journal entirely devoted to reviewing contemporary literature (including scientific and medical publications), was shut down in June 1866. The same month the same fate befell *Medical News*, a weekly newspaper published by Florinskii's younger colleague, gynecologist Martin Gorovits, and, two months later, *Self-Education*, a popular journal issued by Emanuil Khan, yet another physician-turned-publisher.

It looks like none of the journals that — despite mounting government pressure — survived beyond 1866 noticed Florinskii's book. *Book Herald*, another important bibliographic journal that continued publication until July 1867, carried only a one-line bibliographic record of the book in its October 1866 issue, but nothing else.[177] None of the "thick" journals that systematically reviewed new scientific and medical publications lent its pages to a review of Florinskii's treatise. An illustrative example is *Women's Herald*, a new "learned-literary" journal established in September 1866, which provided "refuge" for many of the former contributors to both *Russian Word* and *The Contemporary*. The journal regularly reviewed scientific and medical publications in its bibliographic section. Thus, its third issue that came out in early 1867 carried a critical review of *Hygiene: A Manual for the Preservation of Health* (1860), written by well-known popularizer Stepan Baranovskii.[178] The next issue reviewed *Popular Lectures on Cholera* (1866), delivered and published by a staff physician of the Russian army in the Caucasus, which propagated many ideas closely resonating with Florinskii's, especially regarding the role of poverty in the spread of diseases and the advocacy of social hygiene.[179] But, despite the fact that Florinskii's book had directly addressed many issues dear to the hearts and minds of the journal's contributors and readers — especially "the women's question" — it was never mentioned in the pages of *Women's Herald*. Many other "thick" journals and newspapers during the same time period also regularly reviewed various scientific and medical publications, but not Florinskii's book.[180]

Perhaps, it was the book's explicit critique of existing marriage regulations as both a major potential source of degeneration and a major obstacle to human perfection that deterred contemporary observers from discussing it in newspapers and journals. During the nineteenth century, unlike elsewhere in Europe, marriage in Russia remained the exclusive domain of the church, not the state (the very notions of "civic marriage" and its derivative "fictitious marriage" were just beginning to enter the Russian social conscience at the time).[181] Although the Russian Civic Code had a number of relevant laws on the books, their actual administration (including solemnization, matrimony, divorce, annulment, and so on) was kept firmly in the hands of the clergy that saw marriage as a sacrament and a covenant.[182]

The winds of change that blew over the empire in the aftermath of the Crimean War began to undermine the church's authority, including its grip on the matters of marriage and family.[183] Many educated Russians felt that "the political, social, [and] familial forms of human life had grown old, unsuitable for the present times and are destined to shatter [and] disappear."[184] During the first decade of Alexander II's rule, officially endorsed *glasnost'* stimulated a pointed debate (that unfolded largely on the pages of "thick" journals) over civic laws and the church directives pertaining to the "familial forms of life," including the regulation of "kin" marriages, allowable marital age, age differences between prospective spouses, and "mixed" marriages.[185] Public dissatisfaction with various aspects of what US historian Gregory L. Freeze has described as "a marital order of a rigidity unknown elsewhere in Europe"[186] also found vivid expression in a number of artistic and literary works, such as, to give only two pertinent examples, Vasilii Pukirev's painting *The Unequal Marriage* and Nikolai Chernyshevskii's novel *What is to be Done?*, both of which captivated public attention in 1863. These and other similar works demonstrate that in the first half of the 1860s, a large section of Russian educated society was openly challenging the ecclesiastical cannons, precepts, and mores surrounding marriage.[187] With their explicit commentary on church regulations of "kin" marriages, age differences between prospective spouses, and "mixed" marriages, Florinskii's essays fit quite comfortably into the prevailing atmosphere of critiquing the existing marital order.

The Karakozov Affair changed that atmosphere radically. The imperial edicts and government actions prompted by the affair aimed first and foremost at upholding "the strict basis of religion and morality." This new atmosphere found immediate expression in the fierce defense of traditional church marriage, as Nikolai Cherniavskii's comedy *Civic Marriage*, which premiered in St. Petersburg in November 1866, clearly attested.[188] The play provoked much indignation in the press, which even prompted its author to write a special "foreword" for its second edition justifying his views.[189] These new conditions were hardly conducive to publically discussing Florinskii's treatise. As Blagosvetlov warned Iakobii, after Karakozov's assassination attempt, religion became "a strictly forbidden fruit."[190]

If these "ecclesiastical" dimensions of Florinskii's work could indeed prevent contemporary commentators from discussing it in newspapers and "thick" journals, they did not necessarily preclude a discussion of his "secular" ideas about human variability, heredity, and evolution by Russian scientists. Nor did they automatically exclude debates about the role of social hygiene in human perfection and degeneration or about the more general issues of the applicability of Darwin's evolutionary concept to the understanding of "social ills" by Russian physicians. Yet, neither the scientific nor the medical community seemed to even notice Florinskii's essays. No existing medical periodicals published a review.[191] Neither did any natural science magazines. For instance, *Naturalist*, a popular bi-weekly journal published in St. Petersburg from 1864 through 1867, carried a variety of articles, reviews, and surveys on anthropology, Darwin's theory, human evolution, and other subjects discussed in Florinskii's work, but none of them ever mentioned his book.[192]

It is impossible to estimate how many people actually read Florinskii's treatise, either in the journal or the book versions, but the absence of reviews in popular and specialized periodicals likely decreased its potential readership. We do not know how many copies of the book Blagosvetlov printed in 1866, but he kept it in print for more than a decade and regularly advertised it in his journal.[193] In 1869, he apparently tried to boost sales by reducing the price of the book by a quarter, with an additional twenty per cent discount for the subscribers to *Deed*.[194] Furthermore, there are some indications that sometime in 1871, just as Blagosvetlov published Portugalov's essays "On Degeneration" and released the first volume of his own edition of Darwin's *The Descent of Man*, he also ran an additional printing of Florinskii's book (see fig. 4-8).[195] But, even during the late 1860s and 1870s, with the exception of Portugalov, no one referred to or cited Florinskii's treatise in various publications on any of the subjects it addressed.

A variety of factors might have played a role in the failure of *Human Perfection and Degeneration* to raise any interest in the Russian scholarly community. Florinskii's "thought piece" attempted to synthesize the newest data, ideas, methods, and concepts from a variety of scholarly fields, including anthropology, physiology, and general biology, as well as from a host of medical fields, such as social hygiene, gynecology, psychiatry, and pediatrics. But it failed to find a sympathetic "thought collective," as Ludwik Fleck has perceptively described it, in any of them.[196] In the mid-1860s, nearly all of these fields were in the process of (re)formation. Their practitioners were divided on almost every issue, while their concepts and methods were in flux. In particular, the then current ideas about heredity, reproduction, and evolution were bitterly contested and their application to humans and social issues highly contentious.

Furthermore, Florinskii's treatise came out exactly during the period when the two fields most closely connected and thus, theoretically, most receptive to his ideas — physical anthropology and social hygiene — were becoming fully-fledged disciplines. Their practitioners were in the process of constructing specialized disciplinary communities, building institutional structures, negotiating a consensus over the methods, subjects, and objects of future research, and establishing intellectual and institutional boundaries with competing fields and communities.

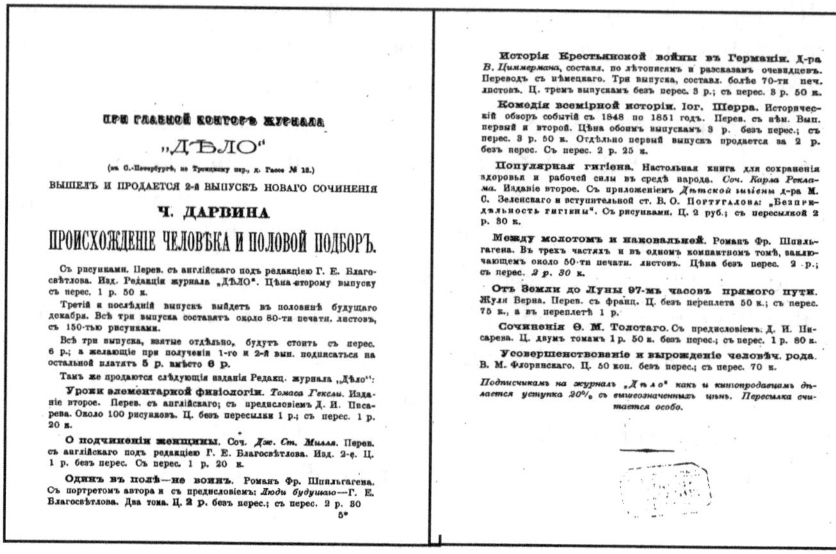

ПРИ ГЛАВНОЙ КОНТОРѢ ЖУРНАЛА
„ДѢЛО"
(въ С.-Петербургѣ, по Троицкому пер., д. Гасое № 18.)
ВЫШЕЛЪ И ПРОДАЕТСЯ 2-й ВЫПУСКЪ НОВАГО СОЧИНЕНІЯ
Ч. ДАРВИНА
ПРОИСХОЖДЕНІЕ ЧЕЛОВѢКА И ПОЛОВОЙ ПОДБОРЪ.

Съ рисунками. Перев. съ англійскаго подъ редакціею Г. Е. Благосвѣтлова. Изд. Редакція журнала „ДѢЛО". Цѣна второму выпуску съ перес. 1 р. 50 к.

Третій и послѣдній выпускъ выйдетъ въ половинѣ будущаго декабря. Всѣ три выпуска составятъ около 80-ти печатн. листовъ, съ 150-тью рисунками.

Всѣ три выпуска, взятые отдѣльно, будутъ стоить съ перес. 6 р.; а желающіе при полученіи 1-го и 2-й вып. подписаться на остальной платятъ 5 р. вмѣсто 6 р.

Тамъ же продаются слѣдующія изданія Редакц. журнала „Дѣло":

**Уроки элементарной физіологіи.** *Томаса Гексли.* Изданіе второе. Перев. съ англійскаго; съ предисловіемъ Д. И. Писарева. Около 100 рисунковъ. Ц. безъ пересылки 1 р.; съ перес. 1 р. 20 к.

**О подчиненіи женщины.** Соч. *Дж. Ст. Милля.* Перев. съ англійскаго подъ редакціею Г. Е. Благосвѣтлова. Изд. 2-е. Ц. 1 р. безъ перес. Съ перес. 1 р. 20 к.

**Одинъ въ полѣ—не воинъ.** Романъ Фр. Шпильгагена. Съ портретомъ автора и съ предисловіемъ: *Люди будущаго*—Г. Е. Благосвѣтлова. Два тома. Ц. 2 р. безъ перес.; съ перес. 2 р. 80

5*

**Исторія Крестьянской войны въ Германіи.** Д-ра *В. Циммермана*, составл. по лѣтописямъ и разсказамъ очевидцевъ. Переводъ съ нѣмецкаго. Три выпуска, составл. болѣе 70-ти печ. листовъ. Ц. тремъ выпускамъ безъ перес. 3 р.; съ перес. 3 р. 50 к.

**Комедія всемірной исторіи.** Іог. Шерра. Историческій обзоръ событій съ 1848 по 1851 годъ. Перев. съ нѣм. Вып. первый и второй. Цѣна обоимъ выпускамъ 3 р. безъ перес.; съ перес. 3 р. 50 к. Отдѣльно первый выпускъ продается за 2 р. безъ перес. Съ перес. 2 р. 25 к.

**Популярная гигіена.** Настольная книга для сохраненія здоровья и рабочей силы въ средѣ народа. *Соч. Карла Реклама.* Изданіе второе. Съ приложеніемъ *Дѣтской гигіены* д-ра М. С. Зеленскаго и вступительной ст. В. О. Португалова: „Безпредѣльность гигіены". Съ рисунками. Ц. 2 руб.; съ пересылкой 2 р. 30 к.

**Между молотомъ и наковальней.** Романъ Фр. Шпильгагена. Въ трехъ частяхъ и въ одномъ компактномъ томѣ, заключающемъ около 50-ти печатн. листовъ. Цѣна безъ перес. 2 р.; съ перес. *2 р. 30* к.

**Отъ Земли до Луны 97-мь часовъ прямого пути.** Жуля Верна. Перев. съ франц. Ц. безъ переплета 50 к.; съ перес. 75 к., а въ переплетѣ 1 р.

**Сочиненія Ѳ. М. Толстаго.** Съ предисловіемъ Д. И. Писарева. Ц. двумъ томамъ 1 р. 50 к. безъ перес.; съ перес. 1 р. 80 к.

**Усовершенствованіе и вырожденіе человѣч. рода.** В. М. Флоринскаго. Ц. 50 коп. безъ перес.; съ перес. 70 к.

*Подписчикамъ на журналъ „Дѣло" какъ и книгопродавцамъ дѣлается уступка 20% съ вышеозначенныхъ цѣнъ. Пересылка считается особо.*

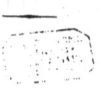

Fig. 4-8. Advertisements for books published by Grigorii Blagosvetlov, available at the editorial office of his journal *Deed*. They open with the announcement of the second instalment of Darwin's *Descent of Man* and end with Vasilii Florinskii's *Human Perfection and Degeneration*. From *Delo*, 1872, 9 (September), backmatter. Courtesy of RNB.

As elsewhere in Europe and the Americas, in Russia discipline building in physical anthropology unfolded during the 1860s and involved first and foremost its separation from a hodgepodge of such fields as ethnography, geography, archeology, and ethnology.[197] In 1864, the Society of Enthusiasts for the Natural Sciences (organized just a year earlier by a group of Moscow University professors) established a special anthropology section that almost immediately began to issue its own journal.[198] Three years later, in 1867, the society also instituted a separate ethnography section that the same year held the empire's first ethnographic exhibition in Moscow.[199] Also that year, the ethnography section of the Imperial Russian Geographical Society began to issue its own *Memoirs*.[200] Amid these events, in the fall of 1866, Anatolii Bogdanov, a leading figure in the institutionalization of both anthropology and ethnography, attempted to articulate the differences between the subjects, objects, and methods of the two newborn disciplines and to draw a distinct boundary between them.[201]

Discipline building in social hygiene occurred during exactly the same period, but took a different form. If the institutionalization of anthropology was largely the result of public initiatives carried out by learned societies, the institutionalization of social hygiene was primarily a state affair. In 1863, Florinskii's former professor Evgenii Pelikan was appointed head of the Medical Department of the Ministry of Internal Affairs.[202] The very next year, the radical reform of local administration (which inaugurated the establishment of *zemstvos* in the European part of the Russian Empire) and court reform (which dramatically reshaped the entire judicial system) necessitated corresponding reorganizations in the provision and administration of medical services to both the state (and its courts)[203] and the populace.[204] Following the precepts of official *glasnost'* that encouraged a limited and carefully controlled debate on state policies, in early 1865 Pelikan created a new journal, the *Archive of Legal Medicine and Social Hygiene*, modeled on the French *Annales d'hygiène publique et de médecine légale*, as a forum for discussion of necessary reforms in both legal medicine and public health. A year later, the Imperial Medical-Surgical Academy split its existing Department of Forensic Medicine, Hygiene, and Medical Police into two separate departments: the Department of Forensic Medicine, and the Department of Social Hygiene.[205] Over the next few years, both the new journal and the new departments were engaged in shaping the agendas of social hygiene and establishing boundaries that would delineate the new discipline from such competing fields as experimental hygiene, community medicine, and epidemiology.[206] It took longer than a decade to consolidate the diverse views of numerous physicians on what constituted the proper domain of social hygiene in Russia. Only in 1877, was the first society of social hygiene established under the name of "the Russian Society for the Protection of People's Health," and it took another seven years for the society to begin publishing its own journal.[207]

Florinskii did not belong to any of the networks responsible for the institutionalization of either anthropology or social hygiene. He never attended their meetings, or published in specialized periodicals organized by the proponents of the two disciplines, and thus took no part in shaping their agendas. Nor did he promote in any way the ideas of his treatise among the members of these newborn specialized communities. Unlike Galton, who used every opportunity to publicize his ideas, methods, and concepts by delivering reports to existing scientific societies in London and elsewhere (especially those of anthropologists, sociologists, statisticians, and physicians), Florinskii never even mentioned the ideas of his treatise at any public gathering he attended.

Furthermore, Florinskii's ideas about human degeneration and perfection appeared largely irrelevant to the agendas of both Russian physical anthropology and Russian social hygiene. As a thought piece based essentially not on his original research, but on published, predominantly western literature, Florinskii's treatise responded to the debates within — and hence reflected the current concerns of — western (especially, French) anthropology and western (mostly French and German) social hygiene. The institutionalization of these two disciplines in Russia, however, required first and foremost the support of domestic patrons and the mobilization of their domestic practitioners and thus prompted the discipline-builders to gear their agendas to local interests and local concerns, quite different from those of their European counterparts. In the 1860s and 1870s, Russian anthropologists were largely engaged in cataloguing the empire's "human diversity" through extensive craniological and anthropometric studies of its various populations, proudly presented at the country's first anthropological exhibition held in 1879 in Moscow.[208] During the same time period, as numerous articles published in the *Archive of Legal Medicine and Social Hygiene* clearly attest, Russian social hygienists focused mostly on the endemic and epidemic diseases that plagued the empire and their relation to such social factors as housing, occupation, sanitation, nutrition, prostitution, and alcoholism.[209] An outbreak of cholera in the summer of 1866, which quickly reached epidemic proportions in Russia, certainly reinforced this focus and perhaps also contributed to social hygienists' neglect of Florinskii's book that came out at the height of the epidemic.[210]

The ideas of human degeneration and perfection were not completely new to the Russian educated public.[211] Florinskii was certainly not the first to write about them in Russian, though he was the first to place them explicitly in the context of Darwin's evolutionary concept and of the fledgling discipline of social hygiene. Yet Florinskii's notions of degeneration and perfection had very little traction with Russian scientific/medical commentators. Consanguineous marriages, which Florinskii presented in his treatise as a major cause of degeneration, as well as "mixed" marriages, which he saw as a major instrument of perfection, were expressly forbidden by both the church rules and the state laws, thus making this part of his argument largely irrelevant to his target audience. His condemnation of slavery, serfdom, and, more generally, the economic and political inequality of various social groups as an important condition conducive to degeneration was, as he himself had readily admitted in his treatise, also clearly beyond the purview of the medical and scientific communities. At best, they might have conceived of any practical implementation of Florinskii's ideas as belonging to a very distant *future*. But busy with debating a mountain of urgent social, political, economic, and cultural issues of their respective vocations, Russian scientists and physicians seemed to follow Chernyshevskii's advice to focus on the present and to "leave the taking care of great-great-children to the great-great-children themselves."[212]

Furthermore, Florinskii himself declared that the Russian people were in no danger of degeneration, thus making the entire subject of a purely "theoretical" concern, in contrast to the very practical and immediate concerns over the provision of medical care to the populace under the new *zemstvo* system, which preoccupied the medical community at the time. Even Portugalov, the only social hygienist who at Blagosvetlov's direction had taken up the further development of Florinskii's ideas, completely abandoned his studies of degeneration after 1876 and devoted his formidable energies to practical work as a physician in the Samara *zemstvo*.

One reason for the marked lack of interest in degeneration concepts in post-Crimean Russia, as well as for Florinskii's optimistic declaration, was the absence of reliable demographic and vital statistics,[213] which underpinned much of the attention such concepts generated in other countries and which Florinskii discussed at length in his treatise.[214] Although "medico-topographical surveys" that usually included some data on demographic and vital statistics of the region under investigation had begun to be conducted in Russia in the first half of the nineteenth century, by the 1860s they covered only a small fraction of the empire's huge territory and population.[215] Moreover, the "raw," *aggregated* data presented in such surveys did not lend itself easily to the *differential* statistical assessments covered by Jean-Marc Boudin's, Johann Ludwig Casper's, Francis Devay's, Adolphe Quetelet's, and Thomas Willis's studies, which Florinskii utilized and cited in his essays.[216] The very first statistical analyses of Russian fertility, mortality, and morbidity figures began to appear exactly at the time Florinskii was writing his essays, largely through the efforts of the country's foremost expert on probability theory, the vice-president of the Imperial Academy of

Sciences, Viktor Buniakovskii.[217] And it was during the exact same time — at the October 1865 meeting of the St. Petersburg Society of Russian Physicians — that Iakov Chistovich voiced a passionate plea to collect "materials for medical geography and medical statistics in Russia" and offered a detailed plan on how to do this.[218] Only in the 1870s and 1880s, however, did the systematic collection of fertility, mortality, and morbidity data become a serious concern of Russian physicians,[219] though it still remained limited largely to the European part of the empire.[220] And only in 1897, did the Russian government conduct the first census of its entire population, providing social hygienists with a baseline for their statistical studies.[221] By that time, however, Florinskii's treatise had long been overshadowed by numerous newer publications and seemed to have been completely forgotten.

From 1865 to 1898, in the course of 33 years since the first publication of *Human Perfection and Degeneration*, Florinskii never returned to the subjects of his treatise — human heredity, variability, and evolution, and various conditions that could lead to either the perfection or the degeneration of the human species — in any of his published works. Only once, in the "preface" to his manuscript *Thoughts and Notes on My Childhood and Education*, considering his own pedigree and life trajectory and noting that he had likely inherited his strong physique and health from his parents, did Florinskii briefly remark:

> It would be interesting to investigate the laws of heredity of human mental and moral characteristics. In this respect, there must exist an anatomical continuity too, but it often gets obscured by the secondary influences of upbringing and environment. Nevertheless, one cannot but admit that the progressive development of civilized nations, estates, and separate families is the result of the gradual perfection of mental organs through exercise and heredity.[222]

But he never pursued this interest. It seems that he simply wished to forget he had ever written his essays on "rational marriage." Indeed, he never included them in any of the lists of publications appended to his official curriculum vitae. Only towards the end of his life, answering requests from the authors of various encyclopedia and biographical lexicons, did he begin to list the 1866 book published by Blagosvetlov (but not the original essays published in *Russian Word*) in the bibliography of his works.[223] Still, when in the mid-1890s he contemplated the possibility of publishing a multi-volume edition of his "Collected Works," Florinskii did not include *Human Perfection and Degeneration* in its projected contents.[224]

With the notable exception of Veniamin Portugalov, it seems that nobody even mentioned in writing *Human Perfection and Degeneration* during its author's lifetime. An illustrative example may be found in the various works by Florinskii's fellow gynecologist Manus Pargamin. In 1891, Pargamin published a "popular-medical essay," *On Degeneration*.[225] Yet even though, like Florinskii, he advocated "mixed" marriages as the major instrument for preventing degeneration, he did not mention his predecessor, basing his essay exclusively on western literature. Five years later, Pargamin produced a voluminous overview, titled *Heredity and Marriage Hygiene*.[226] In the preface, its author claimed that he "had spared neither effort, nor time and had diligently and carefully combed foreign and Russian literature to collect materials for the fullest and all-sided illumination of the subject expressed in its title." Indeed, the volume cited more than 200 English, French, German, and Russian sources. But, though Pargamin addressed many of the issues raised by Florinskii's treatise, he did not mention it at all.

Yet at least once in the course of his post-"Perfection and Degeneration" life Florinskii was forcefully reminded of his treatise. Sometime during the late 1880s, he received a letter from Paris with a request for a copy of his essays from Pavel Iakobii.[227] The correspondent apologized profusely for sending the letter without being personally acquainted with its addressee, but felt that its subject was important enough to grant forgiveness. He introduced himself as a doctor and the author of a book, *Etudes sur la sélection dans les rapports avec l'hérédité chez l'homme*, that had been published in 1881 in Paris. The book had received a special prize from the Madrid Academy of Sciences,[228] Iakobii boasted, as well as favorable reviews in various scientific and medical journals.[229] It had been sold out and he now planned to produce a second, "revised and augmented" edition. It was this project that prompted his request: "I know that you have written an absolutely extraordinary article about the degeneration of the human type. The same theme is examined in my own work and in a published-somewhat-later treatise by [Francis] Galton."[230] "You understand how much I need to thoroughly study your work to reference it in the new edition of my book," he continued, "which is particularly important because you had been the first to articulate this fundamental idea in the literature, and, though both Galton and myself developed the same view independently, not knowing about your work, the *priorité* undoubtedly belongs to you." Alas, he could not find a copy of Florinskii's work anywhere, Iakobii complained, "it has become a bibliographic rarity," and that was why he had decided to trouble the "respected professor" with his request. He concluded:

> I would like very much to produce a *historique* [introduction] to the question of the degeneration of the human type, and I think, not without certain pleasure, that this question — first raised by you, elaborated in detail in my own work, though supplemented by Galton and partially by De Candolle[231] — would remain in science as a result of Russian efforts.

We cannot even guess how Florinskii reacted to Iakobii's letter. Was he flattered? Dismayed? Annoyed? In all likelihood, he did not answer it. Iakobii did publish a second edition of his book. Prefaced by voluble praise from the famous French sociologist, criminologist, and social psychologist Gabriel Tarde, it came out in Paris in 1904, five years after Florinskii's death.[232] But it did not mention even Florinskii's name, to say nothing of his "absolutely extraordinary article." Contrary to what he had promised in his letter to Florinskii, Iakobii did not write a "historical introduction" to his book's subject and did not mention Florinskii's (or Galton's) contributions to its examination. In regard to Florinskii, it seems to be a deliberate omission. Perhaps, Iakobii took offence at Florinskii's refusal to respond to his request for a copy of his work. Iakobii published the new edition of his book almost fifteen years after he had returned to Russia in 1890, and, therefore, he had had ample opportunities to obtain a copy of Florinskii's treatise, if not from its author, then from a public library. After all, his "revised and augmented" volume did include plenty of statistical materials collected by Russian social hygienists over the course of those fifteen years, as well as references to their publications. But not to Florinskii's.

Four years after Florinskii's death, his wife published a large collection of his articles and speeches.[233] The 600-page tome included about forty pieces Florinskii had published during his lifetime in various periodicals on a variety of issues, ranging from women's education to the origins of the word "Siberia." But his treatise on "marriage hygiene" was not mentioned on its pages.

It looked as if with the death of its author, his ideas of human perfection and degeneration died too.

# II. "BOURGEOIS" AND "PROLETARIAN" EUGENICS

> "The principle of the struggle for existence and the survival of the fittest operates not only in the process of plant and animal evolution, but also in ... the book market. Here too, only those books 'survive' — that is gain social recognition — which have proven strong and viable in their own struggle for existence."
>
> Mikhail Volotskoi, 1926

After the death of its publisher Grigorii Blagosvetlov, Vasilii Florinskii's *Human Perfection and Degeneration* completely disappeared from public view. Florinskii outlived his publisher by almost twenty years, but did not do anything to bring his treatise and its major ideas to the attention of his compatriots. Yet, perhaps contrary to its author's wishes and expectations, the book was not irrevocably lost to the merciless currents of time. Less than a quarter of a century after Florinskii's death, his treatise was discovered, actively advertised, and in 1926 reissued in a new, slightly abridged edition. This time its publication did not go unnoticed. In fact, the book became subject of heated public debates over the possibility and necessity of creating a "proletarian" eugenics, radically different from the "bourgeois" eugenics developed by Francis Galton and his numerous followers. But, just a few years later, references to Florinskii's treatise vanished. It seemed that the book and its author would slip into oblivion a second time. Some seventy years later, however, in 1995, the publishing house of the very university he had built in Tomsk more than a century earlier issued a new edition of Florinskii's treatise. Finally, in 2012, yet another edition of the book came out.

In the next four chapters I examine the reasons for, contexts of, and reactions to the repeated appearance and disappearance of Florinskii's treatise, documenting a close intertwining of the fates of Galton's and Florinskii's concepts in Bolshevik, Soviet, and post-Soviet Russia, and considering the implications of the particular historical trajectory of eugenics in Russia for our understanding of the local and global histories of the amalgam of ideas, values, concerns, and actions fused into Galton's eugenics and Florinskii's eugamics.

# 5. Rebirth: Eugenics and Marxism

> "[It] cannot be uniform: every [social] class must create its own eugenics."
>
> Alexander Serebrovskii, 12 January 1926

For sixty years since its first publication, Vasilii Florinskii's *Human Perfection and Degeneration* remained apparently unread and its major ideas dormant. In 1926, however, the book was republished. The main reason for its "resurrection" was the rapid growth during the early twentieth century of a transnational eugenics movement, initiated by, among many others, Francis Galton and advanced by his numerous followers in Britain and elsewhere. The infiltration of eugenics into Russian professional and popular discourse on human variability, heredity, development, and evolution began shortly after Florinskii's death and culminated in the establishment of a Russian Eugenics Society (RES) in 1920. And it was one of the society's founding members and its "scientific secretary" Mikhail Volotskoi who discovered, actively popularized, and eventually reissued Florinskii's book.

Volotskoi portrayed Florinskii as a "precursor" to Galton and his "marriage hygiene" as a "predecessor" of Galtonian eugenics. But for its new publisher and editor, Florinskii's treatise was more than a historical curiosity, and its republication was more than an attempt to demonstrate the long "native roots" of eugenics in Russia. Rather, Volotskoi came to see Florinskii's *Human Perfection and Degeneration* as a model and justification for what, in contrast to Galton's "bourgeois" eugenics, he named a "proletarian," "socialist," "bio-social" eugenics. Volotskoi went beyond a theoretical analysis of what "bio-social" eugenics should be and do. He launched several research projects that would advance its actual development. Indeed, he made explicit the research programme embedded, though not clearly articulated, in Florinskii's notions of "conditions conducive" to human degeneration and perfection. He started detailed investigations on the "eugenic consequences" of various factors, ranging from occupational hazards to women's fashions to such "racial poisons" as syphilis, TB, and narcotics.

## Galton, Eugenics, and Imperial Russia

Unlike Florinskii, who after the 1865 publication of his essays never returned to the issues of human reproduction, variability, heredity, and evolution, Galton made the study of these questions his life's mission and did everything he could to ensure that it would be continued thereafter. For nearly forty years since his first 1865 "eugenic" article, Galton almost single-handedly carried out an extensive programme of quantitative investigations on human variability, heredity, and evolution, ranging from studies of pedigrees, twins, photographic composites, and fingerprints to bio-, psycho-, and anthropo-metry.[1] He reported on his results to various scientific societies and institutions and produced numerous publications, from short letters to newspapers pieces, popular magazine essays, extended articles in professional journals, and voluminous monographs.

Yet unlike the results of his concrete investigations (whether into the principles of biometry, for which the Royal Society awarded him its Darwin Medal, or the classification of fingerprints, which lay a foundation for the personal identification system eventually adopted worldwide), Galton's notion of eugenics and his theorizing on the "possible improvement of the human breed" by means of selective breeding of "men and women of rare and similar talent" had found little support. As he bitterly remarked in his Huxley Lecture delivered to the Anthropological Institute of Great Britain and Ireland on 29 October 1901, eugenics "is smiled at as most desirable in itself and possibly worthy of academic discussion, but absolutely out of the question as a practical problem."[2]

At the time of this remark, Galton was nearing his eightieth birthday and beginning to feel his age. His wife had died a few years earlier and, unlike his cousin Charles Darwin's, his own 43-year-long marriage had borne no children. The approaching milestone seemed to have prodded Galton to contemplate his legacy. He apparently decided to devote his sunset years to putting his "brain-child" on firmer ground. Perhaps the biggest scientific excitement of the previous year — the rediscovery of Gregor Mendel's "laws of hybridization," which led to rapid growth of the new science of heredity, soon to be named "genetics" by his compatriot and critic William Bateson — played a role in Galton's decision. In the opinion of his close collaborator and first biographer Karl Pearson, "Galton had set before himself in the last years of his life a definite plan of eugenics propagandism."[3]

The plan was indeed far-reaching. And it was not limited merely to propaganda. In fact, Galton set out to make his "brain-child" into a proper *scientific discipline* by defining its subjects and goals, outlining its particular methodology, training its practitioners, and establishing its research facilities, teaching programmes, regular forums, and publishing outlets. In 1901 Galton supported the establishment of *Biometrika*, the first journal devoted specifically to the quantitative studies of variability, heredity, and evolution, by both lending his name to the editorial board and underwriting the journal's financial security with his personal funds.[4] Run by his younger disciples and admirers — London mathematician Pearson, Oxford zoologist Walter Frank Raphael Weldon, and Harvard biologist Charles Davenport — *Biometrika* became, for Galton, a convenient vehicle for the dissemination of both his statistical methods and his eugenic ideas.[5] His Huxley Lecture, delivered just a few days after the first issue of *Biometrika* came out and titled "The Possible Improvement of the Human Breed under the Existing Conditions of Law and Sentiment," was another step in his propaganda campaign. But, although its text appeared almost immediately on both sides of the Atlantic in four different periodicals, including *Nature*,[6] the most widely circulated scientific journal of the time, according to Pearson, its main audience — anthropologists — remained deaf to Galton's ideas. This probably induced Galton to intensify and expand his efforts.[7]

In the spring of 1904, he found a new venue for his proselytizing — the London Sociological Society. Established at the beginning of that year by a diverse group of scientists, economists, social reformers, and writers, the newborn society proved receptive to eugenics.[8] At the society's widely publicized and carefully stage-managed meeting on 16

May 1904, Galton delivered a long lecture on "Eugenics, its Definition, Scope and Aims" to an audience overflowing a large hall of the London School of Economics. It was the first time that Galton put the word "eugenics" in the title of a presentation and attempted to define the meanings, goals, and methods of "the science which deals with all influences that improve the inborn qualities of a race; [and] also with those that develop them to the utmost advantage."[9] With Pearson in the chair, the lecture was followed by a lively discussion that featured contributions by the leading intellectuals of the day (including political theorist Leonard T. Hobhouse, sociologist Benjamin Kidd, psychiatrist Henry Maudsley, playwright George Bernard Shaw, philosopher Lady Victoria Welby, and writer H. G. Wells). Galton's lecture was widely covered in the press and published in the society's journal.[10] In Pearson's words, it "got an excellent advertisement for Eugenics,"[11] which Galton quickly proceeded to build upon.

In October, obviously inspired by the reception of his lecture at the Sociological Society, Galton approached "the authorities of the University of London" with an offer to fund "a small establishment for the furtherance of Eugenics."[12] The university's officials readily agreed to lend its name and space to a "Eugenics Record Office," while Galton underwrote its operational costs for the next three years, paying the salaries of a "Research Fellow" and a "Research Scholar in National Eugenics."[13] It was in the course of his negotiations with university officials that Galton redefined eugenics as "the study of agencies under social control that may improve or impair the racial qualities of future generations, either physically or mentally,"[14] a definition that from this time on replaced his earlier, rather vague pronouncements on eugenics' scope and aims.[15]

A few months later, on 14 February 1905, Galton delivered a second lecture to the Sociological Society on what he apparently considered to be the most important among the "agencies under social control": marriage. The lecture again sparked a heated discussion on the floor as well as in the press, which prompted Galton to supplement the publication of its text with several follow-up statements that clarified his views on "Eugenics as a Factor in Religion" and on "Studies in National Eugenics" to be carried out under his direction at the just instituted Eugenics Record Office.[16] It was in this lecture and its supplements that

Galton delineated the two complementary sides of eugenics: the science and the creed of human hereditary improvement, with the former serving as the foundation for the latter and the latter providing the basis for concrete actions aimed at increasing "good" heredity ("positive eugenics") and decreasing "bad" heredity ("negative eugenics") in the British nation.

In 1906, Galton's failing health prompted him to take further steps to assure the future of his "brain-child." He asked Pearson to take over the directorship of the Eugenics Record Office, and amended his will to make provisions for its continuation. He also endowed "a professorship in eugenics" at the University of London and initiated the formation of a Eugenics Society. All of his efforts brought plentiful fruits. Over the course of the next year, the Eugenics Record Office was reconstituted as the Francis Galton Eugenics Laboratory under Pearson's directorship and a "Eugenics Education Society" (EES) established with Galton as its "honorary president."[17] In 1909, the EES issued Galton's *Essays on Eugenics*, a collection of his key reports and publications on the subject over the prior few years.[18] The society also began to publish the *Eugenics Review*, a widely circulated journal that became an oracle of the new creed. To the end of his days on 17 January 1911, Galton continued his proselytizing campaign, delivering lectures, publishing articles, and even writing a "eugenic" novel, titled *Kantsaywhere*.[19] Upon Galton's death, Pearson became the first Galton Professor of Eugenics, while Charles Darwin's son Leonard took over the office of EES president, thus ensuring further development of both the science and the creed of eugenics.

A year later, in July 1912, furthering its major goal of making "more widely known to the public the aims of Eugenists,"[20] the EES hosted in London the First International Eugenics Congress presided over by Leonard Darwin. Although representatives of more than twenty countries took part in the congress's proceedings, its primary target was the *British* public. Nearly ninety per cent of its 300-plus attendees were high-profile British politicians, scientists, writers, social activists and reformers, educators, physicians, socialites, and civil servants. In advance of the congress, the EES published in English all the reports by the participants, providing parallel texts of the originals submitted in the French, Italian, and German languages. The British press extensively covered the congress's week-long proceedings, with *The Times* carrying daily reports on its sessions.

The congress vividly demonstrated that while Galton had been waging his proselytizing campaign in Britain, the ideas of the "physical and mental improvement" of humankind had been developed by numerous adherents in many other countries. Indeed, by the time the congress met in London, a variety of research establishments, periodicals, societies, and legislative initiatives — all aimed at the "improvement of humankind" — had emerged in Belgium, Denmark, France, Germany, Italy, Norway, Switzerland, and the United States.[21] Furthermore, several American states had already passed special laws that targeted the procreation of convicted criminals, mentally-ill, and other "undesirables." As reports delivered at the congress clearly showed, various "national" versions of eugenics had developed in different institutional, intellectual, and social contexts and under different names, such as, for instance, *Rassenhygiene* and *Fortpflanzungshygiene* in Germany, humaniculture, euthenics, and stirpiculture in the United States, and *pédotechnie* and *puériculture* in Belgium and France. The congress thus became a key instrument of, and a pivotal point in, not only disseminating Galton's "eugenic gospel" to various countries, but also congealing the variety of "national" approaches to the issues of human betterment into a transnational eugenics movement that in the next few years rapidly spread around the globe.[22]

Yet the eugenics gospel was not greeted with the same enthusiasm everywhere. One notable exception was Russia. An announcement that the "First international congress on eugenics (racial hygiene)" will be held in London appeared well in advance in *The Physician's Gazette*, the country's most widely circulated medical periodical.[23] Symptomatically, it used Galton's term "eugenics" as a synonym of Ploetz's "racial hygiene." Nevertheless, the announcement failed to induce the formation of a national "consultative committee" similar to those set up in Belgium, France, Germany, Italy, and the United States in order "to nominate a strictly limited number of readers of papers for each Country, representing the country at the congress."[24] No one officially represented Russia at the congress. Although, just a few months before the congress's opening, two Russian physicians did attend a conference on "genealogy, heredity, and racial hygiene," organized in April 1912 by psychiatrist Robert Sommer in Giessen,[25] it seemed that the Russian scholarly community was not interested in taking an active part in the advancement of eugenics.[26]

The silence that enveloped Florinskii's *Human Perfection and Degeneration* during his lifetime suggests that in the second half of the nineteenth century the Russian Empire lacked the socio-economic conditions — from industrialization and urbanization to declining fertility, from a developed civil society to an influential hereditary aristocracy, and from immigration to overpopulation — that fueled interest in eugenic ideas elsewhere. The huge, sparsely populated, predominately agrarian, autocratic, poly-confessional, and multiethnic — on the level of both the population and the ruling elites — empire provided neither sufficient data, nor receptive audiences for "eugenic" concerns about racial degeneration and intermixing, falling birth rates, social degradation, or immigration, the very subjects that would constitute the major themes of the London congress.

Just a few years after Florinskii's death, however, the advent of industrialization, along with the rapid growth of medical, scientific, pedagogical, and legal professions in Russia, stimulated some interest in ideas of "human betterment." During the first two decades of the twentieth century, such ideas started to filter into professional and public discourse. Various publishers began to issue translations of works by known British, Dutch, French, German, Swiss, and US proponents of these ideas, including Georg Buschan, Charles Davenport, Emile Duclaux, Alfons Fischer, August Forel, Kurt Goldstein, Max von Gruber, Elie Perrier, Théodule Ribot, Charles Richet, Johannes Rutgers, and Pearson — but, surprisingly, not Galton.[27]

Although after the 1874 translation of *Hereditary Genius*, no other works of Galton's were published in the country, Russian commentators certainly knew of his role as a "founder" of eugenics. The very word "eugenics" (*evgenika*) and a brief exposition of Galton's views on its meanings appeared in Russian for the first time in a 1902 anthropology textbook, written by Ludwik Krzywicki.[28] In 1911, Russia's own "Darwin's bulldog" Kliment Timiriazev published in a popular encyclopedia a lengthy sketch of Darwin's cousin's biography (along with his portrait), emphasizing Galton's role in developing and promoting eugenics.[29] He also wrote for the same encyclopedia an extensive entry on "Eugenics" that praised Galton as the founder of the "new field of knowledge" and outlined its latest developments in Britain.[30]

But Russian observers were also well informed of varied approaches to the issues of "human betterment" developed in other countries. They picked selectively from the pool of available ideas, liberally mixing Anglo-American eugenics with French *anthropologie sociale*, German *Sozialpathologie* with French *puériculture*, and German *Rassenhygiene* with French *eugénnetique*. They coined a special term, *antropotekhnika* (anthropotechnique), modeled on the Russian word for animal breeding *zootekhnika* (zootechnique). The new word served as a synonym for Russian translations/transliterations of such corresponding English, German, and French terms as eugenics (*evgenika*), Rassenhygiene (*rassovaia gigiena*), Fortpflanzungshygiene (*generativnaia gigiena* or *gigiena razmnozheniia*) and *eugénnetique* (*evgenetika*).[31]

After 1900, Russia's budding professional and disciplinary communities of psychiatrists, jurists, pedagogues, anthropologists, social hygienists, and biologists began to take up the ideas and agendas of their western colleagues under consideration, addressing various facets of eugenic research, creed, and policies in professional and popular periodicals. Although Russian observers were quite eclectic in borrowing from their colleagues elsewhere, their approach to eugenics was decidedly critical. Most commentators criticized the "race" and "class" underpinnings of eugenic ideas and policies espoused by German and British eugenicists. Many placed strong emphasis on environment/education/nurture, as did their French colleagues. They largely rejected "negative measures" (be it sterilization or segregation) promoted by US, German, and Scandinavian eugenicists as a means of remedying such "social ills" as alcoholism, venereal diseases, TB, prostitution, and crime. Instead, they advocated for the improvement of social conditions, re-education, and prophylactic medicine.

These critical attitudes were displayed prominently in Russian responses to the 1912 London congress. Even though the Russian Empire had no official representatives at the congress, at least two imperial subjects did attend its sessions. The eminent naturalist, geographer, and theoretician of anarchism Prince Petr Kropotkin, who at the time was living in exile in Britain, took part in the discussions, while Isaak

Shklovskii, a popular journalist (who wrote under the penname Dioneo), covered the congress for Russian "thick" journals.

Kropotkin delivered a passionate diatribe against the congress's class bias.[32] "Who were unfit?" he exclaimed rhetorically, "the workers or the idlers? The women of the people, who suckled their children themselves, or the ladies who were unfit for maternity because they could not perform all the duties of a mother? Those who produced degenerates in slums, or those who produced degenerates in palaces?" He vehemently opposed proposals repeatedly voiced at the congress to sterilize the "unfit": "Before recommending the sterilization of the feeble-minded, the unsuccessful, the epileptic (Dostoevsky was an epileptic), was it not their [eugenicists'] duty to study the social roots and causes of these diseases?" Kropotkin insisted that such social measures as the creation of healthy housing and the abolition of slums "would improve the germplasm of the next generation more than any amount of sterilization."

Shklovskii echoed Kropotkin's criticism. The subtitle of his correspondence from the congress — "Beastly Philosophy" — speaks for itself. If Kropotkin attacked the "class" underpinnings of eugenic ideas, Shklovskii focused his critique on the "racial" ones: "All those, purportedly scientific, data, upon which the doctrines of higher and lower races are based," he declared, "cannot withstand criticism, for the very simple reason that anthropology knows of no pure races."[33] Indeed, although some Russian anthropologists, particularly among proponents of "criminal anthropology" à-la Cesare Lombroso, did engage in the propaganda of the superiority of the "Great-Russian race,"[34] most of their colleagues rejected the "racialization" of their subjects.[35]

The London congress certainly helped intensify interest in eugenics the world over, including Russia. In its aftermath, a number of Russian physicians, biologists, social hygienists, anthropologists, jurists, and educators got involved in discussions about eugenics (in its various "national" versions). Some Russian anthropologists enthusiastically embraced the eugenic vision of "bettering humankind." Eugenics offered them an opportunity to become not simply the "observateurs de l'homme," but also to play a prominent social role as experts on human diversity and evolution.[36] It was the anthropologist Krzywicki who coined the Russian term *antropotekhnika* and wrote entries on eugenics for various Russian encyclopaedias.³⁷ Yet, as did other Russian commentators, in the aftermath of the London congress, Krzywicki cautioned against too hasty applications of "negative" eugenic measures, which, in his opinion, at the present time "turn into the instrument of narrow class interests."³⁸

Many Russian jurists viewed unfavorably both the ideas of "inborn criminality" and proposals to sterilize prisoners, which were quite popular among many western proponents of eugenics. Indeed, jurists were one of the first professional groups to critically address "eugenic" issues in Russia. Several eminent legal scholars mounted a pointed critique of US sterilization laws, particularly the provisions for coerced sterilization of prisoners.³⁹ This critique was part of a wider debate on justice in the relationship between crime and punishment and, more generally, between the law and the rights of the individual.⁴⁰ That debate flared up in the aftermath of the first Russian Revolution of 1905-1906, which for the first time in the country's history granted (at least on paper) its citizens the rule of law and basic civil liberties, is telling. In the atmosphere of political reaction that followed the revolution's defeat, sterilization laws became the jurists' favorite practice target in their veiled attacks on the tsarist regime's continuing infringements on individual rights and freedoms. In 1912, Pavel Liublinskii, a well-known St. Petersburg jurist, published a detailed and highly critical assessment of "eugenic laws" recently enacted in the United States in *Russian Thought*.⁴¹ The London congress reinforced this trend.⁴²

Similarly, many Russian pedagogues and psychologists critically evaluated the ideas of "hereditary feeble-mindedness," arguing that so-called "defective" children could be brought up to be normal members of the society.⁴³ At the First Russian Congress on Public Education in January 1914, for instance, Kharkov University's professor Isaak Orshanskii delivered a lengthy report on "Heredity and Degeneration." Orshanskii's speech prompted the congress to issue a special "resolution on the struggle against criminality, suicide, defectiveness, and degeneration among children," calling for the foundation of specialized schools for the re-education of "defective children."⁴⁴

As were their colleagues elsewhere, many Russian physicians were sympathetic to eugenics. For many doctors dealing with chronic diseases, psychiatrists and neurologists in particular, eugenics offered new research methodologies (medical family histories, twins studies, and statistical analysis) and a new interpretive framework, replacing older, vague ideas of "inborn constitution" with the newly introduced concepts of heredity (be they Galtonian, Weismannian, Mendelian, or Lamarckian).[45] Some psychiatrists, including Vladimir Bekhterev in St. Petersburg and Tikhon Iudin in Moscow, investigated the ideas of "hereditary degeneration" in their studies of the mentally ill.[46] Others, for example, Orshanskii, focused on "hereditary talents," continuing Galton's research programme.[47] During this period, several doctoral dissertations on "heredity and disease" were defended in Russia.[48] In 1913 Kazan University's psychiatrist Alexander Sholomovich produced a 300-page clinical study on "Heredity and Physical Signs of Degeneration in Mentally Ill and Healthy Patients." Sholomovich discussed his findings in light of various theories of heredity, including those of Jean-Baptiste Lamarck, Galton, August Weismann, and Mendel.[49] That same year, Bekhterev invited Iurii Filipchenko (1882-1930), a founder of Russian genetics, to teach Russia's first course on the subject at the Psycho-Neurological Institute Bekhterev established in St. Petersburg a few years earlier. For many physicians, the use of eugenic language and methods became a means of making their special fields more "objective" and "scientific." After the London congress their principal periodical *The Physician's Gazette* regularly carried articles, reviews, and comments on the subject.[50]

Eugenics garnered a warm reception among Russian public health doctors and social hygienists. A programmatic statement opening the first issue of *Hygiene and Sanitation* (a new journal founded in 1910 by eminent bacteriologist Nikolai Gamaleia) declared that "generative hygiene (eugenics)" must constitute an integral part of Russian public health agendas.[51] In the same issue, the journal began publication of a series of articles on eugenics and introduced a special bibliographic section "on eugenics." As did their western colleagues, many Russian social hygienists focused particularly on questions of alcoholism and heredity.[52] Public health specialists strove to keep abreast of the newest developments in the studies of heredity. In November 1912, for example, the Russian Society for the Protection of People's Health organized a special session and invited Roman Provokhenskii, a well-known animal breeder, to lecture on "modern views on heredity" and their application to humans.[53] In the academic year of 1913-1914, Iur'ev (formerly Dorpat) University's professor Evgenii A. Shepilevskii included "racial hygiene" into his courses and reported on the subject at the meetings of the local society of physicians.[54]

Eugenics also found a receptive audience in the nascent community of Russian experimental biologists. As did their western counterparts, Russian biologists exploited eugenic rhetoric in order to legitimize a new field that captivated their interests — genetics. A popular-science magazine, *Priroda* (*Nature*), that became this community's oracle after its establishment in 1912, regularly featured articles on both genetics and eugenics.[55] Two founders of Russian genetics, Nikolai Kol'tsov (1872-1940; who in 1914 became co-editor of *Priroda*) in Moscow and Filipchenko in St. Petersburg, were particularly active in this endeavor.[56] In early February 1917, Filipchenko delivered a keynote lecture on eugenics to a large gathering marking the tenth anniversary of Bekhterev's Psycho-Neurological Institute.[57]

Yet, despite this flurry of reports and publications on eugenics (in its Anglo-American, French, and German variants), each of the aforementioned professional/disciplinary communities focused almost exclusively on the topics and subjects that resonated with their own professional interests. Indeed, eugenic publications appearing in different specialized periodicals often read as if they were devoted to completely different subjects, without any attempt to find a common ground.[58] The pronouncements on eugenics by Gamaleia and Iudin, leading spokesmen for public health specialists (social hygienists) and psychiatrists, respectively, provide an illuminating example.

Established by Gamaleia in 1910, *Hygiene and Sanitation* became the first Russian periodical to address *systematically* various eugenic issues and attempt to apply eugenic ideas to the Russian context. The editorial opening the journal's first issue clearly described its purpose: "all of our attention will be focused on those sanitary measures that are important for the *ozdorovlenie* (healthification) of Russia."[59] Among the various subjects that the journal planned to cover to make the country healthy, Gamaleia listed infectious diseases, clean water supply and sewage disposal, housing, school, and occupational hygiene, demography and vital statistics, military and naval hygiene, and "generative hygiene (eugenics)."[60] Following this programmatic statement, the same issue carried the first of a series of detailed articles — written by Gamaleia's junior colleague Kazimir Karrafa-Korbut — surveying the goals, methods, and ideas of British eugenics and German racial hygiene.[61] A special bibliographic section "on eugenics" regularly reviewed and summarized recent western books and articles appearing in such European periodicals as the British *Eugenics Review*, the German *Archiv für Rassen- und Gesellschafts-Biologie* and *Zeitschrift für Sexualwissenschaft und Sexual-Politik*, and the French *La Presse Médicale* and *L'Hygiène Populaire*.[62] Among various items on the subject, *Hygiene and Sanitation* carried an essay on the history of eugenics, a lengthy report on the proceedings of the First International Eugenics Congress, and a survey of the current views on heredity.[63]

Gamaleia personally wrote several editorials discussing various facets of eugenic research and policies and reviewed several eugenic publications.[64] Furthermore, in late November 1912 the editor published a long article "On the Conditions Favorable for the Betterment of Humans' Natural Qualities."[65] The London congress (whose contents he had detailed on the pages of his journal just a few months earlier) appeared to be the major stimulus for Gamaleia to take up his pen. His article presented a concise analysis of the basic eugenic ideas of "racial degeneration" and "regeneration," their scientific underpinnings in Darwin's theory of natural selection and in the concepts of heredity advanced by Galton, Weismann, and Mendel, as well as the proposed eugenic actions (both "negative" and "positive") to counter degeneration and to promote regeneration, all of which had been discussed extensively at the congress.

Gamaleia questioned the validity of the main eugenic postulate of "racial degeneration." He saw eugenics simply as an extension of "social hygiene" to the issues of human reproduction: "generative hygiene." For him, the rise of eugenics in Britain represented the culmination of a long process of social reforms, which had started in the mid-nineteenth century with wide sanitary and public health reforms, moved on to reform the legislation pertaining to the conditions of children's and women's labor in factories, and then proceeded to introduce state-mandated education for all children. Now, according to Gamaleia, eugenicists were advocating for expanding these "hygienic" reforms to the pre-school, pre-birth, and even pre-conception stages in human life through extensive "care of the future mother."

He concluded that "Russia, which is as yet in her first period of these social reforms [i.e. sanitary and public health reforms], is incapable of generating a strong eugenic movement, but she needs, nevertheless, to understand the problems that trouble her cultured neighbors."[66] Through his journal he pursued exactly that goal: to inform and educate Russian social hygienists about eugenic debates and actions undertaken by their western colleagues.[67] At the end of 1913, Gamaleia stopped the publication of *Hygiene and Sanitation*. But despite his short-lived involvement with eugenics, the journal proved highly influential in awakening Russian public health specialists' interest in the subject.[68]

Just as Gamaleia left the field, eugenics found another champion in Russia: Tikhon Iudin (1879-1949), a young but well-respected Moscow psychiatrist. In early 1914, Iudin became a co-editor of *Modern Psychiatry*, the discipline's leading periodical. On the pages of this journal Iudin continued Gamaleia's mission of educating Russian physicians about eugenics. In April 1914, he published his first lengthy article "On Eugenics and the Eugenics Movement."[69] Iudin picked up the overview of eugenics exactly where Gamaleia had left it off, surveying the development of the field after the First International Eugenics Congress. But his take on the subject differed considerably from that of Gamaleia, for his personal research interests centered on the role of heredity in psychiatric disorders.[70] According to Iudin, eugenics was an "applied science" and, as such, it "depends on scientific data gathered by other theoretical disciplines, first of all, by genetics — the science of heredity." In his opinion, "genetics had directed and still directs the course of eugenics; the successes of the eugenics movement in the last years to a considerable degree are explained by and depend on the successes in the study of heredity."

Iudin emphasized that current views ascribed an exclusive role in defining the "quality of progeny" to heredity: environmental influences, from hygiene to education, were capable of only modifying what was already present in the heredity of an individual, which s/he had received from parents. Citing Reginald Punnett, the first professor of genetics at Cambridge University and the editor of the *Journal of Genetics*,[71] and Pearson, he stressed that the proponents of both Mendelian and biometric (Galtonian) schools in the study of heredity supported this view, which thus provided the major scientific foundation for eugenic ideas and practices. Iudin pointed out that "this scientific belief in the negligible influence of environment on heredity" prompted "some eugenicists" to advance certain "questionable ideas," such as the ideas of racial superiority and inferiority, and to advocate for "decisive policies," such as the sterilization, segregation, and even euthanasia of individuals with "inferior" heredity.

Iudin noted a strong negative reaction to such ideas among various observers, including Russians. But, he stressed, it would be wrong to judge the entire eugenics movement by these "fanatical" ideas and policies. He approvingly cited the opinion of William Bateson — a leading British advocate of Mendelism who had coined the term *genetics* — that scientific understanding of heredity was at that time still too rudimentary to be applied to humans. Iudin also referred to the opinion of Carl Correns, a leading German geneticist who had played a prominent part in the rediscovery of Mendel's laws, that the majority of "bad" hereditary traits (such as Iudin's own subject, mental disorders) were recessive and remained "invisible" in the progeny, and thus any selection against these traits promoted by eugenicists "cannot lead to their elimination."[72] Iudin noted that some proponents of eugenics focused on "positive" as opposed to "negative" measures, searching for ways to encourage the propagation of "good" heredity in future generations. Through active propaganda campaigns, he stated, they sought to instill basic eugenic ideas in the population to make these ideas part of "unconscious" social mores, or even "religious dogmas," that would guide individuals' decisions regarding marriage partners or the desirable number of children in their family.

Iudin observed that the "eugenics movement has spread in a great wave through the cultured world." He provided a detailed overview of eugenic institutions, societies, journals, activities, and legislative initiatives in Britain, Denmark, France, Germany, Italy, Norway, Sweden, and the United States, identifying leading figures in the national eugenic organizations of each country. He underlined a great variety of approaches to the ultimate goal of eugenics — the improvement of humankind — advocated by different individuals, as well as national particularities in the justifications for and the attempts to attain this goal.

He noted, for instance, that US eugenicists were much more enthusiastic about "negative" eugenics than their British or French colleagues. He described the continuing efforts to unite national eugenic organizations, which had begun at the London congress and which, he was sure, would further invigorate the eugenics movement at the next international congress, scheduled to meet in New York City in September 1915. He concluded his overview with a cautious endorsement:

> Of course, at the present, the theoretical substantiation of, and investigations in, eugenics are at the very beginning, and the time when, on the basis of existing knowledge, we would have a right to intervene in social life on a large scale is still far in the future. But the efforts to advance the very idea of the necessity of greater care regarding the health of future generations, the education of humankind in the spirit of this idea, its propaganda, the creation of common sentiment conducive to eugenics, [and] active support for scientific research in this direction, perhaps will indeed prove very beneficial for all of humankind. In any case, eugenic ideas deserve serious attention and study.

Undoubtedly, Iudin himself was planning to pay "serious attention" to such study in his own field, psychiatry. In June 1914, he published in *Modern Psychiatry* a detailed review of the recently (in April) enacted British Mental Deficiency Act, which British eugenicists hailed as a victory for their campaign to educate the public regarding the dangers of "feeble-mindedness" to the nation's health. The Great War that erupted in August put a stop to Iudin's plans: he was drafted to the army and went to the front.

As the materials presented above demonstrate, to a considerable degree Gamaleia's and Iudin's overviews of eugenics read as if they were written about two different movements. Indeed, each of them even preferred to use a different name in his descriptions. Gamaleia used "eugenics" interchangeably with "racial hygiene" and "generative hygiene." Iudin, however, insisted that not "eugenics," but "eugenetics" (*evgenetika*) was the most appropriate name for the movement, for it emphasized the close connection between eugenics and genetics. Each of these observers focused largely on those components/elements/tenets of eugenics that resonated most with his personal/professional interests. Each of them sought to demonstrate the relevance of certain components of eugenics to his medical specialty, disciplinary agendas, and scientific interests. And the very possibility of using eugenics for these purposes certainly played a role in drawing other Russian observers' attention to the early eugenics movement.

Although they published extensively on eugenics, Russian commentators staunchly maintained their status as observers and critics, not propagandists. None of them called on his/her colleagues either to unify their efforts or to join the fledgling eugenics movement: to take part in an international eugenics conference, to organize a eugenics institution (a society, a journal, or a laboratory), and to lobby for the adoption of eugenic laws and regulations. And none of them ever mentioned Florinskii's *Human Perfection and Degeneration*.

All of this changed after the October 1917 Bolshevik Revolution. In the course of just a few years, despite a bloody civil war, famine, epidemics, and economic deprivation engulfing the entire country, eugenics would boast a nationwide society, several research institutions, and specialized periodicals. It would build close links with the transnational eugenics movement, enter teaching curricula in schools and universities, and find a grassroots following in the new, Soviet Russia. Of all the disciplinary and professional groups concerned with eugenics before the revolution, it was Russian biologists, especially Filipchenko in Petrograd (as St. Petersburg was renamed after the outbreak of World War I) and Kol'tsov in Moscow, who spearheaded the closely intertwined processes of the institutionalization of eugenics and of its "stepsister" genetics. But it was social hygiene that provided eugenics with its first institutional home. And it was within this torrent of discipline-building and propaganda activities that Florinskii's book was "resurrected" by Mikhail Volotskoi, a founding member of the Russian Eugenics Society.

## Mikhail Volotskoi

Volotskoi was born on 13 April 1893, in the beautiful ancient town of Rostov, situated some 200 kilometers northeast of Moscow, to an old, but impoverished noble family: his father made a living as a school teacher.[73] After graduating from high school, in the fall of 1913, Volotskoi enrolled in Moscow University's School of Natural Sciences (see fig. 5-1). He chose to study at the Department of Geography and Anthropology, founded and run by the country's leading expert in these fields Dmitrii Anuchin (1843-1923).[74] Volotskoi graduated from university in May 1918 and his mentor Anuchin slated the talented student for "preparation to a professorial position." It seemed that Volotskoi's career path to becoming an anthropology professor at his alma mater was set. Keeping to this path, however, proved impossible.

Fig. 5-1. Mikhail Volotskoi at the time of his graduation from high school, c.1912. Photographer unknown. Courtesy of TsGAM.

Much like Florinskii, Volotskoi had begun his studies in one epoch and finished them in an entirely different one. His student years coincided with the most dramatic and traumatic period in the life of his homeland, encompassing World War I, the fall of the Romanov dynasty, the Bolshevik Revolution, and the Russian Civil War. The beginning of the Great War in August 1914 did not interrupt Volotskoi's education: as a university student he was exempt from the draft and, unlike some of his classmates, he did not enlist as a volunteer. He spent the war years deeply immersed in his studies. He joined the Society of Enthusiasts for the Natural Sciences, Anthropology, and Ethnography, most likely on the recommendation of Anuchin, who had served as its president since 1890. In the summer of 1916, with the society's funding in hand, he embarked on his first field expedition to collect anthropological, ethnographical, and archeological materials in the Tula region, some 250 kilometers south of Moscow.

Perhaps, as did many university students, in February 1917 Volotskoi celebrated the abdication of Nicholas II and the fall of the Romanov dynasty. Prompted by the military failures and economic hardships of the country's engagement in World War I, the abdication led to the formation of a liberal Provisional Government that was to assure Russia's transition to democratic rule by convening a Constituent Assembly, elected by the representatives of all social estates to work out the country's new constitution. The heated debates on — and active propaganda for — various visions of the country's future put forward by numerous political parties became a major preoccupation of the Russian intelligentsia, particularly students of various institutions of higher learning.[75] But it seems that Volotskoi had little interest in politics and stayed focused on his research: in the summer the budding anthropologist again went on a field expedition to the Tula region.

Just as Volotskoi returned from the field, however, in late October the Bolsheviks — a radical faction of the Russian Social-Democratic Labor Party led by Vladimir Lenin — effected a coup d'état in Petrograd and declared the establishment of a socialist republic. Over the next few months, a new system of government administration — the Soviets of Worker, Peasant, and Soldier Deputies — began to take control over the country. On 3 March 1918, the Bolshevik government concluded a separate peace treaty with Germany, ending Russia's participation in

World War I. But within a few weeks, the country erupted in a bloody civil war. The former empire disintegrated into a patch-work of semi-autonomous regions, with the Bolsheviks holding central industrial provinces, and their various opponents (the "Whites"), aided by British, French, and American troops, encircling the Bolshevik strongholds. In mid-March, threatened by the "White" forces advancing on Petrograd, the new government moved to Moscow, which became the capital of the Russian Soviet Federated Socialist Republic (RSFSR) formally established by the All-Russia Congress of the Soviets of Worker, Peasant, and Soldier Deputies held in July. The Bolsheviks moved quickly to dismantle the old "capitalist" system and to build a new "socialist" one in all facets of life, including the alphabet, calendar, economy, government apparatus, laws, arts, and sciences. Of course, their overriding priority was the construction of a new, "Red" army to fight off their numerous opponents.

The new rulers were set on replacing the old "bourgeois" system with their own "proletarian" one. But they lacked qualified "proletarian" personnel in all the vital spheres of life, from industry and education to medical services and government administration. Thus, although they treated educated professionals as part of the bourgeoisie to be harassed and eventually liquidated, the Bolsheviks had to compromise and convince "bourgeois" specialists to join in the Great Experiment of building socialism in Russia, while, at the same time, putting considerable efforts into the preparation of a new, "proletarian" intelligentsia.[76]

For their part, most educated professionals met the Bolshevik Revolution with distrust and open hostility, for they considered the Bolsheviks to be usurpers of the country's nascent representative government — a long-cherished dream of the Russian intelligentsia. They fled *en masse* from the regions controlled by the Bolsheviks and many of them left the country altogether. Yet, for those who stayed, their very existence as professionals depended first and foremost on a functional state and functioning economy. Not surprisingly, many Russian agronomists, architects, artists, engineers, jurists, military officers, physicians, scientists, teachers, and so on did join the Bolsheviks in rebuilding their homeland. Furthermore, for some of them, the destruction of the old imperial system presented not only a threat, but also an opportunity. Probably no other group among the

Russian intelligentsia seized this opportunity with greater success than scientists.[77] The rapid institutionalization of eugenics during the very first years of the Bolshevik regime exemplifies the mutually beneficent, symbiotic relations between Russian scientists and the new rulers. Volotskoi's involvement in this process illustrates exactly how these relations developed.

At first, Volotskoi's life and work seemed remarkably untouched by the dramatic events unfolding in the aftermath of the Bolshevik Revolution. After receiving his university diploma in May 1918, the young anthropologist went on yet another research expedition to the Tula region (see fig. 5-2). But in the fall, the growing civil war led to

Fig. 5-2. Mikhail Volotskoi at the time of his graduation from Moscow University, c.1918. Photographer unknown. Courtesy of TsGAM.

the rapid deterioration of living conditions in Moscow — already severely compromised by the prior four years of all-out war effort. Faced with military opposition, economic chaos, and political turmoil, the Bolsheviks adopted a policy of "War Communism" based on the abolition of private property, banks, the market, and money, the total nationalization of industry, the forced requisition of agricultural produce from the peasantry, and strict administrative control over the distribution of food and goods.[78]

Although Anuchin had selected Volotskoi for "preparation to a professorial position," after the Bolsheviks had seized power it became an empty title. The university's administration fought a losing battle for preserving its autonomy from the People's Commissariat of Enlightenment (Narkompros), the successor to the Imperial Ministry of People's Enlightenment and the Bolsheviks' top agency in charge of education, science, and culture.[79] Anuchin successfully exploited the situation to accomplish his long-relished plan of establishing an independent anthropology department at Moscow University. But he was unable to secure any additional resources and thus to support Volotskoi's position at the new department. Like many of his contemporaries, the young anthropologist had to find a job, a post, a trade, an affiliation, anything that would give him access to the rations distributed by the country's new rulers, which, though barely sufficient to sustain one's life, were the only readily available source of food, goods, and fuel for urban population. Luckily, he was able to secure a teaching position at a secondary school, which enabled him both to avoid being drafted into the Red Army and to survive the first incredibly harsh winter of Bolshevik rule in Moscow.[80]

In the spring of 1919, Volotskoi obtained an additional source of support. Apparently on the recommendation of his mentor, together with Anuchin's right-hand man at the anthropology department Viktor Bunak,[81] Volotskoi became a member of a "scientific-consultative group on the biological question" set up at the State Museum of Social Hygiene (see fig. 5-3).[82] The museum had been established a few months earlier (in January 1919) by the People's Commissariat of Health Protection (Narkomzdrav), the new government's top agency in charge of medical services and public health activities.[83] Narkomzdrav immediately began to build a new state system of *zdravookhranenie* (health protection) that integrated into a unified whole everything related to health: medical services and institutions, health research and development, epidemiological surveillance and prophylactic measures, manufacturing and distribution of vaccines, pharmaceuticals, and medical equipment, specialized education and training, sanitary infrastructure and propaganda. The agency's head Nikolai Semashko, a Bolshevik physician, was an active proponent of social hygiene. Indeed, with its focus on the role of social factors in health and disease and its prioritizing of prophylactic over curative approaches to disease, social hygiene became the foundational doctrine of the entire system of health protection created by the Bolsheviks. Furthermore, its proponents defined social hygiene as "a science of the future, which studies and shapes the facts that promote the biological well-being of humanity," and saw eugenics as "the ultimate goal of all sanitary-medical activities."[84]

Fig. 5-3. The opening of an exhibition at the State Museum of Social Hygiene, 11 July 1919. In the center at the top of the staircase, three men in dark suits: (left to right) Commissar Nikolai Semashko, Alfred Mol'kov, the museum's director, and Deputy-Commissar Zinovii Solov'ev. Photographer unknown. Courtesy of the Museum of the History of Medicine at the Sechenov First Moscow Medical University.

The establishment of the State Museum of Social Hygiene was but the first step in the field's institutionalization in Soviet Russia: four years later, it would be transformed into the State Institute of Social Hygiene.[85] The museum's "scientific-consultative group on the biological question" included specialists who were to advise Narkomzdrav officials — first of all, the commissar himself — on various issues encompassing "general biology, physiology, anthropology, and racial hygiene."[86] A leading member of this group was Kol'tsov, the country's foremost expert in experimental biology.[87] Volotskoi's participation in the "scientific-consultative group" and his acquaintance with the older scientist proved fateful.

A scion of a large family of Moscow merchants, Kol'tsov was undoubtedly one of the most entrepreneurial scientists of his generation. In addition to doing first-rate research (for which the Imperial Academy of Sciences elected him a corresponding member), he organized new laboratories, journals, and teaching courses, recruited and trained numerous students, and built extensive networks of domestic and international contacts. In late 1916, after nearly a decade of continuous efforts, he finally secured funds for the establishment of an Institute of Experimental Biology (IEB) in Moscow (see fig. 5-4).[88]

Fig. 5-4. Nikolai Kol'tsov (seated in the center) with his students, c.1913. Standing on the far left is Alexander Serebrovskii, sitting on the far left is Mikhail Zavadovskii. Photographer unknown. Courtesy of ARAN.

Within just one year, however, the Bolsheviks expropriated the private endowments that had supported the institute, and Kol'tsov had to work hard to find and court patrons in the new state agencies, such as the People's Commissariat of Agriculture (Narkomzem), Narkompros, and Narkomzdrav. His participation in the "scientific-consultative group" at the State Museum of Social Hygiene was just one among various activities he undertook during the civil war years to ensure the survival and eventual prosperity of a large group of his students and the institutionalization of various sub-fields of experimental biology, from biochemistry, endocrinology, and cytology to biophysics, zoopsychology, and genetics.[89]

Kol'tsov's association with Narkomzdrav proved particularly fruitful. Although at that point his institute had almost nothing to contribute to medicine or public health,[90] in January 1920, the agency took the IEB under its wing. Kol'tsov's institute became part of a recently founded State Institute of People's Health Protection, an ever growing complex of research facilities that was to provide scientific grounding for the Bolshevik system of health protection.[91] Apparently in an attempt to justify the IEB's inclusion in Narkomzdrav's research empire, Kol'tsov created within his institute a "eugenics department" that very summer. At the time, the department existed only on paper — in various reports Kol'tsov presented to his patron. In fact, it had neither personnel, nor a research programme. But its founder certainly had some ideas on how to make the "virtual" department real.[92]

In early October 1920, at a meeting of the group on "the biological question," Kol'tsov aired the idea of creating a eugenics society. The idea found immediate support among the group's members: psychiatrist Iudin, who, as we saw, had studied the heredity of mental illness and had been keenly interested in eugenics long before the Bolshevik Revolution; Al'fred Mol'kov and Aleksei Sysin, Semashko's lieutenants in building Soviet social hygiene; and the anthropologists Bunak and Volotskoi.[93] In a few days, the group met again to discuss a charter for a "Russian Scientific Eugenic Society" drafted by Kol'tsov. Initially, Kol'tsov hoped to organize two separate societies: one professional (modeled after such existing scholarly associations as the Russian Physiological Society), open to specialists and devoted to research; another popular (modeled after the British Eugenic Education Society), open to anyone interested in eugenics and devoted to propaganda and education. The contingencies of the time forced the group to establish only one — the Russian Eugenics Society (RES).

On 19 November 1920, the RES held its inaugural meeting. Thirty participants approved the society's charter and elected its executive council: Kol'tsov became its president, Iudin and Bunak council members, and Volotskoi its "secretary." As had happened to Florinskii when he rejoined the St. Petersburg Society of Russian Physicians after his European tour, the younger man was entrusted with the organizational and technical support of RES operations — arranging its meetings, conducting its correspondence, filing reports to its state patrons, and keeping the minutes of its proceedings.[94] Furthermore, obviously impressed with Volotskoi's abilities, Kol'tsov invited the young anthropologist to join the IEB eugenics department as a researcher.[95] Kol'tsov's invitation afforded Volotskoi his first paid academic position and, from the late fall of 1920, eugenics became his full-time occupation. He embraced it with all the zeal of a new convert.

Apparently, Volotskoi's good command of the English language — still a rarity among Russian scientists at the time — played not the least role in Kol'tsov's invitation. The IEB director immediately utilized Volotskoi's language skills in linking eugenics with another project he had undertaken to justify Narkomzdrav's patronage of his institute: extensive studies of "rejuvenation" in animals and humans purportedly achieved by vasectomy and the transplantation of sex glands.[96] As a first step in the realization of this project, Kol'tsov arranged for the publication of a large volume, seductively titled "Rejuvenation," with translations of major works on the subject by its leading proponents, Austrian physiologist Eugene Steinach and French surgeon (of Russian extraction) Serge Voronoff. Volotskoi, however, translated for Kol'tsov's volume a piece that seemingly had no connection to rejuvenation at all — a letter written by Harry C. Sharp, a surgeon at the Indiana Reformatory (the state's major prison), which had appeared in *Eugenic Review* on the eve of the London congress in 1912.

Sharp had been one of the instigators of the first "sterilization law" promulgated by the state of Indiana in 1907,[97] and its text was amended to his letter.[98] In the 1890s and 1900s, the surgeon had performed numerous vasectomies on the reformatory's inmates, and in the early 1920s his published results were regularly cited in support of the alleged "rejuvenating" effects of the operation. The inclusion of Volotskoi's translation of Sharp's letter (and the text of the 1907 Indiana sterilization law it contained) in Kol'tsov's volume unobtrusively linked the research on "rejuvenation" with eugenics, thus perhaps further strengthening the appeal of the new field to the IEB's patrons among Narkomzdrav officials. Obviously satisfied with Volotskoi's work, Kol'tsov suggested that the young anthropologist translate into Russian the "eugenic Bible" — Galton's 1909 *Essays on Eugenics*. Kol'tsov's efforts to promote eugenics as a means of securing government support for his institute and his numerous students soon paid off.

By the time of the RES inaugural meeting in November 1920, the civil war had largely spent its fury. The Bolshevik Red Army had driven out both the "Whites" and the allied expeditionary forces that supported them.[99] But the new rulers paid a steep price for the victory. The entire country lay in ruins: the economy was shattered and cities depopulated, factories stood still and fields empty, epidemics ran rampant, and famine reigned supreme. Faced with these severe crises, in March 1921, the government abolished War Communism and adopted a "New Economic Policy" — NEP. Although under the NEP the Bolsheviks preserved state control over banking and key industries, they reinstated money and the market, permitted private ownership and initiative in trade, services, and the small-scale production of consumer goods, and replaced the forced requisition of produce from the peasantry with a moderate "food tax." The NEP proved effective in the rapid improvement of living conditions throughout the Union of the Soviet Socialist Republics (USSR) that by the end of 1922 had consolidated most territories of the former Russian Empire.[100]

With the burden of waging the war lifted, the Bolsheviks turned their attention and the considerable resources at their disposal to the previously neglected areas of state building and administration. One such area became the rebuilding of Russian science.[101] The government launched enormous campaigns to combat illiteracy and popularize science,[102] began to expand the entire system of education (from primary schools to universities), and strove to promote the rapid growth of all branches of science. The new rulers did not forget about scientists. In December 1921 the highest government agency, the Council of People's Commissars (SNK) presided over by Lenin, issued a special decree "On

Improvement of Scientists' [Living] Conditions." A few months later, in June 1922, the Central Commission to Improve Scientists' Living Conditions — created to implement the decree — opened a "House of Scientists" in Moscow, which was overseen by Semashko, and became the preferred meeting place for the Russian Eugenics Society. Thanks to Kol'tsov and his fellow members of the RES, including Volotskoi, eugenics quickly became part and parcel of the new policies debated and implemented during the NEP.

For Volotskoi, as for many others, the adoption of the NEP eased considerably the daily struggle for survival and facilitated his scientific work. In the summer of 1921, he took part in two field expeditions (mounted jointly by the Moscow University anthropology department and the IEB eugenics department) to study the ethnic minorities of the Upper Volga region. But these expeditions were not typical of the bulk of research conducted by Russian eugenicists at the time. Given the shortage of necessary resources, most of them followed Galton's lead and focused on compiling and analyzing the pedigrees of eminent cultural figures. Volotskoi also got involved in this kind of research on several families of eminent writers, musicians, and artists, including the genealogy of Fedor Dostoevsky that would become his life-long obsession. In the course of the year, he also delivered two lengthy reports to the RES general meetings. One (coauthored with Bunak) dealt with the application of Mendel's laws to human heredity. Another, titled "The 'Indiana Idea' in Light of the Latest Studies by [Eugene] Steinach," was based on research he had conducted in preparing the translation of Sharp's letter and addressed the sterilization of the "unfit" as an instrument of eugenics.[103]

In October 1921, the RES convened a special meeting to mark its first anniversary. By that time, the society's membership had nearly tripled to include not only biologists, psychiatrists, anthropologists, and social hygienists, but also historians, psychologists, sociologists, physicians, educators, and jurists. As we saw, during the two preceding decades, eugenics had meant many different things to distinct professional audiences in Russia, but had inspired no attempts at a unification of their diverging views. With the establishment of the RES, it became necessary to find common ground for the varying approaches to eugenics popular among the representatives of different professions, medical specialties, and scientific disciplines. In his presidential address to the anniversary meeting, titled "The Betterment of the Human Breed," Kol'tsov attempted to find such common ground and to outline the general contours of "the new biological science of eugenics."[104]

Kol'tsov identified and carefully delineated three key elements amalgamated under the name of eugenics. The first element — "pure science," which he named *anthropogenetics* — was to gather knowledge of human heredity and to investigate the principles of inheritance of various human traits. The second — "applied science," which following his pre-revolutionary predecessors Kol'tsov called *anthropotechnique* — was to apply the knowledge provided by anthropogenetics to finding appropriate methods (from social policies and legislation to modifying individual behaviors) for improving the genetic quality of future generations. And the third — "eugenic religion," comparable, in Kol'tsov's opinion, to nationalism, Christianity, Islam, and socialism[105] — was to propagate the "ideal" that would "give meaning to [human] life and motivate people to sacrifice and self-limitation." The ultimate goal of eugenics was "to create [...] a higher type of human, the powerful king of nature and the creator of life." Echoing Galton, Kol'tsov concluded: "Eugenics is a religion of the future and it awaits its prophets."[106]

If there was a dearth of prophets, there was no shortage of apostles. By the time of the RES anniversary meeting, Filipchenko, a zoology professor at Petrograd University, had already established the country's first department of "genetics and experimental zoology" (see fig. 5-5).[107] He had created a "Eugenics Bureau" under the auspices of the Academy of Sciences and launched its own periodical, *Herald of the Eugenics Bureau*.[108] Filipchenko had also initiated the formation of a eugenics society in Petrograd and waged an extensive campaign to promote both genetics and eugenics amongst members of the city's scholarly community, as well as the general public.[109] Over the next few years, Kol'tsov and Filipchenko, together with their students and co-workers, published nearly 100 articles on eugenics in various popular and specialized journals, such as *Priroda*, *Hygiene and Epidemiology*, *The Proletarian of Communications*, and *Scientific Word*, as well as numerous pamphlets, brochures, and books.[110]

Fig. 5-5. Iurii Filipchenko (seated in the center) with his students, 1923. Standing on the far right is Ivan Kanaev. Photographer unknown. Courtesy of S. Fokin.

The mobilization campaign launched by the champions of eugenics proved very successful. There is little doubt that for many individuals and groups who joined Kol'tsov and Filipchenko in their efforts to develop eugenics in Russia, "the new biological science of eugenics" offered a convenient means to advance their own scholarly interests, disciplinary and professional agendas, expert status, and career ambitions. The Sixth All-Union Congress of Gynecologists and Obstetricians held in June 1924 in Moscow provides an illuminating example. The congress devoted a separate session to "eugenics and biological questions."[111] In his keynote address to the session, titled "Eugennétique and Gynecology," Saratov University's professor Nikolai Kakushkin outlined a broad programme of his specialty's relations to eugenics.[112] He insisted that not only "all the questions of woman's health pertaining to her child-bearing abilities," but also "all the questions of breast-feeding, child hygiene, preschool and school education, marriage and sex hygiene, struggle against venereal diseases and prostitution," should come under the purview of the "eugenicist-gynecologist."[113]

Kakushkin's address vividly demonstrated that eugenics offered gynecologists a suitable tool in the competition with other medical specialists over not just the protection of maternity and infancy, which became a major focus of Narkomzdrav's activities, but also the entire field of health protection. The congress's participants also hotly debated the "eugenic" role of contraceptives and abortion (which had been legalized in Soviet Russia in November 1920). Yet a report by Antonina Shorokhova (1881-1958) on successful artificial inseminations in women as a means to fight infertility did not spur much discussion, attesting clearly to the lack of interest in the subject among Russian gynecologists: relative to other European countries that experienced substantial decline in birth rates in the aftermath of World War I, Russian birth rates remained very high.[114]

In parallel with conducting an extensive propaganda campaign and building their institutional bases, despite the nearly total international isolation of the newborn Soviet republic, the Russian champions of eugenics began to establish contacts with their foreign colleagues.[115] One episode illustrates the aims, scope, and venues of Russian eugenicists' international activities. In September 1921, the head of the US Eugenics Record Office (ERO) Charles Davenport received a long letter from Nikolai Vavilov, Russia's leading plant scientist. One-time student of William Bateson, who alongside Kol'tsov and Filipchenko was a major force in building Russian genetics, Vavilov at the time was attending a congress on phytopathology in San Francisco. His letter informed Davenport about the establishment of "the first Russian Eugenic Society" and the work that was being done by Filipchenko and Kol'tsov. Vavilov expressed his regrets that he would not be able to attend the Second International Eugenics Congress being held at that very time in New York City and asked Davenport to collect the congress's materials and other recent eugenic literature for Russian colleagues.[116] Indeed, after the end of the congress, Vavilov came to visit Davenport at the Eugenics Record Office to reiterate the message in person.[117]

Around the same time, Davenport received a letter from Russia, from the RES president Kol'tsov.[118] Sent through the Narkomzdrav representative in the United States, the letter also notified Davenport of the creation of both the IEB eugenics department and the RES. Kol'tsov noted that he would have liked to attend the International Eugenics Congress, but it seemed impossible at the moment. He also lamented the "intellectual famine" Russian scientists had been and were still experiencing, and asked Davenport for assistance in obtaining recent genetic and eugenic literature.

A few weeks later, Davenport received yet another letter from Russia, sent via the Soviet Trade Delegation to Norway, this one from Filipchenko. Filipchenko apprised his US colleague of the establishment of the Eugenics Bureau in Petrograd: enclosed with his letter was the first issue of its *Herald* dedicated to Galton. He too asked Davenport for help in acquiring eugenic literature.[119] Davenport immediately responded to the Russian requests and arranged for a large shipment of books and periodicals to both Moscow and Petrograd. Furthermore, although no Russian scientist had attended the Second International Eugenics Congress, the RES soon became a member of the International Federation of Eugenic Organizations, with Kol'tsov representing the country on the Federation Council.

Bolshevik Russia appeared the least likely locale for concerns with "national degeneration," the increasing fertility of "lower classes," or "interracial meticization," which held the attention of the Second International Eugenics Congress. Yet the rapid institutionalization, internationalization, as well as active propaganda, of eugenics in the immediate post-revolutionary years was fully funded and enthusiastically endorsed by various agents and agencies of the country's new government. Why would a "proletarian state," which claimed to be building a classless society and vocally denounced racism and nationalism, become a hotbed of eugenic debates, support eugenic research and institutions, and adopt eugenics-inspired policies?

At least in part, the answer to this question lies in the confluence between the eugenic vision of "the self-direction of human evolution," as it was expressed in the motto of the Second International Eugenics Congress, and the Bolsheviks "revolutionary dreams" (in US historian Richard Stites's apt characterization) of creating a "new world," a "new society," and a "new man."[120] As Kol'tsov clearly articulated in his 1921 anniversary address, the major goal of eugenics was "to create […] a higher type of human, the powerful king of nature and the creator of life." The Bolsheviks, in the words of one of their leaders Leon Trotsky, believed that with the victory of the Revolution "humankind, frozen *Homo sapiens*, will enter into radical reconstruction and will become — under its own fingers — an object of most complicated methods of artificial selection and psycho-physical training. [...] Man will put forward a goal [...] to raise himself to a new level — to create a higher socio-biological type, an *Ubermensch*, if you will."[121] Resonance between the eugenic vision of "a higher type of human" and the Bolshevik dreams of "a higher socio-biological type" played an important role in the appeal of eugenics to its state patrons, as well as to the numerous followers the fledgling "biological science of eugenics" attracted in 1920s Soviet Russia.

The *Russian Eugenic Journal* (*REZh*) soon established by Kol'tsov under Narkomzdrav auspices vividly demonstrated that eugenics indeed found an enthusiastic following in Bolshevik Russia. Greeted by a review in *Izvestiia*, the country's most widely circulated newspaper, the journal's first issue came out in the fall of 1922 (see fig. 8-2).[122] Opening with Kol'tsov's 1921 presidential address, the issue included Iudin's overview on "the heredity of mental illness" and Bunak's plan for establishing "eugenic experimental stations" throughout the country.[123] The second issue that came out the following spring carried an article "On the Tasks and Paths of Anthropogenetics" written by Alexander Serebrovskii (1892-1948), Kol'tsov's most talented student in genetics.[124] The article outlined the research methodology and agendas of the new science and was supplemented by Bunak's lengthy "critical analysis" of the existing "methods of investigating human heredity."[125] The issue also contained a revised text of Volotskoi's report "on the sexual sterilization of the hereditary defectives" delivered to the RES meeting on 30 December 1921.[126]

If Iudin's, Serebrovskii's, and Bunak's contributions addressed the first facet of eugenics identified by Kol'tsov — the "pure science" of human heredity, anthropogenetics — Volotskoi's assessed the second one: the "applied science" of anthropotechnique. His presentation surveyed an array of concrete policies and actions that should spring forth from the studies of human heredity and advance the principal goal of eugenics, the hereditary betterment of humanity. But he focused predominantly on the most controversial among eugenic policies

— sterilization. Based on more than fifty publications in English, French, German, and Russian, he outlined the long history of various surgical operations that produced sterility in both men and women, along with their anatomical details and "physiological and psychological consequences," especially the widely touted "rejuvenation" that purportedly resulted from such operations.

Volotskoi described the appropriation of sterilization operations as a tool in eugenicists' attempts to limit the spread of "bad" heredity. He detailed the propaganda campaigns, legislative initiatives, and actual laws aimed at the implementation of "eugenic sterilization" in different countries, particularly in the United States, as well as the extensive critique and objections these efforts had elicited from various quarters. Volotskoi briefly recounted and dismissed as unsubstantiated the unanimous negative reaction of pre-revolutionary Russian commentators to eugenic sterilization. He argued that "sexual sterilization of hereditary defectives" was an efficient, harmless (indeed, often beneficial due to its alleged "rejuvenation" effects), and legitimate way of advancing eugenics' goal of improving humankind and as such ought to be adopted and promoted by the RES. The remarks from the floor that followed Volotskoi's report (and appeared in the *REZh* alongside its text) show that he failed to persuade his audience. Seven commentators, including Iudin and Kol'tsov, pointed out what they all saw as the major deficiency of sterilization policies: the enormous difficulties in, indeed the sheer impossibility of, defining and identifying "hereditary defectives" given the present state of knowledge on human heredity.

Volotskoi, however, remained unconvinced by the arguments of his fellow eugenicists. Over the next few months he reworked his report into a book, issued in 3,000 copies under the provocative title *Elevating the Vital Forces of the Race: A New Path*. As he quipped in the book's introduction: "one either believed in eugenics or didn't."[127] He obviously believed what he preached. Volotskoi expanded his analysis to encompass various legislative initiatives and policies advocated by western eugenicists, from marriage regulations to sterilization and segregation, using more than 150 publications in Russian, English, French, and German to substantiate his view of sterilization as the most effective path to the hereditary betterment of humankind.

Although some social hygienists, for instance, Voronezh University's professor Tikhon Tkachev, came to support Volotskoi's position,[128] the majority of Soviet eugenicists remained highly skeptical of "negative" eugenics. Despite Volotskoi's advocacy, they continued to criticize their western colleagues for advocating restrictive "eugenic laws," particularly the involuntary sterilization of the "unfit."[129] Filipchenko published an extensive denigrating review of Volotskoi's book, which summarised the criticisms of eugenic sterilization by many geneticists.[130] Soviet gynecologists also objected to the sterilization of women on social grounds.[131] Yet Volotskoi did not repent: in 1926 he published an expanded, second edition of this book.[132]

It was in the course of his work on *Elevating the Vital Forces of the Race* that Volotskoi discovered Florinskii's treatise. In an attempt to justify his own views, Volotskoi conducted extensive historical research on marriage regulations in various countries, including Russia. In the course of this research, he found, for instance, a decree issued as early as 1722 by Peter the Great that forbade "fools who are unfit to either education or state service" to marry, because they "will not produce good progeny for the State's benefit."[133] Volotskoi interpreted the decree as based on "the recognition of the hereditary transmission of certain mental and physical deviations from the norm,"[134] and thus prefiguring the ideas of eugenics.

But it was his discovery of Florinskii's *Human Perfection and Degeneration* that most profoundly influenced Volotskoi's views. Volotskoi was clearly unaware that Florinskii's essays had originally been published in *Russian Word*, for he referred only to their book version.[135] It seems likely that he found a copy of the treatise in the large library of more than 2,000 volumes collected by Anuchin at the Moscow University anthropology department. Perhaps, aware of Volotskoi's involvement with eugenics, his mentor actually called the young anthropologist's attention to Florinskii's treatise.[136] Whatever happened, Volotskoi did not discuss in any detail the contents of Florinskii's work in the text of his own book, but in the appended bibliography he pointed out that "in many of his ideas, [its] author is a predecessor of Galton's" and "in certain respects, his views on heredity are remarkably close to the modern ones (Mendelism)."[137] In fact, the RES secretary was so impressed that he used excerpts from Florinskii's text — alongside quotations from Plato, Aristotle, and Galton — as epigraphs to several chapters of his own book.

Volotskoi took it upon himself to reintroduce the "long-forgotten book," as he obliquely referred to it in one of the epigraphs, to his fellow eugenicists. In the spring of 1923, he delivered a lengthy report on Florinskii's *Human Perfection and Degeneration* to an RES meeting.[138] After a brief biographical sketch of its author, Volotskoi recounted the contents of Florinskii's book in considerable detail. "It is not difficult to see," he summarized his presentation, "that Galton's 'Eugenics' and Florinskii's 'Marriage Hygiene' are merely different names for the same thing." "What we now call 'eugenics,' as well as Florinskii's 'Marriage Hygiene'," he elaborated, "is, essentially, the understanding by the species of *Homo sapiens* of the process of its own evolution and its striving to subordinate this process to its own will through the study of all the factors underpinning or even tangentially influencing the evolutionary development of humankind."[139]

Yet, he asserted, there were substantial differences between Galton's and Florinskii's views on how to reach this goal. According to Volotskoi, "prohibitive measures play an essential role in Galton's eugenics," whilst in Florinskii's concept such a role is assigned to "the inculcation in the populace of a healthy, developed taste or fashion that could guide the choice of marital partners." "The unacceptability of marrying for hereditarily defective individuals, in Galton's opinion, should become a religious dogma, a sort of taboo in eugenic religion," he stressed, whilst, "in Florinskii's opinion, [we] should strive to influence [marital] fashions in such a way as to approximate, if not an absolute perfection, then at least such examples that present, aside from a conditional aesthetic value, a certain biological benefit."[140]

Admitting that "much of the book does not correspond to current knowledge," Volotskoi insisted that "in the independence and originality of its ideas, Florinskii's work could rightly claim a place of honor in world literature" on eugenics. "Florinskii's unrecognized and forgotten treatise deserves the serious attention of eugenicists," he declared, "and his name as one of the founders of our discipline must be placed alongside with, to be exact, even ahead of, Galton's." Furthermore, Volotskoi emphasized, in addition to "its purely historical interest, because it had been published significantly earlier than Galton's work,"[141] Florinskii's book "had not lost its scientific importance." At this point, he did not elaborate on what that "scientific importance" might be.

## The Champion of "Proletarian" Eugenics

By the time Volotskoi's report on Florinskii's book appeared in print in the spring 1924 issue of *Russian Eugenics Journal*, his academic affiliations, his attitude to Galtonian eugenics, and, indeed, his entire worldview had undergone a radical transformation. In June 1923, his mentor Anuchin died and the chairmanship of the Moscow University anthropology department went to Bunak. Although he did contribute an article to the special issue of *Russian Anthropological Journal* commemorating his teacher,[142] in the fall, Volotskoi severed all connections to his alma mater.[143] Instead, he became a lecturer at the Narkomzdrav Institute of Physical Culture (IFK), the country's first specialized institution to train instructors in physical culture and education for the "proletarian masses."[144] He began teaching one 24-hour course on eugenics and another 36-hour course on anthropology to the fourth year students.[145] The next spring, just as his report on Florinskii's book came out, Volotskoi also resigned his position at the IEB eugenics department. Instead, he became a researcher at the recently established institution with a cumbersome, but revealing name: the Timiriazev State Scientific-Research Institute for the Study and Propaganda of the Natural-Science Foundations of Dialectical Materialism — the Timiriazev Institute, as it was commonly referred to at the time.[146]

Volotskoi's move from the IEB to the Timiriazev Institute was not a matter of convenience, but a deliberate and meaningful choice: in the course of just a few post-revolutionary years he had become a believer not only in eugenics, but also in Marxism. As its full name makes plain, the new institute was part of the Bolsheviks' extensive efforts to replace "bourgeois" science with their own "proletarian" one.[147] A major tenet of "proletarian science," most fully elaborated by its principal theoretician Alexander Bogdanov and widely popularized by his numerous followers during the early 1920s, was a conviction that, as part of society's superstructure, science reflected the interests of the ruling class.[148] Hence, in a capitalist society, science served only the needs of the oppressor class and itself became an instrument of oppression, while in a proletarian state science was to serve the needs of the previously oppressed proletariat. Accordingly, while bourgeois science was based on "idealistic" bourgeois philosophy, proletarian science should adopt as its guiding philosophy the ideology of the proletariat, Marxism — dialectical and historical materialism. From the very beginning of their rule, the Bolsheviks spared no effort to "infuse" Marxism into science and to build an institutional base for "Marxist" science. As early as 1918, a group of high-ranking Bolsheviks established a Socialist Academy (renamed Communist Academy a few years later) as a counterweight to the "bourgeois" Academy of Sciences they inherited from their imperial past. They also created a number of "Communist Universities" and "Institutes of Red Professors" to prepare cadres for proletarian science.[149]

In their initial labors to remake the "bourgeois" and "idealistic" science into a "proletarian" and "materialist" one, the Bolsheviks targeted mainly the social sciences and the humanities. However, a few naturalists, especially from among the younger generation born circa 1890, appeared receptive to Marxism, embracing both its dialectical-materialist method of studying nature and its class approach to the understanding of human history.[150] Paraphrasing Lenin's famous 1922 call for "militant materialism," the rules of appointment to the Timiriazev Institute stated categorically that its researchers "must follow a strictly materialistic view in the field of natural sciences" and "possess a dialectical-materialist worldview."[151] Named after Kliment Timiriazev, one of the first Russian commentators on Darwin's theory who had attempted to link Darwinism and Marxism, the institute became home to a sizeable cohort of "materialist-biologists," as they called themselves.

Volotskoi's application and subsequent appointment to the Timiriazev Institute signaled unambiguously that by the spring of 1924 he had become a convinced Marxist. He was among the first to apply Marxism to eugenics, not as a rhetorical exercise, but as a genuine attempt to grapple with the Marxist philosophy of nature and history in his own work. And it was his "class" analysis of Galtonian eugenics, much more than his support for eugenic sterilization, that led him to parting ways with Kol'tsov and the IEB eugenics department and to joining the "Department of the Biological Foundations of Social Phenomena" at the Timiriazev Institute.

Available materials are completely silent on exactly when, why, and how Volotskoi "converted" to Marxism. It seems likely that his acquaintance with the RES member Evgenii Radin (1872-1939) played a decisive role in arousing the young anthropologist's interest in Marxism.[152] Radin had developed close ties to the Bolsheviks long before they seized power: his older brother had been an active member of the early Russian labor movement and even authored its most popular "anthem."[153] Trained as a psychiatrist at Berlin University, Radin worked as a school physician before the Bolshevik Revolution.[154] After the revolution, he came to work for Narkomzdrav as an expert on children's health, a subject that became a major focus of the agency's policies from its very birth in July 1918.[155] Indeed, Narkomzdrav created two special large departments: one for "the protection of maternity and infancy" (OMM — *okhrana materinstva i mladenchestva*) and another for "the protection of children's and adolescents' health" (OZDP — *okhrana zdorov'ia detei i podrostkov*). Radin became the head of the OZDP department and a leading force in the development of this field,[156] particularly in the creation of the Bolshevik system of physical culture and physical education, becoming a deputy-head of its flagship institution, the IFK.[157] He also became one of the founders of Soviet pedology (the science of childhood).[158] Radin elaborated a system of "bio-social upbringing" (*biosotsial'noe vospitanie*) that was to combine the biological (such as physical culture) and the social (from general education to specialized psychological testing and training) sides in the upbringing of Soviet children.[159]

Most likely it was Radin who, in the fall of 1923, invited Volotskoi to join the IFK faculty. And it was Radin who the same year published the first, though very brief, Marxist assessment of eugenics that clearly influenced Volotskoi's attitude to his newfound faith and his own programme of "bio-social" eugenics.[160] Whatever were the initial stimuli to Volotskoi's "conversion," he diligently studied Karl Marx's *Das Kapital*, along with many other foundational works of what became the official state ideology of Soviet Russia. He soon attempted to put these studies to good use in his own research and writing on eugenics.

Volotskoi's various reports and publications of 1924-1925 indicate that he came to see Galton's eugenics as a prime example of "bourgeois" science and Florinskii's "marriage hygiene" as a model for "proletarian" eugenics. In 1924, under Radin's editorship, the IFK issued the first volume of its proceedings, titled *Physical Culture in Light of Science*. Prefaced by Commissar Semashko's unequivocal declaration that "physical culture is a powerful means of the healthification of human beings and a foundation of eugenics,"[161] the volume contained two lengthy contributions by Volotskoi. One examined "physical culture in light of eugenics," another analyzed "certain currents in modern eugenics."[162] A few months later, under the auspices of the Timiriazev Institute, Volotskoi also published a fifty-page pamphlet, tellingly titled *Class Interests and Modern Eugenics*.[163] In these three publications, Volotskoi advanced a Marxist critique of "bourgeois" eugenics, taking as his starting point the *Communist Manifesto*'s famous statement that "the ruling ideas of each age have ever been the ideas of its ruling class." And in all three he invoked Florinskii's treatise in support of both his criticism of "bourgeois" eugenics and his vision of what "proletarian" eugenics should be and do.

Volotskoi's critique centered on what he saw as the main programme of "bourgeois" eugenics: the "cultivation of talents" by selective breeding among individuals of privileged classes and "higher" races. He identified three main fallacies that, in his opinion, underpinned this programme and clearly betrayed the "bourgeois nature" of contemporary eugenics: class and race biases, privileging the biological over the social (nature over nurture, heredity over environment) in the understanding of human individual and social development, and rejection of the inheritance of acquired characteristics. In advancing his position, Volotskoi repeatedly juxtaposed Florinskii's views on these issues with those of the well-known proponents of eugenics, taking his examples from Britons Galton and Pearson, Germans Hermann W. Siemens and Fritz Lenz,[164] and Russians Kol'tsov and Filipchenko. He attacked the eugenicists' conviction that the "upper classes" and, especially, the intelligentsia, were the bearers of hereditary talents, while the "proletarianization" of the population constituted the major threat to "national" and "racial" heredity. He was particularly incensed by the eugenicists' attempts to put different monetary values on the children of "higher" and "lower" classes, and thus estimate their respective "hereditary worth," while completely ignoring the role of the environment in the realization of hereditary potentials.

Volotskoi appended his 1925 brochure with a long excerpt from Florinskii's treatise to illustrate the latter's attitude toward the "biological, hereditary worth of the representatives of different classes,"[165] clearly expressed in the professor's musings regarding the "natural mind" of the peasantry and the "artificial mind" of the aristocracy. Volotskoi even attempted to explain the diverging fates of Galton's and Florinskii's concepts of the betterment of humankind by claiming that, unlike the former, the latter had been neglected by contemporary society exactly because it did not correspond to the interests of nineteenth-century Russia's ruling classes.

It was not only his newfound Marxist beliefs that influenced Volotskoi's attitude to Galtonian eugenics and its plans of "breeding talents." His own and his fellow eugenicists' actual research on "hereditary talents" also played an important role. In the early 1920s, this line of research comprised nearly one half of all the studies conducted by Russian eugenicists (including Volotskoi) who analyzed the "inheritance" of literary, musical, mathematical, and artistic talents, along with "hereditary inclinations" to scientific research.[166] To give just one example, the *REZh*'s very first issue carried a joint genealogy of Darwin and Galton constructed by Kol'tsov and a similar genealogy of the Aksakovs family (that included a number of eminent Russian writers) presented by Serebrovskii.[167] Many of these studies considered "talent" to be a simple recessive trait that could be identified by tracing "eugenic pedigrees" of famous scientists, musicians, artists, and writers.[168] Justifying this line of research in his 1921 anniversary address, Kol'tsov stressed: "We cannot experiment. We cannot force [Russia's most famous soprano Antonina] Nezhdanova to marry [Russia's most famous bass Fedor] Chaliapin in order to see what kinds of children they would have."[169] Answering Kol'tsov's challenge, Volotskoi found a "historical" experiment of exactly this kind.

According to Volotskoi's research, in 1838 Osip Petrov (1806-1878), a famous bass of the Imperial Mariinskii Theater, married Anna Vorob'eva (1817-1901), a leading contralto at the same theater.[170] Judging by the testimonies of their contemporaries, including prominent composers Modest Musorgskii and Mikhail Glinka, both were "musical geniuses." Their marriage bore seven children (only one of whom died in infancy), thus, as Volotskoi put it, "realizing those conditions that the talent breeders dream of." Yet, although from their infancy these children grew up surrounded by music and musicians and therefore had a very supportive environment, they did not exhibit even a modicum of their parents' talents. Three of them did attempt to embark on a professional career in music and/or theater, but all ended in failure, while the other three showed no inclination of following in their parents' footsteps. Furthermore, Petrov and Vorob'eva's six children produced only one grandchild between them, who died in infancy, and therefore the entire family line went extinct and the great talents of its founders had been lost.

Volotskoi observed that Petrov himself had come from a very humble background (as a child he was a shepherd): he was raised in the home of his uncle, a cattle dealer who had tried to suppress the boy's attraction to music in every possible way, up to breaking a guitar on his head. Yet, this did not stop the young shepherd from learning music and becoming a star singer. As Volotskoi put it, "Petrov has made a great journey from a shepherd to a creator of the Russian opera. His children went in the opposite direction." According to Volotskoi, his analysis of Petrov and Vorob'eva's marriage demonstrated that the notion of talent as a simple hereditary trait was not supported. Any talent, he argued, was a complex combination of numerous hereditary traits that could be expressed or suppressed under the influence of environment. Hence, he insisted, "we should not breed talents, but [help] realize them." He approvingly cited Florinskii's lament: "We can only regret that not all societal groups have the same opportunity for the development of their natural mind, [and] that many excellent, talented individuals remain hidden in the mass of the people as wasted, unproductive capital that has neither purpose, nor use."[171]

Volotskoi did not limit himself merely to criticizing Galtonian eugenics. He actually attempted to create an alternative, "bio-social" eugenics. On 19 November 1925, he delivered a lengthy report to the Timiriazev Institute, titled "A System of Eugenics as a Bio-Social Discipline."[172] His talk was an extensive commentary on an elaborate chart that presented his vision of the "methods, contents, and scientific foundations" of eugenics (see fig. 5-6). According to this vision, "eugenics strives to study the process of the evolutionary development of humankind in order to learn [how] to guide this process in the desired direction, namely from degeneration to renaissance" with the "ultimate goal of the betterment of the human breed."

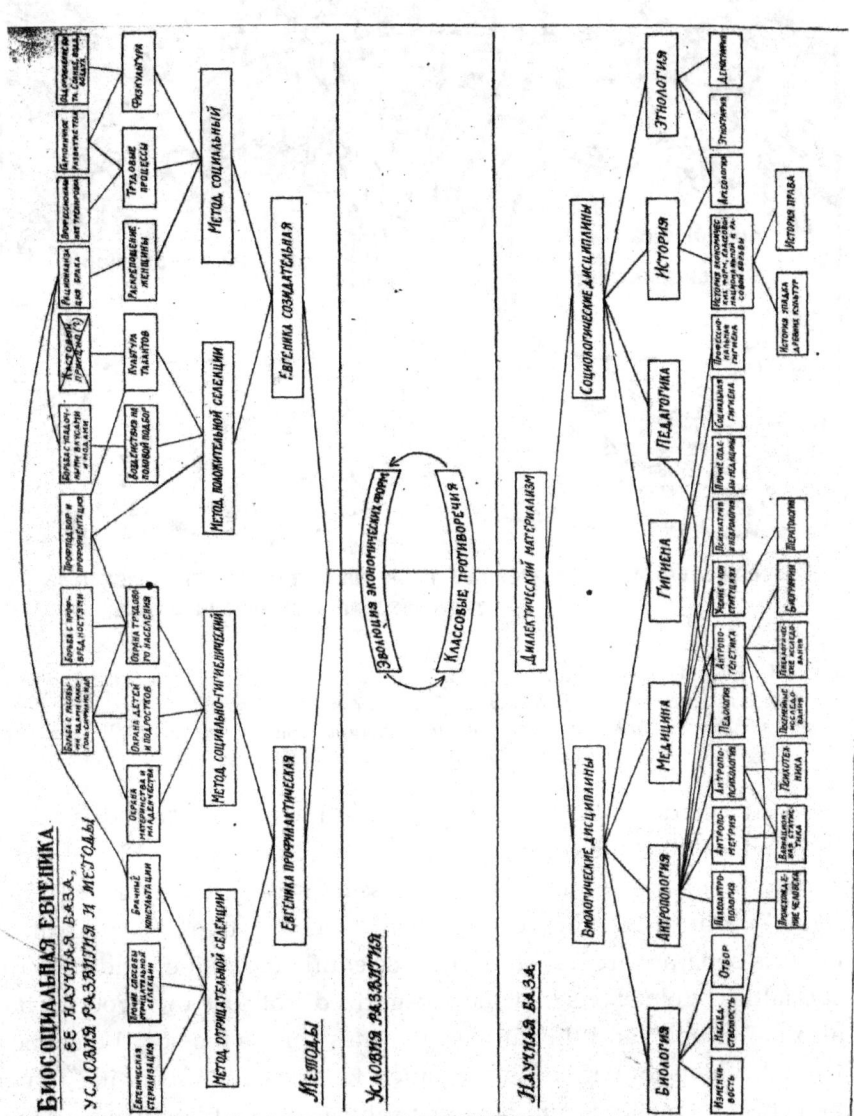

Fig. 5-6. "Bio-social eugenics, its scientific foundation, conditions of development, and methods." From M. V. Volotskoi, *Sistema evgeniki kak biosotsial'noi distsipliny* (M.: Izd. Timiriazevskogo instituta, 1928), insert. Courtesy of INION.

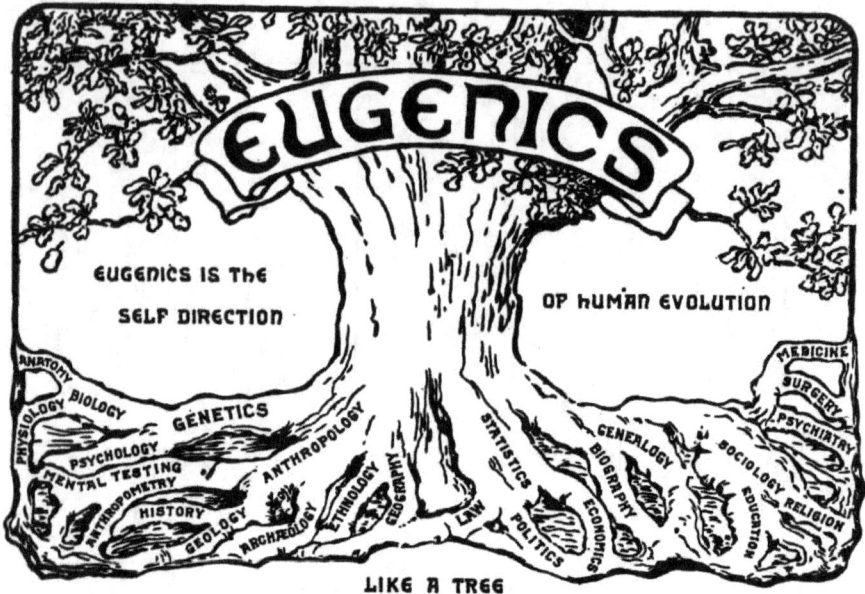

Fig. 5-7. "Eugenics tree." From Harry H. Laughlin, *The Second International Exhibition of Eugenics held September 22 to October 22, 1921, in connection with the Second International Congress of Eugenics in the American Museum of Natural History, New York* (Baltimore, MD: Williams & Wilkins Company, 1923), p. 15.

Volotskoi modeled his chart in part after the famous "eugenics tree" diagram exhibited by the US Eugenics Record Office (ERO) at the Second International Eugenics Congress in New York City in 1921 (see fig. 5-7). But in contrast to the ERO's "eugenics tree" that identified only its separate "roots" in various scientific disciplines and medical specialties, Volotskoi's scheme categorized not just the "roots," but also the "trunk" and the "upper branches" of his bio-social eugenics. Indeed, Volotskoi began his exposition with the analysis of its "upper branches" — methods and policies to advance the ultimate goal of bio-social eugenics. He clearly followed Florinskii's two-pronged approach and suggested that, instead of the customary division of eugenics into "positive" and "negative," all eugenic activities should be split into two categories: "preventive or prophylactic" and "creative." The first should "protect the population (race) from everything that could cause its degeneration," while the second should include "the entire system of measures [that could] actually better the human breed."

Both preventive and creative eugenics, in his view, could be advanced by two complementary methods: positive or negative "selection" (*selektsiia*) and "social eugenic measures," such as "physical culture, protection of mothers', children's, and adolescents' health, and so on." "Supplementing each other," he asserted, "these two methods form that synthetic applied discipline we call bio-social eugenics." Volotskoi described in detail how each of these methods could be used in preventive and creative eugenics. Preventive eugenics employs negative selection, "removing from the production of the offspring" all individuals who are afflicted with such ailments as "hereditary deafness and blindness, feeblemindedness, idiotism, and certain forms of mental and neurological disorders." Since we do not know what produces "such hereditary diseases as hemophilia, schizophrenia, manic-depressive psychosis, feeblemindedness, various physical abnormalities and constitutional anomalies," he explained, we cannot eliminate their actual causes, and hence, "various forms of prophylactic selection (sexual sterilization, marriage prohibition, segregation, and so on) remain so far the only method of protecting the interests of the progeny."

"The social-prophylactic branch of eugenics," according to Volotskoi, should focus on "the healthification and rationalization of all the conditions of life and on the systematic removal of all the factors that could in one way or another [adversely] affect the quality of the race," including "occupational hazards" and "such 'racial poisons' as syphilis and various narcotics (alcohol, cocaine, etc.)." He proclaimed that the existing social institutions devoted to the protection of maternity and infancy (OMM) and the protection of child and adolescent health (OZDP) play an important role in this branch of eugenics. But he suggested that they should be supplemented with a "special department or an institute for the protection of the progeny in the widest sense of the word, that is the protection of future generations."

In contrast to preventive eugenics, creative one, according to Volotskoi, employs not "negative" but "positive" selection (*selektsiia*) through the choosing (*vybor*) and matching (*podbor*) of progenitors. It is in this branch of "selecto-creative eugenics," in Volotskoi's opinion, that the class, race, and other social biases of "bourgeois" eugenics manifest themselves most clearly, especially in its various programmes of "breeding talents." Yet he did not dismiss the method as such.

Rather, he urged his audience to follow Florinskii's ideas of "conscious healthification" of the personal tastes and societal mores that guide people's marital choices and to replace the "breeding of talents" with the "professional orientation of the laboring population," "based on the consideration of all of the organic particularities and abilities of each individual." The last branch of "social-creative eugenics" in Volotskoi's schema covered all social actions, measures, reforms, and revolutions, such as the emancipation of women, the development of physical culture, and the rationalization of marriage, which could produce "positively-eugenic results." He emphasized that his "classification of various eugenic methods has to a considerable degree a conditional character, since there are no strict boundaries among the various branches of eugenics," but it presents a useful tool to demonstrate conveniently how all of them are related to each other.

After the extended discussion of its "upper branches," Volotskoi turned to the "trunk" — the actual contents — of bio-social eugenics. In his opinion, "eugenics is an applied discipline aimed at the betterment, 'ennoblement' of the human breed" and its very name, deriving from the root "good" — "eu- (εὐ-)" — gives it a "certain conditional, normative content." But what is "good" from the viewpoint of the representatives of one class, he elaborated, is evaluated in a completely different way by the representatives of another class, while what is "useful and valuable" under one set of life conditions could become "harmful and excessive" under another one. He proceeded to illustrate this conditionality and normativity of eugenics with examples, taken from Galton and Siemens, of "social evaluations" of certain human qualities in a capitalist society, contrasting them with the possible evaluation of the same qualities in a future communist society. Since "the transition to a communist society naturally involves the re-evaluation of all [previous] social values and the replacement of one set of ideals with another," Volotskoi asserted, "the evolution of socio-economic formations should become the foundation of bio-social eugenics, for it gives the entire eugenics movement its concrete contents and direction."

Clearly following Florinskii's views on health as a universal human ideal, Volotskoi suggested that the social value (whether positive or negative) of certain features does not change, irrespective of the socio-economic conditions of human life. Such human traits as "deafness, blindness, feeblemindedness, and most mental and neurological disorders remain 'bad'," he claimed, "while health [remains] 'good' under all social circumstances." The very existence of such "objective criteria" of good and bad, according to Volotskoi, warrants close attention to "certain advances and methods of bourgeois eugenics." "A complete rejection of the entire modern eugenics as a doctrine totally alien and unsuitable to our conditions" represents "a very narrow and one-sided approach to the issue," he affirmed, and "certain innovations and methods of bourgeois eugenics could be included in our system of bio-social eugenics." Of course, he qualified his statement, any such "borrowings" from bourgeois eugenics require "great caution and criticality."

Volotskoi defined the "roots of eugenics" as "an array of scientific disciplines upon which it is based, using for its own practical goals the achievements of various fields of knowledge." "The main root that should feed our eugenics" and provide "a necessary link between bio-social eugenics and its scientific foundation," according to Volotskoi, "is dialectical materialism," because "thinking dialectically is the first and absolutely essential condition of a truly scientific approach to the problems of eugenics." In his view, the "two main pillars" of this scientific foundation are the theory of biological evolution, Darwinism, and the theory of social evolution, Marxism. The first unifies and synthesizes all biological/anthropological/medical knowledge and the second all sociological/economic/political knowledge pertinent to bio-social eugenics. Volotskoi provided lengthy descriptions of separate disciplines, fields of inquiry, and specialties that contribute to each biological and sociological understanding of humankind, ranging from anthropogenetics, statistics, experimental psychology, and "the theory of constitution (*Konstitutionslehre*)" to social hygiene, "bio-social upbringing," and the history of class struggle, law, and economics. He admitted that in this particular section his scheme had much in common with the ERO's "eugenics tree," but he emphasized such essential differences as the exclusion of religion from and the inclusion of the studies of occupational hazards into bio-social eugenics. Despite its dependence on and close relations to numerous biological and sociological fields of knowledge, Volotskoi insisted, "eugenics does not lose its autonomy." Its main subjects — the process of human evolution, as well as its practical goals of controlling and directing this process — in his opinion are specific enough to insure eugenics' self-sufficiency, self-determination, and independence.

Volotskoi soon went beyond a theoretical analysis of what bio-social eugenics should be and do. He launched several research projects that would advance its actual development. In fact, Volotskoi made explicit the research programme embedded, though not clearly articulated, in Florinskii's notions of "conditions conducive" to either degeneration or perfection. He began a focused study on the influence of such "racial poisons" as alcohol and syphilis on the progeny[173] and initiated a broad research on "eugenics and occupational hazards."[174] He also attempted to expand on Florinskii's idea that "marital tastes and fashions" influence the evolution of certain physical types by investigating "the evolution of the Russian women type in relation to issues in eugenics."[175]

There could be little doubt that Florinskii's book played a key role in both amplifying Volotskoi's critical attitude towards Galtonian eugenics and shaping his own plans for bio-social eugenics. He certainly felt that Florinskii's work deserved much wider circulation beyond the still narrow circle of Russian eugenicists and did everything he could to popularize the writings of his newfound hero. His efforts culminated in a new edition of *Human Perfection and Degeneration*.[176]

This "second edition," as it was described on the title page, appeared in early 1926 in 3,000 copies as the first issue of the Timiriazev Institute's signature series, "The Library of a Materialist" (see fig. 5-8). The 165-page volume contained a shortened text of Florinskii's 1866 book, typeset anew to correspond to the new alphabet and spelling rules introduced by the Bolsheviks. Volotskoi restored the correct initials of its author and placed his portrait on the front page.

Volotskoi took his editorial work very seriously. He divided the text into an "introduction" and four chapters, thus changing considerably the original structure of Florinskii's book. The first chapter was titled "The Changeability of the Human Type," and the second, "The Role of Heredity in the Changeability of the Human Type." Together with the "introduction" the first two chapters corresponded to the text of Florinskii's first essay published in the August 1865 issue of *Russian Word*. The third chapter, titled "Conditions Conducive to the Change of the Human Breed," reproduced the text of the second essay that had

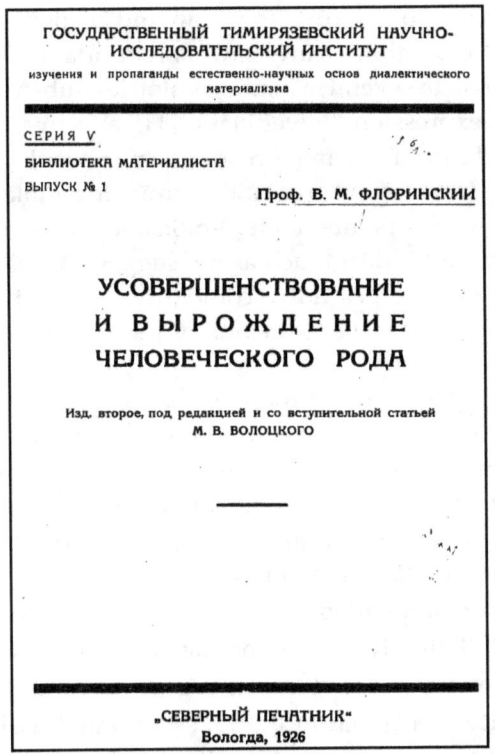

Fig. 5-8. The title page of Vasilii Florinskii's *Human Perfection and Degeneration* published by Volotskoi in 1926. Courtesy of RNB.

appeared in the October issue. It was subdivided into three sections: "Taste and demand for certain qualities," "The influence of external life conditions," and "Rational marriage." The last chapter bore the unwieldy title "The Degeneration of Human Stocks and Conditions that Produce It (Slavery, Poverty, Exploitation, Lack of Stock Renewal, Consanguinity, and so on)," and contained the text of the third essay that had originally been split between the November and December 1865 issues of *Russian Word*.

Although he did not indicate in any way his edits in the published text, Volotskoi substantially edited Florinskii's treatise. Some of the changes were purely stylistic. He divided the long paragraphs of the original into several shorter ones, changed punctuation, and added or removed emphasis indicated by italics. The editor also shortened the original text by nearly ten per cent, excising certain words, phrases, paragraphs, and occasionally whole pages. Some of the cuts look as if they were made to improve the flow and to lighten the prose by removing what the editor clearly saw as redundant. Others, however, were less innocent. In several instances, Volotskoi substituted "peasants" for Florinskii's expression "lower class." He effectively "ungendered" Florinskii's aside to his "female readers" by rewriting it in a gender neutral voice. He removed quotations from the Bible that Florinskii had used to drive his points home. Volotskoi also excised Florinskii's "politically incorrect" ideas, such as his suggestion that "intermixing" of the nobility and the peasantry could "perfect the human type," or his musings on the possible extinction of the "Negroes" in the United States.

It is possible that some of the edits were made not by the editor, but by the censor. The Bolsheviks exercised strict control over all publications, and Florinskii's volume was no exception.[177] The Soviet censorship system was no less vigilant than its imperial predecessor in removing all and any ideas, sentiments, and pronouncements deemed politically and/or ideologically unacceptable.

Volotskoi also appended the book with ten pages of "editor's commentaries." Tellingly, in his commentaries Volotskoi did not use Florinskii's own term *usovershenstvovanie* (perfection) replacing it with *uluchshenie* (betterment) that in the 1920s became Russian eugenicists' preferred word to convey to their audiences the meaning of eugenics' major goal of human hereditary improvement. By translating the "outdated" contents of Florinskii's treatise into the modern languages of both eugenics and genetics, Volotskoi's commentaries "updated" the professor's various statements on human heredity and variability with recent advancements in genetics and anthropology and provided some references to the latest literature on these subjects. For instance, Volotskoi equated Florinskii's expression "hereditary potentials" (*nasledstvennye zadatki*) with "genes" and applied the notion of dominant and recessive genes to the explanation of hereditary diseases offered by Florinskii.[178] In a similar way, he updated Florinskii's characterization of the upper classes' "perverted tastes" in the ideals of women's beauty with an illustration of "bourgeois tastes" in current women's fashion. He actually reproduced a picture of a female fashion model (from a 1922 issue of the popular magazine *Herald of Fashion*), pointing out her narrow pelvis and shoulders, underdeveloped breasts, and hands "unfit for physical labor" (see fig. 5-9). "It is clear," he asserted, "that such a woman is incapable either of labor processes or of childbirth and breastfeeding ... Her entire figure symbolizes the rejection of labor and childbearing."[179]

Volotskoi also wrote an extensive foreword, titled "On the History and Contemporary State of the Eugenics Movement in Relation to V. M. Florinskii's Book."[180] Developing further his claim that Florinskii's work had remained forgotten for so long because it did not suit the

Fig. 5-9. Pictures of female fashion models from *Herald of Fashion*. Unfortunately, I was unable to find the original of the picture reproduced by Volotskoi. The issue of the magazine from which it was taken is absent in all the major research libraries in Moscow and St. Petersburg. From *Vestnik mody*, 1923, 5, insert. Courtesy of BAN.

interests of Russia's ruling classes, Volotskoi opened his foreword with an illuminating metaphor:

> The principle of the struggle for existence and the survival of the fittest operates not only in the process of plant and animal evolution, but also in the whole series of processes that occur in the artificial environment created by civilized humanity. In part, we can see it [operating] in those factors that affect the book market. Here, too, only those books "survive" — that is gain social recognition — which have proven strong and viable in their own struggle for existence, in other words, those which have corresponded to the interests of the ruling class, or at least, of that class which had understood its social role.

According to the editor, Florinskii's book "was too far ahead of its time and thus found no support." "It is sufficient to read the author's merciless assessment of the privileged classes," he continued, "to understand why his treatise did not suit the tastes of the contemporary reading public." "Considering the 'higher' strata of the population as suffering from degeneration and decline," Volotskoi stressed, "he places all his hopes on the country's laboring elements. And it is on this foundation that he builds his system of human perfection." This assessment, of course, implied that Volotskoi's own time — the aftermath of the Bolshevik Revolution that turned the "laboring elements" into the ruling class in Russia — was ripe for appreciating the full import of Florinskii's treatise.

Volotskoi focused on comparing the two "systems of the betterment of humankind": Florinskii's and Galton's. Since he thought that Florinskii's treatise had appeared in 1866, while, as he discovered in the course of his research, Galton's first "eugenic" article had been published a year earlier, Volotskoi revised his earlier claim that Florinskii was "a predecessor" of Galton. Instead, he argued that the general idea of human betterment could be traced as far back as Greek antiquity (to ancient Sparta, Plato, and Aristotle) and had since been supported by various thinkers, "for instance, in the famous utopia *The City of the Sun* by Tommaso Campanella." Galton's 1865 article, Volotskoi maintained, had turned this general idea into a "caste cultivation of talented people, who ... must marry only within their [own] caste and must not mix with the remaining mass of mediocrity." Florinskii, therefore, did have predecessors, Volotskoi conceded, but "his system of the betterment of the human type shows no sign of borrowing [from] or imitating [these predecessors] and has nothing in common with Galton's." In a footnote, he admitted that both Galton's and Florinskii's systems had sprouted from the same root — Darwin's evolutionary theory. But, he insisted, in contrast to Florinskii, Galton and his followers "completely ignore that part of Darwin's doctrine where he speaks of the evolutionary significance of the use [and disuse] of organs and the immediate influence of environment on the organism."

To illustrate further the differences between Florinskii's and Galton's systems, Volotskoi divided several pages of his foreword into two columns. On the left he placed excerpts from Florinskii's text and on the right quotes from Galton and his followers: Pearson, Siemens, and Alfredo Niceforo, a criminologist and statistician who had played a prominent role in founding Italian "national" eugenics.[181] The citations, he stressed, show indubitably that "modern eugenics is imbued with reactionary attitudes towards the proletariat," while Florinskii's treatise is very sympathetic to the "lower" classes.

The editor also critically assessed the works of Kol'tsov and Filipchenko to demonstrate that Soviet eugenics "is almost indistinguishable" from its counterparts abroad. To some, Volotskoi mused, this would seem a good cause to banish eugenics altogether from the land of the victorious proletariat. Yet, he insisted, these "reactionary tendencies [of modern eugenics] ... should not repel us from the extremely valuable idea of bettering humankind by means of conscious actions upon the process of human evolution." This idea of "a conscious, planned betterment of the human breed," he continued, "corresponds completely to the general goals of building a socialist Soviet society." All that is necessary, Volotskoi affirmed, is to create a new, bio-social eugenics that would "correspond to the socio-economic conditions of our great country." His talk on "the system of eugenics as a bio-social discipline" delivered just a few months prior to the book's release clearly shows that he had found a prototype of such new bio-social eugenics in Florinskii's treatise.

Certainly, Volotskoi admitted, any attempt to illuminate the problems of modern eugenics from the viewpoint of a book published sixty years ago would seem unsatisfactory to a modern scientist. Yet, he claimed, the value of Florinskii's work lay not in the "freshness" of its scientific contents, but in its "distinct ideological particularities that give it its own physiognomy." The editor argued that, thanks to his origins in the low-level clergy, Florinskii's "worldview had acquired many features common with the ideology of all the exploited and the oppressed." "Professor Florinskii created the first eugenic system founded not on the caste segregation of the higher, privileged layers of society, as in modern eugenics," he specified, "but, to the contrary, on the abolition of all caste barriers." And this is exactly what, in Volotskoi's opinion, gave Florinskii's book not just historical, but "undoubtable practical importance," for his system could be useful in "finding scientific pathways towards the creation of a new eugenic system." The replacement of Galton's "bourgeois" eugenics with a new system became Volotskoi's avowed goal, as his research projects undertaken during the 1920s clearly demonstrate. "We'll hope," he concluded, "that Florinskii's unrecognized book will finally find its place in the libraries of sociologists, hygienists, anthropologists, and, generally, everyone who is interested in the issue of the conscious betterment of the human breed."

Volotskoi's hope did come true, but not exactly in the way he envisioned it.

# 6. Resonance: Euphenics, Medical Genetics, and *Rassenhygiene*

"True eugenics can only be a product of socialism."

H. J. Muller, May 1936

Volotskoi's reprint gave Florinskii's book a new lease on life. This time, *Human Perfection and Degeneration* found a receptive audience. The book proved influential in shaping not only discussions around eugenics and its "stepsister" genetics, but also the adoption of certain social policies and legislation in the new Soviet Russia. Indeed Volotskoi's use of Florinskii's ideas as a model and justification for his "bio-social eugenics" stimulated heated debates about the place of eugenics in a socialist society and about the necessity and possibility of creating a particular "proletarian," "socialist" eugenics. If Volotskoi and his likeminded colleagues did not succeed in creating a fully-fledged bio-social eugenics out of Florinskii's "marriage hygiene," it was certainly not for lack of trying. But the radical transformations of the country's political, ideological, and institutional landscapes in the course of the Great Break of 1929, the Great Terror of 1936-1938, the Great Patriotic War of 1941-1945, and finally, the rapid escalation of the Cold War during 1947-1948, obstructed their efforts to advance "bio-social" eugenics and to promote the "conscious betterment of the human breed" in the Soviet Union.

## Reception

Volotskoi's re-publication breathed new life into Florinskii's book. This time, its appearance did not go unnoticed, and it garnered attention both at home and abroad. Even before the book came out, Kol'tsov had published in the German *Archiv für Rassen- und Gesellschafts-Biologie* a lengthy report on "the racial-hygienic movement in Russia."[1] He opened his report with a long paragraph on Florinskii and his treatise: Although "the organized racial-hygienic movement has emerged in Russia only a few years ago," the RES president wrote, "its scientific foundations had been laid down" much earlier, with the publication in 1866 of Florinskii's book "On the perfection and degeneration of mankind." Its author, a "broadly-educated physician and biologist, also known for his scientific works in the fields of anthropology, ethnology, and archeology," Kol'tsov claimed, had "formulated clearly and scientifically the fundamental principles of racial biology." In Kol'tsov's opinion, Florinskii had espoused "quite modern views" on a variety of important subjects, including "the non-inheritance of acquired characteristics," "the double-sided effect of consanguineous marriages (in the sense of the accumulation of both good and bad hereditary characteristics)," "selection as the main cause of human evolution," and even "the mathematical laws of the inheritance and distribution of hereditary traits in breeding." Kol'tsov noted that Florinskii had "devoted about half of his book" to "questions of racial biology" and what he had described as "marriage hygiene" would later be "referred to as 'eugenics' or 'racial hygiene'."

In 1927, the same journal carried an overview of "theoretical and practical eugenics in Soviet Russia" as seen not from its centers in Moscow and Leningrad (as Petrograd was renamed after Lenin's death in 1924), but from the periphery. The article was written by Samuil Vaisenberg, a well-known anthropologist, who resided in the city of Elisavetgrad in the Kherson province (today's Ukraine).[2] Obviously impressed by Volotskoi's publicity campaign, Vaisenberg favorably mentioned Florinskii's treatise in his survey. Asserting that "the scientific treatment of eugenic questions" in Russia has begun only after the Great War, he, nevertheless, thought it necessary to point to the "long-forgotten book, whose author was, in many respects, a predecessor of Galton."[3]

A few months later, upon the appearance of the book's new edition, Vaisenberg reviewed it for the same journal.[4] He summarized each of its four chapters, noting Florinskii's "surprising, intuitive grasp" of its complicated subjects, at a time "when Darwinism has just come into existence and no one has even heard of Mendelism at all." "It is most regrettable," the reviewer lamented, "that the estimable editor used the rather long introduction to political ends, to attack what he perceives as the bourgeois tendency of [modern] eugenics." "The unbiased reader," he observed, would find in the book "nothing Marxist," but rather "a romantic glorification of the common man in contrast to the degenerate nobility, which had been characteristic of Russia at that time." Vaisenberg's review prompted the leading German anthropology journal, *Anthropologischer Anzeiger*, to include Florinskii's book in its annual bibliography under the rubric "*Rassenhygiene, Leibesübungen*." The bibliographic record was supplemented with a short explanatory note: "A new edition of the book [originally] published in 1866 that in many respects is regarded as a precursor to contemporary eugenic ideas."[5]

Of course, coupled with Volotskoi's extensive "advertising," the book made a much bigger splash in its homeland. *Pravda*, the mouthpiece of the Bolshevik party, greeted its publication with a stellar review by Vasilii Slepkov, an active "materialist-biologist."[6] Although Slepkov confirmed Volotskoi's opinion that "from the viewpoint of modern scientific data" much of the book looks "mistaken and naïve," he enthusiastically supported its editor's claim that Florinskii's treatise "has far from an exclusively historical interest." In contrast to many modern eugenicists, though "armed with immeasurably greater amounts of scientific materials," the reviewer stressed, "Florinskii poses the question about the betterment of humankind much more broadly and correctly." In Slepkov's opinion, "the author does not reduce all of eugenics, as do modern eugenicists, to heredity and the study of its regularities," but acknowledges the role of the "external social environment," as both an influential factor in determining "human hereditary qualities" and a "means of perfecting human nature." According to Slepkov, Florinskii also correctly approaches the issue of "eugenic worthiness" of different social groups: he neither "hails the eugenic worth of the aristocracy, the bourgeoisie, [and] the intelligentsia," nor "belittles the heredity of the proletariat and the peasantry." In the "lower" classes, Slepkov maintained, Florinskii "sees those social groups which are destined to prevent physical and moral degeneration and to better human heredity." The reviewer praised Volotskoi's "interesting foreword and commentaries" and recommended the book to a "wide circle of readers interested in eugenics."

A month later, Slepkov also published a brief notice on Florinskii's "valuable book" in *Izvestiia*.[7] He again pointed out that the book was a reprint of the 1866 edition, but it "addresses the same questions that nowadays are studied by a special science, eugenics." Although in the actual content of the book "many things are outdated, many simply wrong and naïve," the reviewer reiterated, "this does not preclude it being extremely valuable, for it provides a series of correct general viewpoints."

Another adulatory review appeared in the country's leading popular-science magazine *Priroda*. It was written by Boris Vishnevskii, the curator of the Academy of Sciences Anthropology and Ethnography Museum in Leningrad.[8] A student of Anuchin who had graduated from Moscow University just two years ahead of Volotskoi, Vishnevskii had actually helped his former schoolmate find some biographical materials on Florinskii in Leningrad's libraries. His review, though no less enthusiastic than Slepkov's, paid much more attention to the actual content of *Human Perfection and Degeneration* and the personality of its author. Vishnevskii noted that Florinskii's book "had attracted no attention from his contemporaries and remained undeservedly forgotten" and that "the honor of its discovery" belongs to Volotskoi, who prepared the new edition. The reviewer briefly recounted the contents of each chapter, asserting that "V. M. Florinskii was far ahead of his age in many respects." "In talking about the perfection of humankind, he envisioned a whole system of measures," Vishnevskii stated, which, alas, "turned out to be ill-timed." The reviewer stressed the originality of Florinskii's book that "simultaneous to, but independently from, Galton's work" had begun "the propaganda of the ideas of the perfection of humankind." But the very title of his review, "A forgotten Russian eugenicist," affirmed the equation of Florinskii's "marriage hygiene" with Galton's eugenics.

By far the most detailed and extensive review, published in the Bolsheviks' major bibliographical outlet, *The Press and the Revolution*, however, came from the pen of Kol'tsov. Two years earlier, when Volotskoi's first report on Florinskii's book had appeared in the *REZh*, its editor-in-chief prefaced the publication with a brief introduction:

> On the world scale, the eugenics movement was, of course, created by the works of F. Galton. On the other hand, there is no doubt that he had predecessors. F. Galton's ideas stemmed, first of all, from Ch. Darwin's evolutionary theory. After the publication of the theory of natural selection, expectedly, attempts to apply Darwin's theory to human evolution were [undertaken] in various countries. In Russia, the first such attempt was made by Prof. V. M. Florinskii... Since the book's publication went completely unnoticed and it remained forgotten, we consider useful to remind [our readers] of its [contents] at the present time.[9]

Now, reviewing the actual text of the book, contrary to Slepkov's and Vishnevskii's assessments, Kol'tsov emphasized further the *historical* value of Florinskii's treatise, while completely denying its contemporary import.[10] "As historical material," he readily admitted, "it is, of course, a very remarkable book." He again stressed the role of Darwin's ideas in Florinskii's treatise, hailing it as "one of the first original books on Darwinism in Russia" that had appeared before Darwin published "his books on the variation of domesticated animals and plants (1868) and on the descent of man (1871)." "The young obstetrician had boldly taken on the problem of the application of evolutionary principles to the betterment of the human breed," Kol'tsov continued, which "only several decades later would become the subject of a special discipline, eugenics." According to Kol'tsov, Florinskii "had placed marriage hygiene, marital selection (*brachnyi podbor*) in the foundation of the betterment of mankind" and, "in this respect, he is not only a follower, but also a predecessor of Darwin, who only published his own doctrine of sexual selection (*polovoi podbor*) [in humans] a few years later."[11] "Anticipating Darwin's idea of sexual selection," the reviewer elaborated, Florinskii, "in full accord with the then current views, had demonstrated how the development of unhealthy tastes and fashions, as well as chasing after wealth and nobility, in the choice of a spouse among the higher classes often leads to their degeneration."

Kol'tsov particularly stressed Florinskii's close attention to the role of heredity in the processes of human perfection and degeneration. It is not surprising, the leader of Soviet eugenics pointed out, that Florinskii's views on heredity are outdated, for at the time of his writing all of the major discoveries in the mechanisms of heredity — from Mendel's laws to August Weismann's germplasm concept to Thomas Hunt Morgan's chromosomal theory — were still far in the future. What is surprising, according to Kol'tsov, is "how correctly he had evaluated the relative importance of various hereditary factors, and in most cases, had ascribed the leading role to the selection (*podbor*) of hereditary potentials and not to the inheritance of acquired characteristics."

As did Volotskoi in his editorial commentaries, Kol'tsov effectively "translated" Florinskii's ideas on human heredity, variability, development, and evolution into the language of modern genetics. He praised Florinskii's clear distinction between "hereditary, genotypical human nature and a phenotypical manifestation of that nature dependent on external conditions." "The very notions of genotype and phenotype were introduced [by Danish geneticist Wilhelm Johannsen] only twenty years ago," the reviewer marveled, and "many biologists (including the editor of the book under review) still do not quite understand them." Kol'tsov forcefully refuted Volotskoi's juxtaposition of Florinskii's treatise to modern eugenics: "Instead of emphasizing that the author had by forty years presaged the newest advances of genetics, the editor of the second edition, for some reason, decided to enter into inappropriate and unfounded polemics with modern eugenicists, who [actually] stand much closer to V. M. Florinskii's views than to those of M. V. Volotskoi." "The historical interest of Florinskii's book is not diminished by the fact that its various parts — the doctrine of [human] races, the issue of sex determination, the influence of parental age on the progeny, inbreeding, and so on — are outdated and full of mistakes," Kol'tsov reiterated. Yet, he warned, "numerous commentaries by the book's editor were often incomplete and not always corresponding to the state of modern science" and thus could easily mislead "an unprepared reader." Therefore, he insisted, the book's readership should be limited to "the specialists interested in the history of biological ideas." This suggestion contrasted markedly with Kol'tsov's assessment of several other books on eugenics, which appeared the same year and which he reviewed for the same journal. For instance, in his very favorable review of the translation of R. Ruggles Gates's *Heredity and Eugenics*, prepared by Filipchenko, the RES president recommended the book to "wide circles of Russian readers."[12]

The new edition of *Human Perfection and Degeneration* certainly made the ideas of its author well known in Soviet Russia. Three thousand copies was a huge print run for a long-forgotten sixty-year-old scientific treatise. Reaffirming Volotskoi's suggestion, which was repeated in Slepkov's and Vishnevskii's reviews, a 1927 overview of eugenic literature for school teachers reiterated that "the book could be recommended to anyone interested in the subject."[13] As Volotskoi had hoped, Florinskii's book did "find its place in the libraries of sociologists, hygienists, [and] anthropologists."[14] Copies of the new edition appeared on the library shelves of universities and other schools of higher learning in Russia.

It is impossible to estimate how many people actually read the new edition of Florinskii's book. But "the specialists interested in the history of biological ideas," as Kol'tsov had defined the book's readership, certainly did. Every general overview of eugenics published in Russia after its rediscovery took note of Florinskii's treatise. Following Volotskoi's characterization of its author "as one of the founders of our discipline," they repeated its editor's statement that, even though the book had been unnoticed and forgotten, "in many of his ideas, Florinskii had been a predecessor of Galton's."[15] Florinskii's name thus became firmly incorporated into the history of "Russian eugenics." To give but one example, in a lengthy entry on eugenics published in the *Great Medical Encyclopedia* in 1929, Iudin included the new edition of *Human Perfection and Degeneration* in the list of recommended literature alongside Galton's works.[16]

Slepkov's statement that Florinskii's treatise "has far from an exclusively historical interest" was more than rhetorical praise. Indeed, there are some indications that Volotskoi's active popularization of Florinskii's "marriage hygiene" influenced not only the discussions of the history of eugenics in Russia, but also the contemporary debates surrounding the adoption of new marriage laws and the establishment of "marriage consultations."

The Bolshevik Revolution had separated church and state and placed all matters of marriage and family under the authority of the latter.[17] As early as December 1917, the Bolshevik government introduced the institution of "civic marriage."[18] The newborn state established throughout the country special offices that took over the functions of parish churches in registering births, marriages, divorces, and deaths of its citizens. The old laws and church rules pertaining to marriage and family were abolished and a set of new civic regulations introduced.[19]

In the early 1920s, social hygienists closely associated with the Russian Eugenics Society began to lobby for the adoption of "eugenic legislation."[20] The head of Narkomzdrav's Sanitary-Epidemiological Department, Aleksei Sysin, became the spokesman of this extensive campaign. In September 1923, he published in *Izvestiia* a lengthy article titled "Marriage, Health, and Progeny."[21] Tellingly, to support his proposals, Sysin referred to Volotskoi's pamphlet *Elevating the Vital Forces of the Race*, which, as we saw in the previous chapter, had first introduced Florinskii's ideas of "marriage hygiene" to Soviet readers. A month later, Sysin delivered a lengthy report to a special meeting of the RES on "The Eugenic Evaluation of a Law Proposed by Narkomzdrav on the Protection of Health of the Persons Entering Marriage."[22] The same subject was discussed again in early January 1924 at a joint meeting of the State Institute of Social Hygiene, the Institute for the Protection of Maternity and Infancy (OMM), and the RES.[23] In the summer, Sysin published an extensive summary of his reports in the oracle of social hygienists, edited personally by Commissar Semashko.[24]

Sysin provided a detailed overview of historical and current legislative initiatives and actual laws promulgated in various countries to further the "principles of eugenics." Echoing the opening statement of Florinskii's treatise, Sysin asserted that "no matter how one considers the attempts at bettering the quality of future generations by means of eugenic laws, one thing remains indisputable, namely that the state can no longer go on on being indifferent to this issue." According to Sysin, the first Soviet Civic Code adopted in 1918 "did not reflect in the least degree any eugenic issues" and "did not pursue any eugenic goals." Since the adoption of the NEP necessitated a revision of the Civic Code that the People's Commissariat of Justice was currently conducting, he explained, Narkomzdrav officials took the initiative of introducing in the revised code certain "socio-eugenic" articles aimed at "the protection of the health of prospective spouses and their progeny."

None of the proposed articles even mentioned the coerced sterilization or the segregation of individuals with "bad" heredity. The planned legislation introduced largely voluntary measures, Sysin emphasized, using marriage registration as a suitable moment for "sanitary-eugenic propaganda" that would force prospective spouses, as Florinskii had suggested in his book, "to look at their future marriage from the viewpoint of health." According to the recommended new law, prior to registering their marriage, a prospective couple must inform each other of their personal "medical histories," particularly "in relation to venereal, mental, and tubercular diseases," which at the time were seen as "hereditary." The law required bride and groom to present to the marriage registration agency a written deposition to that effect. If they refused to give such a deposition, the agency would not register their marriage. And if they lied in their deposition, they would face criminal charges.

The lobbying campaign was supported by detailed analyses of the "eugenic consequences" of various pieces of legislation in Russia and abroad published by several jurists with a longstanding interest in eugenics, notably by Pavel Liublinskii, who in 1924 joined Kol'tsov and Filipchenko as a co-editor of the *REZh*.[25] The campaign for "eugenics legislation" proved successful.[26] The new Civic Code promulgated in 1926 put all of the initiatives outlined by Sysin into law.[27] Furthermore, the new code also incorporated certain restrictive measures, such as the prohibition on marriage before eighteen years of age and on marriages between close relatives and between mentally ill persons, which read as if they had been lifted almost wholesale from Florinskii's treatise. Indeed, in its discussion of the new legislation, a popular article published in *Women's Magazine* at the beginning of 1927 referred directly to Florinskii's views on the detrimental effects of "kin marriages."[28]

Moreover, the same year, as if answering Florinskii's suggestion that "in doubtful cases" the prospective spouses should seek a doctor's opinion that "without limiting personal freedom could do more good than any law," the State Institute of Social Hygiene, the cradle of Soviet eugenics, established a "marriage consultation" to advise prospective brides and grooms on "certain eugenic aspects" of marriage (see fig. 6-1).²⁹ The next year, the Moscow Society of Neurologists and Psychiatrists created a special "genetics bureau for the study of hereditary diseases."³⁰ Run by Sergei Davidenkov (1880-1961), a well-known neurologist and RES member, the bureau became a springboard for a "genetic consultation" that Davidenkov soon founded at his neurological clinics.³¹ "Eugenic" advice to prospective spouses became a staple of Soviet eugenicists' propaganda as witnessed by Iudin's 1928 brochure on *Health, Marriage, and Family*, or a 1930 article on "Marriage and Eugenics" by Kol'tsov's former student and one of the architects of population and mathematical genetics in the Soviet Union, Petr Rokitskii, both of which included many passages that sounded exactly like Florinskii's and/or Volotskoi's.³² Rokitskii, for instance, reaffirmed Volotskoi's distinction between "creative" and "preventive" eugenics and strongly endorsed Florinskii's notion of marriage as a matter of social concern and, hence, of state regulation.³³

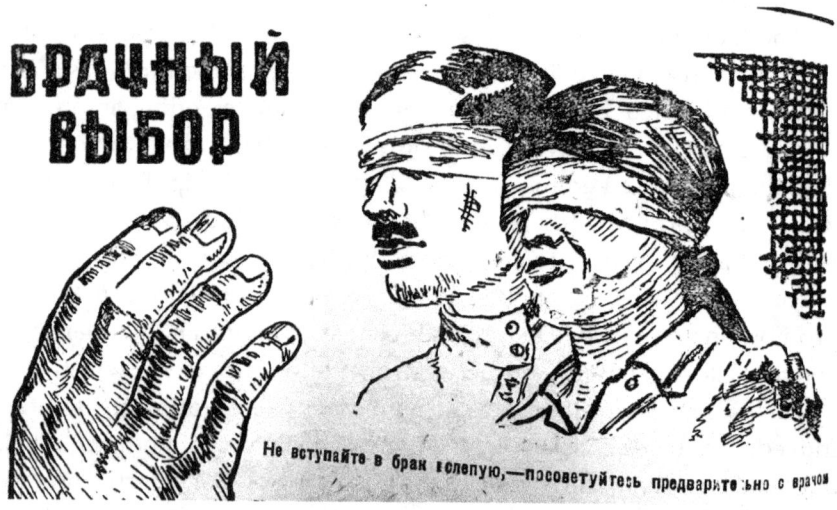

Fig. 6-1. An illustration that accompanied a 1930 article, titled "Marital Choice," in the popular weekly *Hygiene and Health of Worker's and Peasant's Family*. The article opened with a simple statement: "Eugenics is a science about the protection of health of the future generations; its goal is the betterment of the human breed [and] the facilitation of the making of a healthy and strong replacement [of the current generation]." The caption under the picture says: "Do not enter marriage blindfolded — consult a physician beforehand." From *Gigiena i zdorov'e rabochei i krest'ianskoi sem'i*, 1930, 17-18: 12. Courtesy of BAN.

Despite the considerable attention it commanded and contrary to Volotskoi's hopes and despite his extensive efforts, Florinskii's treatise did not become the foundation of "bio-social" or "proletarian" eugenics.

## The Dialectics of Nature and Nurture

There could be little doubt that Florinskii's ideas played a key role in the sharpening of Volotskoi's personal critical attitude towards "bourgeois" eugenics and in the development of his own "bio-social" version. In turn, Volotskoi's active popularization of Florinskii's "marriage hygiene" and his relentless critique of Galtonian eugenics provided an inspiration and a template for many other "Marxist" critics and, thus, contributed considerably to the rising suspicion of and mounting opposition to eugenics. The path of eugenics in Soviet Russia was not all smooth — there were also dangerous potholes. The position of the country's main "State Publishing House" (*Gosudarstvennoe Izdatel'stvo*, GIZ) in regard to issuing books on the subject, most notably, Galton's *Essays on Eugenics*, provides a telling illustration.

In 1922, as part of extensive efforts to popularize both eugenics and genetics, their Soviet proponents enthusiastically celebrated the dual centennial of Galton's and Mendel's births. The RES, for instance, devoted several meetings to reports on various facets of Galton's life and works. Iudin spoke on "Francis Galton, His Life and Scientific Activities," Kol'tsov on "The Genealogy of Galton and Darwin," the eminent psychologist Georgii Chelpanov on "Galton's Role in Modern Scientific Psychology," and Bunak on "Galton as the Founder of the Biometric School."[34] To mark the anniversaries, eugenicists also planned a series of publications, including biographies of Galton and Mendel and Russian translations of their major works. Filipchenko took it upon himself to write a double biography of the "founders of modern genetics," which in 1924 came out under the GIZ trademark in two different printings.[35] For his part, Kol'tsov undertook the task of editing the translations of Mendel's and Galton's key works to be issued by the same publisher in its popular series "The Classics of Natural Sciences."[36] In the summer of 1922, the GIZ board approved contracts for the publication of Galton's *Essays on Eugenics* and Mendel's *Experiments on Plant Hybrids*.[37] In the fall, the translation of Mendel's works was duly submitted to the publisher. It appeared in print about a year later.[38] Galton's, however, did not.

As mentioned previously, at Kol'tsov's suggestion, Volotskoi had begun translating Galton's *Essays* after joining the IEB eugenics department in late 1921. Over the course of the next year or so, he had translated several chapters, but then asked Kol'tsov to be released from the task. Available materials do not reveal his reasons for refusing to finish the job, but the contents of his pamphlet on *Class Interests and Modern Eugenics* written around that very time strongly suggest that Volotskoi came to see Galton's "gospel" as useless, if not outright detrimental, to the development of his own "bio-social" eugenics.[39] Reportedly, Kol'tsov assigned one of his students to finish the translation. It was completed and submitted to the GIZ office in the late spring of 1924. But, although regularly advertised by the publisher as "forthcoming," it never came out (see fig. 6-2).[40]

Surviving documents are completely mute on the reasons for this odd situation.[41] GIZ materials pertaining to the publisher's handling of several other books on eugenics during the same time period, nevertheless, offer certain clues as to why the translation of Galton's *Essays* was not published. In April 1923, the GIZ Petrograd office accepted for publication Filipchenko's manuscript, *Paths to the Betterment of the Human Type (Eugenics)*.[42] It was the first extensive (nearly 200-page long) Russian-language overview of the history, current state, and future prospects of eugenics and human genetics, based on the latest available English, German, and Russian literature. A year later, Filipchenko's volume came out, but not exactly in the form he had written it.[43]

The GIZ board supplied Filipchenko's text with numerous "editorial commentaries" and an unsigned "editorial foreword" that, though praising the book as "valuable and interesting," emphasized its "essential shortcomings."[44] The foreword highlighted the book's "narrowly-biological point of view" and its neglect of a "sociological-Marxist" approach to eugenic ideas, research, and policies. The anonymous author criticized Filipchenko's conviction that the intelligentsia as a social group represented the main reservoir of hereditary talent and recommended that he study the genealogies of "artists, scientists, and sculptors, who had come from the peasants and the proletarians," such as the eminent writer Maxim Gorky and the famous opera singer Fedor Chaliapin.

ГОСУДАРСТВЕННОЕ ИЗДАТЕЛЬСТВО
МОСКВА — ПЕТРОГРАД

СЕРИЯ — „КЛАССИКИ ЕСТЕСТВОЗНАНИЯ"

Под общей редакцией А. Д. Архангельского, Н. К. Кольцова, В. А. Костицына, П. П. Лазарева и Л. А. Тарасевича. При ближайшем участии в редакционной работе В. М. Арнольди, В. Ф. Кагана, Т. К. Молодого, В. В. Шарвина и Э. В. Шпольского.

ВЫШЛИ ИЗ ПЕЧАТИ:

1. И. И. Мечников. — Лекции о сравнительной патологии воспалений. Под. ред. проф. *Л. А. Тарасевича* (2 е изд.).
4. П. Н. Лебедев. — Световое давление. Под ред. академика *П. П. Лазарева* и проф. *Т. П. Кравца*. Ц. 40 к.
5. Г. Гельмгольц. — О сохранении силы. Перев. и ред. академика *П. П. Лазарева*. Ц. 25 к.
6. Г. Гельмгольц. — Скорость распространения нервного возбуждения. Под. ред. акад. *П. П. Лазарева*. Ц. 65 к.
7. Сади Карно. — Размышления о движущей силе огня. Под ред. проф. *В. Р. Бурсиана* и проф. *Ю. А. Крутнова*. Ц. 50 к.
8. М. В. Ломоносов. — Физико-химические работы. Под ред. проф. *Б. Н. Менщуткина*. Ц. 70 к.
9. Космогонические гипотезы. — Кант, Лаплас, Фай, Дж. Дарвин, Пуанкаре. Под ред. проф. *В. А. Костицына*. Ц. 80 к.

ПЕЧАТАЮТСЯ:

Русские классики морфологии растений. Сборник статей. Под ред. проф. *В. М. Арнольди*.
И. П. Павлов — О работе пищеварительных желез (2-е изд.).
К. Э. Бэр. — Избранные работы под. ред. проф. *Ю. А. Филипченко*.
Г. Мендель. — Гибриды у растений. Под ред. проф. *Н. К. Кольцова*.

ГОТОВЯТСЯ К ПЕЧАТИ:

Р. Клаузиус и В. Томсон. — О втором законе термодинамики. Под ред. проф. *В. Р. Бурсиана* и проф. *А. И. Крутнова*.
Эндрус. — Непрерывность газообразного и жидкого состояний. Перев. проф. *Е. Е. Сиротина*.
Гильберт. — О магните и магнитных телах. Перев. под ред. проф. *А. И. Бачинского*.
Ньютон. — Оптика. Перев. проф. *С. И. Вавилова*.
Галилей. — Разговор о двух главных мировых системах — Птоломеевой и Коперниковой. Под ред. проф. *А. И. Бачинского*.
Г. Гельмгольц. — Принцип наименьшего действия. Перев. и редакция, акад. *П. П. Лазарева*.
Б. Риман. — О гипотезах, лежащих в основании геометрии. Под ред. проф. *Д. М. Синцова*.
Ф. Гальтон. — Лекции по евгенике. Под ред. проф. *Т. И. Юдина*.
Н. И. Пирогов. — Работы в области хирургии. Под редакцией проф. *Н. Н. Бурденко*.
Зюсс. — Лик земли. Заключительные главы. Под редакцией проф. *А. Д. Архангельского*.

Fig. 6-2. An advertisement for the GIZ series "Classics of Natural Science." The first section lists nine books that have been published. The second lists four more volumes currently "in production." The last book in this section is Gregor Mendel's *Experiments on Plant Hybrids*. The third section lists ten books "in preparation." The eighth in this list is Francis Galton's *Essays on Eugenics*. This advertisement appeared on the back cover of Elie Metchnikoff's *Lectures on the Comparative Pathology of Inflammation*, published as part of the series in 1923. From I. I. Mechnikov, *Lektsii o sravnitel'noi patologii vospaleniia* (M.: GIZ, 1923), backmatter. Courtesy of RNB.

Just as Filipchenko's book came out, the GIZ Moscow office was preparing another, even more voluminous account of eugenics to be published in its popular series "Modern Issues in Natural Sciences." The publisher had contracted the book, preliminarily titled *Introduction to the Study of Eugenics*, to Iudin a year earlier, and, along with the translation of Galton's, Iudin's volume was supposed to go into production in early 1924.[45] As with Filipchenko's book, the GIZ board planned to preface the volume with an editorial foreword. The board asked Zinovii Solov'ev (1876-1928), a deputy-head of Narkomzdrav, long-time member of the Bolshevik party, and Semashko's right-hand man in building the Bolshevik health care system, to write it.[46]

Admitting that he was not a biologist and even less a "specialist in modern eugenics," Solov'ev nonetheless felt justified in saying "a few words on 'breeding human stocks'," as he titled his lengthy essay.[47] In contrast to the anonymous author of the foreword to Filipchenko's book, Solov'ev focused not so much on the actual contents of Iudin's volume as on Kol'tsov's 1921 presidential address and on a few articles that had appeared in the first issues of the *Russian Eugenics Journal*. But the general thrust of his foreword was very much the same: eugenicists must learn "Marxist materialist sociology." Peppered with long quotes from Marx and Lenin, the foreword accused eugenicists of "equating biological laws with the laws of sociology and transferring [certain] regularities from the field of biology into the field of social relations." According to Solov'ev, Marx's critique of both "the Malthusian law of overpopulation" and the "sociological applications" of Darwin's notion of "struggle for life" had a long time ago demonstrated the futility of biological explanations of social phenomena.

Citing various studies, which had allegedly confirmed the inheritance of acquired characteristics, Solov'ev bitterly criticized eugenicists' "belief in a negligible role of the external environment and in an overwhelming role of heredity." He outright rejected eugenicists' conviction that "human life is predetermined by hereditary features" and was particularly incensed by their claims that "all people are born unalike and unequal." In his view, since eugenicists ignore "environment, economics, [and] social conditions," all their practical proposals "to breed humans" the same way as chickens or pigs were no more than "fantasies." But, Solov'ev emphasized, such fantasies were "deeply reactionary," for they provided "scientific justification" for men-hatred, chauvinism, elitism, and racism. He offered as an example of such openly reactionary attitudes — characteristic of "corrupt, bourgeois science" — a 1923 edition of a textbook on *Applied Eugenics* written by well-known US eugenicists Paul Popenoe and Roswell H. Johnson that had recently been reviewed in the *Russian Eugenics Journal*. Solov'ev concluded his foreword with a forceful statement: "The working class's philosophy demands [both] clarity in the issues of eugenics and essential reconsideration of eugenic theories and practices on the basis of consistent materialism, Marxism."

It is unclear from available materials, whether Solov'ev's foreword prompted the GIZ board to reject Iudin's manuscript, or upon reading the foreword, Iudin himself decided to withdraw it. Eventually, in late 1925, Iudin did publish the volume (without Solov'ev's foreword, of course,) with a different, private press.[48] Galton's *Essays*, however, never saw the light of day.[49]

Although in the case of the translation of Galton's *Essays on Eugenics* I did not find a "smoking gun," it seems unlikely that Volotskoi somehow directly affected the GIZ editorial decisions. But, indirectly, he perhaps did influence the attitude to Galtonian eugenics on the part of the GIZ board members. The "editorial forewords" to both Filipchenko's and Iudin's books sounded remarkably similar to the criticism of Galtonian eugenics advanced by Volotskoi's in his reports and publications of 1923-1924.[50] There is little doubt that Volotskoi's critique — inspired as it was by Florinskii's ideas — set an example and provided a template emulated by many others.[51] The republication of Florinskii's book made this example all the more accessible and appealing. Volotskoi's foreword, together with Slepkov's and Vishnevskii's reviews published in periodicals with country-wide circulation, made this template easily available to any interested individual, and especially to "materialist-biologists."

This situation to a certain extent explains both the contents and publication venue of Kol'tsov's review of Florinskii's treatise. The doyen of Soviet eugenics certainly understood that Volotskoi's commentaries, coupled with Slepkov's and Vishnevskii's accolades, could provide critics of eugenics and its founder Galton with powerful ammunition. His review obviously aimed at disarming them by relegating *Human*

*Perfection and Degeneration* to history and denying its relevance to contemporary Soviet eugenics. Although focused almost exclusively on "the historical value" of Florinskii's book, Kol'tsov's review did not mention Galton at all, instead emphasizing the role of Darwin's ideas in Florinskii's concept of "marital selection." But Kol'tsov did not simply assess the contents of Florinskii's book, he specifically refuted its editor's attempts at presenting it as a model for "bio-social" eugenics. Furthermore, the RES president published his review not in his own *Russian Eugenics Journal*, but in the Bolshevik major bibliographical periodical, *The Press and the Revolution*, addressed to and read by "Marxists" in all fields of arts and sciences.[52]

An article published in 1927 under the revealing title "Eugenics, Ours and Theirs" by no less a figure than Commissar Semashko demonstrates that the RES president had every reason to be apprehensive.[53] As its title makes clear, Semashko's article juxtaposed "theirs," "bourgeois," "biological" eugenics to "ours," "proletarian," "social" eugenics. In his view, "eugenics comprises a part of social hygiene that takes under its purview the issues of the influence of harmful or beneficial factors on progeny." But, the commissar emphasized, the major mistake of western eugenicists, such as Galton and Siemens, as well as their Russian supporters like Kol'tsov, is that "they understand these eugenic factors very narrowly, as biological and hereditary factors," and "want to reduce all of eugenics to zootechnique." However humans are not chickens or cows, Semashko asserted, echoing his deputy Solov'ev's assessment, for "in human life, social factors play a much more important role than the factors of pure heredity." "The eugenicists forget a small difference between humans and other animals formulated by Marx," he continued sarcastically, namely that "humans are social animals." Furthermore, the commissar declared, although the discussion is far from over, "chances are that acquired characteristics are inherited," and hence the role of social environment could be even greater. "Their eugenics is to encourage the propagation of talents among the nobility," he exclaimed, and "to bar the proletariat from entering universities." "Our eugenics," according to its official patron, is just the opposite: "to provide education, and especially higher education, to the people."

As Semashko's article makes evident, in the mid-1920s criticism of eugenics was elaborated along the three main lines clearly identified by Volotskoi: egalitarianism, Marxism, and Lamarckism. Although Soviet eugenicists tried very hard to distance their own research and propaganda from the race and class biases underpinning much of western eugenic research and debates, they still shared with their western counterparts a strong *elitist* bias vividly manifested in their numerous studies on "The Intelligentsia and Talents," as Filipchenko titled one of his programmatic articles.[54] The "editorial foreword" to Filipchenko's 1924 book shows that critics saw this elitist bias of Soviet eugenics as utterly incompatible with the proclaimed egalitarianism of the Bolshevik state. The very title (not to mention the contents) of an article published in 1925 by the Communist Academy's mouthpiece *Under the Banner of Marxism*, "Not From the Upper Ten Thousands, But From the Lower Millions," demonstrates that this attitude to eugenics became a staple of "Marxist" critics.[55] As did Solov'ev in his foreword to Iudin's volume, they branded eugenics a "bourgeois science" and charged its Soviet proponents with advancing "capitalist," "reactionary" ideas about human heredity, variability, development, and evolution, which meant to divert "the revolutionary energy of the proletariat" from its first and foremost task — the struggle against capitalism.[56] In Florinskii's emphasis on the "hereditary potentials" of the "lower" classes, they found substantial support for this line of criticism.

Following Volotskoi's lead, many critics accused Galtonian eugenics of focusing exclusively on heredity (nature) and ignoring environment (nurture), above all, social and economic conditions, which, according to Marx, play a defining role in the formation of the "social human being." They criticized eugenicists for overemphasizing biology to the detriment of sociology, reiterating Marx's dictum that human beings are "the result of social conditions and upbringing."[57] The writings of Grigorii Batkis (1895-1960), a young Bolshevik lecturer at the Department of Social Hygiene established by Semashko at Moscow University, offer an illustrative example.[58] Batkis insisted that there was a conflict between social hygiene aimed at eliminating the social conditions conducive to the spread of disease and degeneration, on the one hand, and "eugenics understood in a narrow sense, i.e. as the improvement of humankind by biological methods," on the other. He disparaged "bourgeois" eugenics for overemphasizing "biological methods, i.e. the selection of procreators" to the detriment of "the protection of the progeny from hereditary venereal diseases [and from] weakening by alcoholism and TB," which he saw as the main task of social hygiene. Batkis concluded, however, by endorsing a "broad understanding" of "socialist eugenics," which, in his opinion, was nothing else but "the very same social hygiene."[59] In this respect, too, with its extensive discussion of various external conditions conducive to either perfection or degeneration, Florinskii's treatise provided critics of eugenics with weighty arguments.

Furthermore, the emphasis on "environment" led many Marxist critics of eugenics (and genetics) to supporting the Lamarckian notion of the inheritance of acquired characteristics in their explanations of human heredity, variability, development, and evolution.[60] The first decades of the twentieth century witnessed a resurgence in the popularity of the Lamarckian explanation of evolution spurred by the rapid development of experimental biology. Various experiments conducted by embryologists, endocrinologists, biochemists, and physiologists all over the world seemed to confirm the validity of its central idea that certain features acquired by an organism during its lifetime could become inherited, thus undermining the main postulates of both Mendelian genetics and the corresponding "mutation theory" (*mutationstheorie*) of evolution advanced by Hugo de Vries, one of the co-discoverers of Mendel's forgotten works.

Many Russian biologists came to support what was now called neo-Lamarckism.[61] They regularly reviewed and abstracted works by western proponents of Lamarckism, particularly the Austrian biologist Paul Kammerer who gained wide notoriety for his experiments on the inheritance of acquired coloration patterns in salamanders.[62] Indeed, in 1925, the GIZ issued Kammerer's voluminous treatise on "general biology" that spelled out his Lamarckian views with a sympathetic (though anonymous) foreword.[63] Lamarckism found a particularly large number of adherents among "materialist-biologists." Indeed in early 1926, the Communist Academy leadership even planned to establish a special laboratory to prove the existence of the inheritance of acquired characteristics and invited Kammerer to head it. Kammerer accepted the invitation, but on the eve of his departure for Moscow, faced with accusations of scientific fraud, he committed suicide, and the plan fell through.[64] Nevertheless, in 1927, two publishers, one in Moscow, another in Leningrad, released almost simultaneously the Russian translations of Kammerer's book, *The Enigma of Heredity*, which reiterated his views on the heritability of acquired characteristics and contained an extensive discussion of eugenics in light of Lamarckian inheritance.⁶⁵

Widely popularized by "materialist-biologists," Lamarckism also attracted under its banners a number of physicians. Many doctors interested in what at that time was construed as "hereditary diseases," from TB and hemophilia to syphilis and schizophrenia, sought to reconcile the old ideas of "inborn constitution" with the recently introduced notions of genetics.⁶⁶ But at the same time, they wanted to defend the principles of prophylactic medicine — a cornerstone of the Bolshevik health protection system — that seemed to be ineffectual in the fight against diseases "rooted in heredity." Lamarckism appeared to offer a way out of this conundrum. The numerous publications of Solomon Levit (1894–1938),⁶⁷ a Bolshevik party member and the founder of the "Circle of Materialist-Physicians" at Moscow University's medical school, who favorably reviewed several publications by Volotskoi, provide an illuminating example.⁶⁸ In his article, expressively titled "The Problem of Constitution in Medicine and Dialectical Materialism," Levit stated categorically that "the reconstruction of Soviet medicine on a prophylactic basis" would be theoretically unthinkable without "recognition of the inheritance of acquired characteristics."⁶⁹ As Kol'tsov's review of Florinskii's treatise indicates, the notions of genotype and phenotype in their application to human heredity and hereditary diseases became a major point of contention and a subject of heated debate. Some thought that "constitution" corresponded exclusively to genotype, others to phenotype, and for many on both sides of the debate, Lamarckian inheritance offered a convenient bridge between the two.⁷⁰ Some physicians even proposed a "synthesis" of genetics and Lamarckism.⁷¹

In a similar way, Lamarckism appealed to a number of Marxist historians and sociologists who sought to unify the notions of human biological (Darwinian) and social (Marxian) evolution. As one of them frankly admitted, "as a person interested in the synthesis of the concepts of sociological and biological progress, I cannot imagine how such synthesis could be accomplished without the Lamarckian idea."⁷² As a result, for many "materialist-biologists," "materialist-historians," and "materialist-physicians," Lamarckism became an integral part of Marxism.⁷³ They insisted that "a Marxist is necessarily obliged to be a Lamarckist,"⁷⁴ and decried as "anti-materialist" and "anti-Marxist" any attempt to criticize the inheritance of acquired characteristics.

Fig. 6-3. The author of "bio-social" eugenics Mikhail Volotskoi. This photo was taken at Volotskoi's apartment in either 1926 or 1927, at the very time that he was actively involved in debates over "socialist" eugenics and was elaborating his own "bio-social" eugenics. Photographer unknown. Courtesy of N. Bogdanov.

Volotskoi's lack of basic training in genetics, as well as his association with the Timiriazev Institute that became a major bastion of Lamarckism, certainly influenced the anthropologist's views on the subject (see fig. 6-3).[75] Upon his transfer from IEB to the Timiriazev Institute, Volotskoi joined the "Lamarck Circle," an informal group organized by several members of the institute for the studies and propaganda of the inheritance of acquired characteristics.[76] He regularly attended its meetings and even delivered a report on "Modern eugenics and its relation to Lamarckism."[77] As we saw, in all of his critical assessments of western eugenics and its Soviet followers, especially their leaders Kol'tsov and Filipchenko, he repeatedly emphasized that, in contrast to Galton and his disciples, Florinskii had acknowledged the role of the inheritance of acquired characteristics in human perfection and degeneration. There is little doubt that Volotskoi's publication and active popularization of Florinskii's treatise added fresh fuel to the heated polemics over Lamarckism in the Soviet Union.

Soviet eugenicists spent considerable efforts answering these "Marxist" criticisms. Kol'tsov, Filipchenko, and their numerous students waged a concerted campaign against Lamarckism.[78] Kol'tsov published several articles on the subject in professional and popular periodicals.[79] Filipchenko, meanwhile, arranged for the publication of a special brochure under the revealing title *Are Acquired Characteristics Inherited?*. The brochure included two pieces: one was the translation of an article by the most influential US geneticist Thomas Hunt Morgan, and the other was written by Filipchenko himself.[80] Both critically examined some recent experiments conducted by various scientists to prove the inheritance of acquired characteristics. Needless to say, both answered the question posed in the brochure's title in the negative. But it is worth noting that the only Russian scientist whom Filipchenko mentioned by name as a supporter of Lamarckism was Volotskoi.

Indeed, Filipchenko wrote a special section on Volotskoi's "biosocial" eugenics and severely criticized the anthropologist's articles that had appeared in *Physical Culture in Light of Science* in the preceding year. As we saw in the previous chapter, in these articles Volotskoi had emphasized the role of the environment, and particularly physical culture, in the betterment and degeneration of humankind and supported his views with extensive excerpts from Florinskii's treatise. Filipchenko did not mention Florinskii's name in his article, but its very language shows that he was well aware of where Volotskoi had found his inspiration and took his fellow eugenicist to task. "If the inheritance of acquired characteristics exists," Filipchenko argued:

> then, obviously, the representatives of this [lower] class would carry the footprints of all those unfavorable influences that during long periods of time had affected their fathers, grandfathers, and a number of more remote ancestors. And because of this, our longsuffering proletarians and peasants must possess much fewer favorable hereditary potentials (*nasledstvennye zachatki*), the genes of the most valuable special abilities, than do [the representatives of] other classes that had lived for so long under very favorable conditions.[81]

If Kol'tsov and Filipchenko countered the propaganda of Lamarckism in print, Serebrovskii confronted its supporters from within their own stronghold — the Communist Academy. Much like Volotskoi, in the post-revolutionary years Serebrovskii had developed a serious interest in Marxism and eventually even joined the Bolshevik party. But, unlike Volotskoi, he was a geneticist by both training and occupation (see fig. 6-4). He put his knowledge of both genetics and Marxism to the vigorous defense of his specialty from the accusations of being "anti-Marxist" and "anti-materialist." He actually succeeded in "converting" several ardent Lamarckists, for instance, the militant "materialist-physician" Levit, into students and supporters of genetics.

In late 1925 Serebrovskii joined the Society of Materialist-Biologists established by the Communist Academy.[82] Just a few weeks later, on 12 January 1926, he delivered a long report on "Morgan's and Mendel's theory of heredity and Marxists" to the "section of natural sciences" recently organized within the Communist Academy.[83] He argued forcefully that it was not Lamarckism, but modern genetics that represented a "truly materialist," "dialectical," "revolutionary," and, hence, "Marxist" view of heredity.

Fig. 6-4. During the Fifth International Genetics Congress in Berlin, September 1927. Left to right: US geneticist Leslie C. Dunn, German poultry breeder Bruno Duringen, and Alexander Serebrovskii at Duringen's Poultry Museum in Berlin. Photographer unknown. Courtesy of ARAN.

Like his teacher Kol'tsov, Serebrovskii critically analyzed recent publications by the proponents of the inheritance of acquired characteristics. And like Filipchenko, he switched almost seamlessly from defending genetics to a lengthy discussion of eugenics. Serebrovskii readily admitted that some western eugenicists, including Galton, did approach the issues of the betterment of humankind from "the viewpoint of their own class." But, he insisted, this is not "the reason to accuse Mendel-Morgan's theory of complicity [in this position]." Echoing Volotskoi's report on "bio-social" eugenics delivered just a few weeks earlier, he stressed that, unlike genetics, "eugenics is not an exact science, but a normative doctrine that uses the modern understanding of heredity as a basis for its own conclusions regarding how one could *better* human society." But the moment we begin to discuss "good" and "bad," norms and values, we leave the ground of exact science and enter a "wide field of class interests," Serebrovskii continued, and "naturally, since it *evaluates* something, [it] cannot be uniform: every class must create its own eugenics." Just because bourgeois eugenicists use modern genetics to substantiate their views, he asserted, it does not mean that "in creating its own eugenics, the proletariat should use Lamarckism." This approach "mixes together two parts of the phenomenon": a biological doctrine (genetics) with a normative doctrine (eugenics), and, as Serebrovskii put it, "with the dirty bourgeois-class bathwater [of eugenics], it throws out the biological baby [of genetics]."

Serebrovskii rejected all attempts at evaluating the "hereditary worth" of any particular social group that so irked the Marxist critics of eugenics. But, he asserted carefully, since we know that the distribution of such human traits as height, skin color, temperament, etc. in a population is not uniform, from a theoretical point of view, it is possible that "elements of talents, elements of giftedness are not spread across humanity in a uniform way." Therefore, he argued, we should consider the country's entire population as "our societal capital, in exactly the same way we consider such assets as the entirety of wheat, cattle, [and] horses, which constitute the economic might of our country," and, hence, the biological, hereditary quality of the population must be our utmost concern.

Serebrovskii suggested that "the entirety of [all] genes, which in a human society create talented, prominent individuals, or, on the contrary, idiots, must be seen as national capital, a *genofond* (gene fund), from which the country draws its people." This *genofond* is constantly changing under the influence of such processes as mutations, differential fecundity, and differential mortality, he continued, and these changes could result in the accumulation of either "bad" or "good" genes in the population, and, hence, could lead to either its hereditary degeneration or its hereditary improvement. We do not yet know the exact causes of mutations, Serebrovskii elaborated, but the mutation process occurs independently of human actions, whilst the processes of differential fecundity and differential mortality could be directly influenced by "government policies." "Nearly every action of every government agency, in one way or another, affects the *genofond*," he concluded.

In his report, Serebrovskii did not mention by name either Volotskoi or Florinskii, but the essence of his arguments indicates that he was familiar with and perhaps even inspired by the works of both. As an active RES member, Serebrovskii had undoubtedly heard Volotskoi's report on Florinskii's treatise and then read it on the pages of the *Russian Eugenics Journal*. As an avowed Marxist himself, he must have paid close attention to the critique of Galtonian eugenics advanced by his fellow Marxist: he might well have attended the November 1925 meeting at the Timiriazev Institute where Volotskoi presented his "system of eugenics as a bio-social discipline." Indeed, Serebrovskii's notion of *genofond* appears remarkably akin to Florinskii's sentiment, regarding "many excellent, talented individuals [who] remain hidden in the mass of the people as wasted, unproductive capital that has neither purpose, nor use,"[84] which Volotskoi had repeatedly quoted and emphasized. Serebrovskii's notion of *genofond* as "the entirety of genes" contained in a population profoundly influenced the development of population genetics, geno-geography,[85] and what would later be called modern evolutionary synthesis.[86] It also proved instrumental in the debates over the directions of Soviet eugenics.

Kol'tsov quickly seized the possibilities presented by the ideas of his former student. In November 1926, he delivered a public lecture that had considerable resonance.[87] In this lecture, soon published in the *REZh* and deliberately crafted to respond to the Marxist critics of eugenics,

Kol'tsov presented the "eugenic genealogies" of several prominent cultural figures, who came not from the intelligentsia, but from the peasantry. Probably with the "editorial foreword" to Filipchenko's 1924 book on eugenics in mind, he devoted more than half of his lecture to the pedigrees of Maxim Gorky and Fedor Chaliapin. Based on these genealogies, Kol'tsov argued that the country's population possessed "a gigantic *genofond*," containing countless genes of creativity, talent, and genius, and that the utilization of this "genetic wealth" was the primary task of Soviet eugenics.[88] Answering the accusations of ignoring the role of social and economic conditions in human development, Kol'tsov echoed both Florinskii's and Volotskoi's claims by emphasizing that this *genofond* had been all but lost under the tsarist regime, since many bearers of the genes of creativity among proletarians and peasants could not realize their "genetic potentials." Only the Bolshevik Revolution had created the conditions, he declared, that allow everyone to develop fully their "hereditary talents."

To be sure, Volotskoi did not remain silent. Just two weeks after Kol'tsov's lecture, on 7 December, he answered his opponents blow by blow in a two-hour-long report on "Issues in Eugenics" delivered to a meeting of the Society of Materialist-Biologists.[89] He touched upon practically all of the arguments, objections, and criticisms advanced by Filipchenko, Serebrovskii, and Kol'tsov: the interrelations among eugenics, genetics, Lamarckism, Darwinism, and Marxism, the race and class biases of modern eugenics, the place of the struggle for existence and selection in eugenic ideas and programmes, and the role of "positive" and "negative" eugenics in the betterment of humanity. To support his views, Volotskoi again invoked Florinskii's treatise, especially, the professor's idea that the removal of all barriers to the intermixing of different races and classes constitutes the main instrument of human perfection.

The report spurred a heated discussion on the floor, with nine commentators, including Serebrovskii, addressing various points in Volotskoi's arguments. This time, Volotskoi evidently managed to persuade at least some of his opponents. Mikhail Mestergazi, a geneticist and Bolshevik party member, agreed with him (and Florinskii) that "Wide intermixing and favorable conditions for development are the foundations for the betterment of the human type."[90] All of the nine discussants concurred with Volotskoi's "class analysis" of the essential differences between "bourgeois" eugenics developed by Galton and his followers in the West and in Russia, on the one hand, and "bio-social," "proletarian," "socialist," "objective" eugenics that must be advanced by "materialist-biologists," on the other. Much of the discussion revolved around the interrelations between heredity and environment, genotype and phenotype, nature and nurture, and their relative role in defining necessary actions and policies that could further eugenics' major goal of the betterment of humankind. Naturally, everyone spoke at length either for or against the Lamarckian inheritance of acquired characteristics.

Eminent physician and RES member Fedor Andreev suggested, however, that the main "issue in eugenics" was not the inheritance of acquired characteristics, but whether interventions aimed at the improvement of humanity should be directed at genotypes or phenotypes. Eugenicists, he argued, echoing numerous Marxist critics, are concerned exclusively with genotypes. But since under different external conditions the same genotype could produce vastly different phenotypes, he declared, "it might be more effective to advance *euphenics*," instead of eugenics.[91]

In his comments Andreev articulated a notion that had already been floating in the air. Indeed, just two weeks prior, Mestergazi had voiced a very similar idea in his report to the same society that also addressed the debate between geneticists and Lamarckists. "No matter the way the genotypes are formed," he stated poetically:

> it is the phenotypes that live, suffer, fight for world revolution, and build socialism. We should not be afraid of truth, but seek to penetrate the mystery of genes, conquer them, and then the field of eugenics would become as momentous as euphenics that should for the moment be in the focus of our attention and should direct all its forces to the betterment of phenotypes.[92]

The discussions spurred by Volotskoi's bio-social eugenics and its underpinnings in Florinskii's *Human Perfection and Degeneration* proved highly stimulating.[93] Following Andreev's line of reasoning, Kol'tsov elaborated further the notion of "euphenics" as "a doctrine that constitutes a necessary supplement to eugenics." According to the RES president, the goal of euphenics was to study the "methods of changing the phenotype, without changing the genotype, in order to obtain the most valuable for us phenotypes of cultivated plants, domesticated animals, and humans."[94] Kol'tsov clearly aimed at separating eugenics, with its goal of the *hereditary* betterment of mankind, from various actions and policies, which Volotskoi and his like-minded colleagues advocated under the names of "bio-social," "proletarian," and "socialist" eugenics, blurring the differences between hereditary and non-hereditary improvements. According to Kol'tsov, such "social measures" as education, prophylactic medicine, and the protection of children's, adolescents', and mothers' health cannot affect the genotype and, thus, have no direct eugenic consequences. But they do affect the phenotype and thus work as "powerful euphenic instruments," facilitating or inhibiting the expression of certain genes. "Euphenics requires," Kol'tsov emphasized, "that every child be accorded such conditions of upbringing and education under which his/her special hereditary abilities can find the fullest and most valuable expression in his/her phenotype."

During the 1924-1928 period, eugenicists managed to fend off "Marxist" critics and to continue the institutional and intellectual development of their discipline. In early 1927, the Communist Academy's section of natural sciences established a genetics laboratory headed by Serebrovskii, and anthropogenetics became its major focus.[95] A year later, the "materialist-physician" Levit, who had converted to genetics under Serebrovskii's tutelage, organized an "Office of Human Heredity and Constitution" at the Medical-Biological Institute (MBI) funded by Narkompros.[96] The same year, Kol'tsov spearheaded the establishment of a new "Society for the Study of Racial Pathology and the Geographical Distribution of Diseases" under Narkomzdrav's patronage.[97] At the very beginning of 1929, both the *Great Soviet Encyclopedia* and the *Great Medical Encyclopedia* published Serebrovskii's extensive articles on Galton.[98] The medical encyclopedia also carried a lengthy article on eugenics written by Iudin, followed by Kol'tsov's entry on euphenics.[99] But just a few months later, Soviet eugenicists found themselves under attack once again.

## The Demise of Eugenics?

The new attack on eugenics reflected profound transformations in the economic, ideological, institutional, and political landscapes of Soviet Russia induced by a new revolution — a "revolution from above." During the late 1920s, the General Secretary of the Bolshevik Party Joseph Stalin began to consolidate his own power over the party and that of the party apparatus over the nation.[100] The infamous 1928 "Shakhty Trial" of "bourgeois specialists" — a highly publicized show-trial of several engineers accused of sabotage and of "wrecking" the coal mines entrusted to their direction — heralded the end to the role played by educated professionals as government advisers and experts in all fields of the country's life.[101] That role was now entrusted to party bureaucrats and ideologues, while professionals were obliged merely to follow the directives of the party apparatus.[102] The year 1929, which Stalin himself named the "Great Break," marked drastic changes in all facets of life, including the abolition of NEP, the collectivization of the peasantry, crash industrialization, extensive militarization, and the all-out launching of the ambitious first Five-Year Plan. It also resulted in the replacement of practically all commissars, including Semashko, with trusted Stalinists. The "revolution from above" greatly diminished the autonomy and authority enjoyed by the scientific community during the 1920s and led to the rapid "Stalinization" of Soviet science: science and scientists were "mobilized to the service of socialist construction" and placed under the watchful gaze of the party-state apparatus and its head, the "Great Teacher" Stalin.[103]

Already the first wave of Marxist criticism during 1924-1928 made many Soviet proponents of eugenics wary of "ideological" dangers inherent to their studies. Some of them began avoiding the very word "eugenics," as well as the name of its founder, Galton. In late 1925, Filipchenko added the word "genetics" to the name of his Bureau of Eugenics, and from that time on, its journal, renamed accordingly *Herald of the Bureau of Eugenics and Genetics*, stopped publishing any work on eugenics and human heredity more generally, focusing exclusively on the genetics of cultivated plants and domesticated animals. In 1928, Filipchenko dropped the word "eugenics" from the names of his bureau and his journal altogether. The same year, Rokitskii published a popular brochure, titled *Can Mankind Be Bettered*, which discussed eugenics at length, but did not even mention Galton's name.[104]

The Great Break exacerbated this trend. To give but one example, Volotskoi had initially titled his 1926 book *Eugenics and Occupational Hazards*.[105] But when it finally appeared in print in late 1929, it bore the title *Occupational Hazards and the Progeny*.[106] Furthermore, the entire subject of human heredity seemed to have become suspect. The First All-Union Congress on Genetics and Breeding, which was held (after several delays) in January 1929 in Leningrad and brought together nearly 2,000 participants, did not have a single session on human genetics.[107] The only report on "hereditary diseases" dealt not with human but animal ailments. The only report mentioning humans as research subjects, titled "New Paths in the Selection of Humans and Mammals," was delivered by the gynecologist Antonina Shorokhova at a session on the genetics of domesticated animals presided over by Serebrovskii. Shorokhova's presentation, based on more than a decade of research on artificial insemination in women as a means to fight infertility, suggested that her experience might prove useful for animal breeders.[108] Just a few months after the congress, in May, Filipchenko rejected the offer to renew the membership of his bureau in the International Federation of Eugenic Organizations. In December, he informed Kol'tsov of his intention to withdraw from the *REZh* editorial board, and only his untimely death from meningitis the following spring prevented Filipchenko from making his intentions public.[109]

Of course, several proponents of eugenics tried to adjust their enterprise to the new situation. In late 1929, the first volume of *Proceedings* issued by the MBI Office of Human Heredity and Constitution opened with two programmatic articles by its editors. One was written by Serebrovskii on "Anthropogenetics and Eugenics in Socialist Society," and another by Levit on "Genetics and Pathology (Regarding the Current Crisis in Medicine)."[110] Following the current party line on "mobilizing science for the needs of building socialism," both articles presented research on human heredity as vital for socialist construction. Serebrovskii even suggested that "probably, it would be possible to fulfill the Five-Year Plan in just two and a half years," if only the country's *genofond* would have been "purged of various forms of hereditary diseases." Perhaps inspired by Shorokhova's report at the genetics congress, he found a "truly socialist" way of achieving this eugenic goal: the "separation of love and reproduction" through the artificial insemination of willing women with "recommended sperm" from a "talented producer," who thus could "father up to 1,000 or even 10,000 children." If this program were adopted, Serebrovskii opined, "human selection would make gigantic leaps forward," and would lead to the increased productivity, efficiency, and creativity of "new forms of human beings" in the USSR. To implement this vision, Serebrovskii noted, the country, of course, needed to expand research in anthropogenetics considerably.

Although much less visionary and more technical that Serebrovskii's, Levit's article also advanced the view that anthropogenetics held the key to solving nearly all of the major problems facing modern medicine, ranging from the etiology and epidemiology of various diseases in human populations to the variability of infectious agents and human susceptibility to pathogens. Following Kol'tsov, Levit emphasized the distinction between eugenic and euphenic consequences of medical and social interventions and insisted, contrary to his own earlier pronouncements, that "it is genetics that provides a scientific foundation for prophylactic medicine."

Serebrovskii's and Levit's panegyrics to "socialist eugenics" proved ill-timed. They appeared amidst two major campaigns of the Great Break aimed at placing trusted Stalinists in positions of power within the entire Soviet science system. One campaign was directed "against bourgeois scientists" and resulted in the "Bolshevization" of the leadership of practically all scientific institutions, beginning with the Academy of Sciences.[111] Another was waged "against mechanistic materialism and menshevizing idealism," which Stalin personally identified with the "left" and "right" deviations from the orthodox party line, and resulted in drastic changes of leadership in the institutions of "communist science."[112] Under these conditions, Serebrovskii's "manifesto" of socialist eugenics, with its assertion of the role of specialists in human heredity as leading experts on the Five-Year Plan (and the county's future more generally), was bound to backfire.

The first volley came in early June 1930, when *Izvestiia* published a lengthy satirical poem, titled "Eugenics," written by Dem'ian Bednyi, a well-known "proletarian" poet.[113] The poem interspersed certain phrases from Serebrovskii's manifesto with a sarcastic commentary by the poet and excerpts from an indignant letter he had allegedly received from an anonymous female correspondent. Bednyi mocked both Serebrovskii's vision of "socialist eugenics" and his claim to authority in questions related to the Five-Year Plan and Soviet reproductive policies. Serebrovskii did not take Bednyi's satire quite seriously and even tried to publish on the pages of the same newspaper a similarly sarcastic response, also written in verse. But his attempt at a public retort proved futile: *Izvestiia* rejected his poem.[114]

A few months later, much heavier guns entered the fray. In September, the "Leninism in Medicine" society, a major stronghold of "materialist-physicians" in Moscow, issued a ten-page exposé, under the telling title "Regarding the Production Plan of 'Socialist Eugenics'," which characterized Serebrovskii's eugenic ideas as "psychotic delusion."[115] The unsigned article did not oust eugenics from building socialism outright. But it stated that the true path of "socialist eugenics" was "the path of prophylactic medicine, regular health check-ups, and making labor and life conditions healthier," thus substituting euphenics for eugenics. Serebrovskii immediately published a repentant letter, admitting that his 1929 manifesto contained a number of "anti-party mistakes" and "mechanistic formulas" and suffered from "abstract theorizing." Yet, he clearly did not see the writing on the wall and defiantly insisted that "these mistaken statements [are] in no way related to the main thoughts developed in the article."[116]

Perhaps this new attack would have proven insufficient to spell the end of eugenics in the Soviet Union. Indeed, several concurrent reviews of the MBI *Proceedings* that appeared in specialized medical and biological periodicals praised the research conducted at the Office of Human Heredity and Constitution and called for "distributing it [the *Proceedings* volume] as widely as possible."[117] But the attack on eugenics in print also coincided with certain institutional actions undertaken by the Soviet authorities. Contrary to the assertions of some later historians, these actions were not directed specifically against eugenics. In early 1930, in its drive for establishing control over the entire system of Soviet science, the party apparatus initiated the "inspection" of all learned societies to investigate their "conformity to the goals of the Five-Year Plan and socialist construction" and their "links with industry and agriculture."[118] In the course of the inspection, usually conducted by a "workers' brigade" from a nearby factory, every scholarly society had to present its charter and membership roll (indicating the percentage of workers, peasants, and Bolshevik party members) for review and approval to the People's Commissariat of Internal Affairs (NKVD) — an arm of the country's security services. At the same time, the party apparatus moved to install trusted party members — recruited mostly from among the members of the Communist Academy and the graduates of the Institutes of Red Professors — on the editorial boards of all scholarly periodicals. Emphasizing "deep contradictions between Marxism and eugenics," its critics immediately singled out the Russian Eugenics Society, together with its president, for a particularly thorough "inspection."[119]

In this tense situation, Kol'tsov apparently decided not to subject the society, its journal, and its members to the scrutiny of Stalin's security services: he simply did not submit the required papers to the NKVD. The society ceased to exist and its journal was discontinued with the last 1930 issue. In a few months, preparing the IEB plan for 1931, Kol'tsov renamed its "eugenics department" as the "department of anthropogenetics." In a clear response to Marxist critics, he reformulated the tasks of the "new" department as "studying the various phenomena of human heredity and variability, defined not only by heredity, but also by the influences of external environment."[120]

If the "pure science" of eugenics could perhaps still be pursued under the name of anthropogenetics, the "applied science" was clearly out. In the spring of 1930, Davidenkov, the head of the Genetics Bureau of the Moscow Society of Psychiatrists and Neurologists, wrote a lengthy article, titled "Our Eugenic Perspectives."[121] It was clearly aimed at adjusting Soviet eugenics to the realities of the Great Break. Echoing Serebrovskii's 1929 manifesto, Davidenkov advanced an elaborate "practical eugenic programme" suitable for a socialist society. The programme included the creation of a "Central Eugenics Institute" to prepare the cadres necessary for its administration and a "Supreme State Eugenic Council" to coordinate all eugenic actions and policies. The first step in the implementation of this programme was "the obligatory eugenic screening of the entire urban population." Such screening would identify the "hereditary endowment" of every individual and thus place every single person in a specific "higher or lower eugenic category." "Medico-eugenic bureaus" established throughout the country would advise prospective couples on "eugenic consequences" and on possible benefits or disadvantages that their marriage could have for their future children. Marriages between individuals of the "highest eugenic category" would be encouraged and supported by a number of subsidies and allowances for each child. Marriages between individuals of the "lowest eugenic category" would be discouraged and such individuals enticed (by education and financial incentives) to undergo "voluntary sterilization." Marriages between all other persons would remain uncontrolled. According to Davidenkov, this long-term programme would eventually result in a substantial increase in the number of "hereditary gifted" individuals in the country's population. Davidenkov probably planned to publish the article in the *Russian Eugenics Journal*, but, expectedly, it never saw the light of day.

Social hygienists, the most active proponents of "applied eugenics" during the preceding years, were hit hard by the dismissal of Semashko from Narkomzdrav and rushed to distance their field from its former "constituent part," as the commissar had defined eugenics just a few years prior. During the 1920s, they had presented social hygiene as "a science of the future, which studies and shapes the facts that promote the biological well-being of humanity," and saw eugenics as "the ultimate goal of all sanitary-medical activities."[122] After the Great Break their attitude changed drastically. In 1931, Batkis published an entry on eugenics in the new edition of the *Great Soviet Encyclopedia*. Obviously trying to forget his own earlier equating of social hygiene with eugenics, he offered an extensive critique of capitalist, western, bourgeois eugenics, peppered with quotations from Marx and Engels. He then turned to the Soviet proponents of this "pernicious" doctrine, characterizing Kol'tsov and Filipchenko as "fascists" and Serebrovskii and Levit as "menshevizing idealists."[123] The next year, Moisei Langis, another member of the State Institute of Social Hygiene, produced a volume of Solov'ev's collected works. Along with numerous articles and speeches on various facets of "building the Soviet health protection system," the volume carried the text of Solov'ev's "foreword" to Iudin's book on eugenics, which appeared in print for the first time and was obviously meant to demonstrate social hygienists' critical attitude towards eugenics.[124]

The labels "fascists" and "menshevising idealists" Batkis attached to the Soviet proponents of eugenics were not merely a reflection of "the sharpening of the class struggle" and the division between "us" and "them" forcibly imposed by the party-state apparatus during the "revolution from above." Although the fierce campaign "against mechanistic materialism and menshevizing idealism" aimed first and foremost at establishing the party apparatus's control over scientific institutions, it was given the form of a "public discussion" and as such had important intellectual consequences.[125]

One "side effect" of the campaign was the expulsion of "the biological" from any substantive discussion of human nature and humanity's future. Darwinism, one of the "two pillars" of Volotskoi's bio-social eugenics, was swallowed whole by the other one — Marxism. Any attempt to consider the role that the biological factors identified by Darwin, such as variability, heredity, and selection, could play in the *future* evolution of humanity, and especially, "its vanguard," the Soviet Union, was now condemned as "pernicious" *biologization* (or *zoologization*) and viciously attacked by the new generation of Marxists nurtured by the Soviet educational system.[126] Marxist critics did not deny the role of biological factors in the *past* evolution of the human species. To the contrary, they hailed Darwinism as the "scientific foundation" of their sacred doctrine, Marxism, as the full name of the Timiriazev Institute clearly manifested. But, for them, the role of biology in human evolution had ended with the emergence of "labor" and "class society," at which point, according to Engels's widely publicized brochure on "Anteil der Arbeit an der Menschwerdung des Affen" (The Role of Labor in the Origin of Humans from Apes), social factors took over the determination of the further development of humanity.[127]

From now on, the future *Homo superior* would be created not by managing reproduction and altering heredity, but by manipulating upbringing and education. This point was forcefully articulated in the fall of 1930 by Lev Vygotskii, one of the leaders of "Marxist" psychology and pedagogy, in an article that appeared under the characteristic title "Socialist Reconstruction of Man" in *VARNITSO*, the mouthpiece of the campaign to subjugate science to the tasks of building socialism.[128] As an article published in the *Advances of Modern Biology* under the revealing title "The Transition from the Leading Role of Natural Selection to the Leading Role of Labor" clearly indicated, in the Marxist view of human evolution, Nurture has overcome Nature.[129] The revolutionary dream of creating a "higher socio-biological type, an Ubermencsh," as Trotsky had put it, was replaced by the task of the "socio-political upbringing of the builders of socialism" put forward by Stalin and his ideologues.[130]

The exclusion of biology from the Marxist vision of humanity's future and, with it, the elimination of any substantive discussion of eugenics from Soviet discourse on human reproduction, heredity, variability, development, and evolution, was not merely a result of the imposition of the new policies by the party-state apparatus during the Great Break. It was facilitated by sharp disagreements over the actual contents of "Soviet" eugenics among its proponents, fueled by their different disciplinary affiliations, institutional positions, and general worldviews.

The materials presented above allow one to distinguish roughly three broad groups involved with eugenics in the Soviet Union. The first was a cohort of established, older scientists, born in the 1870s and early 1880s (such as Davidenkov, Filipchenko, Kol'tsov, and Iudin), with a long-standing interest in the issues of human reproduction, heredity, development, and evolution, whose involvement with eugenics had predated the Bolshevik Revolution and who spearheaded its propaganda and institutionalization during the 1920s. This group was largely indifferent or even hostile to Marxism and the Bolsheviks' efforts to build their own "proletarian" science. But they were willing to reach a compromise with the new rulers and serve as advisors, experts, and consultants to government agents and agencies in exchange for the opportunity to continue and expand their own scientific pursuits.

The second was a group of first generation Bolsheviks, also born in the 1870s (such as Radin, Semashko, Solov'ev, and Sysin) who after the revolution became the patrons and conduits of eugenics in various government agencies, especially Narkomzdrav. All of them were avowed Marxists and saw the creation of "proletarian" science as their major task. Yet they valued the expertise of the older "bourgeois" generation of Russian scientists and followed the scientists' recommendations in promoting the development of particular scientific disciplines and research directions, such as eugenics.

The third was a cohort of younger scholars, born mostly in the 1890s, who were starting their academic careers just after the Bolshevik Revolution (such as Batkis, Levit, Serebrovskii, Slepkov, and Volotskoi). They embraced both Marxism and the Bolshevik efforts to build "proletarian," "communist" science whose institutions provided them with employment and career opportunities. It was this last group that supplied the most vocal critics of "bourgeois" eugenics and proponents of "proletarian" science. Each of these groups had its own vision of what eugenics was and what kind of eugenics should be developed in the Soviet Union, illustrated by their diverging views on both Galton's eugenics and Florinskii's eugamics.

As its editor, reviewers, and commentators all readily agreed, the "scientific contents" of *Human Perfection and Degeneration* were outdated and did not correspond to the current state of knowledge. How, then, could the book serve as a model for "socialist" eugenics? How could Volotskoi use Florinskii's treatise to criticize Kol'tsov's vision of eugenics, while the latter could claim that modern-day eugenicists were actually "much closer to V. M. Florinskii's views than to those of M. V. Volotskoi"? The answers, I believe, lay in the close intertwining of the institutional development of eugenics and genetics and in the nature of eugenics as an *amalgam* of ideas, values, problems, and practices concerning human reproduction, heredity, development, and evolution, fused together into a more or less coherent doctrine. Agreements and disagreements among Soviet eugenicists regarding the importance of both Florinskii's eugamics and Galton's eugenics for their own work derived largely from divergent evaluations of the constituent elements of these fusions, and from "reading-into" the works of both Galton's and Florinskii's their own understandings of various issues addressed by the British and Russian "founders."

The first element — the notion of human nature embodied in the ideas about human reproduction, diversity, heredity, individual and social development, and evolution — became a major source of friction among Soviet eugenicists. For the younger generation of "Marxist" biologists, physicians, historians, psychologists, and so on, Florinskii's tacit acknowledgement of the inheritance of acquired characteristics represented the unquestionable strength of his eugamic concept and became one of the arguments for its suitability as a model for "bio-social" eugenics. The infusion of Lamarckism into Marxism also made the interpretation of Florinskii's ideas as "socialist" eugenics attractive to Bolshevik officials, as Semashko's and Solov'ev's pronouncements readily demonstrate. For geneticists, such as Kol'tsov, Filipchenko, and Serebrovskii, however, it was an understandable (since Florinskii had shared it with most of his contemporaries, including Darwin) but notable weakness. For them, the implicit acceptance of Lamarckian inheritance precluded the use of Florinskii's treatise as a foundation of their own vision of Soviet eugenics, which was based on Mendelian genetics.

At the same time, as Kol'tsov's review made clear, geneticists appreciated Florinskii's pioneering distinction between the two basic processes undergirding hereditary phenomena — *transmission* and *development* of "hereditary potentials." For their opponents, thanks to their Lamarckian convictions, such a distinction appeared immaterial and was consistently blurred in their equation of eugenics with social hygiene and physical culture. In contrast, it is this very distinction that led Kol'tsov to elaborate the notion of euphenics as a set of distinct policies that could supplement, but not replace eugenics. On the other hand, Galton's opposition to the Lamarckian inheritance, exemplified by his vocal critique and experimental refutation of Darwin's pangenesis hypothesis,[131] certainly appealed to geneticists, who even made him one of the "founding fathers" of their discipline, as Filipchenko's dual biography of Mendel and Galton showed. But it was unacceptable to Lamarckists, who largely ignored Galton's actual views on heredity and saw him, first and foremost, as the major proponent of the "class bias" in eugenics.

In a very similar way, the three groups diverged in their attitudes to the value systems that underlay Florinskii's eugamics and Galton's eugenics, respectively. Many ideals, norms, and beliefs advanced by the *raznochintsy* of the 1850s and 1860s, which had informed Florinskii's views, became incorporated into the value system of the pre-revolutionary intelligentsia, whose members introduced and developed eugenics in Bolshevik Russia, including the country's new rulers who funded and promoted this development.[132] Seen in this light, Florinskii's treatise with its focus on the people (*narod*), as opposed to Galton's focus on the ruling and cultural elites, appealed to many Soviet eugenicists.

Similarly, Florinskii's ideal of "physical and mental health" that had infused his notions of human perfection and degeneration was certainly very attractive to physicians (from gynecologists to neurologists) and public health doctors (social hygienists) who together formed arguably the largest disciplinary group among Soviet eugenicists. Furthermore, the same ideal resonated very strongly with the agendas of the main state patron of Soviet eugenics — Narkomzdrav.

The third element — specific social concerns (both hopes and fears) addressed by Florinskii's and Galton's concepts — also evoked far from unanimous responses, though all Soviet eugenicists shared an interest in the "physical and moral health" of the population common to eugenics and eugamics. The older generation of Soviet eugenicists fully embraced Galton's hope of creating "a galaxy of genius" and his anxiety about the purported decrease of "hereditary talents," which focused, especially, on scientists (and the intelligentsia more generally) as the bearers of such talents.

Perhaps one reason why this particular concern became a centerpiece of Soviet eugenics was that during the years of revolution and civil war, proportionately very high emigration and mortality rates had severely depleted the numbers of the Russian intelligentsia, while massive ideological attacks on the "bourgeois" intelligentsia had raised fears about its very survival under "proletarian rule." Certain policies adopted by the Bolsheviks at the very beginning of their regime to create their own "proletarian" intelligentsia, such as the administrative barriers for the children of the old intelligentsia, and the preferential treatment of the children of the proletariat, in entering schools of higher learning, amplified such fears even further. As Kol'tsov stressed in his 1921 address:

> The state must first of all take care of the strong and provide for their families [and] their offspring. The best and only method of eugenics is to identify progenitors who are valuable in their hereditary qualities — physically strong, endowed with exceptional intellectual and moral qualities people — and to put these talents (mandatorily and preferentially as compared to the people who do not exceed the average) in such conditions under which they not only could express their capacities to the fullest, but could feed and nurture a large family.[133]

The numerous publications on "the intelligentsia and talents" produced by Soviet eugenicists during the early 1920s could be seen as a manifestation of this particular concern and an attempt both to mobilize public opinion and to prod the Bolshevik government into adopting measures (from enlarged food rations to social support for procreation and education) aimed at preserving and increasing the nation's "creative capital." Yet this elitist bias of Galtonian eugenics and its promotion by the older generation of Soviet eugenicists provoked sharp critique from both the younger generation of "Marxist" eugenicists and their Bolshevik patrons. This critique eventually led to the disappearance of this concern from the agendas of Soviet eugenics and its replacement with a deep interest in *genofond* as the source of the nation's "human capital."

The fourth element — the array of actions, policies, and practices proposed in the name of Galton's eugenics and Florinskii's eugamics — also proved highly contentious, as Soviet eugenicists' debates on sterilization and marriage laws readily show. Strong opposition to sterilization on the part of the majority of Soviet eugenicists derived not only from the perceived impossibility to define and identify "hereditary defectives." It was also deeply rooted in the ideals of personal liberties and their protection from the government dictate, which had been upheld by the pre-revolutionary intelligentsia (and had clearly been articulated in the early critique of sterilization laws by Russian jurists).

Even when proposing marriage laws that limited such liberties, older eugenicists emphasized their voluntary nature, stressing that the major goal of these laws was "sanitary-eugenic propaganda" that would force prospective spouses, as Florinskii had wanted, "to look at their future marriage from the viewpoint of health." This was why they advocated not for marriage restrictions and sterilization but for the creation of "marriage" or "genetic" consultations as the main tool in the struggle against "hereditary diseases."[134] Yet for the younger generation of Soviet eugenicists, whose worldviews had been profoundly shaped by the horrors of World War I and the civil war, government infringement on personal liberties in the name of eugenics appeared acceptable, as Volotskoi's propaganda of sterilization and Serebrovskii's proposal of mass artificial insemination as the instruments of eugenics clearly show.

The difference between the two particular constellations of ideas, values, concerns, and practices embedded in Galton's eugenics and in Florinskii's eugamics became a key point in the heated discussions over the foundations, research agendas, practices, and ultimate goals of eugenics in the Soviet Union. Indeed, Soviet eugenicists "co-constructed" the two concepts to fit their own agendas. Combined with the absence of Russian translations of Galton's actual texts, Volotskoi's edition of Florinskii's treatise made such "co-construction" heavily tilted in favor of the Russian "founder."

Consider for instance, the key issue in debates on eugenics — the interrelations between, in Galton's terms, nature and nurture or, in the terms of Marxist critics, the biological and the social. Neither supporters, nor critics, ever mentioned Galton's 1904 definition of eugenics as "the science which deals with all influences that improve the inborn qualities of a race; [and] also with those that develop them to the utmost advantage."[135] This definition was very close to Florinskii's notions of "hereditary potentials" and "conditions conducive" to their realization. But this particular similarity between the views of the two "founders" was consistently muted, albeit for different reasons, in Soviet debates. For critics, the implication of the equal importance of nature (inborn qualities) and nurture (environmental influences on the development of such inborn qualities) contained in this definition undermined their favorite accusation of Galtonian eugenics of privileging the biological over the social. For supporters, especially among geneticists, the ambiguity of this definition opened the door to a Lamarckian interpretation of those influences "that develop [the inborn qualities] to the utmost advantage." Both critics and supporters preferred to use Galton's later definition of eugenics as "the study of agencies under social control that may improve or impair the racial qualities of future generations, either physically or mentally,"[136] actively propagated by his western followers such as Pearson, Davenport, or Leonard Darwin, which tacitly excluded "nurture" from Galton's eugenic programme, and thus better suited the interests of both groups.

As a result of the "revolution from above," however, the peculiar amalgam of ideas, values, concerns, and actions regarding human reproduction, heredity, development, and evolution, which together comprised "proletarian" eugenics, was broken apart. Its constituent elements were taken over by other disciplines. The "betterment of humankind" was now considered exclusively within the frameworks of social hygiene, physical culture, the protection of maternal, child, and adolescent health, psychology, and pedagogy. As anthropologist Mikhail Gremiatskii emphasized in his entry on eugenics for the 1936 edition of the *Small Soviet Encyclopedia*, "Instead of eugenics, our task is the development and implementation of social-hygienic measures."[137] In a textbook on social hygiene published the same year, Batkis elaborated the same idea in a special chapter on "Bourgeois Theories of Healthification (Eugenics, Racial Hygiene)," stating categorically that "in their theoretical and practical conclusions eugenicists stand in utter conflict with social hygiene."[138] On the other hand, Marxism (in whatever version currently endorsed by Bolshevik ideologues, starting with Stalin himself) provided a general framework for the evaluation of specific practices and policies that targeted human reproduction and development, such as abortions or psychological tests for school children, both of which were prohibited in the same year: 1936.

## The Birth and "Death" of Medical Genetics

For all intents and purposes, by the end of 1930 eugenics in the Soviet Union was dead. It had lost its patrons, allies, institutions, and journals, while courses on eugenics at various schools of higher learning, such as the course taught at the Institute of Physical Culture by Volotskoi, had been abolished. But members of the defunct Russian Eugenics Society, along with the extensive networks they had built throughout the Soviet science system, remained very much alive and did everything they could to save their enterprise. They "gave up" two of the three elements of eugenics identified by Kol'tsov in 1921: the applied science of anthropotechnique and the religion/ideology of human betterment. But they saved its third element — the pure science of anthropogenetics — by reconstituting it as a new discipline, soon named "medical genetics."[139]

A close-knit network of individuals closely involved in the prior development of eugenics, including Bunak, Davidenkov, Iudin, Kol'tsov, and Levit, quickly mobilized to legitimize and institutionalize their new endeavor. They cultivated new patrons and allies, created new institutional niches and publishing outlets, built new networks of personal contacts with colleagues at home and abroad, and trained a new generation of "clinical" and "medical" geneticists. They elaborated a broad research programme, ranging from the morphological analysis of human chromosomes to the clinical investigation of the heritability of pernicious anemia to the detailed examination of conditional reflexes in twins. Yet, surprisingly, in their extensive efforts to build the new discipline, its champions never once invoked Florinskii and his eugamics.

The "materialist-physician" and Bolshevik Party member Levit became the leading spokesman for medical genetics. After the 1930 rebuke, Serebrovskii stopped all work in human genetics.[140] And it was his student Levit who picked up the fallen banner of "socialist anthropogenetics." With the zeal of a recent convert, Levit put his formidable energy to the advancement of his newly acquired faith. In March 1930, as part of the general move to place trusted party members at the helm of scientific institutions, Levit was appointed director of the Medical-Biological Institute (MBI). He immediately "upgraded" his Office of Human Heredity and Constitution to the status of the institute's major department, expanding its personnel and agendas. In the fall, Levit issued the second volume of the MBI *Proceedings*, which included fourteen research articles introduced by his programmatic editorial, titled "Man as a Genetic Subject and Twins Studies as a Method of Anthropogenetics."[141]

The main purpose of Levit's editorial was to distance anthropogenetics from its stepmother, eugenics. As we saw, during the preceding decade, anthropogenetics had been presented as a key component, a foundation of eugenics. This link had provided a principal justification for the rapid institutional development of both genetics and eugenics, as the close involvement of the leaders of Soviet genetics, Filipchenko, Kol'tsov, Serebrovskii, and even Vavilov vividly demonstrates. With eugenics now deemed unacceptable, the champions of anthropogenetics had to reconfigure its relation to eugenics and to find new justifications for its independent existence and further development. And this is exactly what Levit did in his editorial. He opened with a forceful statement that "sometimes, anthropogenetics is completely erroneously equated with eugenics" and proceeded to rectify this "error." In contrast to his previous article on "Genetics and Pathology" published just a year earlier, Levit now presented anthropogenetics not merely as the practical application of the principles of genetics to human diseases and medicine writ large, but as a separate discipline in its own right. According to Levit, anthropogenetics was more than a subdivision of general genetics, as were plant and animal genetics, for "man as a genetic subject" is in many respects unique. Anthropogenetics, therefore, could investigate hereditary phenomena that were inaccessible to researchers studying the genetics of any other animal (not to mention plant) species.

Among the unique features characterizing the human species Levit noted "the almost complete absence of natural selection," which led to the accumulation in human populations of many "Mendelian characters" that otherwise would have been eliminated. He stressed "the possibility of studying the inheritance of psychiatric features and their anomalies." Levit also emphasized "the much greater knowledge of human physiology and morphology (including histology) than those of any other animal," which presents a great advantage in investigating the interrelations between the genotypical nature and the phenotypical expression of certain characters that became the subject of "phenogenetics," a new area of research that combined genetics, developmental mechanics, and evolutionary theory.[142] He further pointed out the benefits of studying humans for the advancement of population genetics, especially geno-geography — the study of the diffusion and distribution of certain genetic traits in human populations.

At the end of the year, Levit had to interrupt his feverish efforts to institutionalize and promote anthropogenetics: he received a year-long Rockefeller Foundation fellowship to "advance his qualifications in genetics."[143] He chose to do so at the University of Texas under the mentorship of H. J. Muller, a former member of Thomas Hunt Morgan's "fly group," who had visited Russia shortly after the end of the civil war.[144] Judging by extant materials, it was Serebrovskii who had recommended that Levit take his fellowship with Muller (see fig. 6-5).[145] Serebrovskii had first met Muller during the latter's 1923 visit to Moscow. The two geneticists renewed their acquaintance at the 1927 Fifth International Genetics Congress in Berlin where Muller announced his success in obtaining the first artificial mutations in *Drosophila* by X-ray irradiation — an achievement whose import for biology Serebrovskii likened, on the pages of *Pravda*, to that of the Bolshevik Revolution for the world.[146] Muller was not only one of the United States' leading geneticists, he also had a long-standing interest in eugenics and participated in the Second International Eugenics Congress, which, too, probably played a role in Serebrovskii choosing him as Levit's future advisor.[147]

During his fellowship, Levit also spent two months at the Genetics Department of the Carnegie Institution at Cold Spring Harbor headed by US leading eugenicist Charles Davenport.[148] Levit wanted to stay in the United States for the summer of 1932 to take part, along with his mentor Muller, in both the Third International Eugenics Congress planned for July in New York City and the Sixth International Genetics Congress scheduled for August in Ithaca, NY. But unlike the imperial authorities in the case of the similar request by Florinskii, the Soviet authorities refused to extend his stay abroad.

Fig. 6-5. Solomon Levit at Herman J. Muller's laboratory in Texas, 1931. Left to right: S. Levit, H. J. Muller, C. Offerman, I. Agol. Photographer unknown. Courtesy of the Lilly Library, Indiana University, Bloomington, IN.

Upon his return to Moscow in early 1932, Levit found that his temporary replacement as MBI director had curtailed genetics research at the institute. By that time the turmoil of the "revolution from above" had largely subsided, and Levit was able to use his extensive contacts within the party-state apparatus to gather together the scattered pieces of his enterprise and reorient the entire institute to studies of human genetics. As he noted in a foreword to the third volume of the MBI *Proceedings*:

> As of the fall of 1932, the institute ... concentrated on the studies of issues in human biology, pathology, and psychology through the application of the newest achievements of genetics and related fields (cytology, developmental mechanics, [and] evolutionary theory). The institute's main works went in three directions: clinical genetics, twins studies, and cytology.[149]

Expectedly, the distancing of his current research from eugenics remained a major focus of Levit's efforts to legitimize human genetics. In April 1932, on the occasion of a bizarre jubilee campaign to mark the semi-centennial of Darwin's death, he published an article with a telling title, "Darwinism, Race Chauvinism, and Social-Fascism." The article's main purpose was "to cleanse Darwin's theory from its bourgeois mistakes and perversions," including, of course, eugenics.[150]

A year after his return from the United States, Levit acquired a powerful ally in his efforts to justify and advance "socialist anthropogenetics." In the spring of 1933, Muller came to the Soviet Union at the invitation of leading plant geneticist Nikolai Vavilov. After Filipchenko's death in 1930, Vavilov had "inherited" the Bureau of Genetics and quickly reconstituted it into the Academy of Sciences Laboratory of Genetics. Vavilov offered Muller the "scientific directorship" of this institution, which at that very time was expanded even further to become the Academy of Sciences Institute of Genetics. Although, according to Vavilov's plans, Muller's main task was to guide research at the Institute of Genetics and to acquaint its personnel with the latest techniques and concepts of US genetics, Muller immediately extended a helping hand to his former mentee Levit.

Just a few months after his arrival to Leningrad, Muller published a Russian translation of his speech at the Third International Eugenics Congress in the recently created journal *Advances of Modern Biology*, under the title "Eugenics under the Conditions of a Capitalist Society."[151]

When during the next spring Vavilov's Institute of Genetics was relocated from Leningrad to Moscow, Muller also became a "scientific consultant" at Levit's MBI, and published an article on "Eugenics in the Service of the National-Socialism" in the April issue of *Priroda*.[152] Both publications advanced Muller's critique of "bourgeois" eugenics and his ideas of "socialist" eugenics, which resonated strongly with the class analysis of Galtonian eugenics developed a few years earlier by Volotskoi, Serebrovskii, and other "Marxist" critics.[153]

By 1934, former eugenicists had cut their losses and regrouped. On 15 May, Levit organized under the MBI's auspices a "conference on medical genetics," attended by more than 300 participants from Moscow, Leningrad, Kazan, Kharkov, and other provincial centers.[154] He opened the conference with a keynote address on "Anthropogenetics and Medicine." Then Muller delivered a plenary lecture on "Certain basic stages in the development of theoretical genetics and their significance from the viewpoint of medicine." What followed was quite remarkable indeed: four former leaders of Soviet eugenics — Kol'tsov, Davidenkov, Iudin, and Bunak — presented papers on the interrelations of genetics and medicine. All speakers emphasized the profound differences between Soviet and western approaches to human genetics and the pathbreaking nature of many research projects pursued by its practitioners in the land of the victorious proletariat.

Following a general discussion, the conference adopted a resolution that fell nothing short of a manifesto of medical genetics. After the now obligatory "critique" of "bourgeois eugenic perversions," the resolution called upon Narkomzdrav to "create scientific research centers for medical genetics and cytology" in every large city throughout the country, to establish "departments of medical genetics" at all medical institutions, to include genetics on the curricula of medical education (creating corresponding teaching departments at every medical school), to produce necessary textbooks, and to expand the programme of graduate studies in medical genetics at the MBI. The conference also approved the syllabus of a remedial, 52-hour-long lecture "course on genetics for physicians" delivered by Levit at the MBI the previous year as an example to be emulated elsewhere. The course included a special four-hour section on "bourgeois eugenics and its class character" that spelled out the profound differences between socialist and capitalist approaches to various issues in human heredity.

The future of medical genetics seemed bright. The new head of Narkomzdrav Grigorii Kaminskii enthusiastically supported Levit's enterprise. Less than a year later, the labors of the former eugenicists culminated in the reconstitution of the MBI into the world's first Institute of Medical Genetics (IMG) under Levit's directorship. Other laboratories and clinics devoted to research in medical genetics sprung up in Leningrad (under Davidenkov), Kharkov (under Iudin), and several other provincial cities. An important center emerged in Koltushi, a "science village" built in the late 1920s-early 1930s on the outskirts of Leningrad for Ivan Pavlov's research, as well as at several clinics associated with Pavlov's institutional empire.[155] A sizeable cohort of scientists and clinicians became engaged in investigations on human (medical) genetics, focusing particularly on twins-based research.[156]

Soviet eugenicists-turned-medical-geneticists continued to develop and maintain close links with their foreign colleagues,[157] exchanging letters and reprints and publishing research papers in American and British journals, including *Journal of Heredity, Eugenics Review, Nature,* and *Annals of Eugenics*.[158] Given the isolationist policies implemented in the wake of the Great Break, which had radically curtailed the foreign trips of Soviet scientists, the Seventh International Genetics Congress scheduled to convene in Moscow in the summer of 1937 occupied a special place in their efforts.[159] In January 1936, Levit was appointed the "scientific secretary" to the Soviet organizing committee for the congress, while Muller became the head of the congress's programme committee. During the ensuing discussions, Levit, with Muller's support, made sure that his favorite subject figured prominently on the congress's agenda, with Davidenkov slated to deliver a keynote address to a special session on "human genetics and racial theories."

In early May, perhaps hoping to secure support for human genetics from the very top of the Bolshevik party, Muller sent Stalin his recently published book *Out of the Night* that elaborated on his report to the Third International Eugenics Congress and spelled out his vision of "socialist" eugenics.[160] In a letter sent along with the book, Muller urged Stalin to implement his ideas in the Soviet Union.[161] "True eugenics can only be a product of socialism," Muller assured the Soviet leader, "and will, like advances in physical technique, be one of the means used by the latter in the betterment of life." Castigating "the evasions and perversions of this matter ... seen in the futile mouthing about 'Eugenics' current in bourgeois 'democracies', and in the vicious doctrine of 'Race Purity' employed by the Nazis as a weapon in the class war," Muller reiterated Serebrovskii's idea that the well-being of the nation could be radically improved through the artificial insemination of willing women with the sperm of "gifted individuals."[162]

It seemed that the future of eugenics-turned-medical-genetics in the Soviet Union was assured. Reporting on research conducted at Levit's IMG, an editorial in the October 1935 issue of *Eugenics Review* forecasted that "It almost seems as if geneticists in this country will have to add Russian to their already formidable linguistic equipment."[163] A year later Davenport affirmed the impressions of his British colleagues: "I have told many students of human genetics in the United States that Russia is taking the lead away from the United States in this subject, which it formerly held."[164]

But within just a few months, the fortunes of medical genetics in the Soviet Union turned once again. The beginning of the Great Terror in the summer of 1936 inaugurated a new nationwide witch-hunt for "wreckers," "traitors," and "agents of imperialism" in all walks of life. It prompted Levit's expulsion from the Bolshevik party membership for his alleged association with the "opposition," which in turn led to his dismissal from the IMG directorship in December 1936. The rising political tensions between Hitler's Germany and Stalin's Russia (clearly manifested during the Spanish Civil War that flared up in September 1936) apparently sensitized the Soviet leadership to the historical and current links among eugenics, medical genetics, and *Rassenhygiene*.

A variety of factors contributed to the "death" of medical genetics in 1936-1937.[165] But the close connections of the new discipline with eugenics — at the level of ideas, methods, and practitioners — undoubtedly played a major role in its demise. Like eugenics just a few years prior, medical genetics was labeled a "reactionary," "bourgeois" science that had no place in socialist society. But this time, the major accusation leveled at medical genetics and its proponents was that they advanced a "fascist science." Since Hitler's ascent to power the very name "eugenics" (particularly its German variant, *Rassenhygiene*) in the Soviet Union had become strongly associated, if not completely equated, with the explicit racist policies of the Nazis.[166] Despite the protracted efforts of Soviet geneticists — inaugurated by Muller's damning 1934 article on "Eugenics in the Service of National-Socialism" in *Priroda* — to "expose" *Rassenhygiene* and to dissociate, in the words of one of them, "real genetics" from its "perversions" in Nazi propaganda and policies, human genetics, in the minds of many, retained strong fascist connotations.[167] In early December 1936, the mouthpiece of party ideologists, *Under the Banner of Marxism*, published an article under the revealing title "The Black-Guard Nonsense of Fascism and Our Medical-Biological Science." Signed by the head of the Moscow party science department, the article accused Levit and his IMG co-workers of holding "fascist views" on human genetics.[168] The journal's next issue carried a denigrating review of the IMG's latest publications advancing the same accusation.[169] The equating of human genetics with eugenics and Nazi racism figured prominently in practically all pronouncements against Levit and his staff, as happened, for instance, at the All-Union Congress of Neurologists and Psychiatrists held in late December 1936.[170] Furthermore, in his attack on genetics during the same month, December 1936, Trofim Lysenko cleverly exploited these links to discredit the leading Soviet geneticists Kol'tsov and Serebrovskii by accusing them of promoting "bourgeois eugenics."

In early May 1937, a special meeting in Narkomzdrav discussed the future of Levit's IMG.[171] Despite strong advocacy by Davidenkov and the sympathetic attitude of Commissar Kaminskii, most participants repeated the accusations against Levit and his co-workers of promoting a "fascist science." Even the research methods of medical genetics such as twins-based studies came to be labeled "fascist." A few months later, the arrest of Kaminskii as a "member of the Trotskyist conspiracy" sealed the IMG's fate: the institute was closed and its staff dispersed. Muller, its main "scientific consultant," left the Soviet Union for Britain. The next spring, Levit was arrested as an "enemy of the people" and executed. With the dissolution of its main research center and the death of its most active champion, the field of medical genetics in the Soviet Union disintegrated. Around the same time, Kol'tsov's IEB was transferred to the Academy of Sciences, and Soviet genetics completely lost Narkomzdrav's patronage, even though some clinical work on hereditary diseases continued at certain medical institutions, such as Davidenkov's neurological clinic in Leningrad.

The demise of Soviet eugenics-turned-medical-genetics had considerable resonance beyond the borders of the USSR.[172] The perceived links between human genetics, *Rassenhygiene*, and eugenics had played a significant role in the "postponement" by Soviet authorities of Moscow's hosting of the Seventh International Genetics Congress in late 1936. This event, in turn, profoundly shaped the international genetics community's attitude towards their discipline's political and ideological ramifications in both Stalin's Russia and Hitler's Germany and led to the relocation of the congress to Britain. It also deeply affected the community's standing vis-à-vis eugenics in general. One of the highlights of the Moscow congress was supposed to be a "discussion of questions relating to racial and eugenic problems" initiated by a group of US geneticists and aimed at delivering a concerted critique of German *Rassenhygiene*. The discussion was to feature presentations by the leading Soviet medical geneticists Davidenkov and Levit and by several eminent western geneticists, including Britons Julian Huxley and Lancelot Hogben, Norwegian Otto Mohr, and American Herbert S. Jennings. The "postponement" of the Moscow congress dampened the resolve of the international community to face the challenges posed by *Rassenhygiene* head on, and pushed it to distance genetics as a discipline from its stepsister, eugenics. After its withdrawal from Moscow and relocation to Edinburgh, the congress's organizing committee adamantly rejected the offer by the British Eugenics Society to hold an international eugenics congress jointly with the genetics one. Furthermore, the committee made sure that the congress's sessions on human genetics touched on neither *Rassenhygiene*, nor eugenics.

But despite the organizing committee's efforts, the "racial and eugenic problems" would not simply go away. Shortly before the congress's opening in Edinburgh in late August 1939, the committee received a cable from "Science Service," a US-based news outlet specializing in science reports. The cable asked the congress's "representative participants" to provide the news agency with "several hundred words discussing how could [the] world['s] population [be] improve[d] most effectively genetically." Muller, the head of the congress's programme committee, enthusiastically took on the job of answering the cable and wrote a long — nearly 1,300 words — memorandum on the subject.[173] Muller carefully crafted his answer: he did not even use the words "eugenics" and "racial hygiene." Yet its contents largely repeated the pointed critique of "bourgeois eugenics" and "racial hygiene" advanced in numerous articles he and his Soviet colleagues had published in the previous years. Echoing the arguments for "socialist eugenics," Muller stated that "the effective genetic improvement of mankind is dependent upon major changes in social conditions." A "major hindrance to genetic improvement," according to Muller, "lies in the economic and political conditions, which foster antagonism between different people, nations, and 'races'." "Both environment and heredity constitute dominating and inescapable complementary factors in human well-being," he elaborated, "but factors both of which are under the potential control of man and admit of unlimited but interdependent progress."

Muller's memorandum was aimed not only at bourgeois eugenics and racial hygiene — it was also intended to help his Soviet colleagues in their struggles with Lysenko's clique promoting the Lamarckian notion of heredity. "It must... be understood," Muller emphasized,

> that the effect of bettered environment is not a direct one on the germ cells and that the Lamarckian doctrine is fallacious, according to which the children of parents who have had better opportunities for physical and mental development inherit these improvements, biologically, and according to which, in consequence, the dominant classes and peoples would have become genetically superior to the underprivileged ones.

Muller did not even attempt to make his memorandum a subject for public discussion at the congress. Instead he personally asked several participants to add their signatures to his text. Twenty-one geneticists signed on to Muller's statement that soon appeared in print in the oracle of US genetics, *Journal of Heredity*, and became known as the "geneticists' manifesto."[174]

The situation in Soviet genetics, however, could not be remedied by mere manifestos. Although more than forty Soviet geneticists had been scheduled to deliver reports at the Edinburgh congress and their leader Vavilov elected its president, none of them came to Scotland. The Soviet leadership forbade the participation of Soviet scientists in the international meeting. It was a heavy blow to Soviet geneticists who had counted on the congress and the support of their western colleagues as a powerful tool in their attempts to stop Lysenko and his cronies' continuing attack on their discipline. Deprived of international support, Soviet geneticists appealed to the party-state apparatus to halt Lysenko's encroachment on genetics institutions and to permit a "public discussion" of their disagreements with Lysenko.[175] The Central Committee Secretariat — one of the Bolshevik party's top decision-making bodies — permitted the discussion. But, contrary to the expectations of Soviet geneticists who had hoped to hold it under the aegis of their stronghold, the Academy of Sciences, the party bosses put "Marxist" philosophers in charge of adjudicating the disagreements. In October 1939 the editorial board of *Under the Banner of Marxism* conducted a week-long conference that gathered 159 participants and featured 53 presentations by the members of three competing groups: geneticists, Marxist philosophers, and Lysenkoists.[176]

Although its main theme was the "practical achievements" of genetics, especially in agriculture, as one would have expected, the conference could not stay clear of eugenics and medical genetics. Lysenko's supporters repeated their stock accusations of "formal" genetics' close links to eugenics. But, given the new political alliance of the Soviet Union with Nazi Germany embodied in the Molotov-Ribbentrop Pact signed on 23 August 1939, the day the International Genetics Congress opened in Edinburgh, they had to drop their favorite indictments of both human genetics and eugenics as "fascist." Similarly, geneticists had to mute the references to their western colleagues' antifascist stance against *Rassenhygiene*.

Nevertheless, the only representative of medical genetics at the meeting, Davidenkov, felt it necessary to begin his report with a declaration that "Soviet geneticists decisively reject all eugenic and racist theories. These theories do not derive from genetics, they have specific socio-economic roots. In certain countries, genetics is raped as are other sciences, for instance, anthropology and history."[177] Davidenkov also lamented that the unfounded critique by Lysenko's supporters made many physicians stay away from medical genetics as incompatible with building socialism. Even so, in its report to Stalin and the Politburo, the editorial board emphasized: "in capitalist countries, the doctrine of 'genes' is used again and again to substantiate the men-hating theories of racism. In the USSR, academician Serebrovskii, and especially Prof. Kol'tsov, based on the theory of 'genes', develop in their 'works' extremely reactionary views and conclusions."[178]

The discussion of issues in genetics "under the banner of Marxism" ended in an impasse, with the geneticists, Lysenkoists, and philosophers all able to maintain their current positions. But the very next year Soviet genetics suffered heavy losses. Kol'tsov passed away. Vavilov, along with a number of his closest co-workers, was arrested and two years later he died of starvation in prison.[179] Lysenko's followers seized administrative control over both Kol'tsov's IEB and Vavilov's Institute of Genetics, the last two strongholds of genetics under the Academy of Sciences.

By the end of the 1930s, then, the broad research programme conceived by the founders of medical genetics was radically curtailed and limited to just a few disparate studies. The Nazi attack on the Soviet Union in June 1941 broke the Molotov-Ribbentrop Pact and plunged the country into World War II. The Great Patriotic War, as it was called in the Soviet Union, lasted four years and put the studies in medical genetics on the back burner. Doctors now had much more urgent tasks of tending to wounded soldiers and preventing epidemics than collecting clinical genealogies or studying twins. Yet almost immediately after the war, research on human and medical genetics was resumed, particularly in Leningrad, where since the early 1930s Davidenkov and his numerous co-workers had conducted their studies under the umbrella of various medical clinical, research, and educational institutions. Pavlov's institute in Koltushi — "inherited" after its founder's death in 1936 by his oldest and most respected student, Leon Orbeli — became a major hub of research on medical and human genetics, with Davidenkov as its principal "scientific consultant."

During the war, Davidenkov had become a member of the USSR Academy of Medical Sciences (AMN), a new agency established by Narkomzdrav in 1944 to administer and control nearly all medical research in the country.[180] He clearly intended to use this elevated position to promote medical genetics. In 1947, he published (with Orbeli's highly praising foreword) a solid tome on *Evolutionary-Genetic Problems in Neuropathology*, which synthesized his twenty-plus-year experience in clinical studies on human heredity and hereditary diseases.[181] Tellingly, in his text Davidenkov did not mention eugenics at all, though he did cite several studies that had appeared in the *Annals of Eugenics*. He did mention Galton, once, in relation to his discussion of the evolutionary origins of the "imperfection" of the human nervous system manifested in numerous nervous disorders that plagued humanity.[182] Summarizing his findings, Davidenkov called for the expansion of research on the genetics of psychiatric and neurological ailments in the clinic and the laboratory.

Just a year later, however, such investigations were completely abandoned. The rapidly escalating Cold War set up a stage for Lysenko's renewed attack on "western," "bourgeois," "imperialist" genetics, this time endorsed personally by Stalin.[183] In August 1948, at a special session of the Lenin All-Union Academy of Agricultural Sciences (VASKhNIL) the entire discipline of genetics was officially banned in the Soviet Union. Lysenko and his supporters rechristened genetics as "pernicious" Mendelism-Weismannism-Morganism, after its acknowledged western founding fathers Gregor Mendel, August Weismann, and Thomas Hunt Morgan. They seized administrative control over nearly all facilities involved with genetics research. They also replaced the teaching of genetics in schools and universities with Lysenko's own doctrine, named "new genetics" or, more often, "Michurinist biology," after Ivan Michurin, an amateur plant breeder accorded the status of a national hero in the 1930s. Genetics research institutions were closed or reorganized, many geneticists fired, and genetics publications removed from libraries. After 1948, in the Soviet Union "classical" genetics in all its forms, including medical genetics, disappeared.

An article published in March 1949 in the popular illustrated weekly *Flash* (*Ogonek*) by an ardent supporter of Lysenko demonstrates that in the public mind, "classical" genetics became tightly bound to eugenics, racism, fascism, and Anglo-American imperialism. Its very title, "Fly-lovers and Men-haters," equated geneticists (dubbed "fly-lovers" in a clear reference to their favourite research object, the fruit-fly *Drosophila*) with the inhumane ("men-hating" in the current Soviet parlance) practices of racism, colonialism, and fascism. The subtitles of the article's various sections speak for themselves: "Mendelian genetics and fascism," "Mendelism defends racial ideology," and "The science of 'horror and fear'."[184] *Flash* was one of the most widely distributed weekly journals with a print-run of more than half a million copies, and to make sure the readership got the article's message its text was interspersed with cartoons drawn by Boris Efimov (1900-2008), a well-known political cartoonist. To convey the article's contents the artist deployed easily recognizable images — a dollar sign, a swastika, a characteristic outfit of a Klu Klux Klan member, a fat capitalist in the appropriate attire, and a policeman with a machine gun — to which the Soviet readership had long been conditioned (thanks in no small part to Efimov's numerous cartoons that appeared regularly in *Pravda* and *Izvestiia*). These images were accompanied by customary symbols of science/genetics, including a test tube with fruit flies, a textbook (whose title is tellingly written in English, not Russian), a microscope, and a Petri dish (see fig. 6-6). Predictably, a large portion of the article was devoted to eugenics, presented as an extension of "pernicious" Mendelian genetics to the issues of human heredity. As entries on Galton, genetics, and eugenics in the new 1952 edition of the *Great Soviet Encyclopedia* made manifest, in Lysenkoist rhetoric, Galton became a "reactionary racist anthropologist;" eugenics—a "pseudoscience" used by the Nazis and by American imperialists for their nefarious purposes; and genetics — an expression of "mysticism and idealism" characteristic of capitalism and its science.[185]

From 1930 on, then, in the Soviet Union eugenics was no longer "ours," only "theirs." The very word *eugenics* became a pejorative reserved exclusively for "bourgeois science," while its Soviet proponents reconstituted their enterprise into medical genetics. The new discipline took over one part of the extensive programme previously advanced within the framework of eugenics — the study of human heredity and its medical applications.

Surprising as it might seem, in their extensive efforts to build medical genetics on the ruins of eugenics, the champions of the new discipline did not refer to Florinskii's *Human Perfection and Degeneration*. Actively advertised as the foundation of "bio-social," "socialist," and "proletarian" eugenics during the 1920s, Florinskii's treatise might well have served as a suitable instrument to hide the origins of medical genetics in what during the 1930s was invariably called "bourgeois," "fascist," and "racist" eugenics founded by Galton. With its clear focus on "physical and moral health" Florinskii's book would seem an ideal tool for differentiating medical genetics from eugenics. Yet, even the discoverer of Florinskii's treatise, Volotskoi, seemed to have completely forgotten his earlier admiration for the book and its author. Although he actively contributed to the development of medical genetics, Volotskoi never mentioned *Human Perfection and Degeneration* in any of his published works on the subject in the 1930s.[186]

Fig. 6-6. A series of cartoons depicting genetics as a "fascist," "racist," and "imperialist" science, by Boris Efimov. From *Flash*, 1949, 11: 14-16. Courtesy of BAN.

This silence speaks volumes. It suggests that during the 1930s *any* form of eugenics was quite deliberately excluded from the *scientific* discourse on human variability, heredity, development, and evolution. All discussion of eugenics was now confined solely to the political and the ideological. Contrary to the claims of many later commentators, in 1930 eugenics in the Soviet Union was neither forbidden, nor outlawed. Throughout the 1930s, *Journals Chronicle*, the country's major bibliographical periodical, maintained its section on "eugenics," established in 1926.[187] But the publications, which were referenced under this rubric after 1930, were nearly all devoted to the political and ideological attacks on eugenics as the embodiment of bourgeois, racist, fascist, capitalist, and imperialist ideas, values, concerns, and practices. Under these conditions, the founders of medical genetics used every possibility to obscure all and any links between their new enterprise and eugenics. They spared no efforts to distance the new discipline from its stepmother. The total silence that once again enveloped Florinskii's treatise was likely a result of these determined efforts. All the projects of "bio-social," "proletarian," "socialist," "Soviet" eugenics advanced during the 1920s were abandoned. And with them, one of their major inspirations — Florinskii's *Human Perfection and Degeneration* — slipped into oblivion, again.

# 7. Afterlife: Medical Genetics and "Racial" Eugenics

> "Eugenics has to wait for its hour, and nobody knows how soon that hour could arrive."
>
> Vladimir Polynin, 1967

After the 1948 banishment of "classical" genetics in the Soviet Union, eugenics and its founder Galton became a practice target for "Marxist" philosophers and party ideologues, while any mention of Florinskii's eugamics completely vanished from Soviet discourse. It seemed that *Human Perfection and Degeneration* was now permanently consigned to gathering dust in some remote library storage. But, Fate is a fickle mistress. Shortly after Stalin's death, during the de-Stalinization campaign launched by his successor Nikita Khrushchev in the 1950s and popularly known as the "Thaw," medical genetics re-emerged in the Soviet Union. Its "resurrection" culminated in the establishment in Moscow of a brand new Institute of Medical Genetics in 1969. Just two years later, both Galton and eugenics were "rehabilitated," and with them, Florinskii and his book re-entered the contemporary discourse on human reproduction, heredity, variability, development, and evolution. In 1995 a new edition of Florinskii's treatise came out under the auspices of the very university its author had created in Tomsk more than a century earlier. In 2012, the book was republished once more.

A series of dramatic developments — both social and scientific — engendered the long quiescence of *Human Perfection and Degeneration* and inspired its successive revivals in late-Soviet and post-Soviet Russia.

## The "Rebirth" of (Medical) Genetics

After August 1948, "classical" genetics in all its forms was banished in the Soviet Union. Yet, though some western observers mourned the "death" of Soviet genetics, the discipline was not dead.[1] Rather, it went underground. Although by that time, the founders of Soviet genetics — Nikolai Kol'tsov, Iurii Filipchenko, Nikolai Vavilov, and Alexander Serebrovskii — had all passed away, the extensive networks of their colleagues and students survived Trofim Lysenko's takeover of the discipline. Even as Lysenko's "Michurinist biology" reigned supreme and his cronies seized control of nearly all genetics institutions, the discipline's newly found importance in the age of nuclear weapons and nascent space exploration enabled its practitioners to build new institutional bases outside of Lysenko's administrative reach. "Classical" genetics and geneticists survived Lysenko's onslaught in the guise of "radiation biology," "chemistry of bioactive compounds," "physico-chemical biology," "medical radiology," and other similarly cryptic names, under the protective umbrella of physics and chemistry research facilities involved in Soviet nuclear, chemical, and biological weapons and space programmes that became the hallmark of Cold War science. Stalin's death in March 1953 and the subsequent de-Stalinization campaign initiated by his successor Nikita Khrushchev greatly facilitated this process.[2]

The Thaw inaugurated a radical departure from Stalinism, by introducing a series of political, economic, and social reforms, liberating millions of prisoners from the Gulag, and parting (though just slightly) the Iron Curtain that had separated the Soviet Union from the West with the start of the Cold War. As Khrushchev himself defined its main goals in his famous "secret speech" to the Twentieth Party Congress in February 1956, the de-Stalinization campaign aimed "to overcome the negative consequences of Stalin's personality cult" and "to rehabilitate victims of Stalinist repression."[3] The entire discipline of genetics certainly qualified as a victim of Stalinist repression, and its proponents launched a concerted campaign for its "rehabilitation."

Already in 1957, barely a year after Khrushchev's "secret speech," a brand new Institute of Cytology and Genetics directed by Kol'tsov's student Nikolai Dubinin was created as part of *Akademgorodok* — a "science city" built for the just instituted Siberian Branch of the USSR Academy of Sciences in Novosibirsk.[4] Although two years later Khrushchev personally fired Dubinin for the latter's opposition to Lysenko who still exercised his stronghold over Soviet biology and agriculture, the rehabilitation of genetics continued: under the directorship of another geneticist, Dmitrii Beliaev, the Institute of Cytology and Genetics rapidly grew to become a leading center of genetic research in the country.[5]

Furthermore, Soviet geneticists were able to re-establish, albeit on a limited scale, their contacts with western colleagues, which had been completely severed after 1948. Although in August 1958, only adherents of "Michurinist biology" came to the Tenth International Genetics Congress in Montreal, a month later, in September, several Soviet "Mendelists," including Kol'tsov's students Alexandra Prokof'eva-Bel'govskaia and Sos Alikhanian, attended the Second International Conference on the Peaceful Uses of Atomic Energy in Geneva. There, after more than twenty years of separation, they re-united with their one-time mentor H. J. Muller and met a number of other western geneticists.

As part of the overall revival of genetics, medical genetics was rehabilitated too. Indeed, medical applications of genetics — especially research into the harmful effects of chemical and radioactive mutagens — became the major justification for the rebirth of the discipline. In contrast to the resurrection of general genetics under the auspices of the Academy of Sciences, which was marred by institutional rivalries and raging controversies (particularly in evaluating the discipline's troubled history and on devising strategies to isolate it from Lysenko's continuing influence) among several groups of its practitioners,[6] the re-institutionalization of medical genetics appeared largely unproblematic. The USSR Academy of Medical Sciences (AMN) — an important base of biological research that had managed to escape Lysenko's administrative domination while paying lip service to his Michurinist biology — became its main institutional springboard.[7] Sergei Davidenkov, one of the last surviving vocal supporters of eugenics, spearheaded the resurrection of medical genetics. Already in 1957, he managed to establish in Leningrad a "medico-genetic laboratory" under the AMN aegis. The following year, he published a voluminous entry on "medical genetics" in the *Great Medical Encyclopedia*, even though Lysenko's Michurinist biology remained the officially endorsed doctrine of heredity. In establishing a proper "native" genealogy of medical genetics and connecting it with pressing current concerns, Davidenkov referred to the four volumes of the IMG proceedings issued by Solomon Levit in the 1930s, as well as to recent publications on "genetic consequences of radioactive irradiation."[8]

Over the next few years, various institutions devoted specifically to medical genetics popped up throughout the vast network of the AMN's research facilities,[9] specialized courses appeared in the curricula of medical schools,[10] and foreign books on the subject were translated and published.[11] In 1961, together with Vladimir Efroimson, a student of Kol'tsov who had just a few years earlier returned from the Gulag, Davidenkov published an extensive entry on "Human Heredity" in the *Great Medical Encyclopedia*. The nearly fifty-page-long article demonstrated that the champions of medical genetics had managed to dispose of the labels "bourgeois," "racist," and "fascist" attached to their specialty by Stalin's ideologues, "Marxist" philosophers, and Lysenko's disciples.[12] The next year, the AMN official *Herald* devoted an entire issue to medical genetics.[13]

With Khrushchev's ousting from power in October 1964, Lysenko finally lost his administrative grip on Soviet biological and agricultural research.[14] Classical genetics returned to its prominent place on the agendas of Soviet science, medicine, and agriculture. Geneticists reclaimed their lost institutional bases and feverishly built new ones. One of their most important challenges was the restoration of the teaching of genetics in various schools of higher learning, which had been dominated by Lysenko's followers for nearly twenty years. In March 1965, under the auspices of Moscow University, several former students of both Kol'tsov and Filipchenko organized a series of "remedial" lectures for professors of universities, as well as agricultural, medical, and pedagogical schools, on "Current Issues in Modern Genetics" that soon appeared in print as a 600-page volume edited by Alikhanian.[15] As a clear sign that "Mendelism" was no longer a dirty word of Soviet political rhetoric, the same year the Academy of Sciences released a new edition of Mendel's *Experiments on Plant Hybrids*, prepared by another student of Kol'tsov.[16] The academy also launched a new journal, proudly titled *Genetics*. A year later, geneticists established their first disciplinary society, the All-Union Society of Geneticists and Breeders (VOGiS). They defiantly named it after Vavilov, one of the martyrs of Lysenko's anti-genetics campaign, who became one of the first scientists rehabilitated (posthumously) during the Thaw.[17] Boris Astaurov (1904-1974), yet another student of Kol'tsov, became the first president of VOGiS and director of his teacher's Institute of Experimental Biology.[18]

Medical genetics became an integral part of this "genetics renaissance", which was aptly manifested in the virtual explosion in the number of publications, conferences, courses, and institutions in the field during the 1960s. The programme of the discipline's extensive development, which had been elaborated by the participants of the 1934 conference on medical genetics organized by Levit, was finally implemented. As the conference's resolution had envisioned,[19] "scientific-research centers for medical genetics and cytology" were created throughout the country, genetics was included in the curricula of medical education, with "departments of medical genetics" established at several medical institutes and "necessary textbooks" produced.

These concerted efforts culminated in the creation in 1969 under the AMN auspices of a brand new Institute of Medical Genetics in Moscow. Nikolai Bochkov (1931-2011) — a young doctor who had studied genetics just a few years earlier under the tutelage of Kol'tsov's former students Dubinin, Prokof'eva-Bel'govskaia, and Nikolai Timofeev-Ressovskii — was appointed director of the new institute and became the discipline's leading spokesman in the party-state apparatus.[20] Following the lead of its predecessor, Levit's IMG, the new institute quickly organized a graduate programme to train a new generation of "medical geneticists."

By that time, nearly all members of the Russian Eugenics Society involved with the early development of medical genetics, including Davidenkov, Iudin, Kol'tsov, Levit, Serebrovskii, and Volotskoi, had passed away. It was their students, and students of their students, who effectuated the second institutionalization of their teachers' endeavor. Indeed, as a way to legitimize their field, the new generation of medical geneticists traced their genealogy to Levit's IMG, as the world's first institution of its kind.[21] The first original Russian textbook on medical genetics published in 1964 by Efroimson contained a lengthy, laudatory account of Levit's and his co-workers' research on the medical applications of genetics.[22]

Yet the successful revitalization of medical genetics did not mean that Galton and eugenics were also rehabilitated. After all, the first generation of medical geneticists had done their best to obscure the links between their new discipline and its predecessor. Characteristically, the *Great Medical Encyclopedia* that contained Davidenkov's articles on "medical genetics" and "human heredity" had no entry on Galton. It did carry a lengthy article on eugenics, though, written by Evgenii Pavlovskii, an eminent protozoologist and Davidenkov's fellow member of the AMN. In the aftermath of August 1948, Pavlovskii had become a vocal ally of Lysenko.[23] Expectedly, his article pointed to "the energetic struggle against eugenicists by our social hygienists [such as] Z. G. Solov'ev and N. A. Semashko," and repeated Lysenko's stock condemnation of eugenics as "racist," "fascist," and "imperialist," but did not provide any references or suggestions for further readings.[24]

Given that after 1930 in the Soviet Union eugenics had become exclusively "theirs," one might think that the new champions of medical genetics would not be aware of the discipline's deep roots in eugenics, and even less so of Vasilii Florinskii and his treatise. And even if they were, they would not be much interested in exposing these roots. After all, the major justification for the resurrection of medical genetics was not the old concerns with "hereditary diseases" and "degeneration," which had fueled the early interest in eugenics, but the new dangers ushered in by the arms and space race of the Cold War. Yet, just as Mendel was reinstated into Soviet discourse on heredity, so too Galton, eugenics, Florinskii, and "marriage hygiene" were brought back into Soviet discourse on human reproduction, heredity, development, and evolution. In contrast to the first institutionalization of medical genetics in the early 1930s, this time, commentators on Galton's and Florinskii's concepts presented both as the predecessors of "new" medical genetics that superseded and replaced the outdated and flawed "old" eugenics. Indeed, they hailed Florinskii as the founder of the discipline in Russia and his treatise as one of the "first approaches to the development of medical genetics."

## Eugenics, Rehabilitated

In 1967, Vladimir Blanter, a young journalist based at the leading popular-science magazine *Priroda* (who wrote under the pen names of

Polynin and Dolinin), published a voluminous popular book on human heredity, enticingly titled *Mother, Father, and I*.[25] Sketching the short but complicated history of human genetics, Blanter wrote a special section that favorably described Galton's efforts in "creating a new science of eugenics" devoted to "human betterment" and outlined the "perversions" of eugenics by its American and German proponents.[26] The journalist briefly recounted eugenic ideas espoused by Filipchenko, Kol'tsov, and Serebrovskii, without once mentioning any other Russian eugenicists, the Russian Eugenics Society, or the *Russian Eugenics Journal*.[27] In fact, Blanter misrepresented Kol'tsov's eugenic programme and conflated the three facets of eugenics carefully delineated by the RES president. He stated that the leader of Soviet eugenics had sought "to search for practical measures to rid humankind of hereditary ailments [and] to turn away from problematical eugenics to practically important anthropotechnique, or, as it was later named — anthropogenetics, or, as it is called now — medical genetics." "Eugenics as a science has withered away," Blanter declared, "not because it was unnecessary, but because it was impracticable at the time." "Eugenics has to wait for its hour, and nobody knows how soon that hour could arrive," he asserted.

In an enthusiastic foreword to Blanter's book, the VOGiS president Astaurov announced, however, that the "hour of eugenics" had already arrived.[28] Echoing Volotskoi, Serebrovskii, and other 1920s proponents of "socialist" eugenics, Astaurov maintained that "every social formation creates its own eugenics." "The mistakes and misunderstandings of eugenicists in capitalist societies," he declared, do not mean that "the very idea of eugenics is wrong." To the contrary, in his opinion, it is socialist society that offers the possibility of creating "true eugenics." "We must create such a system of the protection of hereditary health," he elaborated, "within which the interests of society would not suppress the rights of the individual, [and] the protection of the health of all would not undermine but support the care for the health of the individual." Astaurov praised Blanter for bringing this issue to the attention of his readers. "Whatever forms the socialist eugenics of the future (will hope a very near future!) will take," he concluded, "there is no doubt that it will be built on the strong foundation of exact knowledge about the general laws of genetics [and] about human genetics (anthropogenetics), and it will consist of the rational applications of recommendations offered by medical genetics."

Five years later, in 1972, the rehabilitation of eugenics and its founder came to a head with the publication of the first Russian-language scholarly biography of Galton.[29] The book came out under the auspices of Nauka, the publishing house of the Academy of Sciences, as part of its renowned series of "Scientific Biographies," with a print run of 10,000 copies. Ivan Kanaev (1893-1984), a student of Filipchenko, finished the job begun by his mentor half a century earlier by producing a detailed account of Galton's life and works.

Fig. 7-1. Iurii Filipchenko (in the center of the first row) with his students: Theodosius Dobzhansky (left in the second row) and Ivan Kanaev (third from the left in the second row), 1925. Photographer unknown. Courtesy of N. Medvedev.

Kanaev was perfectly positioned to do the job. Mostly remembered as a historian of biology and a life-long friend of the prominent philosopher, literary theorist, and semiotician Mikhail Bakhtin (and an influential member of the "Bakhtin circle"),[30] Kanaev was actually one of the leading human geneticists of his time.[31] Born in 1893 in St. Petersburg to the family of a civil servant, Kanaev graduated from Petrograd University's school of natural sciences in the spring of 1918, just as the new Bolshevik government had abandoned the imperial capital.³² Together with hundreds of thousands of the city's inhabitants, the young biologist fled starving Petrograd for better pastures. He settled in a small town, some 300 kilometers south of Petrograd, where he survived the harsh years of the civil war and War Communism by teaching science at a local school. With the end of the civil war and the adoption of the NEP, Petrograd began to come back to life, and in the summer of 1922 Kanaev returned to his home city. He made a living as a science teacher at a secondary school and began his academic career as an unpaid assistant in Filipchenko's department of "genetics and experimental zoology" at his alma mater.³³

As did other students of Filipchenko, in the 1920s Kanaev took an active part in the wide-ranging campaign mounted by his mentor to popularize genetics.³⁴ But his personal interests centered on other areas of experimental biology, especially regeneration. He studied the regeneration of *Hydra*, a genus of small, fresh-water animals that had captivated scientists' attention since Swiss naturalist Abraham Trembley first researched these polyps in the mid-eighteenth century.³⁵ In 1926, Kanaev quit his teaching job and entered graduate studies in Filipchenko's department. He employed new experimental methods to investigate the histology, cytology, and morphology of *Hydra* in order to uncover the mechanisms of their remarkable capacity for regeneration. These studies — presented in several articles in Russian and German scholarly and popular journals — laid a foundation for his dissertation, successfully defended in early 1930.³⁶ After the death of Filipchenko in the spring of 1930, Kanaev joined the biology department at the just established Leningrad Medical Institute as an assistant professor.³⁷ Perhaps the move to the medical school induced Kanaev to radically shift his focus. He turned to the subject that had fascinated his late mentor — human genetics. In the early 1930s, Kanaev began research on the genetics of "higher nervous activity" in twins.

The unwieldy expression "higher nervous activity" had been introduced by Russia's Nobel-winning physiologist Ivan Pavlov in his efforts to create an objective vocabulary for describing what contemporary psychologists called the "psyche" or the "mind."³⁸ His method of conditional reflexes provided a major tool for studies of higher nervous activity, which from 1903 on became the main focus of research conducted by Pavlov and his numerous co-workers.³⁹ In the early 1920s, as did many of his fellow physiologists, Pavlov supported the Lamarckian notion of the inheritance of acquired characteristics and even attempted to prove it experimentally, assigning one of his students to investigate "the inheritance of conditional reflexes" in successive generations of mice. Allegedly these experiments proved that conditional reflexes acquired by one generation become inherited in the subsequent ones, and Pavlov proudly presented the results of these experiments at several international and domestic meetings. Pavlov's reports elicited sharp critique from geneticists, including Thomas Hunt Morgan and Kol'tsov, prompting Russia's premier physiologist to turn his attention to genetics.[40] Indeed, at Kol'tsov's instigation, Pavlov created a special laboratory for studies on "the genetics of higher nervous activity."[41] As a symbol of Pavlov's commitment to genetics, a bronze bust of Mendel was placed in front of the laboratory in Koltushi (see fig. 7-2).

Fig. 7-2. Gregor Mendel's bust in front of Ivan Pavlov's Laboratory for the Experimental Genetics of Higher Nervous Activity in Koltushi. An inscription on the building's façade reproduces Pavlov's motto: "Scrutiny and Scrutiny." After August 1948, the bust was removed and put in storage. It was restored to its prominent place only in the 1960s. Photo by the author, 2016.

In the fall of 1932, Kanaev approached Pavlov with a proposal to organize research on higher nervous activity in twins. Inspired by Galton's pioneering studies of twins to distinguish the role of nature and nurture in the development of human mental characteristics, Kanaev suggested that a careful comparison of various parameters in the formation of conditional reflexes in twins might help disentangle the "inherited" and the "acquired" in their higher nervous activity. Pavlov appeared quite interested. He himself, however, never experimented on humans; his preferred research subjects were dogs. But some of his co-workers did carry out research on humans. Nikolai Krasnogorskii, a well-known pediatrician, had been studying conditional reflexes in children since the early 1920s.[42] Thus, it was in Krasnogorskii's laboratory that Kanaev began his experiments on conditional salivary reflexes in twins.[43] A few years later, he joined Davidenkov's laboratory and greatly expanded his research. Kanaev published numerous articles in academic and popular journals detailing the results of his studies, which formed the basis for the Doctor of Science dissertation he successfully defended in 1939.[44]

The Nazi invasion in 1941 interrupted Kanaev's research. Together with the staff of the Leningrad Medical Institute, he was evacuated from the besieged city to the rear. In summer 1944 he returned to Leningrad and soon resumed his twins-based studies at the institute in Koltushi directed by Leon Orbeli. In early 1948, Kanaev published an overview of his research on the "experimental genetics of higher nervous activity" and outlined its future directions.[45] Just a few months later, the fateful August VASKhNIL session banned genetics in the Soviet Union. Kanaev was fired from his post as the chairman of the biology department at the Leningrad Medical Institute.[46] But Orbeli managed to hide the well-known "Mendelist" at his institute in Koltushi, where Kanaev switched from genetics to physiological research on various motor reactions and time sense in children.[47]

After Stalin's death, Kanaev joined other geneticists in their efforts to rehabilitate their discipline.[48] But with the beginning of the Thaw, he once again radically changed his academic career. In 1957, he joined a branch of the Academy of Sciences Institute for the History of Natural Sciences and Technology (IIET) recently established in Leningrad.[49] By the end of the year he published his first monumental work as a historian: a scholarly edition of Johann Wolfgang Goethe's *Selected*

*Works on Natural Science*, issued by Nauka in its renowned series "Classics of Science." Kanaev not only translated into Russian all the texts included in the 550-page volume, he also supplied them with extensive commentaries and a lengthy analysis of Goethe's work as a naturalist.[50] Two years later he published a sizable monograph on twins, tracing the history of anatomical, physiological, psychological, and genetic studies on twins from Antiquity to the present.[51] In the following decade, Kanaev published numerous articles and half a dozen monographs on the history of biology.[52] He produced detailed historical analyses of the investigations on the physiology of color vision and on comparative anatomy before and after Darwin. He wrote scholarly biographies of such luminaries as Georges-Louis Buffon, George Cuvier, Goethe, Carl-Friedrich Kielmeyer, and Abraham Trembley. In short, during the 1960s, Kanaev became a leading Soviet historian of eighteenth- and nineteenth-century biology. Recognizing his numerous accomplishments, in 1971 the International Academy of the History of Science elected him a corresponding member.

It is unclear from available materials who or what prompted Kanaev to write a biography of Galton. But its publication appears to have been part of an extensive effort to rehabilitate eugenics as a legitimate way to address the issues of humanity's future and the role that human genetics could play in shaping that future. The development of nuclear, chemical, and biological weapons of mass destruction, together with the start of the space era with the launching of Sputnik in 1957 and Iurii Gagarin's flight in 1961, generated acute anxieties about the possible future of humankind. In the tense atmosphere of the Cold War, the apocalyptic visions of humanity's destruction in a possible next (nuclear, chemical, bacteriological) world war — intertwined with dreams of conquering the "final frontier," space, and either joining other sentient beings in intergalactic unions, or fighting off invading aliens — fueled extensive social debates about the future, readily manifested in the virtual explosion of science fiction in both the East and West during the 1950s and 1960s.[53]

The contemporary path-breaking developments in biology — from the deciphering of the genetic code to the synthesis of Darwinism with population genetics and ecology — made biology, and particularly genetics, an integral part of such debates, as witnessed by the 1962 symposium on "Man and his Future" held in London with contributions by the leading lights of contemporary biology, including Francis Crick, J. B. S. Haldane, Julian Huxley, Joshua Lederberg, Fritz A. Lipmann, Peter B. Medawar, and Muller.[54] Indeed in the West, these developments spurred the rise of a "new" eugenics that strove to dispense with the negative connotations and legacies of pre-war "mainline," "old-school" eugenics (engendered by the Nazi atrocities and forced sterilizations of the "unfit" in the United States, Scandinavia, and elsewhere) and to revitalize the eugenic vision of a directed human evolution.[55]

But for Soviet biologists, and especially geneticists, engaging in these debates presented a clear ideological danger. For, as we saw in the previous chapter, after 1930, in the Soviet Union the mere invocation of biology in any discussion of human nature and humanity's future had been condemned as pernicious biologization, and the entire subject had become the exclusive domain of "Marxist" philosophers and party ideologues. Under these conditions, the liberating atmosphere of the Thaw notwithstanding, legitimizing the medical applications of genetics in the eyes of the party-state apparatus was one thing — relatively easily accomplished by references to the harmful genetic effects of nuclear weapons tests and industrial pollution. But justifying geneticists' incursion into issues of humanity's future was an entirely different task, one that required first and foremost dispensing with philosophers' tight control over what in Marxist lingo was elliptically named "methodological problems of biology," but in fact meant the social applications and societal implications of biology, and particularly genetics.

During the late 1950s and early 1960s, Soviet geneticists launched a coordinated campaign aimed at wresting control over the "methodological problems" of their discipline from philosophers.[56] They regularly referred to the debates over the importance of genetics to the future development of humanity by their western colleagues, especially those sympathetic to the Soviet Union, such as J. B. S. Haldane, and even translated some of them into Russian.[57] For instance, in 1966, *Issues in Philosophy* (the mouthpiece of Soviet philosophers that had replaced *Under the Banner of Marxism*) carried a lengthy overview of the polemics on "genetics and eugenics" that had unfolded at the London symposium on "Man and his Future."[58] The campaign reached its apex in a special discussion on "human genetics, its methodological and socio-ethical issues" held in 1970 under the auspices of *Issues in Philosophy*.[59] Eminent Soviet geneticists, including Bochkov, Dubinin, Efroimson, and Alexander Neifakh, a leader of Soviet "molecular" genetics,[60] took part in the spirited defence of their own control over the social implications and possible applications of the latest discoveries in genetics and the role their discipline was to play in the future of humanity.[61] Over the next few years, numerous academic conferences explicitly addressed the interrelations of the biological and the social in the understanding of human reproduction, heredity, variability, individual and social development, and evolution.[62]

The campaign was not limited to debates on the pages of philosophy journals or the proceedings of scientific meetings. Geneticists made sure that their efforts in defending their own right to define the social applications and implications of their discipline reached a much broader audience.[63] One important venue was the *Great Soviet Encyclopedia* that, alongside the multivolume edition of Lenin's *Complete Works*, was an obligatory holding of every public library in the country.[64] In the mid-1960s, scientists began to lobby the party-state apparatus to produce a new edition of the *Great Soviet Encyclopedia* to replace the previous one, which had been issued during the Stalin era and was replete with the "Marxist" lingo and ideological clichés.[65] In 1970, just as the debate between geneticists and philosophers was unfolding, the first volumes of the new, third edition came out under the general editorship of Nobel-prize winning physicist Alexander Prokhorov.[66] The new edition became a convenient instrument for rehabilitating eugenics and its founder.

A brief entry on Galton that appeared in 1971, in the encyclopedia's sixth volume, however, was written not by a geneticist, but by Mikhail Iaroshevskii, a historian of psychology.[67] In contrast to the previous edition, which had characterized Galton as "a reactionary English racist anthropologist, founder of a bourgeois pseudoscience — eugenics,"[68] the new one focused almost exclusively on Galton's contributions to the development of experimental psychology and psychometry. All that Iaroshevskii had to say about eugenics fit into two sentences: "Analyzing hereditary factors, Galton came to the conclusion of the necessity of creating eugenics. The limitations of Galton's psychological views are expressed in his notion of the predetermination of a person's intellectual achievements by genetic endowments, and his political conservatism — in an attempt to present labouring masses as biologically defective."

The encyclopedia's eighth volume published the same year, by contrast, contained an extensive entry on "Eugenics" written by Mikhail Lobashov (1907-1971) — a student of Filipchenko who had "inherited" his teacher's genetics department at Leningrad University — in collaboration with Iurii Vel'tishchev, a well-known Moscow pediatrician.[69] Mistakenly stating that the term eugenics had first been introduced in *Hereditary Genius*, the article identified "English biologist F. Galton" as "the founder of eugenics." "Although progressive scientists put forward humane purposes for eugenics," the authors declared:

> it was often used by reactionaries and racists, who, building on pseudoscientific notions of the inferiority of separate races and peoples and on nationalistic prejudices and conflicts, justified racial and national discrimination, substituted the so-called racial hygiene for eugenics, and legitimized genocide, as [German] Fascism had done in pursuit of its political goals.

As a result, the authors claimed, the very term "eugenics" had become highly contentious. Nowadays, they continued, some scientists consider its use justified, while others suppose that "the main contents of eugenics, including its tasks and goals, as well as the most rational ways to reach them, shall be taken over by such rapidly developing fields as human genetics (or anthropogenetics) and medical genetics." The rest of the article made clear that the authors themselves subscribed to the latter view. They detailed the "preventive role" of medical genetics performed through studies of chemical and radioactive mutagens, the regulation of kin marriages, and medico-genetic consultations. In contrast to such preventive methods, the authors asserted, "so-called positive methods of influencing human nature, which imply the preferential increase of the progeny of persons with exceptional intellectual or physical qualities (artificial insemination, creation of sperm banks, and so on), are, as a rule, future oriented." But, they concluded, "These methods of the betterment of humankind have often been criticized and have not been [commonly] recognized and propagated."

Although Lobashov and Vel'tishchev basically equated eugenics with medical genetics, the entry on the discipline written by Evgeniia

Davidenkova-Kul'kova (Davidenkov's wife and long-time collaborator) and published in the encyclopedia's sixth volume contained no references to either eugenics or Galton at all.[70] Similarly, neither was mentioned in the entries on "Genetics" written by director of the Novosibirsk Institute of Cytology and Genetics Beliaev and on "Human genetics" co-written by Prokof'eva-Bel'govskaia.[71] Only one entry, on "Biometry," co-authored by Timofeev-Ressovskii, did mention Galton — as one of the founders of the discipline.[72]

It seems likely that the sparse and contradictory nature of Galton's portrayal on the pages of the *Great Soviet Encyclopedia* prodded Kanaev into writing a fully-fledged biography of the "English psychologist and anthropologist," as he was identified in Iaroshevskii's entry. Kanaev himself was involved in the work on the new edition of the *Great Soviet Encyclopedia*, having co-authored an entry on "Twins" that appeared in 1970 in its third volume.[73] He also knew Lobashov, the co-author of the entry on "Eugenics," quite well. Aside from their old acquaintance at Filipchenko's laboratory, Lobashov, after having been fired from Leningrad University in the fall of 1948, had also found refuge at Orbeli's institute in Koltushi, alongside Kanaev. Given that by the 1960s Kanaev had a well-established reputation as a foremost historian of nineteenth-century biology, perhaps Lobashov even consulted with his older fellow geneticist on writing the entry.

Although in his encyclopedia article on twins Kanaev did not mention Galton, he was certainly well aware of the latter's role in the introduction of the "twins method" in the studies of heredity, which had inspired his own research on the genetics of higher nervous activity.[74] Indeed, in his 1959 volume, *Twins*, Kanaev stressed the pioneering role "anthropologist Galton" had played in using twins as a "means to study the interrelations of heredity and environment, 'nature and nurture', as Galton himself had phrased it."[75] A decade later, in 1968, Kanaev published a popular book titled *Twins and Genetics*, which provided a thorough historical overview of the studies of heredity in twins.[76] In this account, he again briefly mentioned Galton's contributions to the development of this line of research.

But there is no indication that at that time he even considered writing Galton's biography. During the 1960s, the history of genetics occasionally figured in Kanaev's works: he gave several talks on Mendel and on the history of genetic research in Leningrad spearheaded by his teacher Filipchenko. Available materials show that after publishing the volume on *Twins and Genetics*, Kanaev started thinking about writing "a history of twentieth-century human genetics."[77] In February 1969, in a letter to Theodosius Dobzhansky, a leading US population geneticist and architect of the evolutionary synthesis, Kanaev asked his one-time co-worker at Filipchenko's genetics department (see fig. 7-1) a series of questions on the subject.[78] He was particularly interested in "whether anything of Galton's legacy is still alive, especially of his eugenics," and whether "positive eugenics still really exists and has any future." He asked his US colleague for recommendations on "recent good books and articles on the subject," noting that he had read with great interest Dobzhansky's *Mankind Evolving* that articulated the latter's views on the future evolution of humanity. In his correspondence with Dobzhansky, Kanaev did not mention that he was planning to write Galton's biography. From 1969 through 1971, he was finishing his monumental treatise on *Goethe as a Naturalist*, working on a history of the physiological investigations of color vision, and researching Trembley's biography.[79] Yet, busy as he was with these projects, in late 1971, Kanaev also finished writing his biography of Galton. The next summer, in time for the 150th anniversary of Galton's birth, the book came out.

In the preface, Kanaev declared that his book was merely "an attempt to provide only a brief account of the multifaceted activities of the brilliant man whose theoretical positions had been very contradictory."[80] He had not aimed at creating a fully-fledged, comprehensive study of Galton's "manifold and innovative works," Kanaev asserted, for such a study is impossible for a single individual and would require "a group of specialists in all the major disciplines to which this remarkable scientist contributed." Yet, "since at the present we have neither books, nor articles about Galton at all," he stated, "this work is aimed to fill, if only for a time, this lacuna, and perhaps to inspire [further] interest in this great scientist, whose influence in science is still alive even today." "Paying tribute to Galton's scientific achievements," Kanaev cautioned his readers, "we cannot forget that he was a son of his century and his class, and this inevitably influenced his worldview and framed his thoughts." The biographer underscored "shortcomings and mistakes in his views" and promised "to critically evaluate the negative, even reactionary part of his legacy," especially some of Galton's "completely unacceptable" racist, social-Darwinist, and eugenic ideas and proposals.

Predictably, the book focused mostly on Galton's contributions to the studies of human heredity, variability, and evolution. After the first chapter on "Galton's Youth," Kanaev moved straight to his analysis of *Hereditary Genius*. The third chapter detailed Galton's works on "Heredity and Twins." The next dealt with "Psychological Investigations and Composite Portraits." The fifth sketched Galton's research on "Anthropometry and Fingerprints," while the sixth analyzed his development of statistical methods for studies in heredity. The last chapter addressed directly Galton's eugenics.

In his conclusion, Kanaev very briefly recapitulated Galton's contributions to various fields of science, but focused mainly on eugenics, echoing many of the criticisms leveled at Galtonian eugenics by Volotskoi, Serebrovskii, and other 1920s "Marxist" critics. "Evaluating Galton's works on eugenics, one clearly sees their deep contradictions," the biographer stressed, for, "on the one hand, their goal was profoundly humanistic and scientific — to better human hereditary nature," while, on the other hand, "the realisation of this idea, aside of its natural, for the pre-genetic era, scientific naiveté and faultiness of many conceptions, was grounded in a clearly visible class approach." "Most of Galton's ideas," Kanaev declared, "were shaped by the demands of the growing British imperialism of his time." Furthermore, he stated, "Galton's proposals for the betterment of humankind are based on a purely biological approach to the issue" and ignore "social factors in the development of various human qualities." "In the hands of his followers," the biographer continued, "eugenics acquired monstrous forms absolutely unacceptable in their immorality and scientific unfoundedness." Yet, "the very idea of the betterment of human heredity will find its place in science," Kanaev affirmed, supporting his statement with a long excerpt from Astaurov's foreword to Blanter's book. He finished by declaring that in "rethinking critically Galton's [eugenic] ideas, it is useful to understand the history of this issue, which has been the goal of this book" [129-30].

Kanaev's book effectively rehabilitated both Galton and eugenics in the Soviet Union.[81] A laudatory review of the biography published in *Priroda* under the title "The Founder of Human Genetics" by

Kol'tsov's student and leading Soviet mathematical geneticist Petr Rokitskii particularly praised Kanaev's handling of Galton's eugenics.[82] "Undoubtedly, Galton set before himself humanistic and scientific goals," the reviewer declared, reiterating the book's overall message, "and could not foresee what eugenic proposals would turn into in the future." He confirmed Kanaev's "general conclusion" that "perversions and distortions of eugenics (the demands to introduce coerced sterilization, immigration laws to preserve the purity of the white race, and especially the mass extermination of the representatives of ostensibly 'lower' races by German fascists) could not serve as a reason for a nihilistic attitude towards the scientific content of eugenics." "As the use of the atomic bomb cannot be blamed on physics, and the propaganda of social Darwinism on Darwin," the reviewer elaborated, so, too, Galton is not responsible for the misuses and misinterpretations of his ideas. Rokitskii affirmed Kanaev's opinion that, although eugenics "had to a certain degree stimulated studies in human genetics," "in our times there is very little left of Galton's eugenics." "What remains," according to the reviewer, "is its 'healthy nucleus' — the idea that on the basis of science, humanity should control its own reproduction, paying attention to the betterment of its biological characteristics." "Modern medical genetics, as it is known, not only studies hereditary diseases and anomalies," he explained, "but also searches for the means of their prevention and treatment." "I have no doubts," Rokitskii concluded:

> that acquaintance with I. I. Kanaev's book would be very beneficial to biologists and [more] generally to scientific workers in all fields, because it not only depicts very well the talented personality of Francis Galton, but also shows the deep connections of modern biology and genetics with ideas and conceptions propounded a century ago.

Kanaev not only rehabilitated Galton and his "brain-child," he also brought back to discussions on human heredity, eugenics, and medical genetics Vasilii Florinskii and his treatise. In the last chapter of Galton's biography he mentioned Florinskii in passing as "a Russian professor, a contemporary of Galton," who had written a book on *Human Perfection and Degeneration*. According to Kanaev, unlike Galton who had mostly focused on the "possibility of creating exceptionally talented and healthy people," Florinskii had been interested in one particular form of "eugenic activities," namely "marriage hygiene," as a means of "preventing the birth of sick progeny." "This form of eugenic activities is easier and more accessible [than the one advocated by Galton]," he stated, and "today this important practical part [of eugenic activities] is called 'medical genetics'."[83]

Kanaev had probably been aware of Florinskii's book and its active use by Volotskoi in constructing "socialist" eugenics since his early days at Filipchenko's genetics department. His work on *Twins and Genetics* prompted him to review the extensive earlier literature on both eugenics and human genetics and may well have led him to revisit Volotskoi's edition of Florinskii's treatise.[84] In the postscript to his 1969 letter to Dobzhansky, Kanaev inquired whether his correspondent knew of "V. M. Florinskii's 'Human perfection and degeneration'," asserting that "from a historical viewpoint, it is a remarkable book." A month later, clearly in answer to Dobzhansky's question, Kanaev provided a brief description of the book and its author. He noted that he had written an article on Florinskii and hopes to send it to his US colleague as soon as it came out.[85]

This article, however, appeared only after Kanaev had finished Galton's biography. Published in *Priroda* under the title "On the Path to Medical Genetics," the article provided an extensive summary of *Human Perfection and Degeneration* and a brief biography of its author.[86] "If we imagine that Turgenev's Bazarov — a physician, materialist, and democrat — became a professor and decided to write a tract on the subject," Kanaev mused opening his article, "the resulting book would undoubtedly be very close in spirit to Florinskii's." To support this characterization of Florinskii as "a man of the 1860s," "a physician, materialist, and democrat," embedded in the popular image of the main character of Turgenev's classic, *Fathers and Children*, Kanaev sketched the main facts of the professor's life and work, from his family's "peasant origins" to his role in establishing Tomsk University.[87]

The bulk of the article dealt with Florinskii's "remarkable book" that, as Kanaev put it, echoing Kol'tsov's characterization, "had further developed Darwin's ideas." Like his predecessor Volotskoi, Kanaev clearly did not know about the publication of Florinskii's essays in *Russian Word*. His article was based on the 1866 book edition. He likely found a copy in the Leningrad Public Library and reproduced its title page in his article.[88] According to Kanaev, Florinskii's treatise was "about the principles of the rational reproduction of the human type, ... to be exact, about the first approaches to the development of medical genetics that one hundred years ago had not existed even as a project" [63]. In Kanaev's opinion, Florinskii was a "true pioneer," since he did not know "the first works in the field of genetics" by Mendel and Galton and of Darwin's works was familiar only with the *Origin of Species*. "In 1866, when Florinskii's book was published," the author stressed, "the birth of genetics as a science remained thirty-five years ahead and the paths of future human genetics were completely unknown." "We need to remember," Kanaev underscored:

> that Florinskii believed in the widespread at that time doctrine of the inheritance of acquired characteristics [and] did not know even the basics of genetics, to say nothing of chromosomal heredity, the reduction of chromosomes [number in meiosis], mutations, and many other things from modern genetics, which nowadays are known even to schoolchildren [65].

Kanaev provided a detailed overview of the book's contents illustrated by lengthy quotes. As had Volotskoi, he effectively "translated" Florinskii's book for his readers, supplying each quote with comments (usually placed in brackets) that substituted current genetic terminology for Florinskii's own words. Thus, he clarified, instead of Florinskii's "hereditary potentials," "we would say 'genes' or, even better, 'the norm of reaction'," and instead of "blood mixing" — "hybridization," whilst such expressions as "stability and instability of a breed" apparently meant "homozygous and heterozygous breeds." He especially noted that even when Florinskii's used words familiar to the modern reader, such as "race," "nation," and "tribe," "they are all very imprecise and their contents cannot be equated with their modern meanings" [64]. Kanaev went so far as to portray Florinskii's "advice to prospective couples," "imperfect as it was," as a "pioneering attempt at creating a medico-genetic consultation in our country" [67]. Florinskii's book had gained no recognition among his contemporaries and had not been republished during his lifetime, Kanaev observed, but "in 1926, Moscow geneticist M. V. Volotskoi reissued it," supplying an introductory article and commentaries, which, he put elliptically, "do not always deserve praise" [68].

On the following pages he explicated his disagreements with Volotskoi's understanding of Florinskii's treatise. In contrast to Volotskoi's emphasis on the differences between Galton's and Florinskii's views, throughout his article Kanaev repeatedly drew parallels between Galton's eugenics and Florinskii's "marriage hygiene," stressing that "much like Galton, Florinskii is interested in the protection of human hereditary health." For instance, describing Florinskii's notions of the inheritance of mental qualities, he claimed that the professor's "arguments for the inheritance of mind and talents in essence are identical to those F. Galton used in his book 'Hereditary Genius' (1869), and, despite certain primitiveness deriving from the [low] level of contemporary science, as convincing [as Galton's]" [66]. Furthermore, Kanaev declared that Florinskii's concept of "marriage hygiene" "approximately coincides" with Galton's "negative eugenics and modern medical genetics," but the Russian professor did not write about "what Galton named 'positive' eugenics — the creation of a perfected stock exceeding the average norm" [68].

Indeed, Kanaev directly refuted Volotskoi's contrasting of "Florinskii's system and Galton's eugenics," claiming that his predecessor "had falsely interpreted Galton's 'positive' eugenics as an attempt to defend scientifically the ruling class of England as eugenically most valuable" [68]. "Florinskii's book is important not only as a document from the history of science," Kanaev declared, "It was an original, innovative for its time work, whose main idea (and not only that idea) — marriage hygiene — remains relevant to our times." "Even though its author's knowledge base is obsolete," he concluded, "the progressive, democratic, and humanistic spirit that permeates this book echoes the pursuits of modern medical genetics" [68].

Kanaev's publications played an important role in the "rehabilitation" of eugenics as a legitimate subject in Soviet scientific discourse. They helped remove the stigma that had been attached to Galtonian eugenics and its proponents by labeling them "anti-Marxist," "racist," "fascist," and "bourgeois." They also restored Florinskii and his treatise to their prominent place — assigned to the book and its author by Volotskoi nearly half a century earlier — in the history of eugenics in Russia. Furthermore, Kanaev's article effectively equated Galton's and Florinskii's ideas with those of modern medical genetics, presenting the pair of English and Russian "pioneers" as forerunners of the discipline.

As a result of Kanaev's publications, along with Galton's, Florinskii's name began to figure in accounts of the discipline's history, for instance, in the 1978 textbook on *Human Genetics*, published by IMG director Bochkov.[89] Indeed, the entry on "medical genetics" Bochkov wrote for the new edition of the *Great Medical Encyclopedia* stated unambiguously that "in the middle of the nineteenth century in Russia, V. M. Florinskii worked on the problems of hereditary diseases and human hereditary nature."[90] The spokesman for medical genetics proceeded to describe "English biologist F. Galton's great impact on the development of medical genetics," attributing to Galton the introduction of "a genealogical method, a twins method, and a statistical method into the studies of human heredity." Furthermore, the encyclopedia's entry on "hereditary diseases" declared unequivocally:

> In 1866, V. M. Florinskii in his book "Human perfection and degeneration" gave a correct assessment of the role of external environment in the formation of hereditary characteristics, [as well as] of the detrimental influence of consanguineous marriages on progeny, [and] described the inheritance of a series of pathological traits (deaf-mutism, retinitis pigmentosa, albinism, harelip, and so on).[91]

The entry also asserted that "English biologist F. Galton was the first to put forward the issue of human heredity as a subject of scientific research," and, as did Bochkov's entry, credited Galton with the introduction of modern methods "in the studies of human heredity." Needless to say, special entries on these various methods in the same encyclopedia all duly acknowledged Galton's contributions.[92]

The encyclopedia's entry on Florinskii himself, however, did not mention *Human Perfection and Degeneration*, nor did it list the book among Florinskii's publications, even though it did include Kanaev's 1973 article in the list of "further readings."[93] It focused almost exclusively on the professor's contributions to gynecology and obstetrics and referred readers to his dissertation and textbooks. Florinskii's role in Russian pediatrics as the founder of the country's first specialized pediatrics department at the IMSA was also briefly acknowledged in the corresponding entry.[94]

Kanaev's work on Galton's biography and on the article about Florinskii clearly piqued the geneticist-turned-historian's interest in the history of eugenics, and Florinskii as part of that history. Perhaps, he even thought about writing a biography of Florinskii and/or reissuing Florinskii's treatise. Indeed, he concluded his *Priroda* article with a plea to his readers to send him any information they might have about Florinskii and his work. We do not know whether anyone answered this plea. But, reportedly, during the 1970s, Kanaev did write a voluminous manuscript providing a general outline of, and detailing various episodes in, the history of eugenics and human genetics in Russia,[95] which probably included a chapter on Florinskii and *Human Perfection and Degeneration*.[96] His attempt to publish this manuscript with Nauka, however, proved futile. The "renaissance" atmosphere of the Thaw was by this time long gone, replaced with the deadening "stagnation" of the Brezhnev era.[97] Reportedly, the censor demanded that Kanaev rewrite his "objectivizing" account of still suspect eugenics. He refused and withdrew the manuscript.[98] In 1984, after a long illness Kanaev passed away and his history of eugenics never saw the light of day.

But his *Priroda* article on Florinskii became a spark that reignited the interest in this long-forgotten historical figure and eventually led to a new edition of Florinskii's treatise some twenty years later.

## Florinskii in Post-Soviet Russia

Just a few months after Kanaev's death, his homeland was engulfed in a new revolution. In the spring of 1985, the General Secretary of the Communist Party Mikhail Gorbachev launched a series of political, economic, and social reforms that became known as *perestroika*. A major geo-political result of *perestroika* was the "end" of the Cold War and the disintegration of the so-called Soviet bloc — epitomized by the 1989 fall of the Berlin Wall. Two years later, in 1991, the Union of Soviet Socialist Republics itself fell apart. Its constituent republics became independent states. The largest — the Russian Federation — emerged as the *de jure* heir to the Soviet Union, and its first elected president Boris Yeltsin as the champion of further political and economic reforms. The Communist Party was "banned" and new "democratic" institutions and procedures replaced the party-state apparatus's tight administrative hegemony over all facets of life. Initiated by the Bolsheviks, the "Great Experiment" of building socialism was abandoned. The country rapidly slid into the chaos of rampant capitalism that shredded the very fabric of its economic, political, and social life. The dissolution of the Soviet Union weakened the Moscow-based central institutions' control over the regions. It also reignited long-smoldering inter-ethnic tensions, brutally manifested in the first Chechen War of 1994-1996.[99] The country's economy was shattered (with its GDP falling more than fifty per cent during the 1990s) and its population impoverished, while a small number of oligarchs and organized criminal gangs made gigantic fortunes by "privatizing" its assets and pillaging its resources.[100] On the eve of the new millennium, faced with failing health, economic collapse, and renewed military conflict in Chechnya, Yeltsin resigned, appointing Vladimir Putin as his successor.

The former head of the Federal Security Service (FSB, the successor of the KGB), only a few months earlier, in August 1999, promoted by Yeltsin to the post of Prime Minister, Putin thus became acting president of the Russian Federation. In March 2000, he won the next presidential election. Putin began consolidating the power of central authorities over the country's regions, strengthening the Russian military and security services, reining in the oligarchs and criminal gangs, and establishing control over the State Duma (the country's highest legislative body) and the mass media. It was during these fearful aftershocks of the Soviet system's demise that Florinskii's treatise was republished, first in 1995 and then again in 2012.

The radical transformations of the country's political, economic, ideological, and social landscapes generated a new wave of interest in eugenics — its history, its current issues, and its future promises. During *perestroika* the interest in the history of eugenics had become an integral part of the re-evaluation of the country's historical past. As had Alexander II during the Great Reforms more than a century earlier, Gorbachev actively promoted *glasnost'* as a main instrument in gathering popular support for his reforms and an important feedback mechanism in articulating the country's future trajectory. "Rethinking the country's past" and "filling gaps" in its historical record became the main preoccupation not only of scholarship, but also of literary fiction, cinema, theater, and the media.[101] With the censorship system in disarray,[102] numerous previously unmentionable subjects — from the 1920s purges and the personalities of the country's rulers to the horrors of the Gulag and the country's economic, political, and social development under the tsars — now commanded close attention.[103]

The disintegration of the Soviet Union prompted a "revision" of the uniformly negative portrayal (or simply total silence), which had characterized the official Soviet histories of events, individuals, ideas, and institutions of the imperial past, and its replacement with similarly uncritical and exalted praise. At the same time, the celebratory accounts of the Bolshevik Revolution and its impact on life in the country were reversed and the entire Soviet period uniformly painted in bleak colors. Bringing back "forgotten history" and tracing "historical roots" became a major means of searching for a new identity and a possible future not only for the country as a whole, but also for individuals and social groups — families, generations, professions, ethnicities, religious confessions, and so on.

In the history of Soviet science one such gap waiting to be filled was the history of eugenics. As one would have expected, in the late 1980s and early 1990s, several IIET scholars in both Moscow and St. Petersburg (in 1991 Leningrad regained its original name) began publishing articles on various facets of this history in their professional journals.[104]

But it was not just the history of eugenics that now commanded attention. Eugenics itself became a subject of intense interest to various individuals and groups, to whom *glasnost'* gave an opportunity to publicly express their views. *Perestroika* and the subsequent breakup of the Soviet Union generated a number of dire health and demographic consequences, ranging from sharply increased mortality and lowered life-expectancy (especially among men) to rapidly falling fertility rates and a deteriorating epidemiological situation in regards to both old scourges, like TB, diphtheria, and syphilis, and new diseases, such as AIDS. Spurred by the breakdown of the country's public health and welfare systems, these "social ills" raised the specter of "degeneration" and, with it, interest in eugenics as a suitable instrument of arresting the degeneration and promoting the "revitalization" of the nation.

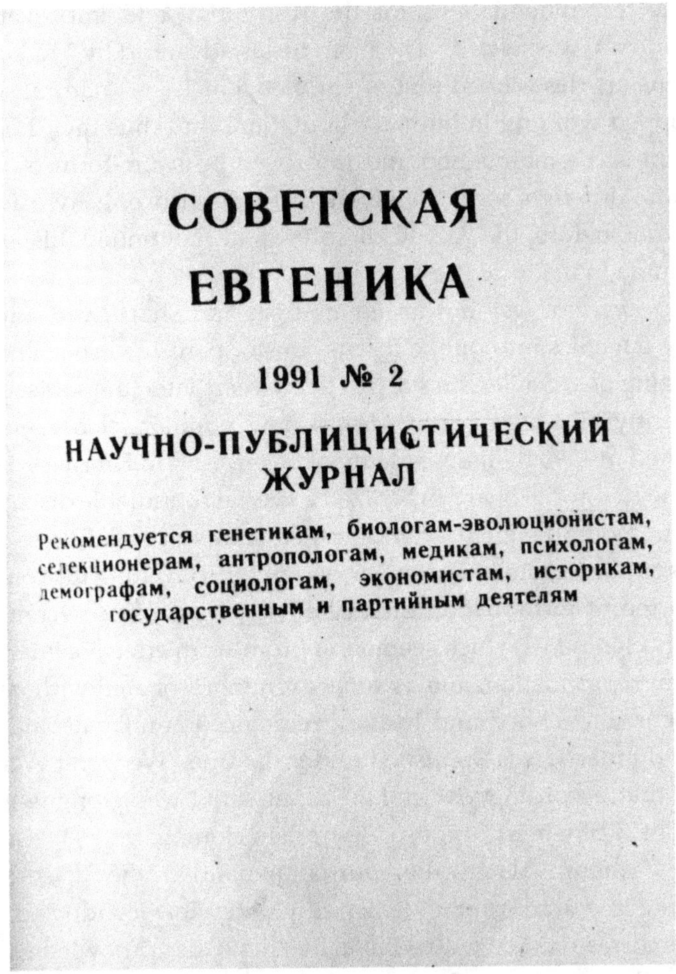

## 7. Afterlife: Medical Genetics and "Racial" Eugenics

### СОВЕТСКАЯ ЕВГЕНИКА

1991 № 2

НАУЧНО-ПУБЛИЦИСТИЧЕСКИЙ ЖУРНАЛ

Рекомендуется генетикам, биологам-эволюционистам, селекционерам, антропологам, медикам, психологам, демографам, социологам, экономистам, историкам, государственным и партийным деятелям

Fig. 7-3. The coverpage of the second issue of *Soviet Eugenics*, 1991. The inscription states that the journal is "recommended to biologists, evolutionary geneticists, breeders, anthropologists, physicians, psychologists, demographers, sociologists, economists, historians, party and state officials." Courtesy of BAN.

A new "scientific-publicist journal," boldly titled *Soviet Eugenics*, that appeared in early 1991 in Kazan offers an illuminating example (see fig. 7-3). The journal was published, edited, and largely written by Stanislav Motkov, an engineer by education and occupation.[105] As he stated in the editorial that opened its first issue, the journal's main goal was "to evaluate economic, political, and social consequences of the genetic degradation of the population of the Soviet Union and to elaborate recommendations for the restoration and improvement of [the country's] *genofond*."[106] The main thesis advanced by Motkov in his numerous articles echoed that of early eugenicists — natural selection had stopped working in human populations and thus their hereditary quality must be maintained and improved by some form of artificial selection. Although Motkov managed to publish only two issues of his journal before the USSR dissolved, he continued his eugenics propaganda in the local press.

*Soviet Eugenics* was but an amateurish and short-lived attempt at bringing eugenics into public discussions of perceived social problems in emerging post-Soviet Russia. Just a few years later, the subject figured prominently on the pages of *Man*, a new popular-science magazine, established in 1990 by the Academy of Sciences.[107] In February 1996, the magazine's editorial board organized a special roundtable discussion on "issues in modern eugenics" and invited leading specialists in medical genetics to participate.[108] The major reason for convening the roundtable was the recent introduction of several new biomedical techniques — primarily cloning — which seemed to promise an effective intervention in human reproduction, and as such strongly resonated with eugenics. The discussion echoed and heavily referenced similar debates in the West, and indeed, was published under the title "We Don't Want to Be Clones" that, according to one of the discussants, was a tongue-in-cheek slogan of US geneticists opposed to human cloning.

A few months later, the journal expanded the discussion by publishing the transcript of a lengthy interview conducted by the journal's correspondent with Nikolai Bochkov, the acknowledged leader of and spokesman for medical genetics, who had been unable to attend the February roundtable at the journal's office. The transcript's title, "A Science That Has Outlived Itself," effectively summarized Bochkov's position. He likened eugenics to alchemy and claimed that medical genetics has surpassed and overtaken eugenics in developing effective methods of both protecting human hereditary health from degeneration and improving humanity's hereditary makeup.[109] The same year, in a clear attempt to capitalize on current interest in the subject, a Moscow publisher issued (without any commentaries) a facsimile edition of the 1874 Russian translation of Galton's *Hereditary Genius*.[110] Three years later, Kanaev's biography of Galton was republished also.[111]

The 1990s discussions of "issues in modern eugenics" stimulated further interest in the history of eugenics in Russia, which rapidly accelerated in the next decade.[112] In addition to historical studies, a number of original works by the Russian proponents of eugenics were republished, along with some relevant archival materials.[113] Thus, in 2008, an 800-page volume, promisingly titled *The Dawn of Human Genetics: The Russian Eugenics Movement and the Beginning of Medical Genetics*, appeared in Moscow. Compiled by Vasilii Babkov, an IIET historian of genetics, the book — despite the promise of its title — was merely a reprint of about forty articles and a few archival documents from the 1920s and 1930s.[114] Supplied with extensive, but not always accurate commentaries by its editor, the volume reproduced the eugenic works by Soviet geneticists (Filipchenko, Kol'tsov, Levit, Serebrovskii and their students) and anthropologists (Volotskoi and Bunak). Works by representatives of other disciplines and specialities involved with eugenics, from jurisprudence to social hygiene to psychiatry (such as Liublinskii, Sysin, and Iudin), found no place on its pages.

The same year, another 400-page volume, titled *The Genealogy of Genius: From the History of 1920s Science*, reprinted 25 articles published by Russian eugenicists. Compiled by Evgenii Pchelov, a historian at the Russian State Humanities University, the volume included only the works that dealt with "eugenic genealogies" collected by RES members. The book was supplemented by an introduction by its editor, which explored the relations between eugenics and genealogy as a historical discipline.[115] Finally, in 2014, the first book-length examination of the history of eugenics in the Soviet Union written by Roman Fando, another IIET historian, came out.[116] All of this literature focused almost exclusively on documenting the development of eugenics during the early Soviet period, its "transformation" into medical genetics in the early 1930s, and the subsequent destruction of genetics. Florinskii and his treatise were at best only briefly (if at all) mentioned in these accounts of the "tragic fate of eugenics" in Soviet Russia, as one of the authors put it in his title.[117]

## The Forefather of Eugenics, a.k.a. Medical Genetics

Yet, the rising wave of new "revisionist" histories did catch Florinskii in its wake. In 1990, the Military-Medical Academy publicly acknowledged his role in establishing the institution's first children's clinics and pediatric department by issuing a special commemorative medal (see fig. 7-4). And just five years later, a new edition of Florinskii's *Human Perfection and Degeneration* was issued under the auspices of the very university he had built in Tomsk more than a century earlier. The new "resurrection" of Florinskii's treatise was prompted by extensive efforts of Evgenii Iastrebov (1923-2003), a retired Moscow geographer, who in the early 1990s produced a whole series of publications on Florinskii, including a complete bibliography of his works, a substantial biography, a collection of his correspondence, and a sketch of his work at the Imperial Medical-Surgical Academy.[118]

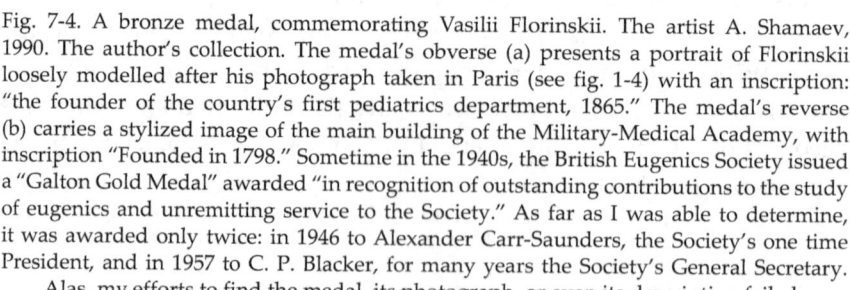

Fig. 7-4. A bronze medal, commemorating Vasilii Florinskii. The artist A. Shamaev, 1990. The author's collection. The medal's obverse (a) presents a portrait of Florinskii loosely modelled after his photograph taken in Paris (see fig. 1-4) with an inscription: "the founder of the country's first pediatrics department, 1865." The medal's reverse (b) carries a stylized image of the main building of the Military-Medical Academy, with inscription "Founded in 1798." Sometime in the 1940s, the British Eugenics Society issued a "Galton Gold Medal" awarded "in recognition of outstanding contributions to the study of eugenics and unremitting service to the Society." As far as I was able to determine, it was awarded only twice: in 1946 to Alexander Carr-Saunders, the Society's one time President, and in 1957 to C. P. Blacker, for many years the Society's General Secretary.
Alas, my efforts to find the medal, its photograph, or even its description failed.

Why would a *geographer* put so much effort into investigating and publicizing the life story of a gynecologist? Iastrebov's initial interest in Florinskii was intensely personal. As did many of his compatriots during *perestroika*, he got deeply involved in uncovering the "forgotten history" of his own family: his older brother had perished in Stalin's Gulag, while many of his relatives had come from the clergy and had been persecuted, facts unmentionable during the Soviet era.[119] But he got particularly interested in a very distant relative — the younger brother of his maternal great-great-grandmother Maria Kokosova (née Florinskaia), Vasilii Florinskii (see fig. 7-5). Iastrebov's interest was apparently spurred not only by the "blood relation" between the two men, but also by the shared affection for their common "little motherland" — the Urals and Western Siberia.

Iastrebov was born in 1923 to a family of school teachers in Ekaterinburg (which would be renamed Sverdlovsk the very next year and regain its original name in 1991), less than 200 kilometers from Florinskii's beloved Peski, virtually in the "same neighbourhood," according to "Siberian standards" that habitually measure distances in thousands of kilometers. Just as he finished high school, the Nazis invaded the Soviet Union. As did millions of his compatriots, the eighteen-year-old joined the army and went to the front. Twice seriously wounded (the last time during the famous battle of Stalingrad) and discharged from the army on medical grounds, the young man returned to his hometown, set on continuing his education. Since childhood he had dreamed of traveling, especially of exploring his homeland — the still uncharted territories of the Urals and Siberia. Unexpectedly, the horrible war helped his childhood dream come true: at the beginning of the war, the renowned geography school of Moscow University (created in the 1930s on the basis of Anuchin's geography department) had been evacuated to Sverdlovsk and merged with the local Urals University.[120] The war veteran became a student of geography. The next year, the geography school returned to Moscow and Iastrebov went with it. But after receiving his diploma in 1947, the young geographer came back to his home city and continued with graduate studies at the Urals University. Four years later, he defended a dissertation on the geomorphology of one of the numerous local rivers and began working at the Urals Branch of the Academy of Sciences and the Urals University, eventually becoming the dean of its geography faculty and one of the leaders of environmental protection in the region.[121]

In 1955, the Urals University geography faculty was relocated to Tomsk and Iastrebov became an associate professor at the very university his "ancestor" had created. Perhaps it was during his time in Tomsk that he learned of Florinskii's role as the founder of Tomsk University and supervisor of the Western Siberia educational region, which by that time had been all but obliterated from "official" historical memory. In 1961, Iastrebov transferred to the geography faculty of the Moscow Regional Pedagogical Institute but continued his research on geography, geomorphology, and environmental protection in the Urals and Western Siberia.[122] Busy with his teaching and research — he spent every summer taking his students on expeditions throughout the country — he had very little time to pursue his "historical hobby." But it seems likely that Kanaev's 1973 article rekindled his early interest in Florinskii. As Iastrebov's biography of Florinskii makes clear, in 1976 he traveled to Kazan for the first time to study the collection of his relative's papers held at the local museum.

In 1986, after 25 years of teaching at the Pedagogical Institute, Iastrebov retired. Now he could devote much more time to investigating his family's history, and especially Florinskii's.[123] Undoubtedly, the general atmosphere of "rethinking the country's past" fueled his interest and deeply influenced his general attitude to the subject of his studies. As Iastrebov pointedly remarked in the foreword to Florinskii's biography, "Vasilii Markovich is one of those individuals whose names under the Soviet regime have deliberately been disremembered and undeservedly erased from the history of Russian science and education." "In 1980, during the centennial celebrations of the Tomsk University founding," he observed indignantly, Florinskii's "name was not mentioned even once."[124] He set out to correct this historical injustice.

Over the next few years, in search of relevant materials, Iastrebov diligently worked in archives and libraries in Moscow, Leningrad, Perm, Tomsk, and Kazan. He compiled an extensive bibliography of Florinskii's publications, which included more than 300 items and, for the first time, identified *Russian Word* as the venue of the original publication of *Human Perfection and Degeneration*. He collected numerous documents and photographs, illustrating various facets of Florinskii's

7. *Afterlife: Medical Genetics and "Racial" Eugenics* 383

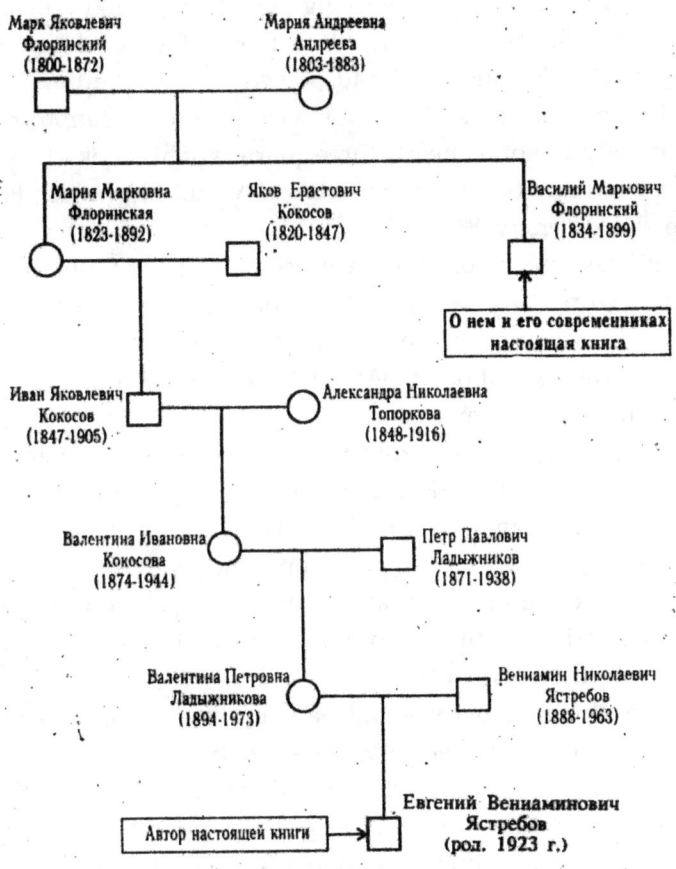

Рис. 1. Фрагмент родословной В. М. Флоринского

Fig. 7-5. Evgenii Iastrebov's "family tree" that demonstrates his relations to Vasilii Florinskii. Following the customary rules of "eugenic" genealogies, the male members of the family are represented by squares, female by circles. The tree starts with Florinskii's parents Mark Florinskii and Maria Andreeva, followed by Vasilii Florinskii and his older sister Maria (omitting all other siblings). The square representing Vasilii Florinskii has a boxed comment: "this book is about him and his contemporaries," while the last square on the diagram, representing Iastrebov, identifies him as "the author of the present book." From E. Iastrebov, ed., *Sto neizvestnykh pisem russkikh uchenykh i gosudarstvennykh deiatelei Vasiliiu Markovichu Florinskomu* (Tomsk: Izd-vo Tomskogo Universiteta, 1995), p. 6. Courtesy of BAN.

life and career, and began working on a book-length manuscript. In describing Florinskii's life story, Iastrebov focused largely on the prominent role his "ancestor" had played in the founding of Tomsk University and the development of science and education in Siberia more generally, devoting nearly three quarters of his book to the subject. Indeed, he initially titled the manuscript "Vasilii Markovich Florinskii and Tomsk University."[125]

In 1992, Iastrebov published a brochure with the bibliography of Florinskii's works and began looking for a publisher for Florinskii's biography. Finding one, however, turned out not to be an easy task. By that time the disintegration of the Soviet Union, combined with Yeltsin's "privatization" campaign, had decimated Russia's economy. Economic ties among the regions had broken down, raw materials were in short supply everywhere, and the old "planned" system of the production and distribution of manufactured goods collapsed. A new "market" economy was geared to selling anything and everything the country could offer to foreign markets at cut-rate prices. Hyperinflation made the money nearly worthless and almost the entire domestic economy reverted to a barter system.[126]

The publishing industry — hit hard by acute shortages of paper, binding materials, and even ink, as well as by the rapidly dwindling purchasing power of its customers and the near total destruction of its distribution networks — barely survived by drastically cutting down both the number and print runs of books it produced. At the same time, the breakdown of the state's monopoly over the printing press spurred the emergence of numerous commercial publishing outlets geared to making quick profit by playing to the demands of the market and the needs of the new political and economic order. Needless to say, scholarly publishing suffered the most. To give but one example, during the 1990s, as compared to the preceding decade, the number of titles issued by Nauka in its famous biographical series (in which Kanaev's biography of Galton had appeared) dropped by more than fifty per cent, while the median print run for each book fell from 15,000 to just 300 to 500 copies.[127] For instance, the biography of Ivan Glebov, Florinskii's mentor at the Institute of Young Doctors, came out in 1995 with a print run of only 180 copies.[128]

Under these conditions who would be interested in producing the biography of an obscure historical figure? Iastrebov published his bibliography of Florinskii's works at his own expense, covering the production costs for 500 copies from his own pocket. The biography, however, was much larger and included more than twenty illustrations, which would have raised its production costs considerably. The retired geographer, whose generous (by Soviet standards) pension had by that time been reduced to virtually nothing by inflation, simply had no means to pay for its publication.

Perhaps, Iastrebov's biography of Florinskii would have remained unpublished and eventually ended up in the Sverdlovsk regional archive along with many other manuscripts he has written during the 1990s, if it were not for Valerii Puzyrev, director of the Tomsk Institute of Medical Genetics. It was Puzyrev who managed to marshal the necessary resources to publish Iastrebov's biography of Florinskii under the trademark of Tomsk University Press in the summer of 1994. He also helped Iastrebov publish a collection of Florinskii's correspondence and wrote a laudatory foreword for the volume that appeared the next year.[129] And it was Puzyrev who spearheaded the publication of a new edition of Florinskii's treatise.[130]

Unlike Iastrebov's, Puzyrev's interest in Florinskii was not personal, but professional. Although Iastrebov did include *Human Perfection and Degeneration* in Florinskii's bibliography, he mentioned the treatise in Florinskii's biography only in passing, merely as an example of his hero's varied activities as a scientist.[131] For medical geneticist Puzyrev, however, this particular book represented the pinnacle of Florinskii's works, for he was well aware of the professor's reputation as a "founder" of his own specialty, which had been promoted by its historian Kanaev and its spokesman Bochkov.

As for other physicians of his generation, Puzyrev's decision to become a geneticist was inspired and enabled by the revival of medical genetics in the Soviet Union.[132] Born in 1947, Puzyrev graduated from Novosibirsk Medical Institute in 1971. The "genetics renaissance" was in full swing, with Novosibirsk emerging as the country's leading center of genetics research, rivaling the old genetics hubs in Moscow and Leningrad, and the young physician decided to specialize in the exciting new field of medical genetics.[133] He started graduate studies at his alma mater, working on a dissertation devoted to the role of heredity in heart disease.[134] According to Puzyrev's recollections, in 1973 he read with great interest Kanaev's article on the history of his chosen discipline and Florinskii as its "founding father." As it happened, that very year, Puzyrev went to Moscow for a few weeks to work and study at the country's leading center of his future specialty — the AMN Institute of Medical Genetics directed by Bochkov. Out of curiosity, he decided to look up Florinskii's book. He found a copy in the Lenin State Library and over two evenings he read it from cover to cover. But, upon return to Novosibirsk, swamped with work, he all but forgot about the "founder of medical genetics" and his treatise. After successfully defending his dissertation, the young geneticist continued his career as an assistant professor at his alma mater. He also came to head a brand new "medico-genetic laboratory" just created at the Institute of Clinical and Experimental Medicine of the AMN Siberian Branch.

In 1981, the Moscow IMG established a branch of the institute in Tomsk, and Puzyrev was invited to head it.[135] The move to Tomsk reawakened his interest in Florinskii. In the Tomsk University Library he found a copy of Florinskii's 1866 book. To his surprise, it has never been read — its pages remained uncut! He made several copies and distributed them among his co-workers. But pursuing the distant history of his profession was not high on the list of his current priorities: he was busy with research and administrative work. In 1986, he managed to reconstitute his "branch" as a separate Institute of Medical Genetics, which became the second institution of its kind in the country, responsible for both research and "medico-genetic consultations" in the whole of Siberia. Puzyrev spearheaded extensive studies on the geno-geography of various populations in Siberia, ranging from the Far East to its polar regions.[136] Part of this enormous project formed the basis of his doctoral dissertation on "medico-genetic study of the populations of the circumpolar regions," defended in 1987.[137] In recognition of his role in developing medical genetics in Siberia, in 1993 the AMN elected him a corresponding member (see fig. 7-6).

It was around this very time that Puzyrev learned about Iastrebov's work on Florinskii's biography, which apparently reignited the geneticist's interest in the history of his profession and its "founder."[138] He spared no effort in helping Iastrebov publish his findings.

## 7. Afterlife: Medical Genetics and "Racial" Eugenics 387

Fig. 7-6. Photograph of Valerii Puzyrev, c.2000. Photographer unknown. Courtesy of V. Puzyrev.

Furthermore, he himself got deeply engaged in popularizing Florinskii's life and works. The next year, the 160th anniversary of Florinskii's birth offered Puzyrev a convenient occasion for waging a veritable campaign for promoting Florinskii as a "founder of medical genetics." He commissioned a local artist to paint a large portrait of Florinskii for the entry hall of the Tomsk IMG. He delivered talks at various conferences and published articles on Florinskii in the popular press and specialized journals.[139] An apex of this campaign was a new edition of Florinskii's book issued in the fall of 1995 by Tomsk University Press with a print run of 500 copies.

A slim paperback printed on cheap paper, the new edition differed from the one prepared by Volotskoi some seventy years earlier not only in its outward appearance (see fig. 7-7). Following his biographer's discovery that Florinskii's book had first come out in *Russian Word*, the new edition reproduced the complete text of Florinskii's essays published in the journal, preserving the structure, paragraphing, and references of the original. It opened with a photograph of its author, taken in 1862 in Paris (see fig. 1-4), which Iastrebov found among Florinskii's papers in Kazan.[140] Typeset anew to correspond to the rules and spellings of current Russian usage, the new edition diverged from the original in only three other minor features. Following the titles and subtitles of the original essays, its editor generated a table of contents for the book. He also inserted two new subtitles: "Introduction" for the first three pages of the treatise and "The Influence of Consanguineous Marriages" for the last 25. Finally, all the italics its author had placed in the original to highlight certain points and thoughts were removed, probably as a result of available typesetting and printing technology rather than deliberate editing.

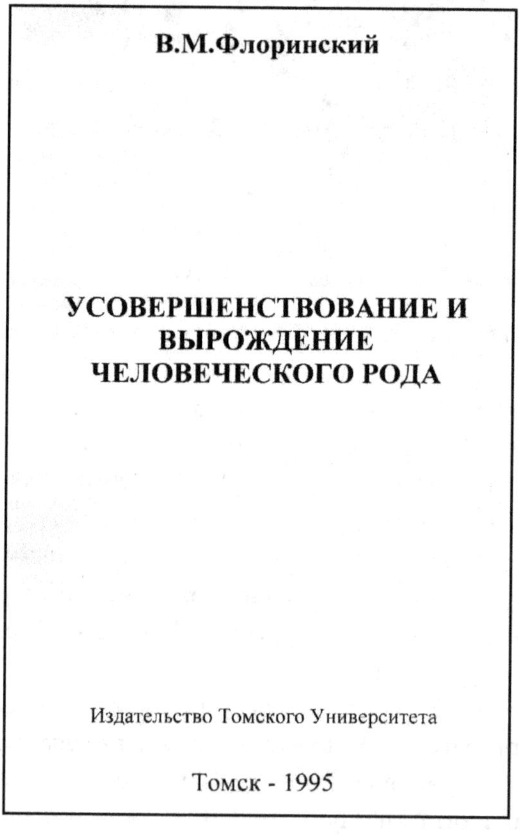

В.М.Флоринский

УСОВЕРШЕНСТВОВАНИЕ И
ВЫРОЖДЕНИЕ
ЧЕЛОВЕЧЕСКОГО РОДА

Издательство Томского Университета

Томск - 1995

Fig. 7-7. The title page of the third edition of Vasilii Florinskii's *Human Perfection and Degeneration*, published by Puzyrev in 1995 with Tomsk University Press. Courtesy of V. Puzyrev.

Unlike his predecessor Volotskoi, Puzyrev did not "translate" the text for its readers: the book contained no editorial comments at all. Instead, he used various extra-textual supplements to illuminate its contemporary import. Thus a "blurb" on the back of the front cover stated that the book's purpose was to acquaint modern readers with "eugenic views of its author, presented simultaneously with the 'father' of eugenics, English scientist F. Galton." "Factual material used by Florinskii to substantiate his concept of 'marriage hygiene'," the blurb stressed, "will be of interest to today's researchers in the fields of medicine, population genetics, anthropology, ethnography, medical genetics, and the history of science." A brief unsigned "afterword" on the last page elaborated: "Many concerns, which had troubled its author in the previous century, are now the subject of scientific research at the laboratories of the [Tomsk] Institute of Medical Genetics". It itemized these putatively "mid-nineteenth century" concerns as "the ethnogenesis and the genetics of various peoples and ethnicities in Siberia, the influence of consanguineous marriages on the population's health, the epidemiology of hereditary diseases, medico-genetic consultations, and so on."

Furthermore, Puzyrev supplemented Florinskii's text with three appendices. The first was his own report, titled "Eugenic Views of F. M. Florinskii on 'The Improvement and Degeneration of Humankind," delivered to a large conference on "The health of the population in Russia" held in Novosibirsk the previous summer.[141] The second was a Russian translation of the article on "The Human Genome Project and Eugenic Concerns" published a year earlier in the *American Journal of Human Genetics* by Pittsburgh University geneticists Kenneth L. Garver and Bettylee Garver.[142] And the third was a four-page English-language abstract of Puzyrev's article on Florinskii's book.[143] In his texts Puzyrev particularly praised Florinskii's innovative approach to "marriage hygiene" and the professor's clear formulation of "the important role heredity plays in human health." He also noted that Mendel, Galton, and Florinskii had all published their path-breaking works in the same year — 1865 — and stated that the trio "had defined the path from eugenics to medical genetics."

The main point of Puzyrev's articles, as well as of the entire project of re-publishing Florinskii's book, however, clearly was not the history, but the present state, of his specialty. Taken together, the three appendices placed Florinskii's treatise squarely within the context of the renewed international debates on eugenics spurred by the rapid development of human genetics during the previous decades, and, especially, the inauguration of the Human Genome Project.[144] The next year, at the Ninth International Congress on Human Genetics held in August 1996 in Rio de Janeiro, Brazil, Puzyrev presented a poster on Florinskii's "physiological laws of heredity."[145] He also distributed among the congress participants an eight-page booklet in English about Florinskii's book[146] and published a brief note on Florinskii's concept of "marriage hygiene" in the *European Journal of Human Genetics*.[147] In December 2001, Puzyrev delivered an inaugural lecture to the Scientific Council of Tomsk Medical University, celebrating the establishment two years earlier of its first medical genetics department.[148] Titled "The Frivolities of Genome and Medical Pathogenetics," the lecture opened with a special section on the "Tomsk footprint in the development of classical genetics." Puzyrev referred to the new edition of Florinskii's treatise, presenting the professor's "physiological laws of heredity" and his concept of "marriage hygiene" as "essential contributions to the development of the doctrine of heredity."

Yet despite all of these efforts, the new edition of Florinskii's treatise went virtually unnoticed, both in his homeland and abroad.[149] Indeed, it seems that the book did not even make it outside of Siberia: I could not find a single copy in any of the major research libraries in Moscow and St. Petersburg. Aside from Puzyrev's articles, the only author who mentioned it was Iastrebov. In 1999, Iastrebov published a 100-page addition to his biography of Florinskii, which detailed the twenty years his ancestor had spent at the IMSA, first as a student and then as a professor.[150] This time he devoted two pages to *Human Perfection and Degeneration*, listing its four separate editions (1865, 1866, 1926, and 1995) and repeating Puzyrev's characterization of the treatise's contemporary import. The brochure was issued, apparently at the author's expense, only in fifty copies. But since it was printed in Moscow, it did make it into the major research libraries. Still, its publication also went completely unnoticed.

Certainly, Iastrebov's and Puzyrev's publications stirred some interest in Florinskii and his works. But this interest appears to have been confined to Siberia and its regional community of historians.

Indeed, over the next decade, publications on Florinskii's role as the founder of Tomsk University and his efforts to promote science and education in Siberia came out in various scholarly journals issued in Tomsk, Novosibirsk, Omsk, and Kurgan.[151] In 1999 Puzyrev himself published in the *Siberian Medical Journal* an article about Florinskii's role in establishing the Tomsk Society of Physicians and Naturalists — the first scientific society in Siberia.[152] In 2003, Tomsk University even named its Archeological Museum after its founder, Vasilii Florinskii.[153] But in all this extensive historical and commemorative activity, Florinskii's *Human Perfection and Degeneration* and his "eugenic views" found no place.

They did find a place in an entirely different domain, however. In 2012, Florinskii's treatise became the opening piece in a massive *Reader on Russian Eugenics*, issued as part of the notorious "Library of Racial Thought," established by the self-styled "racial encyclopedist" and "bio-politician" Vladimir Avdeev (b. 1962).[154]

## The Founder of "Racial" Eugenics

Avdeev's path to (re)publishing Florinskii's treatise was quite circuitous and is illustrative of a general trend that characterized much of the intellectual milieu of post-Soviet Russia. The disintegration of the Soviet Union had not only dire political and economic consequences, but also ideological ones. It led to the flaring up of inter-ethnic and inter-confessional tensions that had long been brewing in the "affirmative action empire" (to use US historian Terry Martin's catchy phrase) under the pressure of its proclaimed internationalism and militant atheism.[155] The crumbling of official "Marxist" ideology triggered a desperate search for a new ideology that could guide the state-building efforts and geopolitical maneuvers of the newborn Russian Federation, on the one hand, and help forge a new collective identity for its diverse populations, on the other. This search enthralled countless individuals and took numerous forms. Vastly divergent doctrines competed for the role of such new ideology, from "eurasianism" and "neo-paganism" to Orthodox Christianity dogmata to the multiple shades of rabid nationalism, as contrary as "national-bolshevism" and "national-capitalism."[156] Forged mostly from the long-forgotten writings of nineteenth- and early twentieth-century Russian philosophers, theologians, scholars, and political thinkers, a wide variety of "new" ideological constructs percolated in the public conscience of post-Soviet Russia.

One such construct was the "science of raciology (*rasologiia*)" concocted by Avdeev.[157] An electrical engineer by training and occupation, during *perestroika* Avdeev felt Calliope's calling and published several poems in newspapers and magazines.[158] In the early 1990s he also published two novels of "intellectual prose,"[159] which went unnoticed by either critics or readers, but earned him membership in the "patriotic" Writers Union of Russia.[160] He thus became a professional writer, but poetry and fiction did not hold his attention for long. In 1994, he published a manifesto of "neo-paganism," pretentiously titled *The Overcoming of Christianity*.[161] During the next few years, the former engineer contributed numerous articles to various right-wing and nationalist periodicals on subjects as varied as "national-hedonism" and "genetic socialism."[162] But his favorite theme became "raciology" that claimed to investigate the "racial nature" of the Russian people and to address the country's current "racial problems," including rapidly falling fertility rates among "Russians" and extensive immigration of "non-Russians" from the former republics of the Soviet Union, especially from Central Asia and the Caucasus.[163]

Avdeev infused the popular nationalist slogan "Russia for the Russians" with explicitly "racial" meaning. In 2000, he joined forces with several outspoken nationalists in creating a "Library of Racial Thought" under the imprint of a private publishing outlet named "White Elves," a reference to the creatures of Norse mythology appropriated by various racist groups as a symbol of white race's superiority.[164] The library was inaugurated with a two-volume set, tellingly titled *The Racial Essence of the Russian Idea*, which Avdeev compiled jointly with Andrei Savel'ev, a physicist-turned-politician, at the time a deputy of the State Duma.[165] Over the next few years, Avdeev produced numerous publications on "raciology" for the "Library of Racial Thought." He compiled two volumes (of more than 1,300 pages in total) of "original works by the Russian classics [of racial theory]," issued under the title *Russian Racial Theory before 1917*.[166] He was instrumental in bringing out Russian translations of writings by such notorious German "racial

Fig. 7-8. Vladimir Avdeev, a portrait by Roman Iashin, oil on canvas, 120 x 100 cm, c.1998. The portrait clearly depicts its subject as a blue-eyed, blond "Aryan," which contrasts markedly with his actual appearance (see fig. 7-9). From Avdeev's personal page at http://racology.ru/galereya. Courtesy of V. Avdeev and R. Iashin.

theorists" as Ludwig Woltmann, Carl H. Stratz, Ernst Krieck, and Hans F. K. Günther.[167] He supplied all of these publications with his own lengthy introductions and commentaries, emphasizing their relevance to Russia's current "racial problems." Avdeev's own "theorizing" on the subject appeared in 2005 as a 500-page monograph, *Raciology: The Science of Hereditary Human Qualities*, which two years later came out in a second, expanded edition.[168]

Needless to say, Avdeev's unabashed propaganda of racism under the guise of the "new science of raciology" provoked an indignant response from the Russian scientific community. Already in 2003, after the appearance of the first volume of Avdeev's anthology on *Russian Racial Theory before 1917*, fourteen well-known anthropologists, ethnographers, and geneticists published in *Priroda* a fierce rebuttal, titled "Recurrences of Chauvinism and Racial Intolerance."[169] A few years later, Viktor Shnirel'man, a leading specialist at the Academy of

Sciences Institute of Anthropology and Ethnology in Moscow, provided a detailed analysis of Avdeev's "raciology" in a series of articles,[170] culminating in a monumental, two-volume damning examination titled *The Ideology and Practice of New Racism*.[171] As Shnirel'man put it, "Like his German teachers, Avdeev does not seek scientific truth at all. He strives as mightily as he can to build anew an edifice of 'Nordic (Aryan) science,' which is supposed to prove the greatness of 'white man' and his 'vanguard' — 'pure-blooded Russian Aryans'." He vividly characterized Avdeev's "scientific" approach to this task:

> The distinctiveness of this book [*Raciology*] lies in that, in essence, it presents merely a large collection of quotations carefully selected and commented upon by the author. He was, of course, attracted only by those quotes and those authors that could, in his view, confirm his own favorite conceptions of racial theory. He is not ashamed of forgery, that is of distorting certain citations taken from well-known authors who had never subscribed to a racial theory. Therefore, all citations provided by Avdeev ought to be checked against the original. Sometimes he uses correct citations, but perverts completely their meaning in his commentaries.[172]

Avdeev was not fazed by this critique.[173] He continued to publish extensively on various facets of "raciology" and to build contacts with like-minded individuals at home and abroad. In 2007, he co-authored a special volume, *Race and Ethnos*, and three years later, published a voluminous *History of English Anthropology*, both of which claimed that "race" was the key to understanding humanity's past history and possible future.[174] In 2011, his *Raciology* came out in English with an enthusiastic foreword by Kevin MacDonald — a retired psychology professor from California State University, Long Beach — who had gained wide notoriety as an outspoken anti-Semite.[175] Although the book has no identifiable publisher, its production was likely backed by another notorious "scientific" racist, Canadian psychologist J. Philippe Rushton, the author of the highly controversial *Race, Evolution, and Behavior* (1996). From 2002 till his death in 2012, Rushton served as president of Wickliffe P. Draper's infamous "Pioneer Fund" and director of its "Charles Darwin Research Institute."[176] According to Avdeev's "scientific-literary biography," it was this institute that funded his "research."[177]

Perhaps as a kind of quid pro quo for producing the English translation of his book, Avdeev, in turn, helped produce a Russian translation of

Rushton's 1996 book that also came out in 2011.[178] The English translation of *Raciology* was warmly greeted on well-known white supremacist websites in the United States and the United Kingdom; while Avdeev maintained close contacts with such notorious western "racial thinkers" as Jared Taylor and Sam Dickson (see fig. 7-9).[179] But, just as its English edition came out, a Russian court decision included the Russian version on the list of "extremist literature" and prohibited its circulation in Russia.[180] It seems likely that this event stimulated Avdeev's turn from "raciology" to racial hygiene and eugenics, manifested in his publication the following year of his *Reader on Russian Eugenics*.

Fig. 7-9. Vladimir Avdeev with Jared Taylor in Moscow, 2016. Photographer unknown. From Avdeev's personal page at http://racology.ru/galereya. Courtesy of V. Avdeev. A similar photo of Avdeev with Sam Dickson (Moscow, 2016) is available on the same website.

Avdeev seems to have been aware of eugenics since his first attempts to develop a theoretical framework for his raciology. Probably his initial interest was spurred by the resonance of German "racial hygiene" with his own racist ideas. In 1997, in the journal *The Ancestral Heritage*,[181] which he helped found, Avdeev published an article on "Personal Freedom and Racial Hygiene." In this incoherent mix of paganism, nationalism, mysticism, racism, and anti-Bolshevism, he claimed that "true personal freedom inevitably leads us to upholding the principles of racial hygiene, and, vice versa, the strict following of eugenic prescriptions maximizes personal freedom."[182] In the next issue of the same journal, Avdeev published a manifesto of "a principally new ideology" he named "Genetic Socialism."[183] The article responded to the sensationalist press coverage of the first successful cloning of a mammal, Dolly the sheep, and presented cloning as the instrument with which to solve Russia's "racial problems." Avdeev proposed "to collectivize the nation's entire *genofond* and on this basis build a society of genetic socialism, with all the ensuing socio-cultural and racio-biological consequences." "A state built on the basis of genetic socialism," he asserted, "will be in its form and essence a eugenic state."

The next year, he published in the same journal a Russian translation of Nikolai Kol'tsov's 1925 overview of the "racial-hygienic movement in Russia" that had appeared in the *Archiv für Rassen- und Gesellschafts-Biologie*.[184] "With this publication we want to tear off the veil from a mystery," Avdeev stated in his preface, "in order to stimulate scientific discussion on a topic, which in the Soviet Union constituted a taboo, and which is little talked about in the modern democratic Russia of today."[185] He misdated Kol'tsov's article, claiming that it had been published in 1935 (i.e. after Hitler's ascent to power), and thus illustrates the collaboration and competition between "Bolshevik and Nazi raciologists." As he put it, "responsibility for the publication by the Soviet geneticist in the racial 'Archiv' of the Third Reich lies entirely with the Gestapo and the NKVD." Demonstrating complete ignorance (or deliberate distortion) of the history of eugenics, Avdeev declared:

> It was in the works by Soviet theoreticians [such as] N. K. Kol'tsov, Iu. A. Filipchenko, M. V. Volotskoi, A. S. Serebrovskii, T. Ia. Tkachev, and T. I. Iudin that many radical doctrines related to the creation of a "new man," to the positive change of the human breed, and to the coerced sterilization of genetically-defective, useless members of society had first been proposed.

In the course of the next decade, Avdeev focused on elaborating his own "racial theory," only occasionally invoking eugenics and racial hygiene in support of his "theorizing."[186] After the 2011 court prohibition of his major opus, however, he apparently decided that it would be safer to propagate the same racist ideas under the guise of "eugenics."

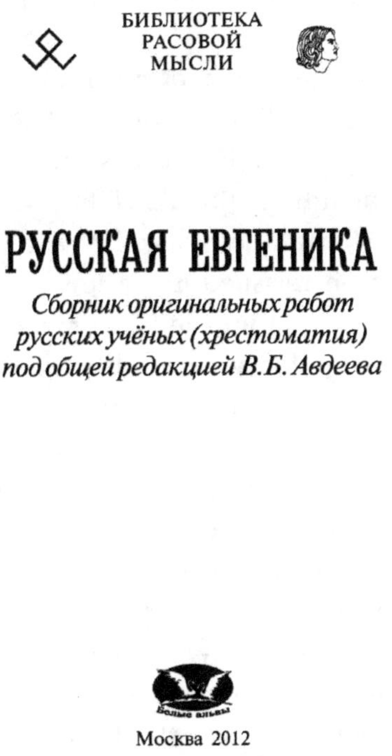

# РУССКАЯ ЕВГЕНИКА

*Сборник оригинальных работ
русских учёных (хрестоматия)
под общей редакцией В. Б. Авдеева*

Москва 2012

Fig. 7-10. The cover page of *Russian Eugenics*, published in 2012 as part of the "Library of Racial Thought" under the "White Elves" trademark, which opened with another edition of Florinskii's treatise. The subtitle identifies the volume as a "collection of original works by Russian scientists (a reader) edited by V. B. Avdeev." The author's collection.

In 2012, Avdeev published in his "Library of Racial Thought" a 570-page anthology of "original works by Russian scientists," titled *Russian Eugenics*.[187] The book's very title carried an explicit nationalist message (see fig. 7-10). The English word "Russian" covers two very different Russian adjectives: *rossiiskii* and *russkii*. The first refers to the country's name, Russia, and thus defines something or someone as belonging to the country. The second refers to the Russians as an ethnic group, and thus defines something or someone as belonging to that group. The title of Avdeev's compilation *"russkaia evgenika"* referred explicitly to the ethnic Russians who had allegedly developed their own particular eugenics, not to the development of eugenics in Russia by scientists of whatever ethnic origins.

The stated goal of this compilation was "to help modern native readers to form their own understanding of such a grandiose and historically important phenomenon as Russian eugenics" [3]. Introduced by its editor's lengthy article on "The Ideology of Russian Eugenics," the volume included eighteen publications by the members of the Russian Eugenics Society, including Bunak, Filipchenko, Iudin, Kol'tsov, Liublinskii, Serebrovskii, and Volotskoi.[188] Although four out of the eighteen items had been reissued just a few years earlier by Babkov, the rest were being reprinted for the first time, thus, indeed providing access to rare, long-forgotten writings by proponents of eugenics in Soviet Russia. In contrast to the previous collections on eugenics by Babkov and Pchelov, Avdeev's was decidedly unscholarly. It simply reproduced the texts, often without even indicating the dates and venues of their original publications. The reason for such sloppy handling of the *Reader's* contents is clear: Avdeev was not interested in the actual history of eugenics in Russia. The main purpose of his compilation was to legitimize its editor's "raciology." "Many Russian eugenicists wrote not merely on eugenics," Avdeev alleged in his introduction, "but on racial eugenics, for their goal was the healthification not of abstract humanity, but of a concrete race — the white race that created Russia" [41].

Echoing, perhaps inadvertently, the Stalin era "patriotic" campaigns for establishing Russian priority in all fields of science and technology,[189] the introduction claimed that eugenics was a Russian creation. As Avdeev put it:

> Despite the fact that from a formal viewpoint Galton is considered to be the founder of eugenics, the most remarkable fact is that, as in many other fields of knowledge, [in eugenics] the priority should belong to Russia. If we set aside the term he introduced into the international usage and look at the principles of organization of this field of natural science themselves, our conclusion would become obvious [6].

It was in order to support this claim that the anthology opened with "a small, but absolutely revolutionary in its essence book" — Florinskii's *Human Perfection and Degeneration.*[190] Avdeev had probably learned about the existence of Florinskii's treatise from Kol'tsov's 1925 article, whose Russian translation he had published in 1998. As we saw in the previous chapter, the RES president opened this overview of the recent development of eugenics in Bolshevik Russia with a long paragraph on

Florinskii's 1866 book. Avdeev did not bother to look for the original and was obviously unaware of its 1995 reprint produced by Puzyrev. Without any acknowledgment, he simply reproduced the text of Volotskoi's 1926 abridged edition. Avdeev, however, removed all extra-textual additions made by his predecessor, including Volotskoi's foreword and commentaries (even though the numbers indicating Volotskoi's footnotes remain in the published text!). But he did not write his own comments on the text. Instead, the editor devoted a substantial part of the volume's introduction to Florinskii and his treatise.[191]

"International Soviet science did everything it could," Avdeev declared, "to drown in obscurity the name of the Russian genius Vasilii Markovich Florinskii who actually should be considered the forefather of eugenics" [6]. According to the editor, Florinskii "analyzed the racial differentiation of humanity and described the population of Russia on the basis of its characteristic traits, along with the causes of the formation of various racial types, taking into account the diverse historical processes [that had unfolded] on the country's gigantic territories" [9]. Avdeev stated that Florinskii was "one of the first in the world's scientific practice" to have provided a "sociobiological interpretation of [human] history" [10] and "established one of the key rules of classical racial theory, long before this [theory] flourished" [11]. According to Avdeev, the main conclusion of Florinskii's book was that "if a state wishes to assure its future prosperity, it must inevitably regulate the [racial] purity and rationality of marriages between its citizens" [13]. "The political vitality of a state, according to the prophetic generalizations of V. M. Florinskii," he asserted, "is defined by the degree of complementarity of [its] ethnic groups, which make it [the state] into a singular historical whole" [39]. "As most Russian pioneers in science," Avdeev lamented, Florinskii "was far ahead of his time, and for this, as is customary here, he was consigned to oblivion" [12].

Compared to the 500 copies of Puzyrev's edition, Avdeev's *Russian Eugenics* came out in a huge (for contemporary Russia) print run of 3,000 copies. Furthermore, very soon after its publication, the book became freely available for download on the Internet. Its editor and publisher advertised the anthology in various venues, including the nationalist web-based channel, "The first Slavic Rusich TV."[192] This active propaganda/marketing campaign certainly made Florinskii's treatise much more widely known and available to contemporary readers.

Indeed, in recent years, references to the latest reprint of Florinskii's treatise began to appear in various publications, whose authors likely do not even suspect that they had read and cited not the original text of Florinskii's essays, but their abridged and edited version produced by Volotskoi.[193] One could surmise, however, that this was not the impact anticipated by the editor and the publisher of *Russian Eugenics*: apart from a few blogs written by supporters of his raciology, "Russian racial eugenics" concocted by Avdeev has found no response, thus far.[194]

## Reading into Florinskii's Book

The repeated resurrection of Florinskii's treatise during the late Soviet and post-Soviet periods was prompted by its utility as a suitable instrument for the justification and legitimization of new approaches to the mix of ideas, values, concerns, and policies related to human reproduction, heredity, development, and evolution, which during the preceding decades had been characterized as eugenics. The diverse uses of Florinskii's treatise by its new publishers and commentators were made possible by reading into its actual contents various meanings and connotations absent in the original text, but resonating strongly with their own specific intellectual, political, economic, institutional, and ideological contexts and aspirations, and by assigning to the mid-nineteenth-century book new significance and importance.

There is little doubt that the "rehabilitation" of both Galton and Florinskii in the late 1960s and early 1970s was an important part of the (re)legitimization of medical genetics in the Soviet Union. At the first stage in the discipline's development during the early 1930s, its champions had sought to distance their enterprise as far as possible from Galton's eugenics and had maintained silence over Florinskii's "marriage hygiene." This time, however, they did exactly the opposite, claiming both as the forerunners of their discipline. This drastic reversal stemmed from two interconnected developments: the decline of the role of "Marxist" philosophy and the rising importance of international disciplinary consensus in the negotiations between Soviet scientists and their patrons in the party-state apparatus.

As we saw in previous chapters, Florinskii's treatise provided Volotskoi and his like-minded colleagues with a suitable template for elaborating "proletarian," "socialist," "bio-social" eugenics, whilst "Marxist" critique played a crucial role in making eugenics "theirs," not "ours" in the Soviet Union during the 1930s. At the same time, certain concurrent international events surrounding issues of eugenics and human genetics — first and foremost, the use of "racial hygiene" in Nazi political rhetoric and actual policies — played a profound role in the legitimization strategy Soviet eugenicists-turned-medical-geneticists devised for their new enterprise in the early 1930s. The essence of this strategy was to separate completely the new discipline of medical genetics from its stepmother, eugenics, including the disremembering of its Russian "founder," Florinskii. But this strategy failed: the actual and perceived links among eugenics, *Rassenhygiene*, and medical genetics (at the level of ideas, methods, institutions, and individuals) became a major target for the new wave of "Marxist" critique leveled at medical genetics and a leading cause of the discipline's decline in the Soviet Union in the late 1930s. Furthermore, the same links were successfully exploited by Lysenko in order to take over the entire field of genetics in 1948.

Apparently the 1960s and 1970s spokesmen for medical genetics learned from the mistakes of their predecessors and developed a new strategy for re-legitimizing their enterprise. This strategy was based on acknowledging the roots of their discipline in eugenics, but separating "true" eugenics — now represented by both Galton's and Florinskii's concepts — from its various later "perversions." Characteristically, they treated as such "perversions" not only the racist interpretations and uses of "racial hygiene" by the Nazis, but also Volotskoi's juxtaposition of Galton's and Florinskii's approaches to their common subject. The "return to the roots" of medical genetics in Florinskii's treatise allowed its new spokesmen to pass over in silence the fierce 1930s "Marxist" attacks on the discipline and to assert their own control over its "methodological issues."

At the same time, in contrast to the 1930s strategy that had emphasized the path-breaking nature of the goals, agendas, and foci of research on medical genetics by its Soviet practitioners and their fundamental differences from those of western eugenicists, the discipline's legitimization in the 1960s and 1970s heavily exploited the concurrent advances in western medical genetics, as well as the rise of "new" eugenics in the West. In his detailed analysis of the "rebirth" of Soviet genetics, Mark B. Adams has convincingly shown that such western advances as the deciphering of the genetic code and the rapid growth of molecular biology became weighty arguments that Soviet scientists used in both restoring the legitimacy of genetics and undermining Lysenko's monopoly over the discipline in the upper echelons of the Soviet power structure — the Central Committee of the Communist Party and its various departments that oversaw science and agriculture.[195]

As with the re-legitimization of genetics writ large, western developments played a key role in the revival of Soviet medical genetics. Undoubtedly, in the Cold War atmosphere of intense competition between the superpowers, the authority of western geneticists became a powerful cultural resource that their Soviet colleagues successfully exploited in asserting the legitimacy of their specialty. From the late 1950s through the early 1970s, Soviet medical geneticists translated into Russian major works on the subject by such well-known western colleagues as Charlotte Auerbach, Robert P. Wagner and Herschel K. Mitchell, Kurt Stern, James V. Neel and William Shull, Victor A. McKusick, and Alan C. Stevenson and B. C. Clare Davison.[196]

Expectedly, the Russian translations were decidedly edited, especially in those sections that touched upon the interrelations of medical genetics and eugenics. For instance, in his preface to Neel and Shull's book on human heredity, Solomon Ardashnikov, Levit's close collaborator and the "scientific secretary" of his IMG, pointedly noted:

> Not all views of the authors could be unquestionably accepted by Soviet readers. A particularly critical approach is required for the chapter devoted to eugenics. As is known, this area of anthropogenetics was subject to numerous perversions and was used to justify racism and colonial wars. Reactionary eugenics especially flourished in Fascist countries. The book's authors criticize many attempts at using human genetics for these purposes. However, even in this sufficiently judicious evaluation of eugenic issues, the authors put forward a series of notions incompatible with our position [on these issues]. It is also important to note that this chapter reflects the views on eugenics by the moderate and progressive representatives of Western scientists, which could be of interest to the Soviet reader.[197]

Neel and Shull's chapter on eugenics was considerably shortened, with sections on eugenic sterilization and differential fertility in various social groups cut out completely.[198] Other translated works on medical genetics underwent similar editing.[199]

Towards the end of the 1970s, the very word "eugenics" began to disappear from literature in both the West and the Soviet Union, whilst "medical genetics" became the preferred descriptor for many ideas, values, concerns, and actions that their proponents had characterized as "eugenic" just a decade or two prior. In August 1978 the Fourteenth International Genetics Congress met in Moscow, bringing together nearly 3,500 participants from some sixty countries. The congress's official motto was "Genetics and the wellbeing of mankind" and medical genetics was its major focus. The IMG director Bochkov chaired the Soviet organizing committee for the congress, and one of the four volumes of its proceedings dealt in its entirety with various issues in medical applications of genetics.[200] Tellingly, there was not a word on eugenics in the volume. Under these circumstances, the acknowledgement of Galton as a founder of "medical genetics" helped establish a common genealogy for western and Soviet medical genetics, whilst Florinskii and his treatise served as a convenient tool in appropriating western advances in medical genetics by demonstrating their alleged "native" roots. By "translating" the mid-nineteenth-century book into the language of modern medical genetics for its late-twentieth-century readers, first Kanaev and then Bochkov effectively "domesticated" concurrent western developments in medical genetics, at the same time distancing the discipline from western "perversions" of eugenics.

In the 1990s, in republishing Florinskii's book, Puzyrev, too, pursued the same goals, emphasizing that "many concerns, which had troubled its author in the previous century, are now the subject of research at the laboratories of the [Tomsk] Institute of Medical Genetics." The inauguration of the Human Genome Project, along with the introduction of cloning techniques, provided a new impetus for the revival of eugenics debates around the world, including Russia, as witnessed by the 1996 roundtable organized by the editorial board of the journal *Man*. In the context of these debates, Florinskii's "marriage hygiene" became a convenient bridge between western and "native" approaches to such contemporary problems identified by Puzyrev as "the ethnogenesis and the genetics of various peoples and ethnicities in Siberia, the influence of consanguineous marriages on the population's health, the epidemiology of hereditary diseases, medico-genetics consultations, and so on."

But for Puzyrev, Florinskii's book apparently also served a different agenda, first and foremost, the preservation of his Institute of Medical Genetics and, indeed, his specialty itself during the rapid deterioration of the Soviet science and health-protection systems triggered by the collapse of the Soviet Union. With its major justification — the Cold War — over and its only patron — the party-state apparatus — gone, the gigantic, centralized, hierarchical system of research institutions that comprised Soviet science fell into disarray.[201] With government coffers dry and science no longer a priority for the new ruling elites, the directors of scientific institutions fought tooth and nail for the survival of their fiefdoms by offering both the public and emerging private businesses whatever services and resources they could scrounge together. Remote from the centers of power (and funding) in Moscow, provincial scientific institutes were hit the hardest. Many academic institutions generated funds necessary for the maintenance of their assets and personnel by renting out their offices, facilities, and equipment. Many strove to find and court new patrons among local individuals and institutions.

In this context, Puzyrev's extensive campaign in the local press to popularize Florinskii's legacy could be seen as an attempt to highlight the deep local connections of his own specialty with Tomsk, and Siberia more generally, and thus raise its profile and importance in the eyes of prospective local patrons. It seems that Puzyrev had successfully accomplished this goal. As indicated by the acknowledgments in his edition of Florinskii's treatise, its publication was made possible by financial support from the Tomsk branch of a certain foundation, named "Human Health," and from a private company, named "DialogSibir'-Tomsk." At the same time, Puzyrev's efforts to advertise Florinskii's treatise on the international scene (at the international congress and in international periodicals) perhaps helped attract the attention of prospective international partners and patrons to his institute and its personnel. In 1994-1995, Puzyrev himself became a recipient of the "Soros Professorship" — an award granted by the International Science Foundation created by billionaire George Soros to support science and scientists in the countries of the former socialist bloc.[202]

The Tomsk IMG, however, was not merely a research, but also a clinical institution, and as such a component of the Soviet health-protection system that also crumbled after the dissolution of the Soviet Union. The impoverishment of the population, the curtailment of the social safety nets, and the disintegration of the healthcare system resulted in deep demographic and health crises in post-Soviet Russia.[203] They provoked a massive rise in alcohol and drug abuse and associated morbidity and mortality.[204] They led to a sharp increase in the number of abortions and a rapid decrease of birth rates, accompanied by a marked decline in women's health status.[205] Deteriorating social conditions worsened considerably the country's epidemiological situation manifested not only in the resurgence of the old "killers" such as TB, diphtheria, and syphilis, but also in the flaring up of new epidemics such as diabetes and HIV/AIDS.[206] The breakdown of the state system of sanitary controls over the quality of air, water, foodstuffs, pharmaceuticals, and other consumer products also contributed substantially to rapidly rising morbidity and mortality rates and dropping life expectancy (especially among men). This dire situation threatened to make the entire specialty of medical genetics irrelevant to the urgent health needs of the population. After all, the morbidity and mortality figures associated with hereditary diseases were negligible compared to those induced by the country's "social ills."

The discipline's spokesmen, then, desperately needed to reassert the importance of their specialty as an integral and indispensable part of a new health-protection system that began to emerge on the ruins of its Soviet predecessor.[207] One might suggest that the renewed interest in Florinskii's treatise (that culminated in its new edition) reflected the search for a suitable means to do just that. Aside from tracing the mutagenic effects of environmental pollution and occupational hazards, medical genetics seemed to have little to offer in alleviating the country's *current* health and demographic crises.[208] But the discipline could perhaps help ameliorate their impact on its *future*. The image of Florinskii as a "pioneer" of medical genetics and of his treatise as the manifesto of the discipline's mission — to guard the health of future generations — could serve as a suitable instrument in the discipline's re-legitimization and expansion in post-Soviet Russia. This probably explains Puzyrev's references to Florinskii in his lecture at the inauguration of the medical genetics department at Tomsk Medical University, which strengthened the position of his specialty by creating a new center for preparing the next generation of its practitioners. The equating of Florinskii's "marriage hygiene" with eugenics apparently served to further underscore the future-oriented contributions medical genetics could make to protecting the country's health, as Puzyrev's report on Florinskii's "eugenic views" delivered to the large conference on "The health of the population in Russia" and reprinted in the new edition of Florinskii's treatise readily demonstrates.

In the 2010s, Avdeev used Florinskii and his treatise for entirely different purposes. But once again, it was the treatise's perceived implications for the country's *future* that apparently attracted the attention of the self-proclaimed "bio-politician." For him, the re-publication of *Human Perfection and Degeneration* served primarily as a means to legitimize his own "raciology" as a truly "Russian" and "scientific" solution to the troubling economic, demographic, political, ideological, and social consequences of the breakup of the Soviet Union and the rise of the Russian Federation as the heir to the past glory of the Russian Empire destroyed by the Bolsheviks. Avdeev's raciology could be seen as a particular nationalist vision of the country's future — clearly expressed in the popular slogan "Russia for the Russians" and built on "racial" and "eugenic" foundations — articulated in his explorations on *The Racial Essence of the Russian Idea* and *Race and Ethnos*. The main "scientific" justification of raciology was the notion of the superiority of the "Great-Russian race" propounded by several anthropologists and psychiatrists in the late imperial period, whose works Avdeev reprinted in his two-volume anthology on *Russian Racial Theory before 1917*.

An extensive critique of this notion by leading Russian ethnographers, anthropologists, and geneticists, which culminated in the prohibition of his magnum opus, *Raciology*, apparently prodded its author to expand the "scientific" foundation of his "theory" by inserting racial hygiene — i.e. eugenics — into its contents. To justify this expansion he claimed that eugenics had first originated in Russia, and he re-published Florinskii's book as the proof of this claim. Furthermore, he presented the early development of eugenics in Soviet Russia as a decisively nationalist and racist project by alleging that its proponents had elaborated a "racial" eugenics. At the same time, the demise of eugenics in the Soviet Union in 1930 gave Avdeev a convenient pretext to indict the Soviet regime's antiracist and internationalist stance as contrary to the interests of the "Russian people."

Moreover, three years after the publication of his anthology on "Russian Eugenics," Avdeev also reprinted Florinskii's monumental treatise on *The Primordial Slavs according to the Monuments of Their Prehistoric Life*. He prefaced the reprint with his own rendering of its author's biography based largely on Iastrebov's publications and ostentatiously titled, "The Restorer of the Russian Worldview, V. M. Florinskii."[209] In recounting Florinskii's life and works, Avdeev again highlighted *Human Perfection and Degeneration* as the foundational work of "Russian" eugenics. Furthermore, he presented the professor's hypothesis on the migration of "prehistoric Slavic tribes" from India to Siberia, and farther on to Europe, as the fundamental truth about the origins and historical development of the "Russian Aryan race."

As with the first revival of Florinskii's treatise in the 1920s, its afterlife from the early 1970s through the 2010s was determined by its commentators' and publishers' reading into the actual text of *Human Perfection and Degeneration* new meanings and new significance drawn from their own specific intellectual, ideological, institutional, and social contexts in the late Soviet and post-Soviet eras.

# 8. Science of the Future: With and Without Galton

> "If blind, opportunistic, and automatic natural selection could conjure man out of a viroid in a couple of thousand million years, what could not man's conscious and purposeful efforts achieve even in a couple of million years, let alone in the thousands of millions to which he can reasonably look forward?"
>
> Julian S. Huxley, 1962

Vasilii Florinskii and Francis Galton, two of the major protagonists of this study, never met and were not aware of one another's efforts to tackle their common subject, the "improvement of humankind." But in Russia their names and their ideas became intimately linked. This peculiar linkage is visualized in the leading current Russian textbook, *Clinical Genetics*: the same page carries portraits of both Florinskii (on the upper left) and Galton (on the lower right). The caption under the first portrait introduces its subject as a "gynecologist-obstetrician and pediatrician," the author of the book "Human Perfection and Degeneration (1865)," and the founder of Tomsk University. The caption under the second portrait introduces its subject as "one of the founders of human genetics and eugenics" and provides the list of his major works as follows: "Hereditary Talent and Character (1865)," "Hereditary Genius: Studies of its Laws and Consequences (1869)," and "Essays on Eugenics (1909)." The textbook's "historical introduction" states unequivocally:

> We can say definitively that by the mid-nineteenth century the notion of pathological heredity was firmly established and accepted by many

medical schools. With this understanding of pathological heredity, the concept of the degeneration of humankind and of the necessity of its betterment was born; what is more, simultaneously and independently from each other, this concept was enunciated by V. M. Florinskii in Russia and F. Galton in England.[1]

In the unlikely event that the Russian medical students for whom Nikolai Bochkov wrote (and Valerii Puzyrev updated) this textbook decide to actually check on this claim and read the original works listed in the captions under the portraits of the two "founders" of clinical genetics, they would be in for a surprise. Today, unlike a century ago, they will have no trouble finding a version of Florinskii's book in their school library or, better still, online. But they will face considerable difficulties in finding Russian translations of Galton's publications. Indeed, they will quickly discover that Galton's works, in which he actually advanced his understanding of eugenics, never appeared in their native language, while *Hereditary Genius*, which was translated into Russian in 1874, and is currently available in a 1996 facsimile edition, does not even contain such words as "degeneration," "pathological heredity," and "eugenics," to say nothing of "genetics."

On the previous pages I have described exactly when, how, by whom and for what purposes the linkage between Florinskii and Galton was constructed and how medical/clinical genetics became "heir" to both Florinskii's eugamics and Galton's eugenics in Russia. In what follows I develop further my arguments about the peculiar historical trajectory of eugenics in Russia, as seen through the biography of Florinskii's treatise, and its implications for the understanding of the history of eugenics, locally and globally.

## An (Unfinished) Biography

Florinskii's *Human Perfection and Degeneration* was published, read, and commented upon during very different periods in Russian history. It was written and first published in the aftermath of the Crimean War and at the height of the Great Reforms that completely changed Russia's historical trajectory. But it attracted almost no attention from its prospective audience. The second time, the treatise was reprinted in the wake of the Bolshevik Revolution and at the pinnacle of NEP, events that profoundly reshaped the country's political, social, cultural, and economic landscape. Yet shortly afterwards, it was disremembered. In the late 1960s and early 1970s, in the last vestiges of the Thaw that inaugurated a radical departure from Stalinism, the book was brought back from oblivion and extensively discussed, even though this time its text was not republished. A new edition did appear in 1995 amidst the aftershocks of the "end" of the Cold War and the dissolution of the Soviet Union, accompanied by the drastic social, political, cultural, and economic reorganizations of the life of the country and its people. Most recently, it was reissued on the tails of the abandonment of *perestroika*'s nascent democratic reforms in favor of a new era of authoritarian rule, the replacement of the economic chaos of the previous decades with oligarchic state capitalism, and the revival in "new" Russia of imperial ambitions and Cold War-like confrontation with the West.

What could explain the historical durability of Florinskii's book? Why and how could the same text be published and read in such radically different times? Some clues to the answers to these questions may be found in the words of the second-century grammarian Terentianus Maurus used as the epigraph at the beginning of this book: "the capacity of the reader sets the fate of books."[2] The dates of successive revivals of Florinskii's treatise point to an interesting regularity: each time the book appeared or was extensively discussed was within approximately a decade from a series of events that redefined the course of Russia's history. Each of these five decades — the 1860s, the 1920s, the 1960s, the 1990s, and the 2010s — represented a break, politically, economically, socially, ideologically, culturally, and so on, with the past and was characterized by heightened social anxieties, both hopes and fears, regarding the future, which thoroughly permeated contemporary society. And each of these breaks was followed by something of a restoration — a return to certain practices, structures, and ideologies of the past and a dampening of concerns about the future (the latest one is still to come and its consequences remain to be seen). It seems that the liberating and future-oriented atmosphere stimulated an appreciation of the book's contents by certain individuals and groups, while the "restoration" essentially led, if only temporarily, to its internment.

The punctuated life of Florinskii's treatise demonstrates that its meanings were defined and repeatedly redefined by — and in fact became almost inseparable from — its particular contexts: both the contexts within which the original text had been created and the contexts within which it was reissued and discussed. Various commentators did not simply read Florinskii's essays. They actually *read into* the same text certain meanings and connotations absent in the original, but important and relevant to their own specific eras, contexts, interests, and agendas. They repeatedly re-interpreted its contents in the terms and concepts reflecting contemporary understandings of the ideas, values, concerns, and actions amalgamated in his eugamics, adapting them to their own ideas, values, concerns, and policies.

Jorge Luis Borges has brilliantly demonstrated this phenomenon of deliberate reinterpretation of any text by its various readers in his celebrated short story "Pier Menard, Author of the Quixote" (1939). The story's title character aspires to write *de novo* Cervantes's classic only to produce a word-for-word copy of several fragments of the original. As the narrator puts it, "The text of Cervantes and that of Menard are verbally identical."[3] Yet, he claims, "The second is almost infinitely richer. (More ambiguous, his detractors will say; but ambiguity is a richness)." In a similar way, he characterizes the stylistic "differences" between the two: "Equally vivid is the contrast in style. The archaic style of Menard — in the last analysis, a foreigner — suffers from a certain affectation. Not so that of his precursor, who handles easily the ordinary Spanish of his time." This "contrast" between the two "verbally identical" texts obviously derives from the different contexts within which the same text is placed and read.

The narrator illustrates this — seemingly paradoxical — interdependence of a text and its contexts through an examination of Cervantes's famous assertion: "... truth, whose mother is history, who is the rival of time, depository of deeds, witness of the past, example and lesson to the present, and warning to the future." In the narrator's opinion, "written by the 'ingenious layman' Cervantes, this enumeration is a mere rhetorical eulogy of history." Yet the same phrase "written" by Menard acquires an entirely different meaning: "History, *mother* of truth; the idea is astounding. Menard, a contemporary of William James, does not define history as an investigation of reality, but as its origin. Historical truth, for him, is not what took place; it is what we think took place." Even more, the narrator feels that "The final clauses — *example and lesson to the present, and warning to the future* — are shamelessly pragmatic." The reference to James, a founder of pragmatism, in the analysis of the work by his contemporary Menard looks justified, but it would look shamelessly anachronistic in the analysis of Cervantes's *Don Quixote*. Yet, as we saw, it was exactly what anthropologist Vishnevskii did in his 1926 review of Florinskii's treatise by calling its author a "eugenicist," and what Bochkov and Kanaev did in the 1970s by characterizing Florinskii as a "pioneer of medical genetics."

In Borges's story, the narrator praises this "technique of deliberate anachronism and erroneous attributions" conceived by Menard as actually enriching "the hesitant and rudimentary art of reading." Obviously, for any (especially fictional) text, the very possibility of its reading/interpretation in a context remote from, if not completely alien to, the contexts of the time and the place within which it was written, constitutes its great feature. "The ambiguity is a richness," as Borges puts it — and, along with many other features, such "ambiguity" defines the text's historical durability and cultural universality. As we saw, it was exactly the possibility of reading into Florinskii's text certain new meanings that prompted its publishers and commentators to resurrect the long-forgotten book.

One could suggest that what secured the continuing appeal of Florinskii's treatise was, first and foremost, its fundamental idea: humankind is the ruler of its own future and science is the instrument of its rule. But in certain times and contexts this idea was reinterpreted by and resonated in different ways with different individuals and groups.

The materials presented in these pages leave little doubt that both the author and the first publisher of the book strongly believed in this idea. It followed logically from Darwin's analysis of the origin, divergence, extinction/degeneration, and progress/improvement of species and varieties in nature and under domestication. And it resonated strongly with the spirit of Imperial Russia's Great Reforms, which convincingly demonstrated that a seemingly eternal, "God-ordained" social order could be changed by human actions informed by scientific knowledge. Florinskii's condemnation of the existing economic, political, and cultural inequalities between sexes, social estates, religious confessions, and ethnicities as a major source of degeneration and his proposal of "rational," "hygienic," "mixed" marriages as a major tool of perfection were based on this very idea, even though it was not stated explicitly in his text.

*Human Perfection and Degeneration* was undoubtedly a child of the Great Reforms, a fact clearly recognized by Ivan Kanaev in his likening of its author to Bazarov, the iconic image of the "man of the 1860s" perpetuated by Ivan Turgenev. The treatise responded to the invitation to the Russian educated public to express their opinions on, as Nikolai Shelgunov put it, "the future fate of the entire country," prompted by the Crimean War fiasco and enabled by *glasnost'* introduced by Emperor Alexander II. For its author Florinskii, this thought piece became an expression of his belief in the obligation of an educated person to put his knowledge to the public good and to respond to the perceived needs and concerns of his contemporary society — in Florinskii's own words, "to bring to my Fatherland as much benefit as possible." It also was a — somewhat accidental — venture into the "social applications" of his scientific/medical expertise and critical abilities, which reflected a new appreciation of the importance of science to the country's future developments.

Just a few months after the appearance of Florinskii's essays, however, Karakozov's assassination attempt dampened the resolve of Alexander II and his "enlightened bureaucrats" to continue and expand the Great Reforms. It also sowed in the minds of many members of the Russian educated public the seeds of doubt in the timeliness and effectiveness of their vocal critique of the existing social order and the various visions of the country's future they advanced. Florinskii apparently was one of those affected. He never promoted or pursued further the main ideas of his treatise. Indeed, it seems as if he tried to forget he had ever written it. He found new outlets for his talents and new ways of bringing benefits to his Fatherland — writing a popular domestic medicine manual, building an education and science system in Siberia, and investigating the artefacts of "prehistoric Slavs."

For its first publisher, Grigorii Blagosvetlov, *Human Perfection and Degeneration* was an integral part of his extensive campaign to popularize the natural sciences as the ultimate remedy for numerous "social ills" that plagued the Russian Empire. He firmly believed in the power of science as a key instrument of human progress in general and the progress of his Fatherland in particular, and he spared no effort in promoting this belief through *Russian Word*. Karakozov's shot that led to his arrest and the prohibition of his journal did not change Blagosvetlov's mind. Less than two months after his release from the Peter-Paul Fortress he issued Florinskii's essays in book format and continued to examine their major ideas on the pages of his new journal, *Deed*.

Blagosvetlov seemed to be the only contemporary who fully grasped the import of Florinskii's essays as a synthesis of Darwinism and social hygiene, which offered a scientific solution — "rational" or "hygienic" marriage — to a number of perceived social problems in post-Crimean Russia and outlined its possible effects on the country's future. This appeared to be the reason why he reprinted them as a book and kept it in print until the end of his life. Indeed, Florinskii's treatise inaugurated a whole series of books issued by Blagosvetlov, which in one way or another explored various questions raised in Florinskii's tract, including translations of Karl H. Reclam's *Popular Hygiene* (1869), Charles Darwin's *The Descent of Man* (1871-1872), and Alfred Russel Wallace's *Contributions to the Theory of Natural Selection* (1878), as well as publication in book format Veniamin Portugalov's *Issues in Social Hygiene* (1873).

Yet, despite its publisher's continuing efforts to popularize its main ideas, Florinskii's treatise failed to stir its prospective audiences into action, whether by creating a research programme that would substantiate and advance its basic conclusions, or by implementing its practical proposals into the life of his compatriots. This failure prompted its subsequent publishers and commentators to claim that the book had been ahead of its time. The nearly simultaneous appearance elsewhere of other works dealing with the same subject along similar lines, including Maximien Rey's *Dégénération de l'espèce humaine et sa régénération* (1863) in France, Galton's "Hereditary Talent and Character" (1865) in Britain, and Eduard Reich's *Ueber die Entartung der Menschen, ihre Ursachen und Verhütung* (1868) in Bavaria, indicates, however, that Florinskii's *Human Perfection and Degeneration* was very much in line with contemporary thinking in Europe. Indeed, it was Florinskii's deep immersion into the problematics and debates of his western colleagues (begun during his European tour and continued afterwards in his regular extensive reviews of nearly everything published on his specialties by western colleagues) that allowed the young gynecologist to address the subject of his essays in a novel way. It was not Florinskii but rather his homeland that was behind the times — neither Russia's scientific/medical community, nor its extensive bureaucracy paid much attention to his ideas.

The silence that met Florinskii's treatise after its first appearance suggests that the Russian Empire lacked the socio-economic conditions — from industrialization and urbanization to immigration and overpopulation — that fueled interest in its subject matter elsewhere during the second half of the nineteenth century. The huge, sparsely populated, agrarian, autocratic, poly-confessional, and multiethnic empire provided neither sufficient data, nor receptive audiences for concerns about "racial" degeneration and intermixing, differential fertility rates, or social degradation (seen in the spread of such social ills as crime, prostitution, pauperism, and alcoholism). In other countries, these issues drove the emergence of a particular amalgam of ideas, values, concerns, and actions aimed at averting the degeneration and advancing the improvement of humanity through deliberate interventions in human reproduction, which Galton named eugenics. Even though various "national" versions of eugenics began to infiltrate Russian professional and popular discourse shortly after Florinskii's death, his book did not become part of the initial debate about the suitability and applicability of eugenics to the country's life. Nor did these "national" variants of eugenics find much support among Russian anthropologists, biologists, physicians, jurists, civil servants, pedagogues, and other disciplinary, professional, and social groups, which at the time actively propagated the "improvement of humankind" elsewhere in Europe and the Americas.

The situation changed radically after the Bolshevik Revolution of 1917. Within just a few years after the Bolsheviks seized power, eugenics boasted a nationwide society, close links with its counterparts abroad, several research institutions, and specialized periodicals in new, Soviet Russia. It entered teaching curricula in various schools of higher learning, instigated a grassroots following, and inspired numerous cultural representations — novels, films, and plays. And it was during this period that Volotskoi discovered, actively popularized, and ultimately republished Florinskii's treatise.

Volotskoi was the first to explicate clearly Florinskii's fundamental idea that humankind is the ruler of its own future (evolution) and science is the instrument of its rule. Furthermore, he pointed out that it was this very idea that underpinned both Florinskii's and Galton's concepts. "What we now call 'eugenics,' as well as Florinskii's 'Marriage Hygiene'," he asserted, "is, essentially, the understanding by the species of *Homo sapiens* of the process of its own evolution and its striving to subordinate this process to its own will through the [scientific] study of all the factors underpinning or even tangentially influencing the evolutionary development of the human type." This shared idea — "the self-direction of human evolution," as it was expressed in the motto of the Second International Eugenics Congress in 1921 — determined to a large degree the subsequent entwined fates of Florinskii's and Galton's concepts in Russia.

The rapid and extensive development of eugenics in Bolshevik Russia was a direct result of the combined efforts of two distinct groups. The first included educated professionals (scientists, jurists, physicians, pedagogues, historians, and so on); the second consisted of their patrons and conduits among the country's new rulers, the Bolsheviks. The rationales and interests of these two groups in promoting eugenics differed considerably. As did their colleagues abroad, many supporters of eugenics among Russian scholars and professionals capitalized on the current popularity of eugenic ideas — manifested in their swift spread through numerous countries during the 1920s — to advance their own, varying scholarly interests, disciplinary agendas, social status, and career ambitions. The Bolsheviks' support of eugenics derived from a close affinity of certain eugenic ideas, values, concerns, and policies with their foundational doctrine, Marxism, and their political programme of building a socialist society. Thus, propounded by Galton in 1865, the eugenic vision of creating "men of a high type" resonated strongly with the Bolsheviks' "revolutionary dreams" of creating, in Leon Trotsky's words, "a higher socio-biological type."

Furthermore, the activist attitude towards human (social) development clearly articulated in the popular Marxist slogan, "the emancipation of the workers must be the act of the working class itself,"[4] predisposed the Bolsheviks to embracing wholeheartedly the fundamental eugenic idea that humanity was the ruler of its own future. Indeed, they established a number of "agencies under social control," which aimed, as Galton had wanted it, at improving the "qualities of future generations, either physically or mentally." At the same time, the Bolsheviks' recognition of science as the essential tool in fulfilling their proclaimed programme of building socialism made them accept unquestionably the notion of science as the instrument of humanity's control over its own future. As Volotskoi put it, "a conscious, planned betterment of the human breed corresponds completely to the general goals of building a socialist Soviet society."

It was the ultimate dependence of Soviet eugenicists on their state patrons, however, that, despite their close connections to foreign colleagues, made the development of eugenics in 1920s Russia so different from that in other countries. If the *science* of eugenics (what Kol'tsov named "anthropogenetics") in Bolshevik Russia was virtually indistinguishable from that elsewhere, eugenics as *ideology* ("religion," in Kol'tsov's own words) and, especially, *policy* ("anthropotechnique," in Kol'tsov's terminology) diverged substantially from counterparts abroad. Indeed, the research methods employed by Soviet eugenicists — collecting pedigrees and medical histories, studying twins, and investigating genetic effects of "racial poisons" — were nearly identical to those practiced by their colleagues everywhere. But the foci of their research differed noticeably. Although the Soviet Union's population offered ample material to study "interracial hybridization," Soviet eugenicists paid scant attention to issues of "meticization" and "mongrelization,"[5] which were a major preoccupation of their colleagues abroad (for instance, in Brazil, Germany, and the United States). Nor did they conduct much research on the "hereditary unfit" similar to such infamous studies as Henry H. Goddard's *The Kallikak Family*, which became a hallmark of eugenics in the West.[6] In contrast, numerous institutions the Bolsheviks set up in the early 1920s "to study the criminal and criminality" took up as their major slogan — "there is no [such thing as] an inborn criminal."[7]

Similarly, the actual policies promoted as "eugenic" by their Soviet advocates looked quite different from those proposed elsewhere. Soviet eugenicists never even attempted to introduce "sterilization laws" that were the preferred eugenic instrument in Germany, Scandinavia, and the United States at the time. To the contrary, with very few notable exceptions, such as Volotskoi and Tkachev, they repeatedly criticized their foreign colleagues for promoting sterilization. Instead, they advocated for extensive popularization of eugenic ideas, marriage regulation, re-education of criminals and so-called defective children, and artificial insemination of "willing women with the sperm of talented producers," as Serebrovskii put it.

These differences reflected a peculiar array of scientific ideas, a specific set of social values, a particular constellation of perceived social problems, and a distinct range of social actions deemed relevant and acceptable by the Bolshevik patrons of eugenics and expressed in a certain vision of humanity's future imbedded in their adopted doctrine, Marxism. It was the rejection of "bourgeois" and "capitalist" ideas, values, concerns, and policies fused into "their" eugenics that inspired calls for creating a different — "proletarian" and "socialist" — "our" eugenics, which, in Volotskoi's formulation, would "correspond to the socio-economic conditions of our great country." And it was the close affinity of the ideas, values, concerns, and actions fused into Florinskii's *Human Perfection and Degeneration* with those endorsed by the Bolsheviks in the 1920s that made the book a suitable model for such "bio-social" eugenics actively propagated by Volotskoi and his like-minded colleagues. Many scientific ideas regarding human reproduction, heredity, diversity, development, and evolution elaborated in the book were now seen as outdated and mistaken. But certain values, concerns, and actions amalgamated with these ideas appeared timely and true.

Florinskii's synthesis of Darwinism and social hygiene became particularly important in this context. During the 1920s, Darwinism and social hygiene came to be seen as the foundational concepts of Soviet biology and Soviet medicine/public health, respectively, and, as such, both were incorporated in the concurrent version of Marxism. This made Florinskii's eugamics all the more appealing to "materialist-biologists," "materialist-physicians," and "materialist-sociologists" who passionately debated the possibility and necessity of creating their own version of eugenics. These debates addressed the foundations, agendas, and instruments of "bio-social" eugenics and garnered numerous innovations, including such theoretical concepts as *genofond* and euphenics and such practical tools as marriage consultations and artificial insemination.

The perceived contradictions between Darwinism and Marxism — the two pillars of "bio-social" eugenics identified by Volotskoi — became a decisive factor in the fate of eugenics in Soviet Russia. The major source of these contradictions rested in the opposing — biological and social — interpretations of human nature and the corresponding visions — biological and social — of humanity's future. Some Marxist proponents of eugenics attempted to resolve these contradictions by resorting to the Lamarckian notion of the inheritance of acquired characteristics. For them, Lamarckism offered a suitable mechanism for an "automatic," as it were, translation of social actions, be they physical education or legislation, into desired biological effects, such as the hereditary betterment of the health, minds, and beauty of future generations.

Florinskii's acceptance of Lamarckian inheritance (which he shared with his contemporaries, including Darwin) became yet another argument for using his eugamics as a suitable model for "socialist" eugenics, while Galton's opposition to Lamarckism became yet another reason for the rejection of his "capitalist" eugenics. The vocal critique of Lamarckian inheritance by supporters of eugenics among Soviet geneticists effectively derailed the 1920s attempts to produce a Lamarckian synthesis of Darwinism and Marxism embodied in the very notion of "bio-social" eugenics. Indeed, they separated the biological (eugenics) from the social (euphenics) and attempted to link the two through the concept of *genofond* as the entirety of all genes contained in a population, which, in the words of its author Serebrovskii, was affected "in one way or another, by nearly every action of every government agency."

The primacy of the social over the biological in the "Marxist" understanding of human nature and humanity's future forcibly imposed during Stalin's "revolution from above" led to the abandonment of the search for a "socialist" eugenics and the splitting of "bio-social" eugenics into its constituent biological and social parts. The biological was taken over by the new discipline of medical genetics that reduced the issues of "degeneration" to the study, diagnosis, and prevention of hereditary diseases, but steered clear of any open discussion of humanity's future. The social was subsumed into general and specialized education, social hygiene, and ideological indoctrination now considered the exclusive tools of both "human improvement" and the realization of the Bolsheviks' futuristic programme of building a socialist society. All the various visions of humanity's *biological* future were condemned as "pernicious" biologization and replaced with a Marxist vision of a future communist society, whilst theoretical discussions of human nature and humanity's future became an exclusive domain of Stalin's ideologues.

At the same time, the extensive use of racial hygiene (which Soviet observers equated with eugenics) as the "scientific foundation" of racist Nazi policies and actions provided new ammunition for the fierce attacks on eugenics as the embodiment of the "capitalist perversion" of science. As a result, any form of eugenics came to be seen as exclusively "theirs." The subsequent condemnation of medical genetics and then general genetics as "fascist" and "imperialist" erased the biological from any discussion of human nature and humanity's future in the Soviet Union. By the late 1940s, in the aftermath of Lysenko's campaign, eugenics in Soviet discourse had been transformed into a "racist and imperialist pseudoscience," its founder Galton into a "reactionary racist anthropologist," whilst Florinskii's *Human Perfection and Degeneration*, a major template for "our" eugenics, became a collateral casualty and was utterly disremembered.

Two decades later, however, the "rehabilitation" of medical genetics, accomplished by the surviving members of the Russian Eugenics Society and the students of its late founders, reignited debates over the interrelations of the biological and the social in human nature and humanity's future. In the liberating atmosphere of the Thaw medical geneticists managed to break the monopoly of "Marxist" philosophers and ideologues in defining the social applications and societal implications of their discipline and to reassert their own control over its "methodological issues." As an amalgam of ideas, values, concerns, and actions regarding human nature and humanity's future, eugenics became an important theme in these debates that echoed the earlier fierce polemics over the suitability and applicability of eugenics to socialist society.

Both Galton's and Florinskii's concepts became key reference points in these debates, as witnessed by Kanaev's publication in the early 1970s of both a biography of Galton that focused almost exclusively on eugenics and a detailed analysis of Florinskii's treatise. Indeed, as we saw, both Galton's eugenics and Florinskii's eugamics became important tools in consolidating the social legitimacy and authority of medical genetics in the discipline's re-institutionalization during the

1960s and 1970s. The discipline's leading spokesman, Bochkov, and its leading historian, Kanaev, popularized the view of eugenics as a regrettable deviation from the true "path to medical genetics," as Kanaev put it in the title of his article about Florinskii's book. According to both Bochkov and Kanaev, this path had been mapped out simultaneously and independently by the discipline's respective British and Russian "pioneers," Galton and Florinskii, but their followers had "strayed away from" it. The "stagnation" of the Brezhnev era (1964-1983) stifled the liberating impulse of the Thaw and both eugenics and eugamics again disappeared from public view.

Both re-emerged in the heady atmosphere of *perestroika* and *glasnost'* ushered in by Mikhail Gorbachev's reforms and further boosted by the subsequent collapse of the Soviet Union. The crumbling of official "Marxist" ideology and the rising of acute anxieties regarding the future of the new, post-Soviet Russia stimulated the interest in, and extensive search for, a suitable vision of the country's future. Eugenics offered one such vision. The 1990s witnessed the flaring up of public discussions on "issues in modern eugenics," the appearance of the first scholarly biography of Florinskii, a new edition of his treatise, a reprint of the Russian translation of Galton's *Hereditary Genius*, and a new edition of Kanaev's biography of Galton.

An important factor in the renewed interest in Galton's eugenics and Florinskii's eugamics was the introduction of several new techniques, above all, cloning and the sequencing of the human genome, which seemed to promise effective tools for intervention into human reproduction, and, hence, human future evolution. But no less important were the actual medical problematics of Florinskii's treatise — his extensive discussion of hereditary diseases and, especially, his view of physicians as the guardians of the health of future generations. Accompanied by the rapid disintegration of the science and health protection systems, the collapse of the Soviet Union threatened the very existence of medical genetics in a new Russia struggling with deep demographic and health crises. For the discipline's new champion Puzyrev, the republication of Florinskii's book, then, became a suitable means to reassert the importance of medical genetics in the eyes of prospective patrons at central and local, domestic and international levels.

8. Science of the Future: With and Without Galton   423

Fig. 8-1. A portrait of Vasilii Florinskii commissioned in 1997 to the Tomsk artist Vasilii Cheremin (1926-2002), oil on canvas, 72 x 88 cm. Modeled in part on the 1888 photograph of Florinskii (see fig. 4-6). Initially, it hanged in the lobby of the Tomsk Branch of the Academy of Medical Sciences. Now it is in Valerii Puzyrev's office at the Tomsk Institute of Medical Genetics. Photo courtesy of V. Puzyrev.

The subsequent 2012 edition of Florinskii's book was also instigated by the search for a suitable vision of the country's future. Its publisher Avdeev had found such a vision in the popular nationalist idea of "Russia for the Russians," which he dressed in explicitly racist clothing. It was the resonance of this racist interpretation with certain conceptions of German racial hygiene that led this self-proclaimed "bio-politician" to eugenics and inspired his (purely fictitious) presentation of Florinskii's treatise as the pioneering articulation of a particular "Russian," "racial" eugenics.

The possibility of reading into Florinskii's text certain new meanings, thus, ensured the book's continuing appeal to its various audiences, be they eugenicists, medical geneticists, or nationalists. The same possibility will perhaps inspire interest in its contents by future publishers and readers of various persuasions, or, as the case might be, will consign the book to oblivion once more.

## The History of Eugenics: Local and Global

The peculiar historical trajectory of eugenics in Russia, as seen through the life and afterlife of Florinskii's treatise, offers certain insights into the history of eugenics writ large, illuminating both its multiple local variations and its common global trends.[8] It suggests that the amalgam of ideas, values, concerns, and actions regarding human nature (i.e. heredity, reproduction, diversity, individual and social development, and evolution), which embodied a certain vision of humanity's future and which Galton named eugenics, had multiple "centers of origin."

In each particular setting and each specific time period, this amalgam acquired a distinct *local* configuration, depending on the exact combination of its constituent parts.[9] In different settings, it was based on different ideas about human nature: Christian, Hobbesian, Lockean, Lamarckian, Darwinian, Galtonian, Mendelian, Marxian, and so on. It incorporated different arrays of values: progressive and conservative, religious and atheistic, democratic and authoritarian, imperialist and anticolonial, capitalist and socialist, etc. It addressed different societal concerns: physical and mental "degeneration," racial and ethnic tensions, gender and class inequalities, "social ills" and endemic diseases, and so forth. It offered different sets of policies, tools, and practices, from sterilization laws to artificial insemination to medico-genetic consultations. It was forged by, and attracted the close attention (both criticism and support) of, numerous individuals representing a variety of professional, occupational, and other social groups. It garnered site-specific institutional arrangements, patronage patterns, and research foci. And the historical trajectory of each version of this eugenic amalgam clearly reflected local political, economic, cultural, and ideological dynamics and imperatives.

The variable contents of this amalgam help explain certain parallelisms, convergences, and divergences in its historical development in different settings. Quite often, the same societal concerns — for instance, rising crime rates— gave birth to divergent "eugenic" solutions, such as sterilization laws in the United States, institutional segregation in the United Kingdom, and re-education in the Soviet Union. Equally often, the same policies — for example, marriage regulations — were proposed and adopted as "eugenic" solutions to different societal concerns, ranging from falling birth rates in post-World War I France, to the perceived growth of "feeblemindedness" among "lower classes" in Britain, to national minorities' claims for equality in Romania. The same policies were at times justified by very different value systems, as reflected in both the defense and critique of "eugenic" sterilization in Germany, Italy, Scandinavia, the US, and the USSR, for instance. Certain policies and practices, such as restrictions on consanguineous marriages or sterilization, had actually predated the emergence of eugenics, and were appropriated by eugenicists. Others, such as medico-genetic consultations, were indeed pioneered by supporters of eugenics, as the history of what today is called genetic counseling in both the Soviet Union and the US readily demonstrates.

The history of eugenics in Russia shows that particular local variants of eugenics easily penetrated national borders, often cross-pollinated, and occasionally hybridized. The migration of such "national" versions from one locale to another depended on the flow of publications, materials, techniques, and people across national borders and often followed the established cultural, political, and economic ties among separate countries. Thus, Ploetz's *Rassenhygiene* shaped the development of eugenics in the German "sphere of influence" that extended from Russia to Japan to various countries in northern, eastern, and central Europe. Galton's eugenics generated most resonance on the territories dominated at one point or another by the British Empire, from the US and Australia to Canada and New Zealand. French *puériculture* exerted an important influence in the development of eugenics in various countries in Latin America, Iran, and Romania. And Soviet "Marxist" eugenics affected many left-leaning eugenicists in Britain and the US, including J. B. S. Haldane, Julian Huxley, and H. J. Muller. But the actual hybridization and the incorporation of certain elements of "foreign" versions into its particular local variant always depended on their "domestication" by local communities.

The extensive cross-national exchange and cross-fertilization did not result in the blending of local variations of eugenics into a universal coherent concept, or in the formulation of a universally accepted programme of actions. Rather, in each particular setting, foreign imports were selected, reconstructed, adopted, and adapted to fit local agendas, interests, cultures, and concerns. Such domestication included, first and foremost, the creation of local eugenic vocabularies.[10] The terms comprising such local vocabularies often had meanings, connotations, referential circles, allusions, synonymic and antonymic associations, and inferences quite different from those of imported terms and concepts. Such local terminology reinterpreted and adapted foreign concepts to a specific local culture embedded in its native tongue. The multiplicity, variety, and interrelations of such terms as *antropogenetika* (anthropogenetics), *antropotekhnika* (anthropotechnique), *evgenika* (eugenics), *evgenetika* (eugennétique), *evfenika* (euphenics), *rasovaia gigiena* (Rassenhygiene), and *generativnaia gigiena* or *gigiena razmnozheniia* (Fortpflanzungshygiene) coined, translated, and transliterated by the Russian proponents of eugenics in distinct eras provide a vivid illustration to the complexities and ambiguities of this process. The alternating synonymic, antonymic, or complementary use and disuse of these and many other eugenic terms by different authors in specific time periods serve as an effective indicator of how and which particular foreign concepts were thus appropriated or rejected.

Such domestication rested on a critical selection of foreign works on the subject to be translated into the local language, summarized in various local compilations and textbooks, and reviewed in local periodicals, as the story of Galton's and Kammerer's publications in Russia clearly manifests. Various factors shaped this selection process: personal contacts of local supporters (and critics) of eugenics with their foreign counterparts, local political and ideological pressures, the recognized authority of certain authors (often related to their institutional positions and international visibility), familiarity with the languages and actual availability of particular original works, diverse disciplinary affiliations of local proponents and critics of eugenics, and many others. Introductions, prefaces, and commentaries written by their translators, editors, publishers, and reviewers highlighted or down played, and sometimes simply ignored, certain elements of foreign doctrines, marking some of them as appropriate and others as inappropriate for emulation and incorporation in a local version.

This domestication process often involved identifying (or inventing) local antecedents of imported concepts, exemplified by Volotskoi's references to Peter the Great's laws on "fool marriages" and his active popularization of Florinskii's book. The proponents of eugenics in particular settings invoked such local "founding fathers" and local "proto-eugenic" conceptions to assess and legitimize the applicability or irrelevance of the imported doctrines and their particular components to local agendas, as the varied uses of Florinskii's treatise by its later publishers and commentators, from Volotskoi to Kanaev, Bochkov, Puzyrev, and Avdeev, readily demonstrate. Indeed, they co-constructed the domestic and the foreign doctrines to fit the latter into local moldings and thus justify their acceptance or rejection. To legitimize their own conceptions, ideas, and disciplines, they created particular "genealogies" that established connections with certain acceptable and accepted "founding fathers" while excluding other existing historical connections to suspect and objectionable ones, as Galton's "familial tree" placed on the cover of the *Russian Eugenics Journal* readily shows (see fig. 8-2).

Fig. 8-2. The title page of the first issue of the *Russian Eugenics Journal* edited by Nikolai Kol'tsov. It carries a stylized "family tree" that depicted the relations of Charles Darwin (left) to Francis Galton (right) via their grandfather, Erasmus Darwin (center). This "genealogy" was clearly meant to emphasize eugenics' deep roots in both Charles Darwin's theory of evolution and the "materialistic philosophy" of his grandfather, whose poem "The Temple of Nature" had been translated into Russian in 1911 by Nikolai Kholodkovskii, a zoology professor at St. Petersburg University. With the very beginning of the Bolshevik regime, "materialism" became the battle cry of Bolshevik ideologues and "Darwinism" was actively incorporated into Marxism, as the materialistic explanation of biological evolution. The genealogy depicted in *Russian Eugenics Journal*, then, implied that eugenics founded by Erasmus Darwin's grandson and Charles Darwin's cousin Galton also was a "materialistic" doctrine, and thus could be accepted and developed in Bolshevik Russia. Courtesy of BAN.

Along with domesticating its foreign versions, the local proponents of eugenics put considerable efforts into spreading a eugenics gospel beyond its "centers of origin." The history of *international* eugenics conferences, organizations, and publications shows that its adherents regularly propagated — and often sought to impose — the ideas, values, concerns, and policies amalgamated into "national" variants of eugenics on the international scene. Initiated by the proponents of *Rassenhygiene*, these efforts were carried out largely by British and US eugenicists who hosted the three consecutive international eugenics congresses in 1912, 1921, and 1932. It was the Anglo-American leadership in the "internationalizing" of this amalgam that resulted in the wide adoption of "eugenics" as a general umbrella term covering its numerous local permutations and in the widespread acknowledgement of Galton as its "founding father."

As a *transnational* phenomenon, however, eugenics was very loosely organized, a fact clearly reflected in the very names of its flagship institutions, the International Federation of Eugenic Organizations, established in 1925, and the Latin International Federation of Eugenic Organizations, established in 1933. It was merely an aggregate of multiple local versions, each of which had its own national roots, antecedents, support base, justifications, patrons, institutions, and agendas. From the first to the last international eugenics congress, their organizers and participants showed little interest in standardizing their ideas, research practices (data collection and analysis), or policies. As Leonard Darwin emphasized in his presidential address to the 1912 London congress: "In so new a field, wide differences of opinion as to the methods to be adopted are certain to exist, and it is only by a tolerant consideration of all these divergent views that the true path of progress will ever be discovered."[11] Such a "true path," however, proved elusive. Although by that time eugenics was not so new anymore, similar "wide differences" persisted through the next two international congresses and even led, in the aftermath of the Third International Eugenics Congress, to the formation of a "break-away" group, the Latin International Federation of Eugenic Organizations. These "differences of opinion" allowed proponents of eugenics in various countries to pick and choose from the large pool of available ideas, values, concerns, and policies covered by the umbrella term "eugenics," liberally mixing certain elements of British or US eugenics with those of German *Rassenhygiene*, French *puériculture*, or Italian *antropologia criminale*.

One could argue that the very lack of cohesiveness gave eugenics a unique strength and durability. The amalgam of widely varying ideas, ideals, concerns, and activities fused together under the name of eugenics offered to a variety of individuals in numerous countries the possibility of choosing, adopting, and adapting its particular elements to their own national, professional, institutional, and disciplinary contexts, interests, and agendas. It was this very looseness of eugenics that attracted anthropologists, educators, doctors, policymakers, public health activists, social reformers, biologists, and jurists in various countries under the banner of eugenics.

This lack of uniformity in the concepts, concerns, ideals, and policies among the proponents of eugenics raises questions about the actual goals of their efforts to convene international congresses, form international organizations, and publish international journals, as well as about their successes and failures in spreading a eugenics gospel locally and globally. In their innovative analysis of the "transnational eugenics movement," sociologists Deborah Barrett and Charles Kurzman have attributed the early success of eugenics on the international scene to a "global culture" that appeared conducive to the movement's basic goals, concerns, and policies.[12] They have identified two fundamental components (frames) of this global culture: the ideology of statehood and the ideology of personhood. In their view, the ideology of statehood in this period involved increasing state interventions into such novel spheres of social and individual life as family, education, health, migration, etc., thus expanding the state's purview far beyond such traditional areas of state authority as defense, law and order, and taxation. The concurrent ideology of personhood limited "full personhood" only to propertied males, and in colonial settings only to those of European ancestry, and included a rigid hierarchization of populations according to "innate" ethnic, racial, class, and gender characteristics. Barrett and Kurzman argue that these particular ideologies permeated the "global culture" of the time and resonated strongly with eugenic ideas, values, concerns, and actions, thus enabling the success of eugenics as a transnational movement.

The history of eugenics in Russia suggests, however, that both the "global" character of these ideologies and their decisive role in the successful spread of eugenics around the world are debatable. One could suggest that these ideologies played a part not only in the "global success" but also in certain "local failures" of eugenics. As we saw, these particular ideologies of statehood and personhood were largely unacceptable to Florinskii, as well as to the Russian observers of eugenics during the late imperial period. Many members of the Russian intelligentsia saw interventions of the autocratic state (and its important component, the Orthodox Church) into civic life as unwarranted, infringing on individual liberties, and threatening to their professional aspirations and agendas.[13]

Indeed, Florinskii was one of the first to propose that, instead of the state's and the church's regulations, the issues of human degeneration and perfection could be better addressed through "rational" and "hygienic" marriage by individuals guided in their marital decisions by a physician. Florinskii also shared with many of his compatriots the belief that various social — class, ethnicity, religion, race, and gender — hierarchies constructed on the basis of "biological," "innate" characteristics were unsubstantiated. Later Russian observers were also quite critical of the class and race biases of eugenics and were equally unsympathetic to the gender bias that occupied such a prominent place in early eugenic ideas and actions.[14] This ideological incompatibility played a significant role in the failure of early eugenics to lure Russian scientists, physicians, and jurists into joining the nascent movement before 1917.

The ideology of statehood certainly contributed to the popularity and growth of eugenics, even though, as several recent studies of interrelations between anarchism and eugenics have shown, its influence varied in different settings.[15] Its impact is particularly visible in the development of eugenics in various countries that emerged in post-World War I Europe on the ruins of the Habsburg, Ottoman, and Russian empires, from Austria and Poland to Czechoslovakia and the Kingdom of Serbs, Croats, and Slovenes to Estonia and the Union of Soviet Socialist Republics. As many historians have convincingly argued, in their efforts to build a modern nation-state, certain individuals and groups in these newborn countries eagerly embraced the "blood and soil" mythology expressed in the perceived "innate" racial and ethnic hierarchies that underpinned much of contemporary eugenics. Such hierarchies, however, were anathema to the Bolsheviks who loudly denounced racism and nationalism and actively promoted internationalism. Indeed, these "innate" hierarchies became a major factor in their rejection of "bourgeois" eugenics.

Yet, although the Bolsheviks were not building a *nation* state, they were nevertheless *building a state*, creating its governing apparatus, laws, institutions, practices, and bureaucracies, and thus establishing and expanding their control over the population.[16] Many historians of eugenics (particularly in other multiethnic states such as Mexico and Brazil) have argued that the extensive medicalization connected to various eugenic projects became one of the "social control" instruments of state-building and modernization. The close involvement of Soviet social hygienists, from Radin to Semashko, in elaborating "bio-social" eugenics and promulgating "eugenic" laws indicates that the ideology of statehood did exert a noticeable influence on the development of eugenics in Bolshevik Russia, albeit in ways different from those in many other countries.

The same could not be said about the "global" ideology of personhood. The Bolsheviks abolished the male-dominated model of gender relations and decreed gender equality in all forms of life: marital, social, political, economic, cultural, familial, and so on.[17] Indeed, they adopted a number of policies, including legalization of abortion, abolition of the very notion of "illegitimate children," protection of maternal health, paid maternity leave, and wife's entitlement to alimony in case of divorce, which actually privileged women. The Bolsheviks also decreed the equality of all races and ethnicities populating the vast territories under their control.[18] They put considerable effort into eliminating economic, social, and cultural disparities among various ethnic groups, which even prompted some historians to characterize Bolshevik Russia as "the affirmative action empire." Not surprisingly, both race and gender biases so prominent in certain "national" versions of eugenics found little support among Soviet eugenicists.[19] But, as we saw in Avdeev's writings, some elements of colonial attitudes and racial hierarchies re-emerged in the post-Soviet renderings of eugenics.

The history of eugenics in Russia suggests that Barrett and Kurzman's analytical scheme also needs one more component (frame): the ideology of *scientific internationalism*.[20] The cosmopolitan, "multi-national" character of early eugenics certainly appeared a major point of attraction for its Russian observers: they repeatedly pointed out that various eugenic programmes had been advanced in numerous western countries and hence deserved serious attention at home. More important, the insistent claim of many eugenicists, beginning with Galton, that eugenics was a science, made this particular ideology highly influential in assuring the success of eugenics in various national, as well as international, settings.[21] Indeed the institutional trappings of a scientific discipline, which Galton had given to his brain-child in Britain in the early 1900s, were quickly emulated elsewhere, for instance, in the United States, where Charles Davenport "borrowed" from Galton even the name for the US first eugenic institution — the Eugenics Record Office.[22]

The power of this ideology was further enhanced by the membership composition of the early eugenics movement, which in general was limited to educated professionals and included many eminent scientists. The decades before World War I were the heyday of scientific internationalism characterized by the growing number of international meetings, journals, awards, expeditions, and societies in practically every scientific field and discipline.[23] This emerging "scientific internationale" rested on the notion of the universality of scientific knowledge and aimed primarily at creating and maintaining a disciplinary consensus regarding research methods, subjects, standards, concepts, terms, boundaries, and objectives in specific fields of knowledge. But, the issues of disciplinary cohesion aside, all of these international organizations and activities also conferred on their participants from various countries shared "transnational" prestige and authority. As numerous studies of "international science" have demonstrated, such prestige and authority associated with participation in international activities and organizations proved highly instrumental in the advancement of scientists' domestic interests and agendas, for instance, in courting local patrons or mobilizing local resources and allies.

One could suggest that, even though many activists of the transnational eugenics movement were not particularly interested in building a disciplinary consensus, they definitely sought to capitalize on the authority and prestige attached to "international science" in the pre-World-War-I and inter-war eras to propagate their views on both domestic and international scenes. They also sought to parlay such transnational authority and prestige into certain domestic advantages and developments. As we saw in Chapter 5, the primary target of the organizers of the First International Eugenics Congress in London was the British public. But a noticeable increase in the number of publications on eugenics in Russia after the congress demonstrates that they also succeeded on an international stage: the congress did attract the close attention of numerous observers, many of whom had never before even heard of eugenics.

This could be further illustrated by the rapid development of "national" eugenic organizations in France, Italy, Scandinavia, and eastern and central European countries in the aftermath of the London congress. For the most part, the very same individuals who had attended the congress and sat on its "Permanent International Eugenics Committee" spearheaded this development. Their attendance at the international gathering became a weighty argument in persuading their domestic patrons and peers of the necessity of organizing "national" eugenics in their home countries. Thus, the first 1913 meeting of the Permanent International Eugenics Committee in Paris helped the organizational efforts of French would-be-eugenicists and the advancement of their national agendas: that very year they established the *Société française d'eugénique*. Similarly, the post-war annual meetings of the committee facilitated the growth of national eugenics in the host countries, as well as the spread of eugenics gospel throughout the world.[24]

Scientific internationalism also enabled the transnational exchange and circulation of certain ideas, techniques, concepts, and research methods developed within particular scientific disciplines regarding human reproduction, heredity, variability, development, and evolution, which became fused into eugenics. Given the wide range of disciplinary affiliations of eugenics proponents, a number of international scientific congresses in various fields of knowledge, from anthropology, mental hygiene, and psychiatry to sexology, psychology, and genetics, became venues for both the propaganda and critique of eugenics. At the same time, certain novel approaches to their specific subject matter developed within these different disciplines, be it IQ testing or artificial insemination, were quickly incorporated into the concurrent versions of eugenics and boosted the debates over their legitimacy the world over.

Of course, ideas and ideologies are not Kantian "things in themselves": they do not exist in a vacuum or in the minds of historians who study them. They originate, come to being, and are effectuated by certain individuals and groups, which form particular *networks*.[25] It was the multitudes of entwined personal networks among scientists involved with eugenics that made all the transnational transfers and exchanges possible and effective.[26] Such disciplinary and cross-disciplinary networks became major channels for domestic and international exchanges of ideas, methods, results, publications, tools, honors, students, and even funding (for instance, in the form of Rockefeller fellowships). As we saw, the active efforts by Soviet scientists to tap into such international networks was an important factor in the rapid development of eugenics in their homeland during the 1920s. Individual members of such networks facilitated the spread of — and spearheaded the negotiations over — ideas, values, concerns, and policies amalgamated under the name of eugenics both across and within national borders. Examples of this international networking include Davenport sending large amounts of US eugenics literature to his Soviet colleagues in 1921-1922 and publishing information on Soviet eugenics in his *Eugenical News*, and Muller's four-year stay in the Soviet Union in 1933-1937. These networks profoundly influenced the domestication of eugenics in specific locales and shaped local eugenics institutions, agendas, and practices. As we saw, it was the personal networks of teachers, their students, and students of their students that assured the temporal continuity of eugenics in Russia, despite its ideological condemnation and the nearly complete takeover of institutions for, first, eugenics, then, medical genetics, and then general genetics by "Marxist" critics and Lysenko's disciples. One could expect to find similar networks that carried the torch of eugenics on both domestic and international scenes in other geographical and temporal settings.

The "global" ideologies of statehood, personhood, and scientific internationalism — mediated through and by the fluid networks of eugenics' adherents and critics — then, do illuminate some features of eugenics as both an inherently local and a transnational phenomenon, but certainly not all of them.

## Science of the Future: A View from the Past

Probably the most important common feature that characterized various local versions and shaped global trends of eugenics was its proponents' explicit preoccupation with *the future*. This focus linked the problematics of eugenics with the fundamental existential questions of human nature, human origins, and human destiny: who are we, where did we come from, and where are we going? Fears and hopes about the future of particular communities and humanity as a whole appear to have been the major stimulus for the development of and debates over eugenics, locally and globally. The same hopes and fears made eugenics a favorite trope of the nascent literary genre of science fiction (SF) that took examination of possible futures as its core theme. Conversely, many self-identified eugenicists (and later historians of eugenics) readily appropriated as "eugenic" almost every excursion into a possible (utopian or dystopian) future that included some form of intervention into human reproduction, be it Plato's *Republic*, Campanella's *City of the Sun*, H. G. Wells's *Modern Utopia*, Aldous Huxley's *Brave New World*, or George Orwell's *1984*.[27] This orientation made eugenics particularly appealing to every social, disciplinary, and occupational group interested in the future, be they social reformers, SF writers, evolutionary biologists, philosophers, revolutionaries, or social hygienists. Indeed, the focus on *prevention* — with its explicit goal to deter *future* diseases and epidemics — made hygiene a natural ally of eugenics, clearly expressed in the close involvement of public health specialists of all kinds with its development everywhere and the very names — "marriage hygiene," "generative hygiene," and "racial hygiene" — it assumed in certain settings.[28]

Seen in this light, the dates of publications and intense discussions of Florinskii's treatise in Russia reflect not merely the internal dynamics and local imperatives of the country's political, cultural, economic, and scientific developments. They also point to certain global changes in the perception, imagination, and anticipation of the future. Welded into all of its local variations, the possibility to affect humanity's future by deliberate intervention in human nature made eugenics repelling or appealing to various audiences especially in times of heightened anxieties about the future.

The active development of various "eugenic" schemes in the 1860s and the subsequent waves of global interest in eugenics — from the 1900s through the 1920s, again from the 1950s through the 1970s, and, lately, from the 1990s through the 2010s — correspond closely to the periods of such heightened anxieties. Although, as we saw in the case of Russia, there certainly were some local variations, this particular chronology strongly suggests that two major groups of factors, which profoundly affected our views of possible futures, also shaped the waning and waxing popularity of eugenics. The first includes certain global scientific developments — from the Darwinian revolution to the Human Genome Project — that have redefined the understanding of life and notions of human nature (reproduction, heredity, diversity, development, and evolution). The second consists of certain global events — from the Industrial Revolution through the two World Wars to the "end" of the Cold War — that have profoundly reformatted the political, cultural, economic, and ideological terrains of the very world we live in.

The nearly simultaneous appearance of various "eugenic" conceptions during the 1860s was undoubtedly stimulated by Darwin's *Origin*, which had not only thrown "light on the origin of man and his history," but also offered a convenient framework to illuminate humanity's future. Almost immediately translated into the major European languages, Darwin's concept of biological evolution ushered a radically new worldview by offering clear-cut *naturalistic* answers to the fundamental existential questions: who we are, where we came from, and where we are going. According to a Darwinian viewpoint, humans are a particular species of the animal kingdom; they evolved over countless millennia from some other "lower" animals; and, as any other species, they will either further evolve/diverge into different species, or become extinct.

Furthermore, like other "laws of nature" — from the laws of celestial mechanics to the laws of thermodynamics — Darwin's "laws of evolution" were seen as inescapable and inexorable. They implied that the factors, mechanisms, stages, and conditions, which had defined the past of the human species, would also define its future. They prompted a re-interpretation of human history (from the rise and fall of civilizations to the evolution of languages) in Darwin's terms of "struggle for existence" and "natural selection." They raised the frightening certainty of inevitable human degeneration and extinction. But, they also offered a comforting hope of the possibility of human "improvement" and further "progress" through a deliberate manipulation of human reproduction.

Underpinning Darwin's analysis of speciation, the close analogy between artificial selection (that produced new and "improved" varieties of plants and animals under domestication) and natural selection (that produced new and "improved" species in nature) provided a major intellectual impulse for elaborating various "eugenic" concepts in Britain, France, Germany, Italy, Russia, the US, and elsewhere. All of these varying schemes responded to the implications of Darwin's theory for humanity's future by proposing certain ways of averting the looming threat of degeneration/extinction and realizing the tantalizing promise of perfection/progress, clearly expressed in the very title of Florinskii's treatise. Not surprisingly, in searching for instruments that would be acceptable, in Galton's formulation, "under the existing conditions of law and sentiment," all of them relied on Darwin's analysis of "sexual selection" and borrowed extensively from the toolbox of plant and animal breeders, including "selective mating," "in-breeding," and "out-breeding."

Another important catalyst for the rise of these various eugenic proposals was "social hygiene" that embodied the conception of health and disease as outcomes of social conditions and, hence, subjects of social control.[29] The Napoleonic wars and the Industrial Revolution of the early nineteenth century directed the attention of physicians, civil servants, scholars, and social reformers in many European countries to differential morbidity and mortality rates among separate well-defined (by location, occupation, income, sex, or age) segments of their populations, especially the military and the "labouring population," as Edwin Chadwick put it in the title of his famous 1842 report.[30] The statistical data they began to systematically collect and analyze led to a growing understanding of the socio-economic underpinnings of health and disease developed in the early studies of Johann Peter Frank in Prussia, Alexandre-Jean-Baptiste Parent-Duchâtelet and Louis René Villermé in France, Chadwick and Friedrich Engels in Britain, and Adolphe Quételet in Belgium.[31]

In 1848, a pandemic of cholera, coupled with the violent uprisings that seized practically all European capitals, helped solidify the notion of the state as a key player in not merely monitoring (i.e. "medical or sanitary police") but actually protecting and promoting the health of its citizens, first articulated in Frank's now famous statement that the most successful physician was in fact "the civil administrator."[32] Rudolf Virchow in his analysis of the 1848 typhus epidemics further elaborated the notion of the social, economic, and political underpinnings of health and disease and actively propagated it in his writings, as well as in his practical work as a member of first the Berlin City Council and then the Prussian National Assembly.

From the 1840s through the 1860s various legislative acts — from the laws on compulsory smallpox vaccination to the sanitary regulations of water supply and sewage disposal — institutionalized the state's responsibility for improving the health of the population and preventing future epidemics. This newfound responsibility also prodded a variety of government agencies and agents to expand collection of relevant data on health and disease status of the population, which could serve both to justify and to measure the effects of the state's interventions. The fundamental idea that health and disease were subjects to social control paved the way to the theoretical possibility of extending such control to human reproduction in order to prevent the degeneration and to promote the improvement of the population, which were articulated in early "eugenic" proposals.[33] However, for nearly forty years, all these "eugenic" schemes lay largely dormant, attracting little attention from their prospective users.

By the 1900s, the situation began to change. Various nations were reaping both the bitter and sweet fruits of industrialization, ranging from the impoverishment, urbanization, and proletarianization of large segments of their populations to the rise of new financial, cultural, and political elites and from the rapid development of new technologies of warfare, mass production, communication, and transportation to the mass migrations to and from metropoles and colonies. All of these processes generated profound health, demographic, and social consequences, such as differential fertility, morbidity, and mortality among different segments of the population, rising rates of social "deviance" (criminality, suicide, prostitution, alcoholism, "feeblemindedness," and so on), and growing economic, political, and cultural inequalities among different social groups. Recorded and interpreted by physicians, civil servants, and scientists of all stripes, these "side effects" of industrialization generated acute anxieties about the future of separate nations (hence, Galton's notion of "national" eugenics) and humanity as a whole.

Some observers attributed these side effects to the advances of civilization. According to their views, civilization slowed down or eliminated completely natural selection in human populations and thus fostered "the survival of the unfit" (as the American eugenicist

David Starr Jordan put it in the title of his 1902 pamphlet),[34] who then propagated their "bad heredity" in their progeny. These anxieties over the detrimental effects of "civilization" were amplified by the notion of inevitable human degeneration, divergence, and eventual extinction "predicted" by Darwin's inexorable laws of evolution. These anxieties were clearly expressed, for instance, in the dark visions of the future in H. G. Wells's *Time Machine* (1895) and *War of the Worlds* (1897), both of which enjoyed enormous popularity with readers around the world. And the very same anxieties fired up interest in eugenic solutions to various social problems engendered by industrialization in Britain, France, Germany, the United States, and elsewhere, from the 1900s through the 1920s.

The Great War that redrew the world's political map facilitated the spread of eugenics through numerous countries. The terrific and terrifying shock of the war incited the fears of the impending doom of western civilization, epitomized in Oswald Spengler's monumental *Der Untergang des Abendlandes* (1918-1923) and vividly reflected in the literary fiction of the post-war decade (from Henri Barbusse to Ernest Hemingway to Erich Maria Remarque). It also fomented the Bolshevik Revolution, which materialized the "spectre of communism" that Marx and Engels had first seen "haunting Europe" amid the 1848 uprisings. The hopes and fears regarding the future fanned by World War I provided a major impetus for debates on the eugenic means to address the war's cataclysmic economic, political, social, demographic, health, and cultural consequences.[35]

These debates were empowered by the concurrent "experimentalist revolution" in what today is called the life sciences (at the time represented by the closely intertwined fields of experimental biology and experimental medicine), which dramatically changed the understanding of life (and death). Begun in the pre-war decades and spurred by the wholesale borrowing by biologists and physicians of the experimentalist spirit and experimental methods from chemistry and physics, this revolution gave birth to a host of new disciplines, from genetics, immunology, and endocrinology to biochemistry, biophysics, and "developmental mechanics" (experimental embryology). These disciplines generated numerous discoveries in the mechanisms of such basic life phenomena as metabolism and reproduction, growth and ageing, heredity and evolution, immunity and behavior, health and disease. The experimentalist revolution also stimulated the advancement of new theoretical conceptions about life and its basic features, including neo-vitalism, the concept of hormonal regulation (of reproduction, embryonic development, and growth), Mendelism, behaviorism, the chromosome theory of heredity, neo-Lamarckism, and many others.

More important, experimental biology and medicine produced a whole new arsenal of practical techniques and instruments, such as organ and tissue transplants, hormones, vaccines and sera, tissue cultures, blood transfusion, and artificial insemination, which allowed scientists to manipulate life processes and which came to be seen as powerful *biotechnologies* that would reshape the future. As early as 1922, Mikhail Zavadovskii — a student of Kol'tsov, leading specialist in "developmental mechanics," and active participant of debates on eugenics in Russia — envisioned "the time, when advances in the study of living nature will create conditions for the flowering of biotechnology (*biotekhnii*) alongside the technology of dead materials, [and when] the biologist's tasks of making new life forms, now seem akin to [H. G.] Wells's fantasy [*The Island of Doctor Moreau* (1897)], will be as mundane as those of a construction engineer."[36]

Both the new theoretical concepts and the new practical tools gave a tremendous boost to belief in science, and especially, new experimental biology and medicine, as the ultimate means of humanity's control over its own future. Powerfully articulated in Haldane's famous 1924 pamphlet, *Daedalus; or, Science and the Future*,[37] this belief inspired numerous futuristic projects of manipulating life phenomena, which Adams has fittingly named "visionary biology."[38] But it also awoke deep-seated fears of potential misuses and abuses of science's newfound powers, clearly expressed in the very title — *Icarus; or, The Future of Science* — of Bertrand Russell's response to Haldane's vision,[39] and visibly manifested in the post-war revival (especially, cinematographic) of the nearly forgotten classic, Mary Shelley's *Frankenstein; or The Modern Prometheus*.[40]

Furthermore, advanced within the framework of experimental biology and medicine, the "engineering ideal" of controlling life, in the apt expression of historian Philip J. Pauly,[41] resonated strongly with the kind of "social engineering" embedded in Galton's conception of "agencies under social control that may improve or impair the racial qualities of future generations."[42] Indeed, as one active proponent of eugenics in the US asserted in 1915, "the work before the true promoter of Eugenics is that of social engineering."[43] Expectedly, as an essentially futuristic project, eugenics found numerous supporters, as well as critics, among experimentalists of all stripes and became a key component of "visionary biology." A number of embryologists, biochemists, immunologists, endocrinologists, and especially geneticists came under the banners of eugenics. The new biological concepts and biotechnologies provided the prophets and apostles of eugenics around the world with new justifications for, and new instruments of, "social control" over human reproduction, heredity, diversity, development, and evolution. They contributed substantially to the growing popularity of eugenics during the 1920s and were scrutinized in such influential cultural productions as Olaf Stapledon's *Last and First Men: A Story of the Near and Far Future* (1930) and Aldous Huxley's *Brave New World* (1932).[44]

During the 1930s, coupled with the Great Depression, the elevation of *Rassenhygiene* to a foundational concept in Nazi political and social programmes had a dual effect on the further development of eugenics. On the one hand, it stimulated the rise in the popularity of various eugenic schemes, especially in the countries that would form the Axis powers. On the other, it resulted in the decline of their popularity, particularly in the countries that would form the anti-fascist coalition. And, it fueled an extensive critique of the scientific foundations, stated goals, moral justifications, and adopted instruments of various (especially German and US) versions of eugenics not only in the Soviet Union but elsewhere, as evidenced by the 1937 Congress of Latin Eugenics held in Paris.

In the aftermath of World War II, the revelation of the Nazi atrocities committed in the name of "racial hygiene" turned the very term "eugenics" into an expletive in many different quarters, not just in Stalin's Russia, but almost everywhere. US geneticists, for instance, made a concerted effort to distance their discipline from its stepsister. This move away from its "historic roots" was manifest during the Golden Jubilee of the rediscovery of Mendel's laws grandly celebrated by the

Genetics Society of America in September 1950.[45] It was also inscribed into the histories of the discipline produced during the next decade by its leading spokesmen, Leslie C. Dunn and Alfred H. Sturtevant, both of which barely mentioned eugenics.[46]

Yet, the attempts to obscure the close historical connections between eugenics and genetics did not result in either disappearance of eugenics or the demise of various schemes to "improve humankind" in the post-World-War-II period. Rather, the amalgam of ideas, values, concerns, and actions aimed at averting the degeneration, and advancing the improvement, of humankind through deliberate interventions in human reproduction, which Galton had named eugenics, evolved/diverged into two new "species." As had happened in the Soviet Union in the 1930s, some ideas, values, concerns, and policies previously construed as eugenic now morphed into "human" and "medical" genetics. This metamorphosis was revealed, for instance, in the re-christening of the oracle of British eugenicists, *Annals of Eugenics*, into the *Annals of Human Genetics* in 1954,[47] in the 1959 reconstitution of the Swedish Institute for Racial Biology (the stronghold of racial hygiene/eugenics established in 1922) into the Department of Medical Genetics at Uppsala University, and in the 1963 renaming of the Galton Laboratory for the Study of National Eugenics into the Galton Laboratory of the Department of Human Genetics and Biometry at University College London. At the same time, some eugenic ideas, values, concerns, and practices informed, and became incorporated into the arsenal of, the population control movement, as witnessed, for example, in the 1960 renaming of the Racial Hygiene Association of New South Wales into the Family Planning Association of Australia.[48]

This "speciation process" was driven by two concurrent sets of developments — social and scientific — that precipitated acute anxieties regarding the future around the world. On the social side, World War II and the ensuing Cold War radically redrew the world's political map, resulting in the rise of the two opposing camps — the "socialist East" and "capitalist West" — and the emergence from the ashes of the old colonial empires of newly independent countries, many of which refused to join either camp and formed the nonaligned movement. The new political order, epitomized in the labels First, Second, and Third Worlds, led to the formation of a number of multinational agencies under the aegis of the

United Nations, such as the Food and Agriculture Organization (FAO), the United Nations Educational, Scientific and Cultural Organization (UNESCO), the United Nations Children's Fund (UNICEF), and the World Health Organization (WHO). It prompted the emergence of new military and economic alliances.[49] It incited a concerted attack on racism and instigated the human rights movement. The new world order also generated numerous economic, political, social, demographic, public health, and cultural challenges, ranging from the arms and space race between the superpowers to the "demographic explosion" in the Third World countries to the rapidly growing environmental pollution and degradation in industrialized nations. These new challenges threatened the survival of not just separate nations, but also humanity as a whole and, indeed, the planet itself. They played a key part in reviving the interest in various "eugenic" solutions to problems and concerns raised by these global threats.

On the scientific side, the post-World-War-II developments were no less dramatic. Science played a critical, though often unsavory, role in both World War II and the Cold War, and was, in turn, transformed by their global impact. It helped produce the most powerful weapons of mass destruction — nuclear, chemical, and biological — which made palpable the fears of humanity's imminent annihilation in the next world war. But it also helped save numerous lives by producing new medicines and technologies (antibiotics and blood transfusions, amphetamines and pacemakers, organ transplants and artificial organs), which raised the hopes of extending human life considerably, if not indefinitely. Furthermore, science opened for humanity "the final frontier," space, and thus "materialized" both the threat of alien invasions and the promise of escape from whatever problems humanity might face on its home planet, which have long been favourite tropes of SF literature. Moreover, the two wars effectively drafted science into state service, which drastically increased its size and funding and enhanced the control of state bureaucracies over scientific activities, leading to the formation of "Big Science" and the "military-industrial-scientific" complex.[50]

More important for this story, however, was the rise in the post-World-War-II decades of a new — populational, molecular, and planetary — biology that once again radically changed the understanding of life phenomena. The joint labors of numerous biologists around the world produced a "modern synthesis," in the words of its major architect Julian Huxley, of Darwin's evolutionary theory with the principles of population genetics and ecology.[51] This synthesis shifted the attention of biologists from *individual* organisms to *populations* as the basic unit of evolutionary processes: adaptation, speciation, and extinction. It thus made the population's *genofond* (gene pool, in its English version) the main arena of evolutionary events. At the same time, the accumulated efforts of experimental biologists (biochemists, endocrinologists, virologists, biophysicists, geneticists, immunologists, etc.) culminated in the deciphering of the genetic code of protein synthesis and the uncovering of the chemical composition and spatial structure of numerous hormones, enzymes, vitamins, and ribonucleic acids that came to be seen as the "molecules of life," as the title of a 1970 book published nearly simultaneously in German, English, and Spanish readily attests.[52] These exciting discoveries led to the consolidation of what historian Lily E. Kay has appropriately named "the molecular vision of life."[53] During the same period, the rapid development of ecology resulted in the formation of a planetary, cosmic vision of life, embodied in the notion of biosphere,[54] as evidenced by a nearly 1,000-page volume that came out in 1970 under the title *Biosphere: A Study of Life*.[55] As a result of these developments, life came to be seen as a molecular, a populational, and a planetary phenomenon.[56]

The new concepts of life prompted the elaboration of various futuristic projects of manipulating life phenomena at a molecular, a populational, and a planetary level, reminiscent of the "visionary biology" of the 1920s. Indeed, during the late 1950s and 1960s nearly every biologist of note (not to mention a host of other science writers) published popular accounts of what the new knowledge of life engendered by new biology meant for the understanding of humanity's past, present, and future.[57] Built upon the earlier euphoric dreams and the "engineering ideal" of humanity's control over life, these new visions of humanity's future found expression in the rapidly growing currency of such terms as "genetic engineering," "population control," "environmental engineering," "terraforming," "biomedical engineering," and "noosphere."[58]

The very titles of several influential collections issued during the 1960s indicate the deep preoccupation with the future that seized the biological community: *Man and his Future* (1963); *The Control of Human Heredity and Evolution* (1965); *Genetics and the Future of Man* (1966); and *Biology and the Future of Man* (1970).[59] A special survey compiled by the Committee on Research in the Life Sciences of the US National Academy of Sciences offers an illuminating example. It was one of several surveys of the "state of the art" in the natural sciences to be used as a foundation for articulating federal science policy in regard to specific disciplines and research directions. The survey results were issued in 1970 by the National Academy Press under the title *Life Sciences: Recent Progress and Application to Human Affairs*.[60] But they also became the basis for a nearly 1,000-page volume for readers "with little previous, directly relevant, scientific background" that summarized "the state of the art" in all of the new areas of biology. Edited by the academy's president, biochemist Philip Handler, it was tellingly titled *Biology and the Future of Man*, even though only its last, twentieth chapter addressed directly the subject indicated in its title.[61]

Once again, both hopes and fears fostered by the seemingly unlimited powers of science (and especially biology) to "control" human life and death became an important impetus to revived discussions of eugenic solutions. These new societal concerns raised by World War II and the Cold War included the anticipated increase of "mutation load" as a result of nuclear weapons testing, the mutagenic and carcinogenic effects of industrial pollution, falling fertility rates in the so-called developed countries and the population explosion in "developing" ones, and the threats of new global epidemics triggered by weaponized microorganisms. And once again, these hopes and fears, threats and promises found powerful expression in the numerous SF writings of the time, from Aldous Huxley's pessimistic *Ape and Essence* (1948) and *Island* (1962) to Robert A. Heinlein's optimistic *Beyond This Horizon* (1948) and *The Moon Is a Harsh Mistress* (1966) to Frank Herbert's epic *Dune* (1965) and its sequels.

New biology became part and parcel of extensive debates on "new eugenics," which unfolded from the late 1950s through the early 1970s. Indeed, one could argue that new eugenics was but an offshoot of new biology: it was experimental/molecular and evolutionary/populational biologists who became most closely involved in both the critique and support of some old and a variety of new "eugenic" tools derived from the advances in the molecular, populational, and planetary studies of life. A fully-fledged analysis of these debates would require a massive volume of its own and is beyond the scope of this brief overview.[62] But the heated polemics that flared up at the 1962 Ciba Foundation symposium on "Man and his Future" and continued during the next decade on the pages of numerous journals and books, as well as at scientific conferences, offer illuminating examples of the new biology — new eugenics nexus, as well as various factors that affected its formation and operation in the 1960s and early 1970s.

## Back to the (1962) Future

The Ciba Foundation was an "international scientific and educational charity" established in 1949 by the Swiss chemical and pharmaceutical company of CIBA Limited with the goal of promoting "international cooperation in medical and chemical research."[63] The foundation sought to accomplish this goal by organizing and funding symposia (up to eight separate meetings annually) on specific, narrowly defined subjects, ranging from "the chemical structure of proteins" (1953) and "bone structure and metabolism" (1955) to "the regulation of cell metabolism" (1958) and "the nature of sleep" (1960). The symposia materials were published in English in the foundation's journal, or in book format, and widely distributed among interested specialists.[64]

As its very name clearly attests, the 1962 symposium on "Man and his Future" was very different from all the previous ones. The Ciba Foundation director Gordon E. W. Wolstenholme clearly articulated its major purpose:

> The world was unprepared socially, politically and ethically for the advent of nuclear power. Now, biological research is in a ferment, creating and promising methods of interference with "natural processes" which could destroy or could transform nearly every aspect of human life which we value. Urgently, it is necessary for men and women of every race and colour and creed, every intelligent individual of our one world, to consider the present and imminent possibilities.[65]

To consider these "present and imminent possibilities" of current biological research, the foundation invited sixteen individuals to deliver reports and eleven more to comment on seven topics: world resources, world population, sociological aspects, health and disease, eugenics and genetics, future of the mind, and ethical considerations.[66]

During the last days of November 1962 — barely a month after the end of the Cuban missile crisis that brought the world to the brink of nuclear war — hand-picked experts gathered at the foundation's headquarters in London. Contrary to Wolstenholme's call to "men and women of every race and colour and creed, every intelligent individual of our one world," the group consisted of 26 white men, all but two of whom came either from Britain and the Commonwealth countries (Canada, India, and South Africa) or the United States, and no one from behind the Iron Curtain.[67] Of all the "creeds" only Christianity and humanism were given voice at the symposium.

The majority of participants represented biomedical sciences, including six Nobel laureates: Francis Crick (who was about to go to Stockholm to receive his prize), Joshua Lederberg, Fritz A. Lipmann, Peter B. Medawar, H. J. Muller, and Albert Szent-Györgyi. The group also featured former and current leaders of the international biomedical establishment: the first secretary-general of WHO, G. Brock Chisholm; the first secretary-general of UNESCO and the outgoing president of the British Eugenics Society, Julian Huxley; the current FAO deputy director-general, Norman C. Wright; and neurologist Russell Brain, the editor-in-chief of the eponymous highly influential international journal. Also in attendance were such luminaries of new biology as gerontologist Alex Comfort, geneticist, biochemist, evolutionist, and long-time supporter/critic of eugenics Haldane, biochemist N. W. Pirie, virologist and immunologist Hilary Koprowski, anatomist and evolutionist J. Z. Young, and endocrinologists Hudson Hoagland, Marc Klein, Alan S. Parkes, and Gregory Pincus.

All other fields of science and scholarship were represented by leading nutrition specialist J. F. Brock; Jacob Bronovskii, a mathematician, historian and philosopher of science, who was one of the first to examine systematically the intersections between science and values; renowned agricultural economist Colin Clark; physical anthropologist Carleton S. Coon; famous city planner and architect Artur Glikson; Donald M.

MacKay, one of the pioneers of research in information theory and artificial intelligence; Derek J. de Solla Price, a historian of science who had pioneered studies of Big Science; and the Reverend Hubert C. Trowell, a specialist in "colonial medicine" and tropical diseases.

Eugenics was not merely one of the seven subjects discussed by the participants. It actually framed the entire symposium. Punctuated by lively discussions of its chosen topics, the five days of the symposium's sessions were bracketed by Huxley's meditation on "evolutionary aspects" of "the future of man" and Haldane's speculation on "biological possibilities for the human species in the next ten thousand years." Furthermore, judging by its published record, the session on "eugenics and genetics" turned out to be the longest and most contentious of them all. Its tone and terms were set by two formal reports: one by Muller on "Genetic Progress by Voluntary Conducted Germinal Choice" and another by Lederberg on "Biological Future of Man." The reports presented two alternative visions of new eugenics: the first articulated the position of the older generation of "classical" geneticists; the second, the views of a new cohort of "molecular" biologists.

Muller had been unable to come to London due to illness, and his paper was read by Ciba Foundation director Wolstenholme. The report was but an abridged version of Muller's recent publications on the subject.[68] Echoing earlier eugenic arguments, Muller described humanity's "genetic predicament," namely that "cultural evolution has undermined the process of genetic selection in man," whereas, according to Muller, what "we need instead, at this juncture, is a means of enhancing genetic selection." He considered various schemes advanced by "eugenists of the old school" for achieving this goal, as well as some new tools "to cause pre-specified changes" by "direct mutagenic operations on the genetic material." He concluded that "all these proposed means of escaping our genetic predicament are impracticable, insufficiently effective, or even positively vicious." What was "the most practical, effective, and satisfying means of genetic therapy," in Muller's opinion, was "eutelegenesis or germinal choice."

Under this new name, Muller largely recapitulated his earlier vision of artificial insemination with a donor's sperm as the most appropriate tool of "socialist eugenics," which he had elaborated in his 1935 eugenic manifesto, *Out of the Night*, and in his 1936 letter to

Stalin. Muller suggested creating "germ-cell banks," where the sperm of "talented producers," as his friend Serebrovskii had put it more than thirty years earlier, would be collected for future use by any interested party. Muller "updated" his earlier scheme with several new techniques derived from the latest advancements in experimental biology, which had been made by several of the very individuals who attended the symposium and which had realized Serebrovskii's vision of "separating sex and reproduction." Thus, he referred to the cryopreservation of the human sperm, pioneered by Parkes, and to the development of oral contraceptives spearheaded by Hoagland and Pincus, which paved the way to the "sexual revolution," actively promoted by its would-be guru, Comfort.

Muller was certain that in addition to artificial insemination with a donor's sperm, "further techniques are in the offing that will radically extend the possibilities of germinal choice," such as the storage of eggs, cloning, and "more delicate methods of manipulating the genetic material itself — what I have termed the use of nano-needles." He concluded that, combined with the use of contraceptives, "germinal choice" "must become increasingly applied in cases of genetic defect, genetic incompatibility, suspected mutagenesis, postponed reproduction, and finally, in serving the ardent aspiration to confer on one's children a highly superior genetic endowment."

Lederberg was 35 years younger than Muller, and his report embodied the aspirations and visions of a new generation of molecular biologists. As he asserted in the opening of his presentation, "Darwin's theory set off the historic debate on man's past. Today, with a new biology we mirror his future." In this "molecular" reflection, "man" was nothing more than "six feet of a particular molecular sequence of carbon, hydrogen, oxygen, nitrogen and phosphorus atoms — the length of DNA tightly coiled in the nucleus of his provenient egg and in the nucleus of every adult cell, 5 thousand million paired nucleotide units long." To assure the future of this "molecular" man Lederberg suggested using "the direct control of nucleotide sequences in human chromosomes, coupled with recognition, selection and integration of the desired genes, of which the existing population furnishes a considerable variety." He named such control "developmental engineering" or "euphenics."

For Lederberg, development was merely "the translation of the genetic instructions of the egg, embodied in its DNA, which direct the unfolding of its substance to form the living, breathing organism." Therefore, according to Lederberg, "Man's control of his own development, 'euphenics', changes the means and also the ends of eugenics, as have all the preceding cultural revolutions that have shaped the species: language, agriculture, political organization, the physical technologies." In his opinion, "Eugenics is aimed at the design of a reaction system (a DNA sequence) that, in a given context, will develop to a defined goal." But for him, the main instrument of eugenics, "somatic selection," was far too "slow in its impact." In order to address "the gravely imminent issues of human numbers and phenotype," Lederberg suggested, "biologists should give the first priority" to euphenics. But, unlike Kol'tsov and his like-minded Russian colleagues who had incorporated in their notion of euphenics all social influences that could affect human development and thus phenotype, Lederberg explicitly excluded the social from his vision. As he put it, "Eugenics and euphenics are the biological counterparts of education, a panacea that has a longer but equally contentious tradition. The troubled history of Utopian education warns us to take care in rebuilding human personality on infirm philosophy."

Lederberg admitted that at the moment molecular biology did not yet have the tools for implementing his vision of "developmental engineering." But he was certain that in the next few years such tools would be ready. As he phrased it: "Embryology is very much in the situation of atomic physics in 1900; having had an honourable and successful tradition it is about to begin! But it will not take long to mature. Most predictions of research progress have proved recently to be far too conservative." He outlined several possible areas of immediate application for the new tools: regulation of the size and metabolism of the human brain, and hence, human intelligence and personality, by prenatal or early postnatal intervention; "development of industrial methodology for synthesis of specific proteins: hormones, enzymes, antigens, structural proteins"; "abolition of immunity to transplants introduced in early life" that would allow "engineering development of artificial organs, e.g. hearts," for transplantations; and "vigorous eugenic programme ... on some non-human species, to produce genetically homogeneous material as sources for spare parts" for the replacement of aged or diseased organs in humans.

The two reports ignited a protracted discussion that revolved largely around the technical feasibility and social implications of the new instruments of control over "man's future" embodied in the two versions of new eugenics. Virtually all participants lamented the low level of biological knowledge among the general public and especially among the political class as the main barrier to the advancement of new biology and new eugenics. As Pirie put it, "I would be very much happier if I thought that those who govern us knew the rudiments of biology." And nearly all felt obliged to voice their positions vis-à-vis new eugenics.

The discussion revealed a deep divide between two groups: one consisted mostly of biologists, another included virtually everybody else. The position of the first group was well expressed by Crick, who opened the discussion by agreeing with both Muller's and Lederberg's versions of new eugenics, "with a few small reservations." Other biologists were divided in their preference for either Muller's or Lederberg's views and articulated such reservations by addressing various issues underpinning Muller's germinal choice and Lederberg's euphenics. Muller's version commanded more attention than Lederberg's, probably because the major tools for implementing Muller's scheme had already been available and tested out,[69] while the actual tools for Lederberg's euphenics remained in the realm of possibilities.

Thus, the participants did not discuss the feasibility of Muller's proposals, but debated the question of whose sperm should be collected in "germ-cell banks" and hence provide the material for germinal choice. As Klein sarcastically pointed out: "twenty years ago, in his book *Out of the Night*, he asked: where is the woman who would not be eager and proud to have in her womb a product of Lenin or Darwin? I don't think Muller would put Lenin and Darwin together now." Since Muller was absent, his friends, Huxley and (to a lesser extent) Haldane, came to his defense. Nearly all biologists agreed with Muller's depiction of humanity's "genetic predicament" and his basic notion that without some form of "genetic control" humanity was doomed. Pincus put it succinctly, "you don't get good genes by breeding in random fashion; you get good genes by selection."

The positions of the objectors were much more diverse but could be well exemplified by a rhetorical question posed by Bronovskii: "What problem are we trying to solve?" He found himself "out of sympathy with much that has been said in Muller's and Lederberg's papers." He saw no reason "to believe that the human population is deteriorating," which underpinned and justified all eugenics schemes, and "no evidence that the present human population is inferior, in any respect that one could quantify, to the human population fifty years ago." To the contrary, in his opinion, certain facts suggested that the population was actually improving. Bronovskii doubted not only the effectiveness but also the very necessity of interfering with "natural means" by which "the human race seems to be improving itself." Brock, MacKay, Glikson, Price, and Trowell expressed similar doubts. MacKay, for instance, noted the absence of clear-cut criteria for what the proponents of new eugenics wanted to accomplish: "to navigate by a landmark tied to your own ship's head is ultimately impossible. If we are ever to make proper use of our growing eugenic powers, we shall need a wisdom greater than our own."

Although quite supportive of Lederberg's discourse regarding the critical role of molecular biology in his own area of research on transplantations, Medawar expressed the same sentiment in much stronger language: "What frightens me about Muller and to some extent Huxley is their extreme self-confidence, their complete conviction not only that they know what ends are desirable but also that they know how to achieve them." Klein, who had survived the Auschwitz death camp, referred to the example of Nazi Germany as a strong argument against any form of eugenics, an argument also invoked by MacKay and Clark. Several discussants (Clark, Klein, and MacKay) supported the notion ardently expressed by Trowell that any eugenics contradicted the fundamentals of Christian ethics. Haldane who had by that time moved from Britain to India, however, remarked that many "eugenic" practices were perfectly compatible with Hindu ethics. Price, on the other hand, pointed out what most biologists completely ignored in their comments — the role of social factors in human development. "We know that a great deal of the performance of man depends as much on social environment as on genetics," he stressed, "and this environment might act in a way completely opposite to that which would be produced by the mechanisms of genetic control which we might introduce."

The chasm between the two groups was further deepened by Crick's inflammatory suggestion that "people" do not have "the right to have children" and only those few licensed by the government should be allowed to procreate and should pay a heavy tax on every child born, which added much fuel to the unfolding discussion. Predictably, the opposing views could not be reconciled and the participants could not reach any agreement regarding "the present and imminent possibilities" of either Muller's germinal choice or Lederberg's euphenics.

The Ciba symposium sparked a far wider debate. Published simultaneously on both sides of the Atlantic, its proceedings were reviewed in the leading biomedical periodicals, from *Science* and *Perspectives in Biology and Medicine* to *JAMA* and *Lancet*, and generated quite a stir, even penetrating the Iron Curtain, as we saw in the previous chapter.[70] But, as one reviewer astutely observed, the symposium merely featured "a group of enlightened but diversely opinioned men who individually spoke each in his accustomed tongue," whilst its "aggregate effect" was utterly insipid: "There is no consensus, and the future of man remains shrouded by the mist of future time. Vague trends are discerned, obvious dangers recognized, but where man goes no one knows."[71]

Yet, the symposium discussion on "eugenics and genetics" did elicit a vocal response. Indeed, since Muller had published versions of his report before the symposium, and Lederberg published a revised and abridged version of his presentation immediately after (in the May 1963 issue of *Nature*),[72] the essence of the debated proposals were known to their prospective audiences even before the proceedings came out. The volume's publication facilitated the spread of the debate that unfolded much along the lines drawn at the symposium. The enthusiasts hailed the advantages of new eugenics, comparing the benefits and shortcomings of Muller's germinal choice and Lederberg's euphenics.[73] The opponents cautioned about the misuse and abuse of the ever growing powers of new biology, denying the urgency and necessity of any form of eugenics and questioning its moral foundations.[74]

The debate continued on the pages of learned journals and books, as well as at academic conferences.[75] Less than a year after the Ciba symposium, in April 1963 eminent US geneticist Tracy M. Sonneborn organized a conference on "The Control of Human Heredity and Evolution" that featured contributions by leading European and

American geneticists, including Salvador E. Luria, Edward L. Tatum, Guido Pontecorvo, and Muller.[76] At the 1965 annual meetings of the American Institute of Biological Sciences, Rockefeller University's leading molecular biologist Rollin D. Hotchkiss delivered a keynote lecture on "The Portents for a Genetic Engineering."[77] The next year, at the annual meeting of the American Association for the Advancement of Science the world's foremost evolutionary geneticist Theodosius Dobzhansky devoted his plenary lecture to "Changing Man."[78] At a symposium co-sponsored by Marymount College, New York, and the *Commonweal* magazine in 1968, well-known human geneticist Kurt Hirschhorn spoke "On Redoing the Man."[79] At the 1968 annual meeting of the Society of American Naturalists, its president, geneticist Jack Schultz devoted his presidential address to "Human Values and Human Genetics."[80] Expectedly, both Muller and Lederberg continued to speak and publish extensively on their respective visions of "Means and Aims in Human Genetic Betterment," as Muller titled one of his articles.[81] James F. Crow, Muller's former student and US leading population geneticist, expressed a general position shared by the advocates of new eugenics in a 1966 article with the telling title "The Quality of People":

> The early eugenics was genetically naive and was connected with various dubious and even tragic political movements. I think the time is here when the subject should be reopened and discussed by everyone — not just biologists — with a serious consideration of the consequences of misjudgments as well as the possibilities for good.[82]

Evident at the Ciba symposium, the critical assessments of the possible consequences of new eugenics found expression in numerous publications by scientists (especially biologists), philosophers, theologians, sociologists, and legal scholars. The essence of their varying concerns is well exemplified by the titles of numerous books and articles that appeared within a decade of the London meeting: *Beyond Morality and the Law*; "On Genetics, Sociology and Politics"; *Fabricated Man: The Ethics of Genetic Control*; "Ethical Aspects of Genetic Controls: Designed Genetic Changes in Man"; "Reservations Concerning Gene Therapy"; "Can Man Shape His Future?"; "The Biologists' Dilemmas"; *Social and Psychological Aspects of Applied Human Genetics*; "Anxiety About Genetic Engineering,"; "Prometheus and Pandora: 1971"; and "The Dilemma of Genetic Engineering."[83] Some of these publications appeared in the same journals that carried the contributions by proponents of new eugenics, *Nature, Perspectives in Biology and Medicine,* and *Science,* others on the pages of legal, medical, and theological periodicals and books.[84]

Journalists and science writers spread the threats and promises of new eugenics. Numerous newspapers and magazines headlined and featured stories with such sensational titles as "Heredity Control: Dream or Nightmare?"; "Man into Superman: The Promise and Peril of the New Genetics"; "Genetics and the Survival of the Unfit"; "Playing God"; and so on.[85] New biology and new eugenics were actively popularized in other media as well, for instance, in the CBS television series *The 21st Century* (1967-1970), summarized by the series science editor Fred Warshofsky in a 1969 book with the telling title, *The Control of Life.*[86]

Lingering debates notwithstanding, the names of the two competing versions of new eugenics — germinal choice and euphenics — did not take root. The first was soon subsumed into the generic term "eugenics." The second, despite Pirie's suggestion to include it in the *Biological Council Dictionary* of new words in biology,[87] was soon swallowed, against Lederberg's vocal protestations,[88] by more general terms such as "gene therapy" and "genetic engineering." But the visions they embodied spurred more than debates between proponents and opponents. They stimulated interest in the historical development of eugenics, especially in the US.[89] More important, they also prompted attempts at practical implementation of both Muller's and Lederberg's versions of new eugenics.

Indeed, shortly after the Ciba symposium, US millionaire Robert Clark Graham approached Muller with an offer to fund his vision. Muller did not live to see it through, as he died in April 1967. But four years later, Graham did establish the Herman J. Muller Repository for Germinal Choice, a sperm bank for Nobel Laureates and other acknowledged "men of genius."[90] Lederberg's prediction that it would not take long to develop technical means for his euphenics proved quite accurate. In fact, by the time of Lederberg's report, several researchers had already demonstrated the technical possibility of transferring genetic material (DNA and RNA) from one organism to another. In less than a decade, various technologies were developed for such transfer by means of viral infections, direct transmission of exogenous or "naked" DNA into a host cell, recombinant DNA and plasmid transformation, as well as by direct transplantation of the nucleus from one cell to another.[91] At the same time, cloning techniques — envisioned by both Muller and Lederberg as a possible tool of future eugenics — were making steady progress from frogs through insects and fishes towards mammals.[92]

As a result of the 1960s debates, in the professional worlds of biology and medicine, new eugenics was *normalized* during the next decade in a number of practices that became widely accepted and construed as standard tools of *medical genetics*.[93] Its new instruments, concerns, and policies — genetic counseling, dietary or drug therapies of certain hereditary diseases (such as phenylketonuria), artificial insemination, sperm banking, in vitro fertilization, embryo transplantation, genetic screening, and prenatal diagnosis — became the subject of special chapters in widely used textbooks on human and medical genetics.[94] Indeed, by the late 1960s, eugenics had acquired yet another new name, as witnessed by the quiet rechristening of *Eugenics Quarterly*, the oracle of US eugenicists, into *Social Biology*.[95] The name change was apparently meant to further distance new eugenics from its suspect ancestry.[96]

At the same time, objections to the very principles of eugenics and the critique of its foundations became *professionalized* within the emerging field of bioethics, a new term coined in 1970 to cover theological, philosophical, sociological, legal, and moral concerns raised by new biological and medical technologies and practices.[97] Expectedly, the hopes and fears associated with new eugenic ideas, technologies, and policies also found vivid expressions in popular journalistic accounts[98] and SF literature, such as Nancy M. Freedman's 1973 novel, *Joshua, Son of None*, whose title character was a clone of John F. Kennedy.[99]

It is hardly necessary to dwell on the fact that the same two sets of factors, social and scientific, propelled a new wave of interest in eugenics (both advocacy and criticism) in the 1990s. The "end" of the Cold War and the collapse of the "socialist camp" once again reshaped the world's political, economic, and ideological maps. It fostered the globalization of neo-liberal capitalism and further deepened economic, political, and social inequalities within and between countries. It stimulated the rise of multinational corporations as a formidable competitor to the state as both a major patron and a prime beneficiary of scientific research, which dramatically reformatted the contours of Big Science, especially in biomedical fields. At the same time, rapidly expanding biomedical research generated new discoveries, concepts, disciplines, and technologies (ranging from polymerase chain reaction to DNA sequencing machines), which made possible such monumental undertakings as the Human Genome Project and geno-geographic atlases of the world's populations. These concurrent social and scientific developments gave rise to new fears and new hopes about humanity's future, associated with "new genetics" and with such products of the new biomedical technologies as GMOs, Dolly the Sheep, and xenografts, to list just a few iconic examples.[100] These hopes and fears, in turn, fanned the embers that had been smoldering during the previous decade, producing yet another "new" eugenics: "liberal," "consumer," "free-market," "libertarian," "homemade," and, as we saw in the case of post-Soviet Russia, "racial."[101] And they fueled extensive debates about the historical development of eugenics and its continuing, though contested, impact on ideas, values, concerns, and policies regarding human nature and humanity's future, which show no sign of abating.[102]

\* \* \*

Given its contrapuntal, polyphonic history, Bochkov's designation of eugenics as "a science that has outlived itself" seems premature. After all, various intellectuals before, simultaneously with, and after Galton had developed versions of "eugenics" as a means to address certain social problems that they saw as threatening the future of their countries by bringing human nature under "social control" and thus arresting its "degeneration" and assuring its "improvement." There could be little doubt that humanity's perpetual anxiety about the future will keep alive the interest in eugenic solutions to numerous new challenges engendered by the inexorable growth of science's powers and compounded by human ambition, bigotry, and greed. No matter what new names the particular amalgam of ideas, values, concerns, and actions regarding human reproduction, heredity, diversity, development, and evolution will assume in different contexts and settings, it will certainly continue to inspire both hopes and fears and, thus, command the attention of various individuals and groups for millennia to come.

Seen in this light, the latest 2012 edition of Vasilii Florinskii's book is probably not the end of its life story. Its author is now firmly inscribed into the historical record of the development of science, medicine, and education in Siberia, with Tomsk University planning to unveil a monument to its founder to mark its 140-year anniversary on 5 September 2018.[103] It seems more than likely that the next wave of heightened anxieties about the future of Russia — spurred by concurrent social and scientific developments — will once again breathe life into his attempt "to bring to my Fatherland as much benefit as possible." At this point, we cannot even fathom what kind of life it will have and what new meanings future commentators will read into *Human Perfection and Degeneration*.

# Apologia: The Historian's Craft

> "…habiendo y debiendo ser los historiadores puntuales, verdaderos y nonada apasionados, y que ni el interés ni el miedo, el rancor ni la afición, no les hagan torcer del camino de la verdad, cuya madre es la historia, émula del tiempo, depósito de las acciones, testigo de lo pasado, ejemplo y aviso de lo presente, advertencia de lo por venir."
>
> Miguel de Cervantes Saavedra, 1605

As Miguel de Cervantes forcefully stated in his immortal book *Don Quixote*: "it is the job and duty of historians to be exact, truthful, and dispassionate."[1] Alas, the author of the first modern novel did not leave us a guide to exactly how historians were to fulfill this tall order. Generations of Clio's worshipers have put considerable efforts into perfecting their craft and honing their tools to ascertain the accuracy of the histories they produce. Yet hardly any other area of scholarship exhibits such a penchant for perpetual "revision" of its products. The high passions history often engenders aside, the origins of its "inherent" revisionism rest on two basic elements of the historian's craft: finding necessary sources and translating them, both literally and figuratively. As for many of my fellow historians, these two foundations of our profession generated numerous challenges in my work on the biography of Vasilii Florinskii's *Human Perfection and Degeneration*. Understanding and meeting these challenges, therefore, became an integral part of my journey on Cervantes's "path of truth" (*camino de la verdad*) to recover the forgotten history of the book, its author, publishers, and readers.

## The Historian as Detective

The historian's craft is very much akin to that of a detective. No one has demonstrated this simple fact more skillfully and more convincingly than the British writer Elizabeth MacKintosh in her novel *The Daughter of Time* published in 1951 under the penname Josephine Tey.[2] The novel's main character, Scotland Yard Inspector Alan Grant, is confined to a hospital bed as the result of an accident he has suffered in the line of duty. Out of boredom, the inspector decides to investigate an old "cold case." The case he picks is not an ordinary one. It had provided the plot for William Shakespeare's famous tragedy: King Richard III's alleged murder of his nephews, Edward, Prince of Wales, and Richard, Duke of York. With the help of a "research assistant" who does all the legwork in searching for the necessary materials in London's libraries, the inspector eventually discovers that Shakespeare's dramatic version of the events surrounding "the mysterious disappearance of the boys from the Tower" has very little to do with historical reality, for it is based on a biased account produced many years after the alleged murder by none other than "St. Thomas More."[3] "I'll never again believe anything I read in a history book, as long as I live," the inspector concludes in disgust. To his dismay, Grant also learns that More's and Shakespeare's portrayals of Richard III as a murderer, in spite of persistent doubts about the accuracy of both expressed by many historians, have been perpetuated in school textbooks and public consciousness. Furthermore, he realizes that numerous historical events exist in public memory only in the form of "a dramatic story with not a word of truth in it." He names this aberration of historical memory "Tonypandy," after the first example of such particular distortion of historic truth he had encountered.

By turning the professional detective Grant into a historian, Tey has vividly depicted the key part in the job of both: gathering all sorts of evidence, from material traces to witness testimonies, relevant to the event under investigation, be it a murder or a publication of a scientific text.[4] Yet, despite the current popularity of TV shows about "cold cases," a real-life detective rarely, if ever, investigates a crime that has previously been "solved" by another detective. The situation is just the opposite for a historian: only in truly exceptional cases, does s/he write about something that has not already been studied in one way or another by numerous colleagues past and present.

As Inspector Grant quickly recognizes, in a historical investigation all relevant evidence falls into the two uneven and vastly different categories of *primary* and *secondary* sources. The former include anything and everything generated at the *time* and the *place* of the event by its active participants and passive observers. The latter consist of everything and anything created *outside* the time and the place of the event by individuals (including historians) who have no direct relation to that event. Unlike a detective who usually deals almost exclusively with primary sources, a historian always has to deal with both — and to distinguish carefully between — primary and secondary sources. As Inspector Grant learns in the course of his investigation of Richard III's purported crime, *who, when, how, for what purpose,* and *for which audience* created a particular source are critical questions that a historian must ask and, whenever possible, answer.

In my own research into the biography of Florinskii's treatise, I had to wear the hats of both a detective and a historian. This book traverses more than 150 years of Russian history, along with the history of medicine, science, journalism, education, and eugenics. It touches upon numerous subjects, events, individuals, institutions, and ideas, nearly all of which and whom have been studied and sometimes hotly debated by several generations of historians. I have greatly benefited from the available secondary sources they produced.

Yet, I could not help but notice the paucity of secondary literature on many important subjects, institutions, and individuals in nineteenth-century Russian history, which hindered substantially my work on this book. To give just a few almost random examples, there is still no fully-fledged scholarly biography in any language of the Grand Duke Konstantin Nikolaevich Romanov, the younger brother of Emperor Alexander II and the main architect of the Great Reforms. Neither has the life story of *Russian Word*'s founder Count Grigorii Kushelev-Bezborodko attracted much attention, despite the fact that he left his footprints in almost every area of nineteenth-century Russian literature, arts, philanthropy, education, and even chess. Compared to the enormous body of historical works on Russian literature, the history of Russian science and medicine remains considerably understudied, with numerous individuals, practices, and institutions still awaiting careful historical investigation.

I have also noticed that much of the Soviet-era historical literature suffers from a dogmatic "Marxist" approach to its various subjects, but nevertheless contains a wealth of factual material, some of which I have happily utilized in my own work. By contrast, much of the pre-1990s western historical literature, especially dealing with the Soviet period, offers a variety of useful analytical schemes and tools, but often contains factual mistakes deriving from lack of access to primary sources. Moreover, this literature not infrequently exhibits a marked Cold War "us versus them" mentality in assessing various events of Soviet history. All told, it was simply impossible to reference all of the historical literature I consulted, and, aside from a few direct quotations, I have kept such references to a bare minimum, only pointing out what I found to be the most relevant and illuminating works. I have also avoided engaging in polemics with divergent interpretations of many events, ranging from the Crimean War to the Bolshevik Revolution to Gorbachev's *perestroika* to the recent rise of rampant nationalism in Putin's Russia, which in one way or another affected the lives of my actors.

The principal foundation of this book is a large array of both archival and published materials related to its main subject: a series of essays entitled "Human Perfection and Degeneration," written by Vasilii Florinskii, and published successively by Grigorii Blagosvetlov in 1865 and 1866, Mikhail Volotskoi in 1926, Valerii Puzyrev in 1995, and Vladimir Avdeev in 2012. Although I faced considerable difficulty in acquiring a copy of the 1995 edition of Florinskii's treatise (it is absent from all the major libraries in both Moscow and St. Petersburg),[5] as one could easily imagine, the foremost challenges arose in finding materials related to its author and its first two publishers.

## Vasilii Florinskii

Luckily for me, in addition to more than 300 published works, ranging from short newspaper notes to voluminous textbooks and monographs, Florinskii left a substantial archival footprint. Materials pertaining to his education and professional career are scattered throughout a number of Russian archives, libraries, and museums that hold the documents of particular institutions he attended as a student or worked for in the course of his career, such as Perm Theological Seminary, the Imperial Medical-Surgical Academy, the War Ministry, the Ministry of People's Enlightenment, as well as Kazan and Tomsk universities.

More important, a large collection of Florinskii's personal papers has miraculously survived not only his continuous movements through the vast expanses of the Russian Empire — from Perm to St. Petersburg to Kazan to Tomsk and back to St. Petersburg — but also the social turmoil, wars, and revolutions that plagued his homeland from the mid-nineteenth through the twenty-first century. Accidentally discovered in 1938,[6] this collection is currently housed at the National Museum of the Tatarstan Republic in Kazan and includes nearly 5,000 items.[7] Yet in this entire collection I found only one (!) document *directly* related to his 1865 essays: an undated letter from Pavel Iakobii that Florinskii received some time in the 1880s. Although towards the end of his life Florinskii began to include the 1866 book (but not its journal version) in the list of his publications,[8] this collection contains no preparatory notes, drafts, manuscripts, indeed, not a single piece of paper that in any way could illuminate his actual work on this lengthy treatise. Nor does it contain any trace of Florinskii's dealings with its editor and publisher. I found no financial records, correspondence, page proofs, indeed no mention of *Russian Word* or *Deed*, Blagosvetlov (the de facto editor and publisher of both), Nikolai Blagoveshchenskii (the de jure editor of *Russian Word*), or Nikolai Shul'gin (the de jure publisher of *Deed*).

Furthermore, this collection includes very few documents dating from before 1870, the period of Florinskii's life most relevant to my project: about thirty letters Florinskii sent to his parents and siblings between 1853 and 1863, several notebooks he kept while attending the IMSA and during his European tour of 1861-1863, and a few photographs. Judging by available materials, Florinskii was very close to his parents and siblings. His own family — his wife Maria and their children, Olga and Sergei — was also a very important part of his life. Yet aside from a few photographs, they are all but invisible in his papers. After her husband's death, Maria Florinskaia took certain steps to preserve his legacy by publishing a portion of his memoirs and a large volume of his selected works. Alas, neither contains a word about *Human Perfection and Degeneration*. Furthermore, she does not seem to have written any personal reminiscences about her husband and the thirty-plus years of their life together.[9]

Altogether Florinskii's personal papers include nearly 3,000 letters, only about fifty of them are written *by* him and nearly all of those are addressed to his or his wife's family members. All others are written *to* Florinskii and thus represent just one side of his extensive professional correspondence with colleagues and superiors. I did find a few of Florinskii's letters among the personal papers of their various addressees, but all of them turned out to be typical business correspondence and contained no information directly relevant to this project. Moreover, the bulk of this correspondence is dated after 1879 and related to his involvement with establishing Tomsk University.[10]

In the 1880s, Florinskii began writing his memoirs that covered three distinct periods of his life. The first part, titled "Thoughts and Recollections of my Childhood and Education," was written in 1882; the second, titled "The Foreign Trip of 1861[-63]," he wrote in 1880-1881; and the last one, titled "Notes and Recollections. 1875-80," was penned on and off from 1881 to 1892.[11] In compiling these memoirs Florinskii seems to have largely relied on his memory, though he also used certain documents many of which have since been lost. These memoirs contain a treasure trove of information, but none of them even mentions *Human Perfection and Degeneration*. Furthermore, in his reminiscences, Florinskii completely left out the dozen years of his life, from 1863 to 1875, during which he had worked at the IMSA and had written his treatise.

These memoirs, of course, must be taken with a grain of salt, and not only because human memory is notoriously unreliable. Unlike diaries, which record their author's impressions of and thoughts about events, peoples, ideas, etc. at the very time s/he encounters them, memoirs, as a rule, reflect the attitudes, thoughts, and values of a much older person, who looks back at their own earlier life from the vantage point of the time, experiences, position, worldview, and so on, much different from those of the younger self. For instance, there is little doubt in my mind that Florinskii heavily romanticized the account of his early life and education, omitting all the hardship and loneliness he undoubtedly experienced. Similarly, his recollections of the 1861-1863 trip to Europe were certainly colored by the mid-1870s conflict with his "German" colleagues, which eventually led to his resignation from the IMSA. Furthermore, along with the memoirs, he wrote an autobiographical novel, *Three Stages of Life*, and writing this fictionalized account probably influenced his reminiscences.[12] It is unclear whether Florinskii ever planned to publish his memoirs or the novel. In the mid-1890s, contemplating the possibility of producing a multi-volume edition of his *Selected Works*, he initially included "the diaries of my foreign trip" in its contents, but then crossed the entry out.[13] In any case, the entire project never went beyond the planning stage. Moreover, none of his memoirs were published during Florinskii's lifetime. Only after his death, did Florinskii's wife publish the last part of his reminiscences devoted to the creation of Tomsk University.[14]

In employing Florinskii's memoirs in the present work, I have sought to verify the information they provide through other sources. Alas, aside from certain events of his career and public engagements, this proved no easy task. Surprisingly, despite his easy-going character, Florinskii does not seem to have made any close friends at any point of his life, though in his later years he did make a few enemies. Equally surprising, despite his nearly thirty-year-long teaching career, he did not have any students who worked closely with him in any of the scholarly fields that held his interest. This perhaps explains, at least in part, the virtually total absence of any memoirs by, diaries of, and correspondence among his contemporaries, which even mention his name, to say nothing of providing more detailed accounts of his life and work. Even the massive unpublished diaries of his long-time supporter, IMSA professor Iakov Chistovich, contain only passing mentions of Florinskii's name, mostly in relation to his involvement in the 1874-1875 "territorial" conflict (described in Chapter 4) between the War Ministry and the Ministry of People's Enlightenment over the administrative control of the IMSA.[15]

In all of the available materials, I also uncovered very little information on one of the most important aspects of Florinskii's life — his religious beliefs and his relationship to the Orthodox Church. Switching from theology to gynecology required the former seminarian to deal somehow with the two very different — Christian and scientific — views of the human body, human reproduction, human evolution, and, ultimately, human destiny. His publications indicate that to the end of his days he believed in God and considered religion the foundation of morality. His treatise also demonstrates that he was quite critical of certain rules and regulations imposed by the church on the life of his compatriots. But neither in his published works, nor in his private diaries, memoirs, and correspondence did he even once address the obvious contradictions between scientific and religious views in relation to the key subjects of his chosen profession.

Although Florinskii finished his career with the third rank in the Imperial Table of Ranks (equal in military terms to a lieutenant-general) and retired from the post of de-facto deputy-minister in charge of all matters related to education for the territory six times the size of France, until very recently his life story has attracted surprisingly little attention. His name does appear in various biographical dictionaries and encyclopaedias,[16] as well as in some historical accounts of nineteenth-century Russian medicine, especially in connection to gynecology and pediatrics.[17] But, as we saw, only in the early 1990s did Evgenii Iastrebov produce a series of publications on his "ancestor," including a bibliography of Florinskii's published works and a biography based on a thorough examination of various published and archival materials. In describing Florinskii's life story, Iastrebov focused predominantly on the preeminent role his ancestor had played in the founding of Tomsk University, the first university in Siberia. In a supplement published a few years later, the biographer did examine Florinskii's studies and work at the IMSA, but paid almost no attention to his 1865 essays. Thanks to Iastrebov's efforts, in the last two decades, various facets of Florinskii's life and career in Siberia have commanded some scholarly attention, but his work on *Human Perfection and Degeneration* has so far received at best only cursory treatment.[18] Although in the 1920s Nikolai Kol'tsov hailed the book as "one of the first original books on Darwinism in Russia," it is absent in the voluminous literature on the reception and development of Darwinism in Russia, which I have examined closely.

As far as I was able to ascertain, the first English-language reference to Florinskii and his 1866 book appeared only a century after its publication in Leslie C. Dunn's presidential address to the American Society of Human Genetics in May 1961.[19] Nearly thirty years later, Mark B. Adams briefly introduced the book to English-speaking audiences in his overview of the history of eugenics in Russia. Since then Florinskii's essays have occasionally been mentioned in other scholarly works (including my own) on the same and related subjects, but have never received an extended treatment. I hope this book has corrected this historical omission.

## Grigorii Blagosvetlov

I have been much less successful in tracing down primary documents related to Blagosvetlov, the first publisher of Florinskii's work. Expectedly, the materials related to his work as the editor of *Russian Word* and *Deed* and publisher of numerous books are preserved in the archives of various censorship agencies. But in the voluminous archival collections of these agencies, I found only one document directly related to the publication of Florinskii's essays — a short report by a censor on their first book edition. Considered a "revolutionary" by the imperial authorities, Blagosvetlov was subject to close surveillance by the secret police since the early 1850s, and some relevant documents can be found in their archives. But, in the spring of 1866, just a few months after *Russian Word* published Florinskii's essays and a few days after an unsuccessful assassination attempt on Alexander II, expecting an imminent arrest, Blagosvetlov reportedly burned most of his personal papers, as well as extensive editorial files relating to his journal.[20] In turn, many of his correspondents also purged his letters from their personal papers. As a result, only his published works and a handful of his correspondence dated from before 1866 have survived.[21] Alas, I did not find any mention of even Florinskii's name, to say nothing of his essays, in the available documents. Much like Florinskii's, Blagosvetlov's wife published after her husband's death a large tome of his selected writings supplemented by a detailed biography written by Nikolai Shelgunov, Blagosvetlov's life-long friend and an active contributor to his journals. Yet Florinskii's name does not appear anywhere in the volume.

Despite the paucity of archival sources, in contrast to Florinskii's, Blagosvetlov's life and works have attracted considerable historical attention, especially from the students of nineteenth-century Russian literature and journalism. They have produced numerous, though contradictory, accounts of his involvement with "revolutionary circles," his role in publishing *Russian Word* and *Deed*, his work as a writer, a literary critic, a publisher, and a mentor to the large cohort of talented litterateurs he had gathered around his journals.[22] Yet in all of this ample secondary literature, Blagosvetlov's efforts to popularize science (and especially Darwin's works) in Russia have received very limited attention, whilst only one author has mentioned Florinskii's name, merely as one among numerous "minor" contributors to *Russian Word*. Thus, one of the aims of this book was to investigate the role Blagosvetlov and his journals played not only in promoting Florinskii's eugamic ideas, but also more broadly in the popularization of a scientific worldview in post-Crimean Russia.

## Mikhail Volotskoi

I failed to find any personal papers of Volotskoi related to his discovery and republication of Florinskii's book. Volotskoi died in 1944 in war-ridden Moscow, and upon his death, his wife apparently decided to save and deposit in a state archive only that portion of his personal documents which dealt with his life-long work on the genealogy of Fedor Dostoevsky, but nothing else.[23] Perhaps, as did many of his fellow members of the Russian Eugenics Society, Volotskoi himself purged his personal archive from all materials related to "bourgeois," "fascist" science, as eugenics came to be called in 1930s Russia. Or perhaps his wife did so in the wake of Trofim Lysenko's 1948 campaign against genetics and eugenics. Whatever happened, very few primary sources shed light on Volotskoi's involvement with Florinskii's treatise, or eugenics more generally, during the 1920s.

Furthermore, available sources (both primary and secondary) even for Volotskoi's biography are scarce. Nowadays he is remembered mostly as the author of a voluminous study of the genealogy of the Dostoevsky family.[24] There is one brief commemorative article hailing his role in the development of Soviet anthropology, especially dermatoglyphics.[25] But his name is absent from the recent extensive bio-bibliographical dictionary of twentieth-century Russian anthropologists and ethnographers.[26] Volotskoi's contributions to several other fields, ranging from physical culture to occupational hygiene, remain completely forgotten. For instance, despite the fact that he worked for nearly a decade as a lecturer and researcher at the Institute of Physical Culture, his works (including those published under the institute's auspices) are not even mentioned in the sizeable bibliography of publications produced by the institute's staff.[27] Furthermore, his name appears only once in the voluminous memoirs about the early years of the institute recently published by its museum.[28]

My reconstruction of Volotskoi's life and career is largely based on materials preserved in the archives of several institutions where he worked, especially his "personnel files" that contain his curriculum vitae from different years. These necessarily brief and formulaic documents are understandably mute on many aspects of his life and provide very little insight as to his motives, inspirations, and aspirations in reissuing Florinskii's book. Volotskoi's published works, then, constitute the main source for my interpretations. Finding these works, however, turned out to be a major undertaking in itself, for they have become a rarity.

In the aftermath of Lysenko's anti-genetics campaign, Volotskoi's eugenics publications were removed from many libraries. Only major research libraries in Moscow and St. Petersburg, such as the Russian National Library, the Russian State Library, the Library of the Academy of Sciences (BAN), and the Library of the Institute of Scientific Information on Social Sciences (INION), have preserved some, but not all, of his publications.[29] To give just one example, the Russian National Library in St. Petersburg, the oldest and second largest public library in the country, does not have a copy of either the first 1923, or the second 1926, edition of his pamphlet *Elevating the Vital Forces of the Race*, in which he first introduced Florinskii's treatise to Soviet readers. Furthermore, for many years, Volotskoi's articles and books on eugenics, as well as many other publications on the subject, were held in the *spetskhran* — a special "closed" section that required "security clearance" to access its holdings. Only after the dissolution of the Soviet Union in 1991 did these publications eventually become available to the general reader. But in some cases, they are still excluded from the libraries' general catalogues. Thus, in BAN, if one attempts to order the aforementioned editions of Volotskoi's pamphlet through the library's main reading room, the response would state that the library does not have them. But in fact, and one needs to know it, they could be ordered in the "reading room for the literature of Russia abroad," as the former *spetskhran* section is now misleadingly called.[30] So, it took considerable effort and the help of many people to collect copies of Volotskoi's published works for this project.

Although his name as one of the founders of the Russian Eugenics Society and the only one among its members who vocally supported the sterilization of the "unfit" appears in practically all historical accounts of eugenics in the Soviet Union, there is no analysis of Volotskoi's actual contributions to its development.[31] The present work, then, brings back to the historical record not only his extensive efforts as a publisher and promoter of Florinskii's treatise, but also his involvement in creating what he named "bio-social" eugenics, which profoundly influenced the debates around and developments of eugenics in Soviet and post-Soviet Russia.

## Is History Bunk?

The above brief description of my hunt for relevant materials shows that I have been able to find very few primary sources illuminating *directly* some of the key episodes in the biography of Florinskii's essays, a situation not uncommon in historical research. No matter how much material pertaining to the event, person, or artifact under investigation a historian collects, there always remain gaping holes in the record, which cannot be filled whatever the effort. Certain events might have left no traces or never even been recorded at all. Some materials might have been lost forever to the merciless passage of time or remain hidden in some private holdings.

How then might a historian avoid creating just another "Tonypandy," so vividly portrayed by Tey and condemned by Inspector Grant as "a dramatic story with not a word of truth in it"? Although we still cannot completely escape the temptation to judge historical actors according to the ideals, values, and mores of our own times, writing a history differs substantially from conducting a criminal trial. The simplest solution to the absence of sources illuminating this or that episode in the life of our actors is to admit that we do not know and might never know exactly what happened and move on to something we do know. But, in certain cases, I personally find this solution utterly unsatisfactory.

History is not an exact science (some argue it is not a science at all) in the sense that, barring the availability of a time machine, any "truth" about a particular historical event cannot be confirmed and reconfirmed, as can be done in physics or chemistry, by direct, so-called control experiments that reproduce all the circumstances and participants of the event in question. The dream of any historian to be "a fly on the wall" during a past event under investigation, and thus be able to witness and record it personally, is just that — a dream. More often than not, a historian has to rely on circumstantial evidence that would be rejected out of hand by a detective and, if not, thrown out by a court of law. Yet like any good detective, in the course of research, a historian gains insight into the minds and lives of his/her "characters" by learning everything and anything possible about them: their words and deeds, their expressed feelings and thoughts, their teachers and pupils, their tastes and beliefs, their friends and foes, their parents and children, and on, and on. S/he "gets to know" the actors and learns to "envisage" the way they might have felt, thought, and acted in particular circumstances, even in the absence of any documentary materials (be they manuscripts, photos, publications, films, and so on) with exact information on a specific event of their lives.

My reconstruction of Blagosvetlov's role in the crafting of Florinskii's treatise, for instance, is based not on direct evidence (for I did not find any), but on conjuncture, guesswork, analogy, and sometimes pure speculation, deriving from the mass of archival and published materials that I have examined. These materials, I believe, did provide me with certain insight into the minds and actions of both the author and the first publisher of *Human Perfection and Degeneration*. To give but one example, thanks to the secret police that monitored and copied parts of his correspondence, I was able to dig out from police files several letters that, though never once mentioning Florinskii's name, clarified for me Blagosvetlov's interests and motivations in waging a wide campaign to popularize Darwin's works and "social hygiene" in Russia. This appeared to have been the main reason for his publication of Florinskii's treatise in the first place and for his continuing efforts to explore further its main ideas and to keep it in print for more than a decade.

Of course, some purists deny the validity of historical extrapolations made on the basis of circumstantial evidence and the legitimacy of reconstructions supported not by archival documents but by the historian's "knowledge" of his/her actors. They see history as nothing more than a dry list of established "facts," denying its actors any semblance of having ever been live human beings and its writers any insight into the thoughts, feelings, and actions of their "characters."[32] The old proverb that provided Tey with the title of her novel posits that "truth is the daughter of time, not of authority." Perhaps, some later historian will find documents that will confirm or disprove some of my reconstructions. I leave it to my readers to decide whether I have taken my "poetic license" too far and whether the insights and interpretations presented on the pages of this book are convincing, believable, and justifiable by the evidence I have managed to assemble.

## The Historian as Translator

The past *is* a foreign country, in more than one sense.[33] Since safe travels through a foreign land require at least some understanding of its native language, the historian's job becomes akin to that of a translator. Investigating a historical event and studying an unfamiliar language are very much alike, while writing a history of that event is very similar to translating (both literally and figuratively) a text written in one language into another. The further from the present day the past event we study, the more difficult and more important this translation process becomes.[34] A language is a living thing and, in the course of time, meanings of many words change, sometimes quite dramatically, which occasionally even earns them the designation of "archaic" in modern dictionaries and often makes them incomprehensible to or misunderstood by modern readers.

In writing the history of a *scientific* event, this translation process is further complicated by the fact that many words have migrated into science lexicons from everyday language, but have acquired meanings very different from those of everyday vocabularies. It is with the explicit goal of avoiding this sort of possible confusion that scientists have developed their own specialized languages (mathematical and chemical formulas, for example) and have regularly utilized the so-called dead languages (Latin and Greek in the European tradition) to create their own vocabularies, as did Galton in inventing the word "eugenics." But of course, even coining a new term does not guarantee the preservation of its original meaning across time and space, as the history of eugenics readily demonstrates.

Furthermore, the multiplicity of meanings in scientific vocabularies is exacerbated by the *multilingual* nature of the unending pursuit of knowledge called science. The entire history of this pursuit as we know it — from Ancient Babylonia to the present day — could be (and often has been) seen as a history of continuous translations extending through both time and space.[35] The "Republic of Knowledge" has always spoken in multiple tongues, even though, at certain times and in particular places, its citizens used some languages (such as Greek, Syriac, Latin, Arabic, French, German, and most recently English) more widely and more often than others.[36] Indeed, in contrast to the parochial views habitually expressed in various national histories, in the actual history of European science only very rare events unfolded exclusively within *one language zone*.

The historian of science, then, regularly faces a situation of constant interchange among multiple languages used by practitioners of science in different settings, which has frequently resulted in the same thing being given different names and the same name being given to different things, creating a wide spread confusion about the exact meaning of particular terms. As Dmitrii Pisarev astutely observed in his interpretation of Darwin's *Origin*:

> In the languages of all educated nations there exist certain words that every intelligent person should use with extreme circumspection. It would be even better not to use them at all, but, alas, it is nearly impossible. ... They obscure actual facts and nobody knows with certainty what they mean, while everybody utters them incessantly and always strives through these unintelligible words to express and explain something or other.[37]

Obviously, when we study the history of a certain event that happened outside of our own language zone, the issues of its correct translation become even more pressing and ever more complicated.

## Science in Translation

In this particular project, I faced three different kinds of translation. Many of my actors, first of all, Florinskii, regularly translated contemporary foreign texts into their native languages. Others, for instance, Volotskoi and Kanaev, "translated" certain historical texts written in their native tongue, such as Florinskii's treatise, for their own contemporary readers. And I myself had to translate whatever my actors said and wrote in whatever language (mostly Russian, but occasionally also German and French) about their specific subjects into current English.

For me, the difficulties in these various translations were substantially eased by the existence of several Russian dictionaries published at the exact time Florinskii was writing his essays, thus representing the contemporary usage of many words in his vocabulary. The first and most important one is Vladimir Dal's massive *Explanatory Dictionary of the Living Russian Language* that first appeared in installments in 1863-1866.[38] The second is the monumental, nearly 3,000-page long *Table Dictionary for Inquiries in All Fields of Knowledge* compiled by Feliks Toll', one of *Russian Word*'s authors, and also published in installments during exactly the same time, 1863-1866.[39] The third is *General Terminological Medical Dictionary in Latin, German, and Russian Languages* (with detailed explanations of the etymology of numerous Greek medical terms) first published in four volumes in 1840-1842 by Lev Grinberg, and updated (with the addition of French medical terminology) in 1862-1864 by Florinskii's classmate Pavel Ol'khin.[40] Then, in figuring out the possible rendering of various mid-nineteenth-century Russian words into English (and vice versa), I relied on the 1838 two-volume *English-Russian Dictionary* and the 1840 two-volume *Russian-English Dictionary* compiled by Jacob Banks. I also used *New Parallel Dictionaries of the Russian, French, German and English Languages* compiled by Charles P. Reiff, which were published multiple times from the 1840s through the 1860s.[41] Finally, I consulted Aleksei Mikhel'son's dictionary of Russian neologisms that from its first 1861 edition of 7,000 words grew to 30,000 by its 1866 edition.[42]

Still, even with all of these dictionaries, writing for a present-day English-language audience about Florinskii's treatise presented a series of challenges. To begin with, though he was fluent in French and German, knew Latin, English, and Italian quite well, and, thanks to his early theological education, also read Greek and Old Church Slavonic, Florinskii never wrote in any language other than Russian. Indeed, he was one of the first IMSA graduates who chose to write his doctoral dissertation in Russian instead of the traditional Latin. Unlike his famous predecessors at the IMSA — Karl von Baer, Russia's foremost embryologist and anthropologist, and Nikolai Pirogov, the country's most celebrated surgeon and anatomist, as well as many of his less illustrious contemporaries — Florinskii published exclusively in Russian. By contrast, von Baer published most of his works in his native German, with very few of them appearing in Russian translations (which were made by his students) during his lifetime, while Pirogov, as a rule, published all of his works in either Russian or Latin *and* nearly simultaneously in either German or French.

As far as I was able to ascertain, during his lifetime, none of Florinskii's works appeared in any of the common science languages of the day. Thus, I did not have the benefit of Florinskii's own (or his contemporary translators') representation of his thoughts and ideas in any other language, no matter how inaccurate it might have been. Furthermore, his writings bear a clear stamp of the flowery style, long-winded sentences, and pages-long paragraphs characteristic of Russian ecclesiastical literature to which he had been exposed during his formative years as a student at theological schools and which is often difficult to translate into readable English. Yet, his stylistic idiosyncrasies were the least of my problems.

More difficult turned out to be the intricacies of translating the nineteenth-century discussions on scientific, medical, and public health issues underlying Florinskii's treatise and its interpretations. For starters, the Russian equivalent of the English word "science" is *nauka* (plural *nauki*). The English word, however, refers exclusively to the natural sciences, while the Russian one means *scholarship* in any field of knowledge, including the humanities, the arts, and what today in English are called social, behavioral, and human sciences. Thus *nauka* is much closer in its meaning to the German *Wissenschaft* than to the English *science*.

Its derivative, the word *scientist* was introduced into the English language only after the establishment in 1831 of the British Association for the Advancement of Science by one of its founding members William Whewell. Whewell's explicit purpose was to create at least verbal, if not yet professional and social, boundaries between science and scientist, on the one hand, and everybody else involved in the pursuit of every other kind of knowledge, on the other.[43] *Scientist*, then, refers only to someone who studies and practices the natural sciences. Its Russian analogue, *uchenyi*, however, means, first and foremost, "a learned person," and thus refers not only to any kind of scholar, but often more generally to any educated person. It is worth noting that it was in the early 1860s that Russian state bureaucracy began working on a legal definition of *uchenyi* to fit the growing number of individuals engaged in the scholarly pursuit of knowledge into the country's rigid estate structures and the Table of Ranks that governed the hierarchy of the military, civic, and court service.[44]

This "linguistic" difference between science and *nauka*, scientist and *uchenyi* was in fact embedded in the very structures of many nineteenth-century Russian scholarly institutions (such as the St. Petersburg Imperial Academy of Sciences that united under one roof the pursuit of both the natural sciences and the humanities) and, as we saw, profoundly shaped mid-nineteenth-century Russian debates on the interrelations between natural and social sciences. Due to this linguistic difference, any English translation of the mid-nineteenth-century Russian discussions of *nauka* and *uchenyi* tends to accentuate a boundary between science and scholarship and between scientist and learned person, even though such distinctions are completely absent in the original Russian texts. This is especially important to the understanding of discussions conducted on the pages of popular magazines and literary journals concerned with the role that the natural sciences were thought to play in the fledgling social sciences. Indeed it was exactly the absence of linguistic, social, and epistemological boundaries between the natural and the social *nauki* that, in my opinion, facilitated the transfer of ideas, methods, agendas, and concepts from the former to the latter (and vice versa) and drove much of the Russian debate on their interrelations, which, as we saw, provided an important stimulus and a specific context for Florinskii's treatise.

Even more important, in his 1865 essays, Florinskii sought to synthesize data, ideas, concepts, and issues from a number of scholarly fields, including physiology/psychology, anthropology/ethnography/ethnology, demography/statistics/sociology, and general biology (particularly, the concepts of evolution, reproduction, variability, embryonic development, and heredity), not to mention a host of such established and emerging public health and medical specialties as anatomy, epidemiology, hygiene, gynecology, pediatrics, psychiatry, and vital statistics. At the time of his writing, nearly all of these fields and specialties were in the process of (re)formation and of becoming *scientific disciplines*.[45] Each of them was undergoing a series of intertwined developments, including the delineation of a specific subject matter, the erection of epistemological and institutional boundaries, the establishment of consensus over objects, methods, and basic concepts, and, most important for my story, the creation of a specialized language/terminology that embodied all of these developments.

All of these processes took place predominantly in one geographical zone — Europe — but in different linguistic zones of the three most widely used science languages — French, German, and English — with a smattering of Italian, Russian, Swedish, and (particularly in medical and biological fields) Latin, occasionally thrown into the mix. Individual practitioners of these fields everywhere, therefore, faced two related tasks. First, each sought to understand and assimilate the contributions that their colleagues published not only in their native tongue, but also in foreign languages. Second, each had to make their own contributions understood and appreciated by colleagues not only in their homeland, but also abroad. The formation of a *trans*national consensus based on the translatability and transferability of a field's specialized vocabulary (its terms, classifications, nomenclatures, formulae, etc.) became a major task of scholarly communities in all of these fields. Tellingly, it was exactly at this time that the first international congresses in various scientific disciplines (with chemists leading the way in 1860 and anthropologists following the next year) began to be held with an explicit goal of unifying and standardizing their methods, ideas, and lexicons.

At the time of Florinskii's writing, however, most fields he dealt with in his treatise had not yet created such established, unified lexicons. In every country, practitioners in these fields were making up the language of their descriptions and explanations on the go, as it were. Since most of them wrote in their native tongues, the same word quite often had different, sometimes overlapping, and only occasionally the identical meaning in different languages, while many different words were used synonymously. Not unexpectedly, the translation of a particular scholarly work into another language often included a "glossary" that explained the meaning of certain words introduced and/or utilized by its author.[46]

It was exactly during the middle decades of the nineteenth century that the genre of specialized multilingual dictionaries, encyclopaedias, and lexicons, which explained the meaning of particular scientific and medical terms in more than one language, began to develop and became a growing business.[47]

Moreover, with very few exceptions, in the mid-nineteenth century, major developments in all of these fields occurred outside of Russia, both geographically and linguistically. Given the relatively undeveloped state of Russian natural sciences at the time, and a limited number of specialized scientific/medical periodicals published in their native tongue, Russian scholars, including Florinskii, read, as a matter of course, original literature appearing in European languages in their particular fields of interest. They regularly published reviews, surveys, and abstracts of, as well as commentaries on, foreign publications in Russian periodicals. Indeed, during the 1850s and 1860s, some specialized journals devoted more space to translations of foreign publications than to original Russian works.[48]

This sort of "bibliographic/translating" service to the professional community constituted a major part of Florinskii's literary output during the years leading up to the publication of his treatise. As we saw, he regularly reviewed and surveyed for Russian medical journals the newest literature in his own specialties (gynecology, obstetrics, and pediatrics) that had appeared in French, German, English, and Latin. As many others among his colleagues, Florinskii was also involved with translating (and editing such translations of) important foreign publications, especially textbooks, in his own specialties. Indeed, during the second half of the nineteenth century the publication of translations, compilations, and extracts of foreign scientific and medical texts was a booming business in Russia.[49]

One side effect of this ever growing translating and publishing enterprise was that many of the foundational works in all of the scholarly fields synthesized by Florinskii in his treatise often appeared in Russian nearly simultaneously in several different translations. And though more often than not, these translations were made from a foreign work's original language, sometimes they were produced from and influenced by its translation into any of the three major languages of science. Since in nineteenth-century Russia, English was much less popular (and hence less known) than French or German, this situation was particularly prominent in translations of the works originally published in English. For instance, the only Russian edition of Robert Chambers's *Vestiges of the Natural History of Creation* was made from a German translation, and its German translator was even ascribed authorship of the book.[50] One Russian translation of Thomas Henry Huxley's *On the Place of Man in Nature* was made from the original 1863 English edition, while another, which appeared the same year, from a German translation.[51] As we saw in Chapter 4, Francis Galton's essay "On Men of Science, their Nature and Nurture" was translated into Russian from its French version. The number of such examples could easily be multiplied.

This situation was further exacerbated by the absence of specialized natural-sciences and medicine English-Russian (or Russian-English) dictionaries, as witnessed by Grinberg's *Terminological Medical Dictionary*, which included explanations and translations of numerous terms in Latin, German, Greek, French, and Russian, but not English. In search of a suitable Russian terminology, a translator of English-language works quite often consulted their French and/or German translations. The translator of Herbert Spencer's *Classification of Sciences*, for instance, openly admitted: "Translating the book posed considerable difficulties exacerbated by the fact that Spencer's works have not been translated into either German, or French; therefore, [I] had no additional interpretation of the original [that is] written in a brusque and dense style."[52]

## Translating Darwin

Consider, for instance, the case of Darwin's *Origin* that provided a major inspiration and a major source for Florinskii's own synthesis. Darwin's volume was published in London in late 1859, under the title *On the Origin of Species by Means of Natural Selection, Or the Preservation of Favoured Races in the Struggle for Life*.[53] Since it sold almost immediately, the second unchanged edition was issued just a few weeks later, at the beginning of the next year. But as we saw in Chapter 2, Russian readers were first introduced to Darwin's work via its French and German reviews and translations by Heinrich Georg Bronn, Édouard Claparède, and Clémence Royer. The participants in the initial Russian debates over Darwin's concept utilized all three language versions of *Origin* then available, but since few of them knew English, most used the French and German translations rather than the English original. To compound the problem, Russian booksellers offered their customers very few English-language books, and quite often a particular book was simply unavailable in Russia, as happened with the first editions of *Origin*.[54] Tellingly, the IMSA library — a major collection of science publications that Florinskii regularly used in his daily work — had copies of the German, French, Russian, and even Polish translations of *Origin*, but not the original English editions.

Only in January 1864, did Sergei Rachinskii, a botany professor at Moscow University, publish the first Russian translation of *Origin*. Although Rachinskii used the second "unchanged" 1860 edition of Darwin's book to produce its Russian version, it bore clear signs that in his work with the original he also consulted its 1860 German translation. For instance, he changed Darwin's title in exactly the same way as the German translator had, by adding several words (marked in italics): "On the origin of species *in the animal and plant kingdoms*," which were absent in the original English publication.[55]

Rachinskii's translation made Darwin's work much more widely available and accessible to a Russian readership. But, just as it appeared, two publishers — one in Moscow, another in St. Petersburg — released two different translations of Darwin's theory as interpreted by Friedrich Rolle, a well-known German geologist and paleontologist. The same year, Pisarev published his reading of Darwin's concept in a lengthy treatise, "Progress in the Plant and Animal Kingdom," while Kliment Timiriazev, a botany student at Moscow University, published his own interpretation of Darwin's *Origin*, which a few months later appeared in book format. In 1873, Rachinskii released the third, "corrected" version of his translation, in which he fixed some of the inaccuracies of the previous editions. Even so, Nikolai Strakhov pointedly noted in his review of the new edition:

> Among all the translations and editions, we do not know a single work by Darwin that could be comfortably read in Russian. ... We are loath to enervate the reader with the list of all the inaccuracies, missing words, incorrect rendering of terms, and so on. ... The bitter experience has

convinced us that, in general, it is impossible to study Darwin using Russian translations and, quite often, one has to turn to the original.[56]

As one might easily imagine, in all of these renderings of Darwin's *Origin* in French, German, and Russian, the same word was often translated in several different ways. The best example is Darwin's key term "selection," which figured prominently in the very title of his book. The word itself (and its various modifiers, such as "natural," "sexual," "artificial," "unconscious," "methodical," etc.) apparently did not pose any *linguistic* difficulties to English native speakers, though, of course, it caused a number of philosophical, theological, and epistemic problems.[57] But it became a major stumbling block for its translators, for the word had no direct analogues in any other language. In his first 1860 German edition, Bronn translated it as *Züchtung* (breeding), while the third, 1867 edition, "updated and corrected" by eminent zoologist J. Victor Carus, rendered it as *Zuchtwahl* (choice).[58] French translators and commentators faced similar challenges. In his review, Claparède used the word *élection* (choice, election) to convey the meaning of Darwin's term. As he candidly explained:

> I hate to use such a paradoxical phrase. It is difficult, indeed, to accept that an election (choice) could be unconscious. The expression used by the English naturalist has the advantage of not containing any contradiction in terms. Unfortunately, our language has no word that renders the term selection exactly. Rather than using a too colorful foreign neologism, I chose [to use the word] election, despite its shortcomings.[59]

In her 1862 translation, Royer followed Claparède's lead and used *élection* to convey the meaning of Darwin's "selection" in her text, while excising the expression "natural selection" from its title altogether. Only in the subsequent 1866 edition did Royer defy Claparède's dislike of "colorful foreign neologisms": she used *sélection* as the "translation" of Darwin's term and restored it to the book's title.[60] Thereafter, all French editions of *Origin* adopted the same "borrowed translation" of selection as *sélection*.[61]

Russian translators and commentators, thus, found themselves in a "linguistic soup" concocted from this mix of English, German, and French terms. In the first Russian edition, perhaps inspired by the German translation of the word as *Züchtung*, Rachinskii rendered "selection" as *podbor rodichei*, literally "matching of kin." In the second edition, he dropped the word *rodichei* (kin) and left only the word *podbor* — matching — as the equivalent of "selection." Others, however, used such words as *vybor* (choice), *otbornost'* (selectivity), *izbranie* (election), and *izbrannost'* (electivity) — the last two variants clearly reflecting the initial French translation of the term as *élection*.

Florinskii, therefore, had a choice of which particular terms to use in his own rendering of Darwin's ideas. As his text makes quite clear, he followed Rachinskii's first translation and used *podbor rodichei* as the Russian equivalent of Darwin's *selection*. As we saw in Chapter 3, this choice profoundly shaped Florinskii's ideas and the contents of his treatise. Furthermore, different renderings of Darwin's term may well have also played a role in later reinterpretations of Florinskii's views after his book was discovered and republished by Volotskoi.

In the late nineteenth century, to distinguish between the different forms of selection described by Darwin, some Russian breeders and scientists began to use a "loan word" and adopted the word *selektsiia* — a Russian transliteration of "selection" — as the preferred rendering of what Darwin had termed "artificial selection" in plant and animal breeding.[62] Around the same time, in his own new translation of *Origin*, Timiriazev introduced the word *otbor* (literally, the process of choosing something and taking it out of a group of similar things) as the proper translation of Darwin's *selection* irrespective of whatever modifiers one attached to it.[63]

Until the 1930s, however, Russian commentators used both *podbor* and *otbor* interchangeably in their discussions of Darwin's *selection*. Only towards the end of that decade, did the latter word become commonly, but not exclusively, adopted and thus acquired at least something of an agreed-upon terminological status and meaning. But, in the 1920s, in his discussion of "bio-social" eugenics Volotskoi used *selektsiia* and *podbor* synonymously. This usage was, perhaps, meant to emphasize the similarities between artificial selection in plant and animal breeding (for which the typical Russian word was *selektsiia*) with its goals of bettering plant and animal stocks, on the one hand, and the methods of eugenics based on sexual selection (*polovoi podbor*) with its goals of bettering human stocks, on the other.

## Translating Florinskii

Just as did Florinskii in his interpretations of various foreign-language texts in *Human Perfection and Degeneration*, in my writing about Florinskii's treatise in English I had to make certain choices on how best to represent his ideas to my readers. In cases when Florinskii discussed certain texts originally published in English, for instance, Darwin's *Origin*, such choices might seem unproblematic — just use the original English words. Alas, the situation is not that straightforward. For, as we saw in the previous section, the translations of Darwin's vocabulary by his numerous Russian commentators varied substantially. One example would suffice. To translate back into English a variety of Russian words and phrases employed to render Darwin's "selection" by using his own term would completely obscure the difficulties and differences in the interpretations of Darwin's concept by its Russian commentators. Thus, I have elected to use a literal translation of Florinskii's phrase *podbor rodichei* as "matching of kin," because the seemingly logical choice of using the word "selection" as the correct rendering of this expression would result in misrepresenting its meaning and connotations in his own thinking and writing.

The situation gets even murkier with English translations of Florinskii's vocabularies, which he used to explain to his prospective audience numerous ideas and concepts from a wide variety of scientific and medical fields, especially anthropology and social hygiene. To begin with, in the early 1860s, the fields variously described as anthropology, ethnology, and ethnography overlapped to a very considerable degree and lacked clearly defined boundaries. The terminology employed by their practitioners might best be described by a single word — chaos. In the 1860s the meaning of nearly all classificatory categories employed in these fields to describe various groupings of humans, such as "race," "tribe," "nation," "nationality," "family," "type," "stock," "breed," "generation," "mankind," and "human type," was very fluid. None of these words had an established terminological meaning in any language. Quite often, many of these words were used interchangeably and inconsistently. Equally often, different authors utilized different words to describe the same group, or the same word to describe different groupings.

Not surprisingly, Florinskii's use of such words as *rod* (genus, clan, or kind), *plemia* (tribe), *rasa* (race), *natsiia* (nation), *natsional'nost'* (nationality), *tip* (type), *poroda* (breed, stock), *generatsiia* (generation), and *familiia* (family) was quite inconsistent. It obviously reflected their contemporary usage recorded in Toll's *Table Dictionary for Inquiries in All Fields of Knowledge*. To give but one example, according to Toll's dictionary, the word *rasa* (race) came into the Russian language from French and was used as a synonym to *plemia* (tribe) or *poroda* (stock or breed). But the preferred Russian word for describing major human groupings, for which French and English authors actually used the word *race*, was *plemia* (tribe).[64] Expectedly, at times, Florinskii employed many of these words synonymously, at other times, he juxtaposed them. What is more, in the early to mid-1860s, there was no foundational work on anthropology in English (similar to Darwin's *Origin*) that had been translated into Russian and that might have served as a lexical model for my interpretations and English translations of Florinskii's anthropological ideas.[65] In his own writing Florinskii mostly relied on French- and German-language works in these fields. So in my translations I have sought to follow the meaning, rather than the form, in choosing an English equivalent for a particular Russian word used by Florinskii, giving the Russian word in brackets.

The nascent field of what today is commonly called public health was in a very similar "terminological limbo." In the course of the nineteenth century this rapidly growing field was variously named *Medizinsche-* or *Sanitäts-Polizei, öffentlichen Gesundheitspflege, Sozialmedizin,* and *Sozialhygiene* in German,[66] *sante publique, salubrité publique, médicine social,* and *hygiène publique* in French,[67] and *social hygiene, state medicine,* and *public health* in English.[68] Russian specialists followed very closely the field's development abroad and its Russian names were as diverse as their foreign counterparts, ranging from *meditsinskaia politsiia* (medical police) to *okhranenie narodnogo zdraviia* (protection of people's health) to *obshchestvennaia gigiena* (social hygiene).[69] All of these names meant to emphasize the field's purview over the issues of health and disease prevention in particular groups of people, as opposed to the focus of clinical medicine on health and disease treatments in an individual. The very multitude of its names reflected the different national roots, traditions, and trajectories of this composite field, the complete lack of an international consensus on its purview and methods,[70] as well as its shifting boundaries and vocabularies.[71]

As we saw, it was Florinskii's close familiarity with this "social hygiene" and its problematics (especially its French and German variants) that profoundly affected his views on human perfection and degeneration and informed his vision of "hygienic" marriage. And it was this very field — with its first successful legislative interventions aimed at the promotion of health and the prevention of communicable and occupational diseases in certain defined groups of people (populations, professions, and so on) — that provided Florinskii with a model for his own suggestions regarding the legal regulation of marriage to promote perfection and to prevent degeneration.

Searching for adequate English translations of Florinskii's vocabulary, I have sought to identify the sources from which he might have borrowed particular terms and expressions. Alas, it turned out to be no mean feat, for he rarely provided exact references, especially to what he considered "common knowledge." Often he simply gave an author's name, in Russian transliteration or transcription. Plus, he avoided polemics with established authorities in any of the fields he dealt with, only occasionally noting his divergence from the opinions of this or that author. Florinskii wrote his essays for a "general public" and published them not in a specialized scientific/medical journal, but in the "literary-political" *Russian Word*. Occasionally he had to use certain foreign terms and expressions, for instance, "breeding in and in," in their original language, for they had no analogues in Russian. But for the most part, he used plain everyday language and, wherever he could, avoided professional jargon in explaining the complicated issues he addressed.

The very title of his treatise provides a good example of such usage and of the challenges inherent to its correct representation in English (or, for that matter, in any other language). In Russian the title reads: *"usovershenstvovanie i vyrozhdenie chelovecheskogo roda,"* which literally means "perfection and degeneration of the human kind." Obviously, Florinskii chose the title with care to present the three core concepts — *usovershenstvovanie* (perfection), *vyrozhdenie* (degeneration), and *chelovecheskii rod* (human kind) — which underpinned it. None of these words (or their German, French, and English equivalents) had any defined terminological status in any of the fields Florinskii sought to synthesize in his treatise. He took all of them from everyday vocabulary.

As noted above, none of Florinskii's publications were mentioned or translated in any of the foreign languages during his lifetime. As we saw in Chapter 6, the first time his book was referenced and reviewed outside Russia was after Volotskoi had discovered it in the 1920s, and the first foreign language the book's title was ever translated into was German. In his report on the "racial-hygienic movement in Russia" for the *Archiv für Rassen- und Gesellschafts-Biologie*, the RES president Kol'tsov translated its title as "Über die Vervollkommnung und Entartung der Menschheit" ("On the perfection and degeneration of mankind"). Anthropologist Samuil Vaisenberg in an overview of "theoretical and practical eugenics in Soviet Russia" also mentioned the book in the next issue of the same journal, but he translated its title in a slightly different way: "Die Verbesserung und Entartung des menschlichen Geschlechts" ("The improvement and degeneration of human stocks"). Vaisenberg used the same translation in his review of the book published a year later. *Anthropologischer Anzeiger*, the leading German anthropology journal, included the book in its annual 1927 bibliography, under the rubric "*Rassenhygiene, Leibesübungen*," using Vaisenberg's translation of its title. However, in 1961, when Leslie C. Dunn mentioned the book in his presidential address to the American Society of Human Genetics, he referenced its title in German, using Kol'tsov's translation, which, as we saw, is actually closer to the original Russian title than Vaisenberg's.[72]

Introducing Florinskii's work in his 1990 history of eugenics in Russia, Adams has translated its title as "The Improvement and Degeneration of the Human Race."[73] Ever since, all English-language works that mention Florinskii's treatise (including my own) have used Adams's translation of its title.[74] Although correctly conveying to its late-twentieth-century audiences the general import of Florinskii's work by emphasizing its similarities to Galtonian eugenics, this translation obscures the historical origins and distinct contemporary connotations of Florinskii's ideas. In this respect, two words in this translation — *improvement* and *race* — seem particularly problematic.

In English, the word *improvement* means, first and foremost, "the act or process of making something better"[75] and it has an exact Russian equivalent in the word *uluchshenie* (betterment). Characteristically, it is this latter word that most Soviet eugenicists employed to convey to their audiences the meaning of eugenics in the 1920s.[76] Although Florinskii did occasionally utilize *uluchshenie* as a synonym of *usovershenstvovanie* in the text of his treatise, he used the latter word much more frequently, and with a good reason. The word *usovershenstvovanie* had several distinct connotations and synonymic associations, which could not have been lost on the mid-nineteenth-century readers of Florinskii's essays.

According to Dal's *Explanatory Dictionary*, two closely connected nouns *usovershenstvovanie* and *sovershenstvo* (perfection), with the first referring to the process and the second to its result, derived from the verb *sovershat'* and meant, first of all, "to make something perfect."[77] *Usovershenstvovanie* did have a connotation of "making something better," but it also included such meanings as "to correct, to dignify, to elevate to a qualitatively higher place, materially or morally," and "to perfect to the highest degree." The word had definite ecclesiastical overtones, since, in the Bible, perfection is considered to be an attribute of God, as, for instance, in the following Gospel dictum cited by Dal': "Be perfect, therefore, as your heavenly Father is perfect" (Matthew, 5:48). These overtones would have been particularly obvious to Florinskii who had been exposed to theological literature throughout his childhood and adolescence.[78] Outside of the theological discourses on the perfection of God and the imperfection of everything earthly, however, the word was often applied to various human trades and activities, such as agriculture, engineering, and manufacturing; and it is in this area that the meaning of *usovershenstvovanie* overlaps with the English *improvement* most closely.[79]

Yet Dal' also recorded that in some texts one could encounter the notion of "perfecting humans" used in two different ways. One is the perfection of moral and mental qualities through the knowledge of "God's word"[80] or education more generally, as in the following phrase that Dal' used to illustrate such usage: "He went abroad to perfect his knowledge." This particular meaning was certainly influential in Florinskii's thinking, for he himself was sent on a two-year trip abroad for exactly that purpose — to perfect his knowledge of his chosen specialties. Another is the perfection of physical qualities (such as beauty or special abilities, like singing) by exercise and cosmetic/medical procedures that became highly popular in the mid-nineteenth century.[81]

It seems very likely that Florinskii's choice of the word *usovershenstvovanie* was prompted by the expression *perfectionnement physique et moral de l'homme* that became a battle cry of early French public health writings. Developing further the perfectionist ideas espoused by such famous French naturalists as Georges-Louis Buffon and Jean-Baptiste Lamarck, during the first half of the nineteenth century numerous French physicians actively promoted the notion of *perfectionnement physique et moral de l'homme* by means of social, medical, and hygienic interventions.[82] Florinskii was undoubtedly familiar with at least a subset of this literature, especially with the writings of Francis Devay, a professor at the Ecole de Médicine de Lyon, whose works he used extensively in his own treatise. Florinskii probably read Devay's 1844 monumental, nearly 1,000-page, two-volume study on *Hygiène des familles,* subtitled "Du perfectionnement physique et moral de l'homme," as well as its second, one-volume 1858 edition that focused specifically on the issues of marriage and included a detailed discussion of heredity and hereditary diseases.[83] He also knew (and was quite critical) of a popular book on *Hygiène et physiologie du mariage* written by Auguste Debay, a well-known French popularizer. From its first appearance in 1848 the book went through more than thirty printings and editions in French and came out in several different Russian translations in the early 1860s.[84] Florinskii was also likely familiar with Debay's popular treatise on *Hygiène et perfectionnement de la beauté humaine* also translated into Russian.[85] What is more important, in various Russian publications the French *perfectionnement* was typically rendered as *usovershenstvovanie*.

In addition, given the Darwinian context of Florinskii's writings, his choice of the word *usovershenstvovanie* might have also been influenced by the use of the same word (in adjective form) in Rachinskii's translation of the subtitle of Darwin's *Origin*. The English subtitle reads: "The Preservation of *Favoured Races* in the Struggle for Life." Rachinskii translated it as "sokhranenie *usovershenstvovannykh porod* v bor'be za sushchestvovanie," literally "the preservation of *perfected breeds* in the struggle for existence." Darwin's subtitle emphasized his fundamental idea that, in the struggle for life, natural selection *favors* those individuums, varieties, subspecies, races, and breeds of animals and plants that have any advantage over their competitors. Rachinskii's translation, however, slightly changed the emphasis by implying that natural selection leads to the *perfection* (in the sense of improvement) of those varieties, subspecies, races, and breeds of animals and plants that are preserved (survive) in the struggle for existence.

Such a reading was not inconsistent with the text of Darwin's work. In describing the action of various forms of selection, Darwin himself regularly employed such expressions as "modification and improvement" and "modification with improvement." It was exactly this subtle inconsistency (emphasized in Rachinskii's translation) in Darwin's own description of natural selection as the motive force of such basic evolutionary processes as species divergence, extinction, and adaptation that led some commentators, such as Royer and Pisarev,[86] to the equation of biological evolution with "progress," which became highly popular among nineteenth-century Darwinists and to which, as we saw, Florinskii's treatise responded.

Thus, in my translations of Florinskii's ideas I chose to use the word *perfection* to convey and emphasize all of these various meanings of Florinskii's term *usovershenstvovanie*, rather than to follow its rendering as "improvement," which impoverishes considerably the rich connotations and allusions Florinskii must have hoped to evoke in crafting his treatise.

For similar reasons, I do not follow Adams's translation of Florinskii's phrase "*chelovecheskii rod*" as "the human race." Florinskii used the phrase to denote humanity as a whole, for his treatise dealt not with a particular group of human beings, identified by anthropologists, ethnographers, or sociologists, but rather with the human species as it had been (re)defined by Darwin's evolutionary concept. In the mid-nineteenth century the word "race" in the singular was often applied to humans with a qualifier, such as black, white, yellow, red, Aryan, Mongolian, etc., to define a particular *subset* of humanity and frequently had explicitly racist undertones (higher and lower, civilized and barbarian race). Although some nineteenth-century authors — for instance, Galton — did use the expression "the human race" to denote humanity as a whole, more often they did so by using the plural form: "human races."[87] The notion of mankind as "the human race" became much more popular in the mid- to late-twentieth century and came into wide use only after a bitter controversy over UNESCO's declarations on race in the early 1950s.[88] Since Florinskii's treatise was written in

1865, his own usage of the very word *rasa* (race) was limited to meaning nothing more than *plemia* (tribe) or *poroda* (stock or breed).[89] Thus, substituting "race" for "kind" in Florinskii's title gives it a distinctly racist bent that, as we saw, it utterly lacked and actually tried to dispel. Therefore, I chose to translate the title of Florinskii's treatise simply as *Human Perfection and Degeneration*, which, I believe, accurately conveys its general meaning.

Translating another word in Florinskii's title, *vyrozhdenie*, seemingly poses no difficulties, for it has an exact English equivalent in the word "degeneration." Here, however, we encounter a different kind of problem: the meaning of these words in both Russian and English (as well as their French, German, and Italian analogues: *dégénérescence*, *Entartung*, *degenerazione*) have changed considerably, and to a twenty-first-century reader they mean something very different from what they meant some 150 years ago. To begin with, at the time of Florinskii's writing, these words did not have the terminological status that they would acquire just a few decades later in any of the major science languages. Although the concept of *dégénérescence* had been presented by French physician Bénédict Morel in 1857, in the mid-1860s it had not yet attained the strong "medico-psychiatric," "medico-judicial," or "socio-pathological" connotations engendered by its later use in the writings of the French Valentin Magnan, the German Max Nordau, the British Henry Maudsley, the Italian Cesare Lombroso, and their numerous followers.[90] Indeed, in 1857, reviewing Morel's treatise for the Paris Medico-Psychological Society, his friend Philippe Buchez pointed out: "The word *dégénérescence* itself is a new word."[91] As we saw in Chapter 3, Florinskii was familiar with, but highly critical of Morel's concept, and his use of the word *vyrozhdenie* had very little in common with Morel's *dégénérescence* that so heavily "infected" the later English uses of the word degeneration.[92] Rather Florinskii utilized it more in accord with the earlier understanding of *dégénération* by the French naturalists Buffon and Lamarck as *déviation naturelle de l'espèce*.

In mid-nineteenth-century Russian, according to Dal's *Explanatory Dictionary*, the verb *vyrozhdat'* (from which the noun *vyrozhdenie* derives) meant "to give birth to, or to generate, somebody or something lower, worse than one's/its own breed, or something completely dissimilar to the breed." It also had such meanings as "to be born unlike father and mother, in a bodily or spiritual image," and "to change from one generation to the next."[93] Florinskii certainly utilized the word in all of these meanings, but he also gave it new connotations obviously inspired by Darwin's evolutionary ideas. His text makes clear that, for Florinskii, *vyrozhdenie* came to signify first and foremost Darwin's *extinction*, with its attended processes of the "retrogression" and "decrease in numbers" of a species. And it is in this sense that he claimed that the Russian people were in no danger of degeneration (*vyrozhdenie*)!

This rather lengthy digression into the various challenges the historian faces in translating, both literally and figuratively, a past text for the modern reader indicates that any translation represents not only, and not even primarily, the retelling of a certain text in another language, but also its transformation into another, new text that acquires its own history and generates its own consequences, as evidenced, for instance, by the convoluted history of Darwin's evolutionary concept and "Darwinism" in Russia. Furthermore, as we saw in the biography of Florinskii's treatise, the re-interpretations of the same text by successive generations of its publishers and commentators generated numerous anachronisms and erroneous attributions. Although in "Pier Menard, Author of the Quixote" Jorge Luis Borges praised such anachronisms and attributions as enriching "the hesitant and rudimentary art of reading," to a historian, especially a historian of science, such anachronisms (to say nothing of erroneous attributions) is an abomination that has long been condemned as "presentism" or "whiggish history."[94] Thus, one of the most important challenges I faced in writing this book was to separate the actual contents and contexts of Florinskii's original text from the various meanings read into it by its later publishers, reviewers, and commentators. And, hence, the question of how exactly to translate its archaic vocabulary into language understandable to the modern reader, without introducing anachronistic terms, concepts, and notions, became critical. I leave it to my readers to judge whether I have succeeded.

# Notes

## The Faces of Eugenics

1 The literature on the history of eugenics is vast and varied. For a sample of books published in English just since 2010, see Alison Bashford and Philippa Levine, eds., *The Oxford Handbook on the History of Eugenics* (New York: Oxford University Press, 2010); Sheila Faith Weiss, *The Nazi Symbiosis: Human Genetics and Politics in the Third Reich* (Chicago, IL: University of Chicago Press, 2010); Marius Turda, *Modernism and Eugenics* (Basingstoke: Palgrave Macmillan, 2010); Francesco Cassata, *Building the New Man: Eugenics, Racial Science and Genetics in Twentieth-century Italy* (Budapest: CEU Press, 2011); Paul A. Lombardo, ed., *A Century of Eugenics in America: From the Indiana Experiment to the Human Genome Era* (Bloomington, IN: Indiana University Press, 2011); Angela M. Smith, *Hideous Progeny: Disability, Eugenics, and Classic Horror Cinema* (New York: Columba University Press, 2011); Christian Promitzer, Sevasti Trubeta, and Marius Turda, eds., *Health, Hygiene and Eugenics in South-Eastern Europe to 1945* (Budapest: CEU Press, 2011); Nathaniel Comfort, *The Science of Human Perfection* (New Haven, CT: Yale University Press, 2012); Alexandra Minna Stern, *Telling Genes: The Story of Genetic Counseling in America* (Baltimore, MD: Johns Hopkins University Press, 2012); Christine Ferguson, *Determined Spirits: Eugenics, Heredity and Racial Regeneration in Anglo-American Spiritualist Writing, 1848-1939* (Edinburgh: Edinburgh University Press, 2012); Stefan Kühl, *For the Betterment of the Race: The Rise and Fall of the International Movement for Eugenics and Racial Hygiene* (New York: Palgrave Macmillan, 2013); Debbie Challis, *The Archaeology of Race: The Eugenic Ideas of Francis Galton and Flinders Petrie* (London: Bloomsbury Academic, 2013); Sharon M. Leon, *An Image of God: The Catholic Struggle with Eugenics* (Chicago, IL: University of Chicago Press, 2013); Clare Hanson, *Eugenics, Literature, and Culture in Post-war Britain* (London: Routledge, 2013); Randall Hansen and Desmond King, *Sterilized by the State: Eugenics, Race, and the Population Scare in Twentieth Century North America* (Cambridge: Cambridge University Press,

2013); Erika Dyck, *Facing Eugenics: Reproduction, Sterilization, and the Politics of Choice* (Toronto: University of Toronto Press, 2013); Richard Cleminson, *Catholicism, Race and Empire: Eugenics in Portugal, 1900-1950* (Budapest: CEU Press, 2014); Fae Brauer and Serena Keshavjee, eds., *Picturing Evolution and Extinction: Regeneration and Degeneration in Modern Visual Culture* (Cambridge: Cambridge Scholars Publishing, 2015); Ewa Barbara Luczak, *Breeding and Eugenics in the American Literary Imagination: Heredity Rules in the Twentieth Century* (New York: Palgrave Macmillan, 2015); Thomas C. Leonard, *Illiberal Reformers: Race, Eugenics, and American Economics in the Progressive Era* (Princeton, NJ: Princeton University Press, 2016); Maurizio Meloni, *Political Biology: Science and Social Values in Human Heredity from Eugenics to Epigenetics* (New York: Palgrave Macmillan, 2016); Adam Cohen, *Imbeciles: The Supreme Court, American Eugenics, and the Sterilization of Carrie Buck* (New York: Penguin, 2016); Shantella Y. Sherman, *In Search of Purity: Popular Eugenics and Racial Uplift Among New Negroes* (CreateSpace Independent Publishing Platform, 2016); Gerald V. O'Brien, *Framing the Moron: The Social Construction of Feeble-Mindedness in the American Eugenic Era* (Manchester: Manchester University Press, 2016); Heike I. Petermann, Peter S. Harper, Susanne Doetz, eds., *History of Human Genetics: Aspects of Its Development and Global Perspectives* (Cham: Springer, 2017); Diane B. Paul, John Stenhouse, and Hamish G. Spencer, eds., *Eugenics at the Edges of Empire: New Zealand, Australia, Canada and South Africa* (Cham: Palgrave Macmillan, 2018); and many others.

2 Of course, some scholars have contested this linear genealogy, see, for instance, John Waller, "Ideas of Heredity, Reproduction and Eugenics in Britain, 1800-1875," *Studies in History and Philosophy of Science*, Part C, 2001, 32(3): 457-89; Diane B. Paul and Benjamin Day, "John Stuart Mill, Innate Differences, and the Regulation of Reproduction," *Studies in History and Philosophy of Biological and Biomedical Science*, 2008, 39: 222-31; and Mark B. Adams, "Eugenics," in V. Ravitsky, A. Fiester, and A. Caplan, eds., *The Penn Center Guide to Bioethics* (Cham: Springer, 2008), pp. 371-82.

3 See the early classic studies by Charles P. Blacker, *Eugenics: Galton and After* (Cambridge, MA: Harvard University Press, 1952); Mark Haller, *Eugenics: Hereditarian Attitudes in American Thought* (New Brunswick, NJ: Rutgers University Press, 1963); Donald K. Pickens, *Eugenics and the Progressives* (Nashville, TN: Vanderbilt University Press, 1968); Kenneth M. Ludmerer, *Genetics and American Society: A Historical Appraisal* (Baltimore, MD: Johns Hopkins University Press, 1972); Geoffrey R. Searle, *Eugenics and Politics in Britain, 1900-1914* (Leyden: Noordhoff, 1976); Lyndsay A. Farrall, *The Origins and Growth of the English Eugenics Movement, 1865–1912* (New York: Garland Press, 1985); and Daniel J. Kevles, *In the Name of Eugenics: Genetics and the Uses of Human Heredity* (Berkeley, CA: University of California Press, 1985). The emphasis on Anglo-American eugenics has been continued in many

later studies, see, for instance, Diana B. Paul, *Controlling Human Heredity: 1865 to the Present* (Atlantic Highlands, NJ: Humanities Press International, 1995); idem, *The Politics of Heredity: Essays on Eugenics, Biomedicine, and the Nature-Nurture Debate* (Albany, NY: SUNY Press, 1998); Robert A. Peel, ed., *Essays in the History of Eugenics* (London: Galton Institute, 1998); Ian R. Dowbiggin, *Keeping America Sane: Psychiatry and Eugenics in the United States and Canada, 1880-1940* (Ithaca, NY: Cornell University Press, 1997); Alexandra Minna Stern, *Eugenic Nation: Faults and Frontiers of Better Breeding in Modern America* (Berkeley, CA: University of California Press, 2005); and many others. Compare also the three overviews of relevant literature published in roughly fifteen-year intervals: Lyndsay A. Farrall, "The History of Eugenics: A Bibliographical Review," *Annals of Science*, 1979, 36(2): 111-23; Philip J. Pauly, "Essay Review: The Eugenics Industry — Growth or Restructuring?" *Journal of the History of Biology* (hereafter *JHB*), 1993, 26(1): 131-45; and Philippa Levine and Alison Bashford, "Eugenics and the Modern World," in Bashford and Levine, eds., *Oxford Handbook*, pp. 3-24.

4 *Problems in Eugenics: Vol. II. Report on Proceedings of the First International Eugenics Congress held at the University of London, July 24th to 30th, 1912* (Kingsway: Eugenics Education Society, 1913), p. 44.

5 On Ploetz and *Rassenhygiene*, see Peter Weingart, Jürgen Kroll, and Kurt Bayertz, *Rasse, Blut und Gene. Geschichte der Eugenik und Rassenhygiene in Deutschland* (Frankfurt am Main: Suhrkamp, 1988); Paul Weindling, *Health, Race, and German Politics between National Unification and Nazism, 1870-1945* (Cambridge: Cambridge University Press, 1989); and Robert N. Proctor, *Racial Hygiene: Medicine under the Nazis* (Cambridge, MA: Harvard University Press, 1988).

6 Karl Pearson, *The Academic Aspect of the Science of National Eugenics* (London: Dulau, 1911), p. 4.

7 Since the publication of Kevles's classic study many historians have emphasized the close similarity and interconnectedness of British and US eugenics, seen particularly in the prominence of Galton's version in both countries. A closer look at the early manifestations of eugenic thought in the United States, however, indicates that American eugenics also had distinct "national" roots and traditions. To give but one example, at the "first national conference on race betterment" held at Battle Creek, Michigan, in January 1914, Galton's name was barely even mentioned (nine times on 600-plus pages of the published proceedings). But participants did hail several local "founding fathers" (and "mothers"!) of the science of "race betterment." See Emily F. Robbins, ed., *Proceedings of the First National Conference on Race Betterment* (Battle Creek, MI: Race Betterment Foundation, 1914).

8 See, for example, Mark B. Adams, ed., *The Wellborn Science: Eugenics in Germany, France, Brazil, and Russia* (New York: Oxford University Press, 1990); Anne Carol, *Histoire de l'eugénisme en France: les médecins et la procréation, XIXe-XXe siècle* (Paris: Éditions du Seuil, 1995); Gunar Broberg and Nils Roll-Hansen, eds., *Eugenics and the Welfare State: Sterilization Policy in Denmark, Sweden, Norway, and Finland* (East Lansing, MI: Michigan State University Press, 1996); Nancy L. Stepan, *'The Hour of Eugenics': Race, Gender, and Nation in Latin America* (Ithaca, NY: Cornell University Press, 1996); R. A. Soloway, "From Mainline to Reform Eugenics: Leonard Darwin and C. P. Blacker," in R. A. Peel, ed., *Essays in the History of Eugenics* (London: Galton Institute, 1997), pp. 52-80; Reinhard Mocek, "The Program of Proletarian Rassenhygiene," *Science in Context*, 1998, 11(3-4): 609-17; Jean-Noël Missa and Charles Susanne, eds., *De l'eugénisme d'état à l'eugénisme privé* (Paris and Brussels: De Boeck University, 1999); Michael Schwartz, *Sozialistische Eugenik: Sozialtechnologien in Debatten und Politik der deutschen Sozialdemokratie, 1890-1933* (Bonn: Nachf, 2000); Pauline M. H. Mazumdar, "'Reform' Eugenics and the Decline of Mendelism," *Trends in Genetics*, 2002, 18(1): 48-52; Nicholas Agar, *Liberal Eugenics: In Defense of Human Enhancement* (Boston, MA: Blackwell, 2004); Marius Turda and Paul J. Weindling, eds.,*"Blood and Homeland": Eugenics and Racial Nationalism in Central and Southeast Europe, 1900-1940* (Budapest: CEU Press, 2007); John Glad, *Jewish Eugenics* (Washington, DC: Wooden Shore, 2011); Bjorn M. Felder and Paul J. Weindling, eds, *Baltic Eugenics* (Amsterdam: Rodopi, 2013); Marius Turda, ed., *The History of East-Central European Eugenics, 1900-1945* (London: Bloomsbury, 2015); Marius Turda and Aaron Gillette, *Latin Eugenics in Comparative Perspective* (London: Bloomsbury, 2016); and Judith Daar, *The New Eugenics: Selective Breeding in an Era of Reproductive Technologies* (New Haven, CT: Yale University Press, 2017).

9 See, for instance, a detailed discussion of the intimate relations between "reproductive culture" and "eugenic thinking" in Maria A. Wolf, *Eugenische Vernunft: Eingriffe in die reproduktive Kultur durch die Medizin, 1900-2000* (Vienna: Böhlau Verlag, 2008).

10 Francis Galton, "Hereditary Talent and Character," *Macmillan's Magazine*, 1865, 12(68): 157-66; (70): 318-27.

11 Francis Galton, *Inquiries into Human Faculty and Its Development* (London: Macmillan, 1883), pp. 24-25 (fn. 1) and p. 44.

12 Galton, "Hereditary Talent and Character," p. 165.

13 *Ibid*, p. 325.

14 V. Florinskii, "Usovershenstvovanie i vyrozhdenie chelovecheskogo roda," *Russkoe slovo* (hereafter *RS*), 1865, 8 (August): 1-57. Hereafter the references to this edition will be given as Florinskii, 1865, 8: 1-57.

15 V. Florinskii, "Usovershenstvovanie i vyrozhdenie chelovecheskogo roda," *RS*, 1865, 10 (October): 1-43; 11 (November): 1-25; 12 (December): 27-43. Hereafter the references to this edition will be given as Florinskii, 1865, 10: 1-43; 11: 1-25; 12: 27-43.

16 F. [V. M.] Florinskii, *Usovershenstvovanie i vyrozhdenie chelovecheskogo roda* (SPb.: n. p., 1866). The title page had a typo that gave the author a wrong initial: "F.," instead of "V." Hereafter the references to this edition will be given as Florinskii, 1866.

17 I have coined the term "eugamics" to distinguish Florinskii's concept with its focus on "rational marriage" from Galton's notion of "eugenics" that focused on "being well-born." To my surprise, in the course of my research, I found that the term had already been proposed to highlight this very distinction in the monumental, two-volume "outlines of general biology" by J. Arthur Thomson and Patrick Geddes, *Life: Outlines of General Biology* (New York: Harper, 1931). In Chapter 12 of its second volume, the authors included a special section on "Eugenics and Eugamics," comparing and contrasting the two approaches to the improvement of humankind (see *ibid.*, pp. 1327-33). But the term obviously did not take root — I did not encounter it in any other publications. So, I have decided to use it in the present volume as a shorthand for Florinskii's ideas of "hygienic" or "rational" marriage.

18 Francis Galton, *English Men of Science: Their Nature and Nurture* (London: Macmillan, 1874), p. 7.

19 For analyses of this process in Britain, see J. Morrell and A. Thackray, *Gentlemen of Science: Early Years of the British Association for the Advancement of Science* (Oxford: Clarendon Press, 1981); R. Barton, "'Huxley, Lubbock, and a Half a Dozen Others': Professionals and Gentlemen in the Formation of the X Club, 1851-1864," *Isis*, 1998, 89: 409-44; more specifically on the role of Galton, see John C. Waller, "Gentlemanly Men of Science: Sir Francis Galton and the Professionalization of the British Life-Sciences," *JHB*, 2001, 34: 83-114. Unfortunately, the literature on the professionalization of science and scientists in nineteenth century Russia is virtually nonexistent; for some observations, see E. V. Soboleva, *Organizatsiia nauki v poreformennoi Rossii* (L.: Nauka, 1983); A. E. Ivanov, *Uchenye stepeni v Rossiiskoi imperii, XVIII v-1917* (M.: RAN, 1994); *idem*, *Uchenoe dostoinstvo v Rossiiskoi imperii, XVIII-nachalo XX veka* (M.: Novyi Khronograf, 2016); Nathan M. Brooks, "Alexander Butlerov and the Professionalization of Science in Russia," *Russian Review*, 1998, 57(1): 10-24; *idem*, "The Science of a Lost Empire and its Internal Colonies: The Case of Russia," in George N. Vlahakis et al., *Imperialism and Science: Social Impact and Interaction* (Santa Barbara, CA: ABC-Clio, 2006), pp.154-77; Elizabeth A. Hachten, "In Service to Science and Society: Scientists and the Public in Late-Nineteenth-Century Russia," *Osiris*, 2002, 17: 171-209; Michael D. Gordin,

"The Heidelberg Circle: German Inflections on the Professionalization of Russian Chemistry in the 1860s," *Osiris*, 2008, 23: 23-49. For a more general treatment of professionalization in other areas such as medicine, psychiatry, civil service, and education, see Harley D. Balzer, ed., *Russia's Missing Middle Class: The Professions in Russian History* (Armonk, NY: Sharpe, 1996).

20 See, for instance, Maximien Rey, *Dégénération de l'espèce humaine et sa régénération* (Paris: Germer-Baillière, 1863); Eduard Reich, *Ueber die Entartung der Menschen, ihre Ursachen und Verhütung* (Erlangen: Enke, 1868); and many others. As John Waller has convincingly demonstrated, Galton was far from the first or the only British scientist to advance "eugenic" ideas at the time. See Waller, "Ideas of Heredity, Reproduction and Eugenics." For similar "eugenic" notions in the United States, see Charles E. Rosenberg, "Bitter Fruit: Heredity, Disease, and Social Thought in Nineteenth-Century America," *Perspectives in American History*, 1974, 8: 189-235. A careful study of "proto-eugenic" and "proto-eugamic" ideas in Austria, France, Germany, and Italy still awaits its champions, but as many detailed studies of the development of "national" eugenics, along with a recent workshop on "proto-eugenic thinking before Galton" (see *Bulletin of the GHI*, 2009, 44 (Spring): 83-88), have indicated, such ideas were indeed advanced and debated in these and other countries as well.

21 See John C. Gunn, *New Domestic Physician* (Cincinnati, OH: Moore, Wilstach, Keys & Co., 1863), p. 120. I am indebted to my student Riiko Bedford for finding this citation. A digital copy of this particular edition is available at https://archive.org/details/63570990R.nlm.nih.gov. On Gunn and his manual see, Charles E. Rosenberg, "John Gunn: Everyman's Physician," in *idem*, *Explaining Epidemics* (Cambridge: Cambridge University Press, 1992), pp. 57-73.

22 For a detailed description of the church's regulations of marriage in Russia, see Chapter 4.

23 Gregory L. Freeze, "Bringing Order to the Russian Family: Marriage and Divorce in Imperial Russia, 1760-1860," *Journal of Modern History*, 1990, 62: 709-46 (p. 711).

24 The portrayal of Florinskii as a "precursor" appeared in the publisher's foreword, see M. V. Volotskoi, "K istorii i sovremennomy sostoianiiu evgenicheskogo dvizheniia v sviazi s knigoi V. M. Florinskogo," in V. M. Florinskii, *Usovershenstvovanie i vyrozhdenie chelovecheskogo roda* (Vologda: Severnyi pechatnik, 1926), pp. vii-xix. Hereafter the references to this edition will be given as Florinskii, 1926. A very brief assessment of this episode has appeared in Mark B. Adams, "Eugenics in Russia," in *idem*, ed., *The Wellborn Science*, p. 170.

25 See, for instance, I. I. Kanaev, "Na puti k meditsinskoi genetike," *Priroda*, 1973, 1: 52-68; and N. P. Bochkov, *Genetika cheloveka* (M.: Meditsina, 1978), pp. 11-12.

26 V. P. Puzyrev, "Evgenicheskie vzgliady V. M. Florinskogo na 'Usovershenstvovanie i vyrozhdenie chelovecheskogo roda' (k 160-letiiu so dnia rozhdeniia)," in V. M. Florinskii, *Usovershenstvovanie i vyrozhdenie chelovecheskogo roda* (Tomsk: Izd-vo Tomskogo universiteta, 1995), pp. 120-26. Hereafter the references to this edition will be given as Florinskii, 1995.

27 V. M. Florinskii, "Usovershenstvovanie i vyrozhdenie chelovecheskogo roda," in V. B. Avdeev, ed., *Russkaia evgenika* (M.: Belye Al'vy, 2012), pp. 44-134.

28 See, for instance, E. N. Gnatik, *Genetika cheloveka: Byloe i griadushchee* (M.: LKI, 2010); on Florinskii, see Chapter 2, http://www.irbis.vegu.ru/repos/11864/ HTML/007.htm; and R. A. Fando, *Proshloe nauki budushchego: Istoriia evgeniki v Rossii* (Poltava: OOO "ASMI", 2014), pp. 82-83. The latter book was produced in Ukraine in 300 copies and is not available in Russia's main research libraries. I am grateful to the author for providing me with a copy. It is also available in a German translation, see R. A. Fando, *Die Anfänge der Eugenik in Russland. Kognitive und soziokulturelle Aspekte*, transl. by Elena Paschkowa (Berlin: Logos, 2014).

29 For a highly readable account of some of the challenges in writing the biography of a book, see Owen Gingerich, *The Book Nobody Read: Chasing the Revolutions of Nicolaus Copernicus* (New York: Walker, 2004).

30 To give a few, almost random examples, see James H. Sledd and Gwin J. Kolb, *Dr. Johnson's Dictionary: Essays in the Biography of a Book* (Chicago, IL: University of Chicago Press, 1955); Paul Eggert, *Biography of a Book: Henry Lawson's While the Billy Boils* (University Park, PA: Pennsylvania State University Press, 2013); and Alice Kaplan, *Looking For The Stranger: Albert Camus and the Life of a Literary Classic* (Chicago, IL: University of Chicago Press, 2016).

31 See, for instance, James A. Secord, *Victorian Sensation: The Extraordinary Publication, Reception, and Secret Authorship of Vestiges of the Natural History of Creation* (Chicago, IL: University of Chicago Press, 2001); idem, *Visions of Science: Books and Readers at the Dawn of the Victorian Age* (Oxford: Oxford University Press, 2014); Gingerich, *The Book Nobody Read*; and Janet Browne, *Darwin's Origin of Species: A Biography* (London: Atlantic, 2006).

32 Such as, for instance, J. B. de C. M. Saunders and C. D. O'Malley, *The Illustrations from the Works of Andreas Vesalius of Brussels: With Annotations*

*and Translations, a Discussion of the Plates and their Background, Authorship and Influence, and a Biographical Sketch of Vesalius* (Cleveland, OH: The World Publishing Company, 1950; repr. New York: Dover, 1973); Koen Huigen, "Biography of a Book: Paratext in all Dutch editions of Louis Couperus' *De stille kracht*." MA thesis, Leiden University, 2015; or Steven van Impe and Mari-Liisa Varila, "The Biography of a Book: The Turku Copy of the 1613 Mercator-Hondius Atlas," *Approaching Religion*, 2016, 6(1): 24-34.

33 Compare, for example, Michael Reynolds, "Hemingway's In Our Time: The Biography of a Book," in J. Gerald Kennedy, ed., *Modern American Short Story Sequences: Composite Fictions and Fictive Communities* (Cambridge: Cambridge University Press, 1995), pp. 35-51; and Helen P. Liepman, "The Six Editions of the Origin of Species: A Comparative Study," *Acta Biotheoretica*, 1981, 30(3): 199-214.

34 See, for example, Julie Bates Dock, ed., *Charlotte Perkins Gilman's The Yellow Wall-paper and the History of its Publication and Reception* (University Park, PA: Pennsylvania State University Press, 1998).

35 Adams, "Eugenics in Russia," in *idem*, ed., *Wellborn Science*, pp. 153-229; *idem*, "The Politics of Human Heredity in the USSR, 1920-40," *Genome*, 1989, 31(2): 879-84; *idem*, "Eugenics as Social Medicine in Revolutionary Russia," in S. G. Solomon and J. F. Hutchison, eds., *Health and Society in Revolutionary Russia* (Bloomington: Indiana University Press, 1990), pp. 200-23; *idem*, "Soviet Nature-Nurture Debate," in Loren R. Graham, ed., *Science and the Soviet Social Order* (Cambridge, MA: Harvard University Press, 1990), pp. 94-138. Adams has also published short biographies of many leaders of Russian eugenics in the *Dictionary of Scientific Biography*.

36 See, for instance, H.-W. Schmuhl, "Rassenhygiene in Deutschland — Eugenik in der Sowjetunion: Ein Vergleich," in Dietrich Beyrau, ed., *Im Dschungel der Macht: Intellektuelle Professionen unter Stalin und Hitler* (Göttingen: Vandenhoeck und Ruprecht, 2000), pp. 360-77; Michael Flitner, "Genetic Geographies: A Historical Comparison of Agrarian Modernization and Eugenic Thought in Germany, the Soviet Union, and the United States," *Geoforum*, 2003, 34: 175-85; and A. Spektorowski, "The Eugenic Temptation in Socialism: Sweden, Germany, and the Soviet Union," *Comparative Studies in Society and History*, 2004, 46: 84-106.

37 Loren R. Graham, "Science and Values: The Eugenics Movement in Germany and Russia in the 1920s," *American Historical Review*, 1977, 82(5): 1133-64; Paul Weindling, "German-Soviet Medical Co-operation and the Institute for Racial Research, 1927-c.1935," *German History*, 1992, 10(2): 177-206; Mark B. Adams, Garland E. Allen, and Sheila F. Weiss, "Human Heredity and Politics," *Osiris*, 2005, 20: 232-62; Nikolai Krementsov, "Eugenics, *Rassenhygiene*, and Human

Genetics in the Late 1930s," in Susan G. Solomon, ed., *Doing Medicine Together: Germany and Russia between the Wars* (Toronto: University of Toronto Press, 2006), pp. 369-404; and Per Anders Rudling, "Eugenics and Racial Biology in Sweden and the USSR: Contacts Across the Baltic Sea," *CBMH/BCHM*, 2014, 31(1): 41-75.

38 See, for instance, Iu. V. Khen, *Evgenicheskii proekt* (M.: IFRAN, 2003); Fando, *Proshloe nauki budushchego*; and Kevin Liggieri, "'[A]n der Front des Kampfes um den Menschen selbst.' Anthropogenetik und Anthropotechnik im sowjetischen Diskurs der 1920er Jahre," *Berichte zur Wissenschaftsgeschichte*, 2016, 39: 165-84.

39 See, for instance, Pat Simpson, "Bolshevism and 'Sexual Revolution': Visualising New Soviet Woman as the Eugenic Ideal," in Fae Brauer and Anthea Callen, eds.,*"Art, Sex and Eugenics": Corpus Delecti* (Burlington, VT: Ashgate, 2008), pp. 209-38; Birte Kohtz, "Gute Gene, schlechte Gene. Eugenik in der Sowjetunion zwischen Begabungsforschung undgenetischer Familienberatung," *Jahrbücher für Geschichte Osteuropas*, 2013, 61(4): 591-610; Vsevolod Bashkuev, "Soviet Eugenics and National Minorities: Eradication of Syphilis in Buryat-Mongolia as an Element of Social Modernization of a Frontier Region, 1923-1928," in Felder and Weindling, eds., *Baltic Eugenics*, pp. 261-87; *idem*, "An Outpost of Socialism in the Buddhist Orient: Geopolitical and Eugenic Implications of Medical and Anthropological Research on Buryat-Mongols in the 1920s," *Études mongoles et sibériennes, centrasiatiques et tibétaines*, 2015, 46, http://emscat.revues.org/2523; and Anne Finger, "Left Hand of Stalin: Eugenics in the Soviet Union," in Ravi Malhotra, ed., *Disability Politics in a Global Economy: Essays in Honour of Marta Russell* (New York: Routledge, 2016), pp. 199-219.

40 I have attempted to address some of these gaps in several articles, see Nikolai Krementsov, "Eugenics in Russia and the Soviet Union," in Bashford and Levine, eds., *Oxford Handbook*, pp. 413-29; *idem*, "From 'Beastly Philosophy' to Medical Genetics: Eugenics in Russia and the Soviet Union," *Annals of Science*, 2011, 68(1): 61-92; *idem*, "The Strength of a Loosely Defined Movement: Eugenics and Medicine in Imperial Russia," *Medical History*, 2015, 59(1): 6-31.

# Chapter 1

1 For the most voluminous accounts of Galton's life and works, see, Charles P. Blacker, *Eugenics: Galton and After* (Cambridge, MA: Harvard University Press, 1952); Ruth S. Cowan, *Sir Francis Galton and the Study of Heredity in the Nineteenth Century* (New York: Garland, 1985); Milo Keynes, ed., *Sir Francis Galton, FRS: The Legacy of His Ideas* (London: Macmillan, 1993); Nicholas W.

Gillham, *A Life of Sir Francis Galton: From African Exploration to the Birth of Eugenics* (New York: Oxford University Press, 2001); and Martin Brookes, *Extreme Measures: The Dark Visions and Bright Ideas of Francis Galton* (London: Bloomsbury, 2004).

2 F. Galton, *Memories of My Life* (London: Methuen, 1909), pp. 287-88.

3 This is the date on Florinskii's birth certificate. However, until the Bolshevik Revolution, Russia used the Julian calendar that was thirteen days behind the Gregorian calendar used in Britain, so in absolute "astronomical" terms Galton and Florinskii were born twelve years and thirteen days apart. A copy of Florinskii's birth certificate is preserved in the Russian State Military Historical Archive (hereafter RGVIA), *fond* (collection) 316, *opis'* (inventory) 63, *delo* (file) 6257, *list* (page) 3; hereafter such references will be given as f. 316, op. 63, d. 6257, l. 3.

4 For a detailed analysis of the social history of the Russian clergy, see Gregory L. Freeze, *The Russian Levites: Parish Clergy in the Eighteenth Century* (Cambridge, MA: Harvard University Press, 1977), and *idem*, *The Parish Clergy in Nineteenth-Century Russia: Crisis, Reform, Counter-Reform* (Princeton, NJ: Princeton University Press, 1983). For a more general analysis of the social estate system in Russia, see *idem*, "The Soslovie (Estate) Paradigm and Russian Social History," *American Historical Review*, 1986, 91(1): 11-36; N. A. Ivanova and V. P. Zheltova, *Soslovnoe obshchestvo Rossiiskoi imperii (XVIII — nachalo XX veka)* (M.: Novyi Khronograf, 2010); and Alison K. Smith, *For the Common Good and Their Own Well-Being: Social Estates in Imperial Russia* (Oxford: Oxford University Press, 2014).

5 For a detailed, three-volume history of this seminary, see N. V. Malitskii, *Istoriia Vladimirskoi dukhovnoi seminarii* (M.: A. I. Snegireva, 1900-02). However, for some reason, Mark Iakovlev's name is not included in the list of the seminary graduates appended to the last volume.

6 My reconstruction of Florinskii's early life is largely based on his unpublished memoirs, "Thoughts and Reminiscences about my Education and Upbringing," held at the National Museum of the Tatarstan Republic (hereafter NRMT), see NRMT, V. M. Florinskii's collection, № 117959, #204, 205; hereafter such references will be given as NRMT, #204, 205. Direct quotations from this manuscript are indicated in the text as [ZV, followed by the page number]. Alas, the collection has been processed according to the rules of the museum curation (whereby each individual item, be it a letter, a book, a photograph, or an artifact is merely given a number in a general inventory), and not according to the established principles of archival preservation (where similar types of documents, such as correspondence, manuscripts, biographical materials, photographs and so on, are gathered under the same heading), which would have made it much easier to identify and locate a particular document.

7 See O. Penezhko, *Gorod Iur'ev-Pol'skii, khramy Iur'ev-Pol'skogo raiona* (Vladimir: n. p., 2005), p. 59.

8 Florinskii relayed the story of his family name in his "Thoughts and Reminiscences." See NRMT, #204.

9 Mikhail's wife was the younger sister of Mikhail Speranskii, a graduate of the same Vladimir Seminary, who became one of the most prominent statesmen during the reigns of Alexander I and Nicholas I. For his biography, see Marc Raeff, *Michael Speransky: Statesman of Imperial Russia, 1772–1839* (Westport, CT: Hyperion, 1979).

10 For detailed biographical information on Grigorii Fedorov, see E. A. Popov, *Velikopermskaia i Permskaia eparkhiia (1379-1879)* (Perm: Nikifirova, 1879), pp. 158-261.

11 Even at the end of the nineteenth century (when the first country-wide census was taken), its population still counted less than two million people.

12 See, for instance, a description of the travel on that road, though in the opposite direction, some fifteen years after the Florinskiis had made their journey in V. P. Parshin, *Opisanie puti ot Irkutska do Moskvy, sostavlennoe v 1849 g.* (M.: A. Semen, 1851). For a historical analysis of the Trans-Siberian highway and its role in Russian history, see O. N. Kationov, *Moskovsko-Sibirskii trakt i ego zhiteli v XVII-XIX vv.* (Novosibirsk: Izd-vo NGPU, 2004).

13 On the 1837 trip of the future Emperor Alexander II through his domains, see L. G. Zakharova and L. I. Tiutiunnik, eds., *Venchanie s Rossiei* (M.: MGU, 1999).

14 On the contemporary views of Siberia as a "promised land," see Mark Bassin, "Inventing Siberia: Visions of the Russian East in the Early Nineteenth Century," *American Historical Review*, 1991, 96(3): 763-94; and *idem*, *Imperial Visions. Nationalist Imagination and Geographical Expansion in the Russian Far East, 1840-1865* (Cambridge: Cambridge University Press, 2004).

15 The remnants of this garden have survived to this day, but only the foundation of the old church remains standing. See N. P. Shusharina, "Rod Betevykh v istorii zemli Kataiskoi," *Soiuznaia mysl'*, 2011, 2 (24 January): 2-4, http://kaz2.docdat.com/docs/index-122554.html.

16 At the Church of the Holy Virgin he had built, Father Mark served his parishioners as best he could until 1867, when he retired and was succeeded by his youngest son Semen. For a very brief description of the church and its history, see *Prikhody i tserkvi Ekaterinburgskoi eparkhii. Istoricheskii ocherk* (Ekaterinburg, 1902), p. 571. For a detailed description of the Florinskii family based on the material from the local archives, see N. P. Shusharina, "Sviashchenicheskii rod

Florinskikh," *Zaural'skaia genealogiia* (Kurgan), 2009, 3, http://www.kurgangen. ru/religion/pravoslavnoe/Florinsky/Florinsky_Rod; and L. A. Biakova, "O dinastii sviashchenosluzhitelei Florinskikh-Kokosovykh," http://www. fnperm.ru.

17 It is preserved among his personal papers kept in the NMRT, "Psaltyr'."

18 For an English translation, see N. G. Pomyalovsky, *Seminary Sketches*, transl. by Alfred Kuhn (Ithaca, NY: Cornell University Press, 1973). The translation's title is misleading, for seminary is a secondary theological school, while Pomialovskii's *Sketches* described *bursa* — that is a primary school.

19 For a detailed, three-volume history of this seminary, see [I. E. Lagovskii], *Istoriia Permskoi dukhovnoi seminarii arkhimandrita Ieronima*, 3 vols. (Ekaterinburg: I. V. Shestakov, 1900-01), parts 1-4.

20 P. A. Kropotkin, *Zapiski revoliutsionera* (London: Izdanie vol'noi russkoi pressy, 1902), p. 82.

21 For a brief biographical sketch of Vishniakov, see [Lagovskii], *Istoriia Permskoi dukhovnoi seminarii*, part 3, p. 677. See also, N. A. Skorobotov, *Pamiatnaia knizhka okonchivshikh kurs S.-Peterburgskoi seminarii, s 1811 po 1895 g.* (SPb.: I. A. Frolov, 1896).

22 *Moskovitianin* (M., 1841-56), *Biblioteka dlia chteniia* (SPb., 1834-65), *Otechestvennye zapiski* (SPb., 1839-84), *Sovremennik* (SPb., 1836-66). For a classic, detailed analysis of the nineteenth-century "thick" journals and Russian journalism more generally, see the latest edition of B. I. Esin, *Istoriia russkoi zhurnalistiki XIX veka* (M.: Moskovskii universitet, 2008).

23 For Makarii's biography, see G. A. Polisadov, *Vysokopreosviashchennyi Makarii* (Nizhnii Novgorod: Gubernskoe pravlenie, 1895).

24 On Florinskii's later contributions to these fields, see A. V. Zhuk, "Vasilii Markovich Florinskii kak arkheolog," *AB ORIGINE: Problemy genezisa kul'tur Sibiri* (Tiumen': TiumGU, 2013), 5: 5-22; and idem, "Vasilii Markovich Florinskii, ego mesto i znachenie v otechestvennoi arkheologii," *Vestnik Omskogo universiteta. Seriia "Istoricheskie nauki"*, 2015, 1(5): 100-15. Florinskii's involvement with "historical scholarship" will be discussed in Chapter 4.

25 For a brief sketch of Morigerovskii's work at the seminary, see [Lagovskii], *Istoriia Permskoi dukhovnoi seminarii*, part 3, pp. 670-71. For his brief biography, see Skorobotov, *Pamiatnaia knizhka okonchivshikh kurs S.-Peterburgskoi seminarii*; and A. S. Rodosskii, *Biograficheskii slovar' studentov pervykh XXVIII-mi kursov S. Peterburgskoi dukhovnoi akademii* (SPb.: I. V. Leont'ev, 1907), p. 281.

26 For a detailed history of the Russian system of learned degrees in theology, see N. Iu. Sukhova, *Sistema nauchno-bogoslovskoi attestatsii v Rossii v XIX – nachale XX vv.* (M.: PSTGU, 2009).

27 See V. N. Sazhin, "Aleksandr Nikiforovich Morigerovskii – obshchestvennyi deiatel' i izdatel'," in *Knizhnoe delo v Rossii vo vtoroi polovine XIX – nachale XX veka* (L.: GPB, 1986), vol. 2, pp. 10-22.

28 Probably this is why Vasilii Florinskii is absent in the published list of the Perm Seminary graduates, though both of his brothers are duly listed. See *Spravochnaia kniga vsekh okonchivshikh kurs Permskoi dukhovnoi seminarii* (Perm: I. Shestakov, 1900).

29 Florinskii's graduation certificate and travel permit are preserved in RGVIA, f. 316, op. 63, d. 6257, ll. 2-2 rev.

30 NMRT, V. Florinskii to his parents, 17 July 1853.

31 For the history of the fair, see Anne L. Fitzpatrick. *The Great Russian Fair: Nizhnii Novgorod, 1840-90* (New York: St. Martin's Press, 1990); and N. F. Filatov, *Tri veka Makar'evsko-Nizhegorodskoi iarmarki* (Nizhnii Novgorod: Knigi, 2003).

32 On the development of highways and stagecoach services, see Alexandra Bekasova, "The Making of Passengers in the Russian Empire: Coach Transport Companies, Guidebooks, and National Identity in Russia, 1820-1860s," in John Randolph and Eugene M. Avrutin, eds., *Russia in Motion: Cultures of Human Mobility since 1850* (Urbana, IL: University of Illinois Press, 2012), pp. 199-217.

33 See *Postroika i ekspluatatsiia Nikolaevskoi zheleznoi dorogi (1842-1851-1901). Kratkii istoricheskii ocherk* (SPb.: Upravlenie dorogi, 1901).

34 Dobroliubov (1836-1861) would enter the Teachers Institute and soon become the leading literary critic and, together with Nikolai Chernyshevskii, the lead author of *The Contemporary*, the most influential journal of the time. Markov (183?-1861) would graduate with the highest honors — a gold medal — from the IMSA in 1858 and would be slated for a professorship at his alma mater, but would die from meningitis just after completing his doctoral dissertation. In his recollections, Florinskii claims that he struck a lasting friendship with Dobroliubov but I could not verify this claim: none of Dobroliubov's published materials (correspondence, diaries, etc.) contains any indication of his acquaintance with Florinskii. Furthermore, Dobroliubov's account of his attempt to enter St. Petersburg Theological Academy recorded in his letters home differs substantially from Florinskii's story.

35 According to the Orthodox Church's rules, only married ordained clerics were allowed to serve as priests. Widowed priests therefore were discharged from their parishes. They had two options: to leave the clergy and try to make a living in the secular world (most often as teachers in low-level secular schools); or to continue their ecclesiastical careers in the Black Clergy, take monastic vows and either become monks or continue their education at a theological academy in their diocese, hoping to get a position in the church administration.

36 Introduced by Peter the Great, the Table of Ranks defined the hierarchy of the court, military, and civil service in the empire. For a complete list of all the ranks included in the Table see D. V. Liventsev, *Kratkii slovar' chinov i zvanii gosudarstvennoi sluzhby Moskovskogo gosudarstva i Rossiiskoi imperii v XV — nachale XX vv. Chast' 1* (Voronezh: RAGS, 2006). For a detailed historical analysis of ranks and associated privileges, see L. E. Shepelev, *Tituly, mundiry, ordena v Rossiiskoi imperii* (M.: Nauka, 1991).

37 Possibly professors overlooked Florinskii's failure in the mathematics exam because the academy had made it a requirement for the first time that year, and many candidates were poorly prepared.

38 The tuition of fifty rubles per annum for "self-supported" students was reinstated two years later, in 1855; see RGVIA, f. 316, op. 26, d. 57. For a detailed account of the history of the academy, including admission rules, tuition, the professoriate, and so on, see G. G. Skorichenko, *Imperatorskaia Voenno-meditsinskaia akademiia*, 2 vols. (SPb.: Sinodal'naia tipographiia, 1902-10).

39 Unfortunately, Florinskii's collection in the NMTR contains none of the letters he must have sent home explaining the situation and announcing his decision to enter the IMSA.

40 Indeed, his great uncle, Archbishop Arkadii, was renowned as one of the leading relentless pursuers of a prominent group of *inovertsy* — the "old believers," a large section of the Russian Orthodox Christians who had refused to accept the reforms introduced by Patriarch Nikon of Moscow in the mid-seventeenth century.

41 On Zinin's very brief biography and his work in the IMSA, see I. S. Ioffe, *N. N. Zinin. Deiatel'nost' v Mediko-Khirurgicheskoi akademii. K 150-letiiu so dnia rozhdeniia* (L.: VMA, 1963).

42 Useful introductions to the history of the Crimean War in English could be found in David Wetzel, *The Crimean War: A Diplomatic History* (Boulder, CO: East European Monographs, 1985); Trevor Royle, *Crimea: The Great Crimean*

*War, 1854–1856* (New York: Palgrave Macmillan, 2000); and Orlando Figes, *The Crimean War: A History* (New York: Metropolitan, 2010).

43 For useful introductions to the history of the Great Reforms, see a recent reprint of the Russian language classic, G. A. Dzhanshiev, *Epokha velikikh reform*, 2 vols. (M.: Territoriia budushchego, 2008); W. Bruce Lincoln, *In the Vanguard of Reform: Russia's Enlightened Bureaucrats, 1825-1861* (DeKalb, IL: Northern Illinois University Press, 1982); *idem*, *The Great Reforms: Autocracy, Bureaucracy, and the Politics of Change in Imperial Russia* (DeKalb, IL: Northern Illinois University Press, 1990); and Larissa Zakharova, Ben Eklof, and John Bushnell, eds., *Velikie reformy v Rossii, 1856-1874* (M.: MGU, 1992); an English version of the latter book was published in 1994 by Indiana University Press.

44 The word *glasnost'*, which entered the English vocabulary only during Mikhail Gorgachev's *perestroika* of the mid-1980s, had come into wide use in the Russian language more than a century earlier. See W. Bruce Lincoln, "The Problem of *Glasnost'* in Mid-Nineteenth Century Russian Politics," *European Studies Review*, 1981, 11: 171-88; and W. E. Mosse, *Perestroika under the Tsars* (New York: St. Martin's, 1992).

45 See P. S. Platonov, *Ob amputatsiiakh chlenov i rezektsiiakh kostei* (SPb.: Ia. Trei, 1855), and *idem*, *Opisatel'naia anatomiia*, 3 vols. (SPb.: Ia. Trei, 1856-1858).

46 See V. Florinskii, "Pronitsaiushchaia rana kolennogo sustava (Vulnus gaesum penetrans genu sinistri)," *Voenno-meditsinskii zhurnal* (hereafter *VMZh*), 1857, 70: 1-12.

47 For a biography of Kiter, see Iu. V. Tsvelev, *Akademik Aleksandr Aleksandrovich Kiter (ocherk zhizni i deiatel'nosti)* (SPb.: Voenno-Meditsinskaia Akademiia, 2004).

48 See A. A. Kiter, *Rukovosdtvo k izucheniiu akusherskoi nauki*, 2 vols. (SPb.: Ia. Trei, 1857-1858).

49 Unfortunately, "due to its poor physical conditions" I was refused access to a special file on Florinskii's appointment held in RGVIA, f. 316, op. 27, d. 55, ll. 1-2.

50 See A. A. Kiter, *Rukovodstvo k izucheniiy zhenskikh boleznei* (SPb.: Ia. Trei, 1858). On the prize, see RGVIA, f. 316, op. 60, d. 112.

51 For an English-language analysis of the IMSA reforms, focused mostly on changes in the academy as a research institution, see Galina Kichigina, *The Imperial Laboratory: Experimental Physiology and Clinical Medicine in Post-Crimean Russia* (Amsterdam: Rodopi, 2009).

52 For a short biography of Dubovitskii, see V. N. Beliakov, *Petr Aleksandrovich Dubovitskii* (Leningrad: VMA, 1976).

53 For a biography of Glebov, see V. A. Makarov, *Ivan Timofeevich Glebov* (M.: Nauka, 1995).

54 See Glebov's description of the IMSA reforms undertaken during the three years, 1857-1859, in I. Glebov, *Kratkii obzor deistvii Imperatorskoi Sanktpeterburgskoi Mediko-khirurgicheskoi akademii za 1857, 1858 i 1859 gody v vidakh uluchsheniia etogo zavedeniia* (SPb.: Imperatorskaia akademiia nauk, 1860). See also a historical account of the academy transformation in Skorichenko, *Imperatorskaia Voenno-meditsinskaia akademiia*; especially Part 1, "Rebirth of the Academy," in vol. 2, pp. 1-130.

55 See RGVIA, f. 316, op. 28, d. 40.

56 For the statute of the new institute, see RGVIA, f. 316, op. 60, d. 118, ll. 23-27 rev.; see also Florinskii's reminiscences in ZV, 135-37.

57 See RGVIA, f. 316, op. 60, d. 131.

58 In the spring of 1860, Krasovskii published a voluminous report on the work accomplished in his clinics during the 1858-1859 academic year. See A. Ia. Krasovskii, "Otchet o sostoianii akusherskoi, zhenskoi i detskoi klinik i gospital'nogo zhenskogo otdeleniia," *VMZh*, 1860, 77 (March): 229-74; (April): 327-35. Even though Florinskii is not listed as an author, it is clear that he had produced the lion's share of this report, which Krasovskii acknowledged in a footnote on p. 229.

59 See the protocol of Florinskii's examination in RGVIA, f. 316, op. 30, d. 129, ll. 3-4 rev.

60 RGVIA, f. 316, op. 60, d. 382, l. 2.

61 See *Protokoly zasedanii Obshchestva russkikh vrachei v S. Peterburge* (hereafter PZORV), 1859-1860: 427. For his reports to the society meetings, see V. Florinskii, "O vtorichnom obremenenii (Superfoetatio)," *ibid.*, 335-57; and *idem*, "O terapevticheskom upotreblenii matochnykh dushei," *ibid.*, 429-60.

62 See V. Florinskii, "Mozhno li dopustit' kefalotripsiiu nad zhivymi mladentsami?" *PZORV*, 1860-1861: 13-56; *idem*, "Zachatiia pri otsutstvii mesiachnykh ochishchenii," *ibid.*, 248-51; *idem*, "Sroki rodov v sviazi so srokom menstruatsii i sutochnym vremenem," *ibid.*, 251-53; *idem*, "O molochnoi likhoradke," *ibid.*, 253-54; *idem*, "O zhirovom pererozhdenii posleda," *ibid.*, 254-57; *idem*, "Sertsebienie mladentsa do i posle rozhdeniia na svet," *ibid.*, 257-59.

63 V. S. Gruzdev, *Istoricheskii ocherk kafedry akusherstva i zhenskikh boleznei Imperatorskoi voenno-meditsinskoi akademii i soedinennoi s neiu akademicheskoi akushersko-ginekologicheskoi kliniki* (SPb.: Imperatorskaia akademiia nauk, 1898), p. 219.

64 This gradation of professorial positions roughly corresponds to the present day titles of full, associate, and assistant professors.

65 On Florinskii's appointment, see RGVIA, f. 316, op. 69, d. 195, ll. 187-187 rev.

66 See V. Florinskii, *O razryvakh promezhnosti vo vremia rodov* (SPb.: I. Markov, 1861). For official documents related to Florinskii's dissertation, see RGVIA, f. 316, op. 69, d. 196, ll. 102-05 rev.

67 Earning the degree automatically moved Florinskii up to the eighth rank (equal in military terms to major) in the Table of Ranks; more on this subject below.

68 See RGVIA, f. 316, op. 60, d. 256, ll. 21-21 rev.

69 See RGVIA, f. 316, op. 69, d. 196, ll. 125 rev.-126.

70 It appeared in the October issue of the journal. See V. Florinskii, "Otchet akusherskoi, zhenskoi i detskoi klinik i gospital'nogo zhenskogo otdeleniia Imperatorskoi S. Peterburgskoi Mediko-khirurgicheskoi akademii za 1859-60 uchebnyi akademicheskii god. 1. Akusherskaia klinika," *VMZh*, 1861, 72 (October): 111-47.

71 See *Meditsinskii otchet Imperatorskogo S. Peterburgskogo vospitatel'nogo doma za 1857 god* (SPb.: n. p., 1860). On the founding and early history of this institution, see A. P. Piatkovskii, "S.-Peterburgskii vospitatel'nyi dom pod upravleniem I. I. Betskogo," *Russkaia starina*, 1875, 12(1); 146-59; (2): 369-80; (4): 665-80; 13(5): 177-99; (8): 532-53; 14(11): 421-43; (12): 618-38. On its role in medicine and medical education, see T. G. Frumenkova, "Peterburgskii vospitatel'nyi dom i podgotovka medikov doreformennoi Rossii," *Universum: Vestnik Gertsenovskogo universiteta*, 2012, 1: 166-74.

72 See RGVIA, f. 316, op. 60, d. 246.

73 V. M. [Florinskii], "Meditsinskii otchet Imperatorskogo S. Peterburgskogo vospitatel'nogo doma za 1857 god. SPb., 1860," *Meditsinskii vestnik* (hereafter *MV*), 1861 (24 June): 125-28.

74 This statement comes from Florinskii's article published in the same newspaper two months later. See V. M. [Florinskii], "Doktorskii ekzamen," *MV*, 1861 (26 August): 205-08 (p. 206).

75 Florinskii's criticism provoked an indignant response from the institution's head physician, which led to a polemic lasting nearly a year. See K. Raukhfus, "Otvet na stat'iu V. M. O meditsinskom otchete Imperatorskogo S-Peterburgskogo vospitatel'nogo doma za 1857g," *MV*, 1861 (5 August): 177-81; V. M. [Florinskii], "Otvet g. Raukhfusu, po povodu ego antikritiki," *ibid.*, 1861 (30 December): 385-89; K. Raukhfus, "Otvet g. V. M.," *ibid.*, 1862 (26 May): 203-06.

76 My reconstruction of Florinskii's "grand tour" is largely based on his letters and unpublished diaries and memoirs, "Diaries of My Foreign Trip," held at the NMRT, #99, 100. Direct quotations from this manuscript are indicated in the text as [DZP, followed by page number].

77 This sum included his regular annual salary (700 rubles) and a special subsidy (1,000 rubles per year) for travel and living expenses. See RGVIA, f. 316, op. 69, d. 196, ll. 119 rev.-120.

78 For Krasovskii's instructions on what Florinskii should do during his tour, see RGVIA, f. 316, op. 30, d. 19, ll. 7-8 rev.

79 He was granted permission with no extra subsidy; for details, see RGVIA, f. 316, op. 30, d. 19, ll. 34-50; op. 60, d. 256, ll. 108-13.

80 See Florinskii's letters to his parents and siblings in NMRT; and to his superiors in RGVIA, f. 316, op. 32, d. 28, ll. 10-11 rev.

81 V. Florinskii to his parents, Prague, 25 September 1861, in NMRT.

82 See I. Glebov, "O zaniatiiakh medikov vrachebnogo instituta," *VMZh*, 1863, 88: 338-64.

83 See, V. M. [Florinskii], "Venskaia akusherskaia klinika," *MV*, 1862 (10 March): 81-87 (17 March): 93-96; (25 March): 105-9; (31 March): 121-25; *idem*, "Miunkhenskaia akusherskaia klinika," *ibid.*, 1862 (15 September): 314-16; *idem*, "Miunkhenskaia i Erlagenskaia akusherskie kliniki," *ibid.*, 1862 (22 September): 367-70.

84 See V. Florinskii, "Obzor trudov [na nemetskom, frantsuzskom, angliiskom i latinskom iazykakh] po chasti akusherstva i zhenskikh boleznei za 1861 i 1862 gg.," *VMZh*, 1862, 87 (August): 171-248; 88 (September): 37-62; *idem*, "Referat [inostrannoi literatury na nemetskom, frantsuzskom, angliiskom i latinskom iazykakh] po chasti patologii i terapii zhenskikh boleznei za 1861 i 1862 gg.," *ibid.*, 1863, 88 (October): 197-219; (December): 341-70.

85 See, for instance, V. M. Florinskii, "Neskol'ko slov o dekapitatsii utrobnogo mladentsa," *MV*, 1862 (11 August): 314-16.

86 V. Florinskii, "Ob izmenenii matki v poslerodovom periode," *MV*, 1862 (21 July): 283-86; (4 August): 299-304; (18 August): 319-23; (1 September): 335-39; (15 September): 351-53.

87 See V. Florinskii, "Ovariotomiia," *MV*, 1863 (9 March): 78-82; (16 March): 91-92; (23 March): 101-03; (30 March): 111-14.

88 See the list of books Florinskii donated to the Tomsk University Library upon his retirement, in NMRT, #262, 31 ll.

89 See, for instance, V. Florinskii to his parents, London, 10 August 1862, NMRT.

90 For a colorful depiction of Russian physicians' sojourns abroad during this period, see the memoirs of Florinskii's older colleague Ivan Sechenov, I. M. Sechenov, *Avtobiograficheskie zapiski* (M.: Izd-vo AN SSSR, 1945), pp. 73-113.

91 On "Russian circles" abroad, see A. Iu. Andreev, *Russkie studenty v nemetskikh universitetakh XVIII-pervoi poloviny XIX veka* (M.: Znak, 2005); and, especially, Michael D. Gordin, "The Heidelberg Circle: German Inflections on the Professionalization of Russian Chemistry in the 1860s," *Osiris*, 2008, 23(1): 23-49.

92 V. Florinskii to his parents, Vienna, 6 October 1861, NMRT.

93 See V. M. [Florinskii], "Doktorskii ekzamen," *MV*, 1861 (12 August): 189-92; (19 August): 197-201; (26 August): 205-208; (2 September): 213-16.

94 See V. M. [Florinskii], "Mnenie russkogo inostrantsa o nashem meditsinskom byte," *MV*, 1862 (20 January): 17-21.

95 *Ibid.*, p. 21.

96 For useful introductions to the controversy between the Slavophiles and the Westernizers, see Nicholas Riasanovsky's classic, *Russia and the West in the Teaching of the Slavophiles* (Cambridge, MA: Harvard University Press, 1952); Andrzej Walicki, *The Slavophile Controversy: History of a Conservative Utopia in Nineteenth-Century Russia* (Oxford: Clarendon, 1975); and Susanna Rabow-Edling, *Slavophile Thought and the Politics of Cultural Nationalism* (Albany, NY: SUNY Press, 2006).

97 See, for instance, V. Florinskii to his parents, Prague, 11 August 1861, NMRT.

98 V. Florinskii to his parents, Vienna, 25 September 1861, NMRT.

99 For details, see RGVIA, f. 316, op. 30, d. 26, ll. 1-4; d. 36, l. 1; op. 32, dd. 28-29; and op. 60, d. 382, l. 5.

100 See V. Florinskii, "Otchet o sostoianii i deiatel'nosti Obshchestva russkikh vrachei v S. Peterburge za 1863-64 akademicheskii god v godovom sobranii obshchestva 12 sentiabria 1864 g.," *PZORV*, 1864-1865: 4-10. The next year, he was re-elected as the society's secretary, see *ibid.*, 1864-65: 11.

101 See V. Florinskii, "Ocherk sovremenngo ucheniia ob anatomii i fiziologii detskogo mesta," *PZORV*, 1863-1864: 71-80; *idem*, "O matochnykh zhelezakh (*Glandulae utriculares*)," *ibid.*, 139-46; *idem*, "O tak nazyvaemom zakonnom isskustvennom vykidyshe," *ibid.*, 223-36; *idem*, "Difterit," *ibid.*, 321-38.

102 V. Florinskii, "Referat [inostrannoi literatury na nemetskom, frantsuzskom, angliiskom i latinskom iazykakh] po chasti patologii i terapii zhenskikh boleznei za 1861 i 1862 gg.," *VMZh*, 1863, 88 (October): 197-219; (December): 341-70; *idem*, "Obzor trudov po chasti akusherstva i zhenskikh boleznei za 1863 i 1864 gg.," *VMZh*, 1865, 94 (October): 23-58; (November) 59-102; (December): 103-39.

103 [V. M. Florinskii], "Povival'noe iskusstvo. Rukovodstvo dlia krest'ianskikh uchenits rodovspomogatel'nogo zavedeniia pri imperatorskom S. Petersburgskom vospitatel'nom dome. Sostavil I. Geppener. SPb, 1862," *MV*, 1863 (7 December): 513-15; (14 December): 525-27; *idem*, "O rasshirenii matochnoi sheiki posredstvom *Laminaria diqitata* Vil'sona," *ibid.*, 1864 (2 May): 172.

104 See RGVIA, f. 316, op. 60, d. 382, ll. 12-14.

105 The exact address was Ital'ianskaia Street, Solomko's House (there were two streets named Ital'ianskaia in 1863), nowadays 22 Zhukovskogo Street (see fig. 1-7). For the history of this house see A. S. Dubin, *Ulitsa Zhukovskogo* (SPb.: Aleteia, 2012), pp. 236-49.

106 V. Florinskii to A. I. Fufaevskaia, 13 February 1865, NMRT.

107 V. Florinskii to A. I. Fufaevskaia, 23 February 1865, NMRT.

108 RGVIA, f. 316, op. 60, d. 382, ll. 15-17.

109 V. Florinskii, "Razbor khronicheskogo vospaleniia ili zavala matki (*Metritis parenchymatosa chronica s infarctus uteri chronicus*)," *VMZh*, 1865, 93 (August): 337-400.

110 V. Florinskii, "Ob ozhivlenii mnimoumershikh novorozhdennykh detei," *MV*, 1865 (28 August): 325-27; (4 September): 333-35.

111 [G. A. Braun], *Kurs operativnogo akusherstva prof. G. Brauna* (SPb.: Ia. Trei, 1865). For the original edition, see G. A. Braun, *Compendium der Geburtshilfe* (Vienna: Braumiiller, 1864).

112 [V. M. Florinskii], "A. Krasovskii, Kurs prakticheskogo akusherstva, Vyp. 1. SPb., 1865 – xx+490 pp.," *MV*, 1865 (21 August): 315-20; V. M. Florinskii, "Obzor trudov [na nemetskom, frantsuzskom, angliiskom, latinskom iazykakh] po chasti akusherstva i zhenskikh boleznei za 1863 i 1864 gg." *VMZh*, 1865 (October): 23-58; (November): 59-102; (December): 103-39.

# Chapter 2

1 Charles Darwin, *On the Origin of Species by Means of Natural Selection, Or the Preservation of Favoured Races in the Struggle for Life* (London: John Murray, 1859), p, 488. The complete text of this edition is available at https://www.biodiversitylibrary.org/item/135954#page/11/mode/1up. All subsequent references to this edition will be given as Darwin, *Origin*.

2 In comparison, only four new periodicals had appeared in the preceding five years. For detailed quantitative analyses of Russian periodicals during the imperial era, see N. M. Lisovskii, *Bibliografiia russkoi periodicheskoi pechati, 1703-1900 gg.*, 2 vols. (M.: Literaturnoe obozrenie, [1915] 1995); and A. G. Dement'ev et. al., *Russkaia periodicheskaia pechat' (1702-1894). Spravochnik* (M.: Gospolitizdat, 1959).

3 N. V. Shelgunov, "Vospominaniia," in N. V. Shelgunov, L. P. Shelgunova, M. L. Mikhailov, *Vospominaniia v 2-kh tomakh* (M.: Khudozhestvennaia literatura, 1967), vol. 1, pp. 93-94.

4 See W. Bruce Lincoln, *In the Vanguard of Reform: Russia's Enlightened Bureaucrats, 1825-1861* (DeKalb, IL: Northern Illinois University Press, 1982).

5 See Esin, *Istoriia russkoi zhurnalistiki*. For an English language account that focuses exclusively on "literary journals," see Deborah A. Martinsen, ed., *Literary Journals in Imperial Russia* (Cambridge: Cambridge University Press, 1997), especially the chapter by Robert L. Belknap, "Survey of Russian Journals, 1840-1880," pp. 91-116, which unfortunately contains a number of factual mistakes. For a more general analysis of the role of "thick" journals in the second half of the nineteenth century, see Anton Fedyashin, *Liberals under Autocracy: Modernization and Civil Society in Russia, 1866-1904* (Madison, WI: University of Wisconsin Press, 2012).

6 For detailed histories of the journal, see L. E. Varustin, *Zhurnal "Russkoe slovo." 1859-1866* (L.: Izd. Leningradskogo universiteta, 1966), and F. Kuznetsov, *Nigilisty? D. I. Pisarev i zhurnal "Russkoe slovo"* (M.: Khudozhestvennaia literatura, 1983). Although the analytical framework of these books bears the unmistakable stamp of Soviet "Marxist" historiography, both contain a wealth of factual information about the journal and its contributors.

7 Although the count left his footprints in almost every area of Russian literature, arts, philanthropy, education, and even chess, surprisingly, there is no fully-fledged biography of this remarkable man in any language, only brief notices in various encyclopedias and biographical lexicons. See, for instance, "Kushelev-Bezborodko, Grigorii Aleksandrovich," in A. A. Polovtsov, *Russkii biograficheskii slovar'* (SPb.: Russkoe istoricheskoe obshchestvo, 1903), vol. 9, p. 703.

8 G. A. Kushelev to F. M. Dostoevskii, 5 July 1858. Cited in Varustin, *Zhurnal*, p. 13.

9 See *Sankt-Peterburgskie vedomosti*, 25 October 1858, p. 1365.

10 This is how Apollon Grigor'ev, the journal's leading critic during its first year of publication, characterized it in his autobiography, cited in Kuznetsov, *Nigilisty*, p. 23.

11 See, for instance, V. P. "Literaturnaia zametka," *Russkii invalid*, 5 September 1859, p. 1.

12 For biographical details, see P. S. Shelgunov, "G. Blagosvetlov," in G. E. Blagosvetlov, *Sochineniia* (SPb.: E. A. Blagosvetlova, 1882), iii- xxviii; and S. A. Vengerov, "Blagosvetlov, G. E.," in *idem*, *Kritiko-biograficheskii slovar' russkikh pisatelei i uchenykh* (SPb.: I. A. Efron, 1892), vol. 3, pp. 345-62.

13 The early years of Blagosvetlov's life are described in detail in A. Lebedev, "K biograficheskomu ocherku G. E. Blagosvetlova," *Russkaia starina*, 1913, 153(1): 166-82; (2): 359-66; (3): 629-43.

14 The Russian State Archive for Literature and Arts (hereafter RGALI), f. 1027, op. 1, d. 3, cited in Kuznetsov, *Nigilisty*, p. 43.

15 See the Russian State Historical Archive (hereafter RGIA), f. 1268, op. 8 (1855), d. 155, ll. 1-10; and f. 733, op. 88, d. 123, ll. 1-3.

16 See G. R. Prokhorov, "G. E. Blagosvetlov, Ia. P. Polonskii, G. A. Kushelev-Bezborodko. Pis'ma i dokumenty," *Zven'ia*, 1932, 1: 293-344.

17 On Herzen, see Martin Malia, *Alexander Herzen and the Birth of Russian Socialism, 1812-1855* (Cambridge, MA: Harvard University Press, 1961); and Edward Acton, *Alexander Herzen and the Role of the Intellectual Revolutionary* (Cambridge: Cambridge University Press, 1979).

18 For a detailed description of Blagosvetlov's involvement with "revolutionary circles," see T. I. Grazhdanova, "G. E. Blagosvetlov v russkom osvoboditel'nom dvizhenii 40-60-kh gg. XIX veka," doctoral dissertation, Leningrad, 1985.

19 Blagosvetlov described the details of his meeting with the count and its

results in a nine-page article that opened the September 1865 issue of the journal. It was clearly added to the issue at the last minute and had no pagination. See G. Blagosvetlov, "Otvet g. Podpischiku 'Russkogo slova' na ego neobkhodimye voprosy," *RS*, 1865, 9: [iii-xii].

20 For an analysis of Blagosvetlov's work as a writer and a critic, see E. K. Murenina, "Literaturno-kriticheskaia deiatel'nost' G. E. Blagosvetlova," doctoral dissertation, Sverdlovsk, 1989.

21 Blagosvetlov, "Otvet g. Podpischiku," *RS*, 1865, 9: [iii-xii].

22 At the peak of its popularity in 1865, the number exceeded 4,000. See Kuznetsov, *Nigilisty*; and Varustin, *Zhurnal*.

23 On the history of *The Contemporary*, see V. Evgen'ev-Maksimov, "*Sovremennik*" *v 40-50 gg.: ot Belinskogo do Chernyshevskogo* (L.: Izd-vo pisatelei, 1934); idem, "*Sovremennik*" *pri Chernyshevskom i Dobroliubove* (L.: Khudozh. lit-ra, 1936); and V. E. Bograd, *Zhurnal "Sovremennik", 1847-1866* (M.: Khudozhestvennaia literatura, 1959). As is the case with the histories of *Russian Word*, these books contain a mass of useful factual information, but their analytical frameworks are shaped by the official Soviet ideology.

24 P. N. Tkachev, "Izdatel'skaia i literaturnaia deiatel'nost' G. E. Blagosvetlova," in *idem, Kladezi mudrosti rossiiskikh filosofov* (M.: Pravda, 1990), pp. 550-76 (p. 554). Written in response to the 1882 publication of Blagosvetlov's biography and "selected works" and slated to appear in Blagosvetlov's journal *Delo* (Deed), this article had been banned by the censorship and was first published only in 1940.

25 G. Blagosvetlov to Ia. Polonskii, Paris, [April] 1859, *Zven'ia*, 1932, 1: 328-29 (p. 329).

26 For details on certain members of this remarkable group, see F. Kuznetsov, *Publitsisty 1860-kh godov. G. Blagosvetlov, V. Zaitsev, N. Sokolov* (M.: Molodaia gvardiia, 1969); and *idem, Krug D. I. Pisareva* (M.: Khudozhestvennaia literatura, 1990).

27 There were very few women writers in 1860s Russia, and Blagosvetlov took special care to recruit several of them to his journal.

28 This admission appeared in a letter Pisarev's mother published at her son's direction in *The Contemporary*, see Varvara Pisareva, "Pis'mo v redaktsiiu 'Sovremennika'," *Sovremennik*, 1865, 3: 218-20.

29 D. I. Stakheev, "Gruppy i portrety. Listochki vospominanii," *Istoricheskii vestnik*, 1907, 109(7-9): 424-37 (pp. 424-25).

30 G. N. Potanin, "Vospominaniia o N. A. Nekrasove," *Istoricheskii vestnik*, 1905, 99(2): 458-89 (p. 477).

31 See Blagosvetlov's complaint that in several issues of *Delo* that came out while he had been abroad, the articles in the first section of the journal had been poorly connected to those in the second section, cited in Varustin, *Zhurnal*, p. 60.

32 The existing accounts often disagree on, and occasionally downplay, Blagosvetlov's actual role in shaping the journal, see Varustin, *Zhurnal*; Kuznetsov, *Publitsisty*; and *idem, Nigilisty*.

33 RGIA, f. 772, op. 8, d. 374. Cited in Varustin, p. 103.

34 All the citations are from Blagosvetlov's letters to D. L. Mordovtsev, which had been published in G. Prokhorov, "Shestidesiatye gody v pis'makh sovremennika," in N. K. Piksanov and O. V. Tsekhnovitser, eds., *Shestidesiatye gody* (M.-L.: AN SSSR, 1940), pp. 432-36.

35 See, for instance, RGIA, f. 772, op. 8, d. 105.

36 The State Archive of the Russian Federation (hereafter GARF), f. 109, op. 1, d. 2044, l. 2.

37 For details on the Russian censorship during the period, see Mikhail Lemke, *Ocherki po istorii russkoi tsenzury i zhurnalistiki XIX stoletiia* (SPb.: Trud, 1904); and, particularly, *idem, Epokha tsenzurnyh reform 1859-1865 godov* (SPb.: Gerol'd, 1904); on the "temporary rules," see the latter volume, pp. 166-78. In the last twenty years, the history of censorship in Russia has become a lively field of historical research, with hundreds of publications ranging from short notes to full monographs. See, for instance, the proceedings of several conferences held in St. Petersburg, M. B. Konashev, ed., *Svoboda nauchnoi informatsii i okhrana gosudarstvennoi tainy* (L.: BAN, 1991); *idem*, ed., *Tsenzura v tsarskoi Rossii i Sovetskom Soiuze* (M.: Rudomino, 1995); *idem*, ed., *Tsenzura i dostup k informatsii: istoriia i sovremennost'* (SPb.: Russion National Library, 2005), as well as seven volumes compiled by Mikhail Konashev and issued during the last ten years under the general title "Censorship in Russia: History and the Present," see *Tsenzura v Rossii: Istoriia i sovremennost'* (SPb.: RNB, 2001-2015), vol. 1-7. For an English language analysis, see Charles A. Ruud, *Fighting Words: Imperial Censorship and the Russian Press, 1804-1906* (Toronto: University of Toronto Press, 2009).

38 See RGIA, f. 777, op. 2, d. 1.

39 Another regular contributor, Afanasii Shchapov, was exiled to Siberia in the summer of 1864. For a voluminous biography of Shchapov, see A. S. Madzharov, *Afanasii Prokop'evich Shchapov: Istoriia zhizni (1831-1876) i zhizn'*

*"Istorii"* (Irkutsk: Tipografiia No. 1, 2005); for an English language assessment of his works, see Andrew M. McGreevy, "A. P. Shchapov's Scientific Analysis of the Nature of Russian Society," doctoral dissertation, Ohio State University, 1973.

40 For details of this peculiar deal, see Kuznetsov, *Nigilisty*, pp. 198-204.

41 At first, this role was played by Alexander Afanas'ev-Chuzhbinskii, a well-known litterateur, and later, by one of the journal's regular contributors, novelist Nikolai Blagoveshchenskii.

42 See RGIA, f. 775, op. 1 (1863), d. 288. Emphasis added.

43 D. I. Pisarev, "Nasha universitetskaia nauka," *RS*, 1863, 7(I): 1-75; 8(I): 1-54.

44 "Ob"iavlenie ob izdanii Russkogo slova v 1863 godu," *RS*, 1863, 1: front matter.

45 A. N. Gertsen, "Byloe i dumy," in *idem*, *Sobranie sochinenii* (M.: Izd-vo Akademii nauk SSSR, 1954-65), vol. 9, p. 168.

46 Piksanov and Tsekhnovitser, eds, *Shestidesiatye gody*, p. 225.

47 D. I. Pisarev, "Shkola i zhizn'," *RS*, 1865, 7(I): 132.

48 These various labels, which defined a particular journal's "direction," have been widely used in contemporary polemics. On polemics between, for instance, *Russian Word* and *The Contemporary*, see Pereira, "Challenging the Principle of Authority."

49 D. I. Pisarev, "Razrushenie estetiki," *RS*, 1865, 5(II): 1-32 (p. 3). Emphasis in original.

50 D. Pisarev, "Pushkin i Belinskii (stat'ia vtoraia)," *RS*, 1865, 6(II): 1-60 (p. 3). Emphasis in original.

51 *Ibid.*, p. 19. Historians of the journal have spilled much ink on this aspect of *Russian Word*'s publications. See, for instance, Chapter 8, titled "The Destruction of Aesthetics?" in Kuznetsov, *Nigilisty*, pp. 467-502; and Varustin, *Zhurnal*. For English language analyses of the debate among Russian writers on utilitarianism in aesthetics, see Robert L. Jackson, "Dostoevsky's Critique of the Aesthetics of Dobroliubov," *Slavic Review*, 1964, 23(2): 258-74; and Charles A. Moser, *Esthetics as Nightmare: Russian Literary Theory, 1855-1870* (Princeton, NJ: Princeton University Press, 1989).

52 D. Pisarev, "Nereshennyi vopros (stat'ia tret'ia i posledniaia)," *RS*, 1864, 11(II): 1-64 (p. 34).

53 See, for instance, V. Samoilovich, "Bezvykhodnoe polozhenie," *RS*, 1864, 2(I): 119-58; 3(I): 29-79; N. Kholodov [N. Bazhin], "Stepan Rulev," *RS*, 1864, 11(I): 301-34; 12(I): 85-104; N. A. Blagoveshchenskii, "Pered rassvetom," *RS*, 1865, 1(I): 39-74; 2(I): 244-92; 4(I): 175-215; and many others.

54 Such as, for instance, Aleksei F. Pisemskii's novel *A Troubled Sea* (*Vzbalomuchennoe more*) (1863) that was viciously criticized in Zaitsev's review titled "A Troubled Novelist," see V. Zaitsev, "Vzbalomuchennyi romanist," *RS*, 1863, 10(II): 23-44; on the general issue of the "anti-nihilist trend" in the literature of the period, see Charles A. Moser, *Antinihilism in the Russian Novel of the 1860s* (The Hague: Mouton, 1964).

55 See "Zapiska A. V. Nikitenki o napravlenii zhurnala 'Russkoe slovo'," *Russkii arkhiv*, 1895, 2: 225-28 (p. 226).

56 Cited in Daniel P. Todes, *Ivan Pavlov* (New York: Oxford University Press, 2014), p. 30. The students to which Pavlov refers would have been only fourteen-eighteen years old.

57 GARF, f. 109, op. 1, d. 1632, l. 2, cited in Kuznetsov, *Nigilisty*, p. 491.

58 After the "Whistle" (*Svistok*) — a satirical appendix to *The Contemporary* established by Dobroliubov in 1858.

59 For a detailed analysis of this characterization and the very label "nihilist," see Kuznetsov, *Nigilisty*, pp. 537-84; Richard Peace, "Nihilism," in William Leatherbarrow and Derek Offord, eds., *A History of Russian Thought* (Cambridge: Cambridge University Press, 2010), pp. 116-40; and Olga Vishnyakova, "Russian Nihilism: The Cultural Legacy of the Conflict Between Fathers and Sons," *Comparative and Continental Philosophy*, 2011, 3(1): 99-111.

60 Compare, for instance, D. Pisarev, "Realisty," *RS*, 1864, 9-11; and M. Antonovich, "Lzhe-realisty," *Sovremennik*, 1865, 7: 53-93.

61 D. I. Pisarev, "Progulka po sadam rossiiskoi slovesnosti," *RS*, 1865, 3(II): 1-68 (p. 35).

62 See a monumental, though dated, two-volume exploration of the place of science in Imperial Russia, Alexander Vucinich, *Science in Russian Culture, a History to 1860* (Stanford, CA: Stanford University Press, 1963), and *idem*, *Science in Russian Culture, 1861-1917* (Stanford, CA: Stanford University Press, 1970). For more recent accounts, though much narrower in their scope, see Michael D. Gordin, *A Well-Ordered Thing: Dmitrii Mendeleev and the Shadow of the Periodic Table* (New York: Basic, 2004); and Charles Eliss, "Natural Science," in Leatherbarrow and Offord, eds., *A History of Russian Thought*, pp. 286-307.

63 Norov's statement is cited in K. A. Timiriazev, "Probuzhdenie estestvoznaniia v tret'ei chetverti veka," in *Istoriia Rossii v XIX veke, v 9 tomakh* (SPb.: Granat, 1909), vol. 7, pp. 1-30 (p. 2). Emphasis in original.

64 N. Shelgunov, "Vospitanie kharaktera," *RS*, 1865, 12(II): 175-210 (p. 175).

65 For the attitude to science by "enlightened bureaucrats," see Lincoln, *In the Vanguard of Reform*.

66 For a general analysis of the Enlightenment ideas and ideals in post-Crimean Russia, see R. G. Eimontova, *Idei prosveshcheniia v obnovliaiushcheisia Rossii v 50-60-e gody XIX v.* (M.: IRI RAN, 1998).

67 N. Serno-Solov'evich, "Ne trebuet li nyneshnee sostoianie znanii novoi nauki," *RS*, 1865, 1(I), p. 127.

68 See RGIA, f. 774, op. 1 (1864), d. 12, ll. 117-18.

69 For a general overview of the changes in the institutional structures and organizational principles of Russian science after the 1850s, see E. V. Soboleva, *Organizatsiia nauki v poreformennoi Rossii* (L.: Nauka. 1983); some useful information on the development of academia in the same period could be found in James C. McClelland, *Autocrats and Academics: Education, Culture, and Society in Tsarist Russia* (Chicago, IL: University of Chicago Press, 1979); and Vucinich, *Science in Russian Culture*, 1970.

70 For a general analysis of scientific societies in Imperial Russia, see Joseph Bradley, *Voluntary Associations in Tsarist Russia: Science, Patriotism, and Civil Society* (Cambridge, MA: Harvard University Press, 2009). Russian physicians were particularly active in organizing various local societies. From 1858 to 1864, they established nearly thirty new societies. See M. M. Levit, *Stanovlenie obshchestvennoi meditsiny v Rossii* (M.: Meditsina, 1974), pp. 102-05.

71 Numerous memoirs record that movement; see, for instance, E. N. Vodovozova, *Na zare zhizni. Vospominaniia* (SPb.: 1-ia trudovaia artel', 1911), especially Chapters 15 and 16: "Among the Petersburg Youth of the 'sixtieths," pp. 442-502.

72 See, for instance, A. P. Katolinskii, *Lektsii v Passazhe* (SPb.: Imperatorskaia akademiia nauk, 1860); and [A. I. Il'inskii], *Piat' populiarno-gigienicheskikh lektsii doktora Il'inskogo, chitannykh v zale Entomologicheskogo obshchestva 26, 29 marta, 1,9, 12, aprelia 1864 goda* (SPb.: A. Ia. Isakov, 1864).

73 On the 1864 educational reform see, for instance, a classic account by N. V. Chekhov, *Narodnoe obrazovanie Rossii s 60-kh godov XIX veka* (M.: Pol'za, 1912). For a recent re-examination of the topic, see A. N. Shevelev, *Shkola*.

*Gosudarstvo. Obshchestvo. Ocherki sotsial'no-politicheskoi istorii obshchego shkol'nogo obrazovaniia v Rossii vtoroi poloviny XIX veka* (SPb.: SPbGUPM, 2001).

74 For examples of the use of this particular expression, see, for instance, M. Antonovich, "Piat' populiarno-gigienicheskikh lektsii d-ra Il'inskogo," *Sovremennik*, 1864, 8: 207-15 (p. 207).

75 For a general overview of the popularization of science in Russia, see E. A. Lazarevich, *Populiarizatsiia nauki v Rossii* (M.: MGU, 1981); and *idem, S vekom naravne: populiarizatsiia nauki v Rossii* (M.: Kniga, 1984). Overviews of "scientific popular" periodicals in Imperial Russia could be found in Jeffrey Brooks, *When Russia Learned to Read: Literacy and Popular Culture, 1861-1917* (Princeton, NJ: Princeton University Press, 1985); V. A. Parafonova, "Nauchno-populiarnye zhurnaly v dorevoliutsionnyi period," *Mediaskop*, 2011, 3, http://www.mediascope.ru/node/897; *idem*, "Stanovlenie nauchno-populiarnykh zhurnalov v Rossii," *Vestnik MGU, Zhurnalistika*, 2011, 6: 61-72; and A. G. Vaganov, "Rozhdenie termina 'nauchno-populiarnaia literatura' v Rossii poslednei treti XIX-pervoi cheverti XX v.," *VIET*, 2014, 4: 75-89.

76 See *Vokrug sveta: Zhurnal zemlevedeniia, estestvennykh nauk, noveishikh otkrytii, izobretenii i nabliudenii* (SPb.: M. O. Vol'f, 1861-68) and *Priroda i zemlevedenie: Zhurnal zemlevedeniia, estestvennykh nauk, noveishikh otkrytii, izobretenii i nabliudenii* (SPb.: M. O. Vol'f, 1862-68). On M. Vol'f and his publishing empire, see S. F. Librovich, *Na kniznom postu: vospominaniia, zapiski, dokumenty* (M.: Gos. Istoricheskaia biblioteka, 2005).

77 *Russian Word* regularly carried this advertisement. See, for instance, the back matter of the October 1863 issue.

78 See RGIA, f. 774, op. 1 (1864), d. 12, l. 117.

79 D. Pisarev, "Nereshennyi vopros (stat'ia tret'ia i posledniaia)," *RS*, 1864, 11(II): 1-64 (p. 52).

80 In late 1865, when Zaitsev left the journal, Tkachev took over this task.

81 For details of Zaitsev's biography, see Kuznetsov, *Nigilisty*.

82 [V. Zaitsev], "Bibliograficheskii listok," *RS*, 1864, 10(II): 91.

83 See Varustin, *Zhurnal*, and Kuznetsov, *Nigilisty*.

84 N. Shelgunov, "Nachala obshchestvennogo byta," *RS*, 1863, 11-12(III): 1-51 (p. 41).

85 For a rather superficial analysis of this view in Pisarev's writings, see N.

N. Zabelina, "Osmyslenie roli nauki kak faktora istoricheskogo progressa v vozreniiakh D. I. Pisareva," *Vestnik MGTU*, 2007, 10(3): 382-87.

86 A useful introduction to the concept of "scientific revolution" could be found in Steven Shapin, *The Scientific Revolution* (Chicago, IL: University of Chicago Press, 1996), especially, in his 44-page "Bibliographic Essay."

87 Blagosvetlov even invited Vogt to become a "staff science writer" for *Russian Word*. See "Ob"iavlenie ob izdanii literaturno-politicheskogo zhurnala 'Russkoe slovo' v 1866 godu," *RS*, 1865, 12: front matter.

88 Zaitsev's statement is cited in the memoirs of Sergei Kravchinskii (1851-1895), an active participant of the revolutionary movement in the 1870s and 1880s, published under the telling title, "Underground Russia." See S. Stepniak-Kravchinskii, "Podpol'naia Rossiia," in idem, *Sochineniia* (M.: Khudozhestvennia literatura, 1958), p. 359.

89 [G. Z. Eliseev], "Vnutrennee obozrenie," *Sovremennik*, 1864, 7(II), p. 90.

90 For a dated, but still useful overview of the historical development of "social science" in Russia, see Alexander Vucinich, *Social Thought in Tsarist Russia: The Quest for a General Science of Society, 1861-1917* (Chicago, IL: University of Chicago Press, 1976).

91 N. S. Kashkin, "Idealisticheskii i pozitivnyi metody v sotsiologii," in P. E. Shchegolev, ed., *Petrashevtsy v vospominaniiakh sovremennikov v 3-kh tomakh* (M.-L.: GIZ, 1926-28), vol. 2, pp. 167-73 (p. 173).

92 For a history of the Petrashevskii circle and the trial in English, see J. H. Seddon, *The Petrashevtsy: A Study of the Russian Revolutionaries of 1848* (London: Manchester University Press, 1985).

93 N. G. Chernyshevskii, "Antropologicheskii printsip v filosofii," in *idem, Sochineniia v 2-kh tomakh* (M.: Mysl', 1987), vol. 2, pp. 146-328 (p. 187).

94 N. Serno-Solov'evich, "Ne trebuet li nyneshnee sostoianie znanii novoi nauki," *RS*, 1865, 1(I): 125-37 (pp. 127, 131).

95 At the time, Pisarev was writing reviews and editing a bibliographic section in *Sunrise* (*Rassvet*), an obscure magazine "for young ladies."

96 See G. Geine [H. Heine], "Atta Troll, perevod. D. I. Pisarev," *RS*, 1860, 12(I): 1-62; and D. I. Pisarev, "Sbornik stikhotvorenii inostrannykh poetov," *ibid.*, 1860, 12(II): 32-42.

97 See D. I. Pisarev, "Ulichnye tipy. A. Golitsynskogo, s 20-iu risunkami Pikkolo. Izd. K. Rikhau. M. 1860," *RS*, 1861, 2(II): 58-70.

98 This long-winded essay was Pisarev's "candidate" dissertation written in the early spring. See D. I. Pisarev, "Appolonii Tianskii ili agoniia rimskogo obshchestva," *RS*, 1861, 6(I): 1-51; 7(I): 1-86.

99 D. I. Pisarev, "Fiziologicheskie eskizy Moleshota," *RS*, 1861, 7(II): 23-53. For the German original, see Jacob Moleschot, *Physiologisches Skizzenbuch* (Giesen: Emil Roth, 1860).

100 Apparently, Viktor L. Khankin, a practicing doctor and the journal's "staff" reviewer of scientific and medical literature, quit at the beginning of 1861. His last review appeared in February, and from March until July *Russian Word* carried no reviews of this sort of publications. Perhaps, knowing Pisarev's proficiency in German, Blagosvetlov offered him Khankin's role reviewing the latest achievements of German physiology.

101 D. I. Pisarev, "Protsess zhizni (Physiologische Briefe, von Carl Vogt)," *RS*, 1861, 9(II): 1-26. Further direct quotations from this article are indicated in the text by the issue and page numbers in square brackets. For the German original, see Carl Vogt, *Physiologische Briefe für Gebildete aller Stände* (Geisen: Rider, 1861).

102 D. I. Pisarev, "Fiziologicheskie kartiny (po Biukhneru)," *RS*, 1862, 2(I): 1-51. For the German original, see Louis Buchner, *Physiologische Bilder* (Leipzig: Thomas, 1861), Bd. 1.

103 Of course, the analogy between social and biological "organisms" has a very long history, but in the nineteenth century it became increasingly popular, largely as a result of the "romantic philosophies" of F. W. J. Shelling and G. F. Hegel. See, for instance, a classic historical analysis in F. W. Coker, *Organismic Theories of the State: Nineteenth Century Interpretations of the State as Organism or as Person* (New York: Columbia University, 1910); for a more sophisticated and nuanced recent analysis, see Robert J. Richards, *The Romantic Conception of Life: Science and Philosophy in the Age of Goethe* (Chicago, IL: University of Chicago Press, 2002); and Joshua Lambier, "The Organismic State Against Itself: Schelling, Hegel and the Life of Right," *European Romantic Review*, 2008, 19(2): 131-37.

104 N. Shelgunov, "Ubytochnost' neznaniia," *RS*, 1863, 4(I): 1-59; 5(I): 1-40. Further direct quotations from this article are indicated in the text by the issue and page numbers in sqare brackets.

105 [N. Shelgunov], "Usloviia progressa," *RS*, 1863, 6(I): 1-26.

106 The literature on the reception of Darwin and his works in Russia is quite voluminous. For English-language accounts, see Daniel P. Todes, *Darwin without Malthus: The Struggle for Existence in Russian Evolutionary Thought*

(Oxford: Oxford University Press, 1989); and Alexander Vucinich, *Darwin in Russian Thought* (Berkeley, CA: University of California Press, 1988).

107 On Bronn and his translation of Darwin's works, see Sander Gliboff, *H. G. Bronn, Ernst Haeckel, and the Origins of German Darwinism: A Study in Translation and Transformation* (Cambridge, MA: MIT Press, 2008), especially Chapter 4, "Bronn's Origin," pp. 123-54.

108 On Royer and her involvement with translating Darwin's works, see Geneviève Fraisse, *Clémence Royer: philosophe et femme de science* (Paris: La Découverte, 1985); and Joy Harvey, *Almost a Man of Genius: Clémence Royer, Feminism and Nineteenth-Century Science* (New Brunswick, NJ: Rutgers University Press, 1997), pp. 62-80.

109 See Charles Darwin, *Über die Entstehung der Arten im Thier- und Pflanzen-Reich durch natürliche Züchtung, oder Erhaltung der vervollkommneten Rassen im Kampfe um's Daseyn*. Uebers. und mit Anmerk. v. Dr. H. G. Bronn (Stuttgart: Schweizerbart, 1860); and [Ch. Darwin], *De l'origine des espèces, ou des lois du progrès chez les êtres organisés par Ch. Darwin*. Trad. par Clémence–Aug. Royer, avec préf. et notes du traducteur (Paris: Guillaumin, 1862).

110 Anon. [Ed. Claparède], "Darvin i ego teoriia proiskhozhdeniia vidov," *Biblioteka dlia chteniia*, 1861, 11: 21-40; 12: 25-36. The authorship of this anonymous publication had been a subject of intense speculation for nearly a century until the Moscow historian of biology Iurii Chaikovskii proved that the paper was in fact a translation of Claparède's review, see Iu. V. Chaikovskii, "O Darvine mezhdu strok," *VIET*, 1983, 2: 108-19; and *idem*, "Pervye shagi darvinizma v Rossii," *Istoriko-biologicheskie issledovaniia*, 1989, 10: 121-41. For the original review, see Ed. Claparède, "M. Darwin et sa théorie de la formation des espèces," *Revue Germanique, Française et Étrangère*, 1861, 16: 523-59; 17: 232-63.

111 See, Charl's Darvin [Charles Darwin], *O proiskhozhdenii vidov v tsarstvakh zhivotnom i rastitel'nom putem estestvennogo podbora rodichei, ili sokhranenie usovershenstvovannykh porod v bor'be za sushchestvovanie*, transl. by S. A. Rachinskii (SPb.: A. I. Glazunov, 1864).

112 Darwin, *Origin*, p. 488.

113 See, for example, Charles Lyell, *Geological Evidences of the Antiquity of Man* (London: John Murray, 1863); T. H. Huxley, *Evidence as to Man's Place in Nature* (London: Williams & Norgate, 1863); M. J. Schleiden, *Das alter des menschengeschlechts* (Leipzig: W. Englemann, 1863); and Carl Vogt, *Vorlesungen über den Menschen, seine Stellung in der Schöpfung und in der Geschichte der Erde* (Giessen: Ricker, 1863).

114 See V. Z[aitsev], "Bibliograficheskii listok," *RS*, 1864, 1(II): 15-21. The book came out in the first week of January 1864, while the censorship permitted the distribution of the journal's first issue that carried the announcement only on 12 January.

115 Indeed in the preceding issue, Zaitsev had reviewed the first installment of Carl Vogt's *Lectures on Man*, see V. Z[aitsev], "Bibliograficheskii listok: Chelovek i mesto ego v prirode. Publichnye lektsii Karla Fogta. Izd. P. A. Gaideburova. T. 1. Vyp. 1. SPb. 1863," *RS*, 1863, 11-12(II): 1-7.

116 V. Z[aitsev], "Bibliograficheskii listok. Chelovek i ego mesto v prirode. Lektsii K. Fogta. Vyp. 2. Izd. Gaideburova, SPb., 1864; Chelovek. Mesto ego v mirozdanii i v istorii zemli. Sochinenie K. Fogta. Perevod d-ra Kanshina, vyp. 1 i 2. Izd. Vol'fa, SPb. 1864," *RS*, 1864, 3(II): 55-57. For the German original, see Carl Vogt, *Vorlesungen über den Menschen, seine Stellung in der Schöpfung und in der Geschichte der Erde* (Giessen: Ricker, 1863).

117 D. I. Pisarev, "Progress v mire zhivotnykh i rastenii," *RS*, 1864, 4(I): 1-52; 5(I): 43-70; 6(I): 233-74; 7(I): 1-46; 9(I): 1-46.

118 Pisarev did not have access to the English original but, even though he was incarcerated in the Peter-Paul Fortress, he apparently had a copy of its French translation. He did not use Rachinskii's rendering of "natural selection" as *podbor rodichei* (matching of kin). Instead, he chose the word *vybor*, which Rachinskii used only occasionally. Most likely Pisarev's choice derived from Royer's translation of Darwin's term as "élection," which literally means "choice," and, thus, has a clear connotation of something *being chosen, selected*. This conveyed Darwin's meaning much closer than Rachinskii's "matching of kin." More on Royer's influence on Pisarev's interpretation of Darwin's work below. For details on the importance and intricacies of translations in this book, see the last chapter, "Apologia."

119 Pisarev most certainly was not alone in popularizing Darwin's book. After the appearance of Rachinskii's translation, the leading "thick" journals *The Contemporary, Reading Library*, and *Annals of the Fatherland* also published substantial articles on the subject. But none of these articles was as extensive and detailed as Pisarev's. See M. A. Antonovich, "Teoriia proiskhozhdeniia vidov," *Sovremennik*, 1864, 3: 63-107; E. E[del'so]n, "O proiskhozhdenii vidov," *Biblioteka dlia chteniia*, 1864, 6: 1-31; 7: 1-14; 8: 1-34; and K. T[imiriazev], "Kniga Darvina, ee kritiki i komentatory," *Otechestvennye zapiski*, 1864, 155: 880-912; 156: 650-85; 157: 859-82; a few months later, Timiriazev's essays were issued in book format, see K. Timiriazev, *Kratkii ocherk teorii Darvina* (M.: A. A. Kraevskii, 1865).

120 The available literature on the notion of progress in history is vast. For a

helpful introduction, see Robert Nisbet, *History of the Idea of Progress* (New York: Basic, 1980).

121 See [Ch. Darwin], *De l'origine des espèces, ou des lois du progrès chez les êtres organisés par Ch. Darwin*.

122 For a detailed examination of the notion of progress in *biology*, see Michael Ruse, *Monad to Man: The Concept of Progress in Evolutionary Biology* (Cambridge, MA: Harvard University Press, 1996); on Darwin's attitude to the notion, see pp. 136-77. On the intersections of the notions of biological and social progress in Britain, see Peter J. Bowler, *The Invention of Progress: The Victorians and the Past* (Oxford: Blackwell, 1989).

123 In the first instance, Darwin stated: "The inhabitants of each successive period in the world's history have beaten their predecessors in the race for life, and are, in so far, higher in the scale of nature; and this may account for that vague yet ill-defined sentiment, felt by many paleontologists, that organisation on the whole *has progressed*" (Darwin, *Origin*, p. 345; emphasis added). This statement in no way suggests that Darwin himself shared in "that vague yet ill-defined sentiment," least of all, that he considered it the essence of his concept. The second time he used the word as a verb in the following passage: "Judging from the past, we may safely infer that not one living species will transmit its unaltered likeness to a distant futurity. And of the species now living very few will transmit progeny of any kind to a far distant futurity; for the manner in which all organic beings are grouped, shows that the greater number of species of each genus, and all the species of many genera, have left no descendants, but have become utterly extinct. We can so far take a prophetic glance into futurity as to foretell that it will be the common and widely-spread species, belonging to the larger and dominant groups, which will ultimately prevail and procreate new and dominant species. As all the living forms of life are the lineal descendants of those which lived long before the Silurian epoch, we may feel certain that the ordinary succession by generation has never once been broken, and that no cataclysm has desolated the whole world. Hence we may look with some confidence to a secure future of equally inappreciable length. And as natural selection works solely by and for the good of each being, all corporeal and mental endowments will tend to *progress towards perfection*" (Darwin, *Origin*, p. 489; emphasis added). Here again, Darwin speaks not of the essence of his theory, but of one possible outcome of natural selection — "perfection" — the other one being "utter extinction."

124 On anthropomorphic explanations of animal behavior, see N. L. Krementsov, "Chelovek i zhivotnoe: k istorii povedencheskikh sopostavlenii," in E. N. Panov, ed. *Povedenie cheloveka i zhivotnykh: skhodstvo i razlichiia* (Pushchino:

n. p., 1989), pp. 6-28; and Nikolai Krementsov and Daniel Todes, "On Metaphors, Animals and Us," *Journal of Social Issues*, 1991, 47(3): 67-82.

125 See Pisarev, "Ocherki iz istorii truda," *RS*, 1863, 9(I): 69-134; 11-12(I): 55-117. The article critically discussed American economist Henry Charles Carey's book *The Harmony of Interests: Agricultural, Manufacturing & Commercial* (1851).

126 Pisarev explains that the label had been introduced as a pejorative by the German opponents of Darwin's theory, but that he, of course, used it without any negative connotations.

127 See, Fridrikh Rolle, *Karla Darvina uchenie o proiskhozhdenii vidov v tsarstve rastenii i zhivotnykh, primenennoe k istorii mirotvoreniia*, transl. by M. Vladimirskii (SPb.: M. O. Vol'f, 1864); and Fridrikh Rolle, *Uchenie Darvina o proiskhozhdenii vidov, obshcheponiatno izlozhennoe*, transl. by S. A. Usov (SPb.: A. I. Glazunov, 1865). For the German original, see [Friedrich Rolle], *Chs. Darwin's Lehre von der Entstehung der Arten im Pflanzen- und Tierreich in ihrer Anwendung auf die Schöpfungsgeschichte dargestellt und erläutert von Dr. Friedr. Rolle* (Frankfurt am Main: J. C. Hermann, 1863). For Schleicher's work, see [August Schleicher], "Teoriia Darvina i iazykoznanie (Pis'mo Avgusta Shleikhera k d-ru Ernestu Gekkeliu)," *Zagranichnyi vestnik*, 1864, 2(4-6): 241-63. It was also issued as a brochure, see A. Shleikher, *Teoriia Darvina v primenenii k nauke o iazyke* (SPb.: P. A. Kulish, 1864). For the German original, see August Schleicher, *Die Darwinsche Theorie und die Sprachwissenschaft* (Weimar: H. Boehlau, 1863).

128 G. B[lagosvetlov], "Bibliograficheskii listok: Edinstvo roda chelovecheskogo, Katrfazha. Perevod A. D. Mi-khna. Moskva. 1864. –Metamorfozy chelovecheskogo roda, Katrfazha. Perev. A. M-na. Moskva. 1864," *RS*, 1864, 4(II): 52-53; for the French originals, see A. de Quatrefages, *Physiologie comparée, métamorphoses de l'homme et des animaux* (Paris: Baillière, 1862), and idem, *Unité de l'espèce humaine* (Paris: L. Hachette, 1861). The first book also appeared in English translation in the same year, see A. de Quatrefages, *Metamorphoses of Man and the Lower Animals*, transl. by Henry Lawson (London: Hardwicke, 1864).

129 G. B[lagosvetlov], "Bibliograficheskii listok: T. Geksli. O prichinakh iavlenii v organicheskoi prirode. Shest' lektsii, chitannykh rabochim v Muzee prakticheskoi geologii v Londone. SPb. 1864," *RS*, 1864, 7(II): 53-58. For the English original, see T. H. Huxley, *On Our Knowledge of the Causes of the Phenomena of Organic Nature (Six Lectures to Working Men)* (London: Harwicke, 1862), https://archive.org/details/onourknowledgeof62prof.

130 V. Z[aitsev], "Bibliografichekii listok: Karla Darvina uchenie o

proiskhozhdenii vidov v tsarstve rastenii i zhivotnykh, primenennoe k isorii mirotvoreniia. Izlozheno i ob"iasneno Fridrikhom Rolle. S prilozheniem biografii Darvina, sostanvlennoi S. Shenemanom. Perevod starshago uchitelia 5-i gimnazii M. Vladimirskogo. SPb. 1864," *RS*, 1864, 8(II): 100-02.

131 [V. Zaitsev], "Bibliograficheskii listok. Mesto cheloveka v tsarstve zhivotnom. Sochinenie Tomasa Genrikha Guksleia. Perevel s nemetskogo izdaniia d-ra V. Karusa N. Gol'denbakh. M. 1864," *RS*, 1864, 10(II): 97-98.

132 N. V. Shelgunov, "Razvitie chelovecheskogo tipa v geologicheskom otnoshenii," *RS*, 1865, 3(I): 218-54.

133 The literature on the history of "social Darwinism" is vast. Useful introductions to the topic can be found in Mike Hawkins, *Social Darwinism in European and American Thought, 1860-1945: Nature as Model and Nature as Threat* (Cambridge: Cambridge University Press, 1997); and Paul Crook, *Darwin's Coat-Tails: Essays on Social Darwinism* (New York: Peter Lang, 2007).

134 See [Ch. Darwin], *De l'origine des espèces, ou des lois du progrès chez les êtres organisés par Ch. Darwin*, transl. by Clémence Royer (Paris: Guillaumin, 1862). On the role and place of Royer in French social Darwinism, see Linda L. Clark, *Social Darwinism in France* (Tuscaloosa, AL: University of Alabama Press, 1984).

135 Clémence Royer, "Préface," in *De l'origine des espèces, ou des lois du progrès chez les êtres organisés par Ch. Darwin*, p. lvi.

136 *Ibid.*, p. lxi. A few years later, she would elaborate and further develop her ideas in a nearly 600-page treatise, see Clémence Royer, *Origine de l'homme et des sociétés* (Paris: Victor Masson et fils, 1870).

137 See Thomas F. Glick, "The Impact of Darwin on Ideas about Race in the US, UK, France, Germany, and Brazil," in Henrika Kuklick, ed., *New History of Anthropology* (Malden, MA: Blackwell, 2009), pp. 225-40.

138 The literature on the history of racial thought and racial debates is immense. Useful introductions to the topic could be found in Nancy Krieger, "Shades of Difference: Theoretical Underpinnings of the Medical Controversy on Black/White Differences in the United States, 1830-1870," *International Journal of Health Services*, 1987, 17(2): 259-78; David N. Livingstone, *Adam's Ancestors: Race, Religion, and the Politics of Human Origins* (Baltimore, MD: Johns Hopkins University Press, 2008); B. Ricardo Brown, *Until Darwin: Science, Human Variety and the Origins of Race* (London: Pickering & Chatto, 2010); Nicolas Bancel, Thomas David, and Dominic Thomas, eds., *The Invention of Race: Scientific and Popular Representations* (New York: Routledge,

2014). For an analysis of Russian anthropologists' attitudes to the issue of race, see Marina Mogilner, *Homo Imperii: A History of Physical Anthropology in Russia* (Lincoln, NE: University of Nebraska Press, 2013).

139 See extensive historical examinations of this subject in I. K. Mal'kova, "Istoriia i politika SShA na stranitsakh russkikh demokraticheskikh zhurnalov 'Delo' i 'Slovo'," *Amerikanskii ezhegodnik*, 1971: 273-94; R. F. Ivanov and I. Ia. Levitas, "N. G. Chernyshevskii o rabstve negrov v SShA i problema grazhdanskikh svobod," *idem*, 1980: 118-38; O. Iu. Kazakova, "Amerika i amerikantsy v otsenke russkogo obshchestva (konets 1850-kh — 1867 g.): sotsiokul'turnye aspekty vospriiatiia", doctoral dissertation, Orel, 2000; and A. A. Arustamova, *Russko-amerikanskii dialog: istoriko-literaturnyi aspect* (Perm': Permskii universitet, 2008), http://litda.ru/images/2017-3/LDA-2017-3_all.pdf.

140 See G. Bicher-Stou, "Khizhina diadi Toma ili Zhizn' negrov v nevol'nich'ikh shtatakh Severnoi Ameriki," *Russkii vestnik*, 1857, 11-12; 1858, 1-4; and *Sovremennik*, 1857, 66(11-12), 1858, 67(1-2).

141 See A. Pal'mer, *Khizhina diadi Toma. Povest' gospozhi Stove, raskazannaia detiam* (SPb.: M. O. Vol'f, 1857). Vol'f reissued the book again in 1865.

142 See M. L. Mikhailov, "Iz Longfello. Pesni o negrakh," *Sovremennik*, 1861, 3: 267-78; and V. A. Obruchev, "Nevol'nichestvo v Severnoi Amerike," *ibid.*: 279-308. For the original of the book excerpted in Obruchev's digest, see John S. C. Abbott, *South and North* (New York: Abbey and Abbott, 1860).

143 A. Toporov, "Nevol'nichestvo v Iuzhno-Amerikanskikh Shtatakh," *RS*, 1861, 4(II): 1-28 (p. 26).

144 N. Strakhov, "Durnye priznaki," *Vremia*, 1862, 11: 158-72. All subsequent quotations are from this source.

145 Here I use Darwin's own language to convey the meaning of Strakhov's praise. Strakhov himself actually used the Russian expressions "*estestvennoe izbranie*" and "*zhiznennaia konkurentsiia*," which were literal translations of the French "*l'élection naturelle*" and "*concurrence vitale*" used by Royer. Perhaps, Strakhov did not even have the English original handy. He did not try to find Russian analogues to Darwin's own "natural selection" and "struggle for existence," as had the translator of Claparede's review of Darwin's work, who had translated them literally as "*estestvennaia otbornost'*" and "*bor'ba za zhizn'*," respectively. See Anon., "Darvin i ego teoriia proiskhozhdeniia vidov," *Biblioteka dlia chteniia*, 1861, 11: 21-40; 12: 25-36. For details on the importance and intricacies of translations in this book, see the last chapter, "Apologia."

146 G. E. Blagosvetlov, "Politika," *RS*, 1861, 2(III): 15.

147 E. Rekliu [E. Reclus], "Politika," *RS*, 1863, 11-12(III): 42.

148 V. Z[aitsev], "Bibliografichekii listok: Edinstvo roda chelovecheskogo. Kartfazha. Perevod A. D. Mikh...na. M. 1864," *RS*, 1864, 8(II): 93-100.

149 M. A. Antonovich, "Russkomu Slovu," *Sovremennik*, 1864, 105(11-12): 164.

150 V. Zaitsev, "Bibliograficheskii listok. Otvet moim obviniteliam po povodu moego mneniia o tsvetnykh plemenakh," *RS*, 1864, 12(II): 20-26.

151 Anon., "Po povodu statei 'Russkogo slova' o nevol'nichestve," *Iskra*, 1865, 8: 114-17. All the subsequent quotations are from this source. Emphasis in original.

152 On Nozhin, see A. E. Gaisinovich, "Biolog-shestidesiatnik N. D. Nozhin i ego rol' v razvitii embriologii i darvinizma v Rossii," *Zhurnal obshchei biologii*, 1952, 13(5): 377-92.

153 According to Gaisinovich, Elie Metchnikoff refers to such a translation in his memoirs, but I was unable to find any indication that it had ever been published. Perhaps, as happened with the majority of Nozhin's works, it perished after his untimely death in April 1866. The first Russian translation of Muller's book appeared only in 1932, see Frits Miuller, *Za Darvina* (M.: Gos. Med. izd-vo, 1932). For the German original, see Fritz Müller, *Für Darwin* (Leipzig: Wilhelm Engelmann, 1864). For an English translation, see Fritz Müller, *Facts and Arguments for Darwin: With Additions by The Author*, transl. by W. S. Dallas (London: John Murray, 1869).

154 A few months after the review was published, the book came out in two different Russian translations, see Karl Fokht, *Chteniia o poleznykh i vrednykh zhivotnykh* (SPb.: L. P. Shelgunova, 1865); idem, *Chteniia o mnimovrednykh i mnimopoleznykh zhivotnykh* (SPb.: M. O. Vol'f, 1865). For the German original, see Carl Vogt, *Vorlesungen über nützliche und schädliche verkannte und verläumdete Thiere* (Leipzig: Ernst Keil, 1864).

155 Nozhin gives the full translation of the title, indicating that he had used the 1864 London edition of Darwin's *Journal*. Just a few months after the publication of Nozhin's article, a complete Russian translation of the book in two volumes was published in St. Petersburg. See Ch. Darvin, *Puteshestvie vokrug sveta na korable Bigl'*, 2 vols. (SPb.: A. S. Golitsyn, 1865).

156 See V. A. Zaitsev, "Bibliograficheskii listok: Charl'z Darvin, Puteshestvie vokrug sveta na korable Bigl'. Perevod pod redaktsiei A. Beketova. Tom 1. Izd. Golitsina. SPb. 1865," *RS*, 1865, 5(II): 31-32.

157 See N. V. Shelgunov, L. P. Shelgunova, M. L. Mikhailov, *Vospominaniia v 2-kh tomakh* (M.: Khudozhestvennaia literatura, 1967).

158 See V. N. Sazhin, "Aleksandr Nikiforovich Morigerovskii – obshchestvennyi deiatel' i izdatel'," in *Knizhnoe delo v Rossii vo vtoroi polovine XIX – nachale XX veka* (L.: GPB, 1986), vol. 2, pp. 10-22.

159 See, for instance, the memoirs of *Russian Word*'s author, D. I. Stakheev, "Gruppy i portrety. Listochki vospominanii," *Istoricheskii vestnik*, 1907, 109(7-9): 424-37, especially, pp. 433-36.

160 See I. E. Barenbaum, "Tipografiia zhurnalov 'Russkoe slovo,' 'Delo' (izdatel'skaia deiatel'nost' Blagosvetlova, Riumina, Morigerovskogo, Tushnova)," in *idem*, ed., *Knizhnoe delo v kul'tunoi i obshchestvennoi zhizni Peterburga-Petrograda-Leningrada* (L.: LGIK, 1984), pp. 30-43.

161 [Florinskii], "Doktorskii ekzamen," p. 201.

162 For the English-language history of the publication and censorship troubles, see Daniel P. Todes, "Biological Psychology and the Tsarist Censor: The Dilemma of Scientific Development," *Bulletin of the History of Medicine*, 1984, 58: 529-44.

163 I. Sechenov, "Refleksy golovnogo mozga," *MV*, 1863, 47: 461-84; 48: 493-512. For historical assessments of this work and its author, see M. G. Iaroshevskii, *Sechenov i mirovaia psikhologicheskaia mysl'* (M.: Nauka, 1981); and David Joravsky, *Russian Psychology: A Critical History* (Cambridge, MA: Blackwell, 1989).

164 Their daughter Olga would be born on 23 January 1866.

165 For publications in verse, the basic unit for calculating the honorarium was one line.

166 For details of financial arrangements in *Russian Word*, see Kuznetsov, *Nigilisty*, pp. 93-96.

167 On the censorship of medical publications, see Anon., "Zakon 8 marta 1862 goda," *MV*, 1862 (21 April): 153-56.

168 See Lemke, *Ocherki*; *idem*, *Epokha*; and Ruud, *Fighting Words*, especially, Chapter 9, "The Reform of 6 April 1865," and Chapter 10, "The First Year of the Reformed System, 1865-66," pp. 137-67.

169 See Todes, "Biological Psychology and the Tsarist Censor."

170 I will discuss these steps in the next chapter.

171 For details on how the three "warnings" were issued and the journal's suspension, see S. S. Konkin, "Zhurnal 'Russkoe slovo' i tsenzura v 1863-66

godakh," *Uchenye zapiski Sterlitamakskogo gosudarstvennogo pedagogicheskogo instituta*, 1962, 8: 231-54.

172 See RGIA, f. 776, op. 3, d. 161, ll. 22-25.

# Chapter 3

1 See D. Bairon, "Don Zhuan. Roman v stikhakh. Per. D. Minaev," *Sovremennik*, 1865, 1: 5-72; 2: 447-512; 3: 169-208; 4: 245-74; 5: 163-203; 7: 207-37; 8: 453-76; 10: 433-68; 1866, 1: 259-74; 4: 385-406.

2 See V. Florinskii, 1865, 8(I): 1-57. The exact citations from this essay are indicated in the text by specific page numbers in square brackets. For the announcement of the release of the August issue, see *Golos*, 7 October 1865, p. 4.

3 Unless noted otherwise, the emphasis in all quotations is Florinskii's.

4 Florinskii actually used the word "*izmeniaemost'*," which is more accurately translated as "changeability," although the contents of the section make clear that he followed closely Darwin's notion of variability. Tellingly, Rachinskii in his translation used the word "*izmenchivost'*," which is much closer to Darwin's meaning of variability. Of course, neither word had a defined terminological status in any language yet.

5 In his discussion of variability, Florinskii borrowed his main analytical category "type" (*tip*) directly from comparative anatomy, where it was employed to denote a particular pattern of organization that distinguishes one "type" from another. On the emergence of the notion of "type" and its uses, see F. J. Cole, *A History of Comparative Anatomy, from Aristotle to the Eighteenth Century* (New York: Dover, 1975); on the historical development of this notion in zoology before and after Darwin, see I. I. Kanaev, *Ocherki iz istorii sravnitel'noi anatomii do Darvina: Razvitie problemy morfologicheskogo tipa v zoologii* (M.-L.: Izd-vo AN SSSR, 1963); and idem, *Ocherki iz istorii problemy morfologicheskogo tipa ot Darvina do nashikh dnei* (M.-L.: Nauka, 1966).

6 Although Florinskii did not refer directly to Lyell's and/or Vogt's books, it is clear that he borrowed much data from their works. As we saw, both Lyell's and Vogt's books were translated into Russian almost immediately after their publications in the original English and German, respectively, see Charl'z Liaiel, *Drevnost' cheloveka. Geologicheskie dokazatel'stva drevnosti cheloveka s nekotorymi zamechaniiami o teoriiakh proiskhozhdeniia vidov* (SPb.: O. I. Bakst, 1864); for the English original, see Charles Lyell, *Geological Evidences of the Antiquity of Man* (London: John Murray, 1863); and Karl Fogt, *Chelovek. Mesto ego v mirozdanii i v istorii zemli* (SPb.: M. O. Volf, 1865); and idem, *Chelovek i mesto ego v prirode*

(SPb.: P. A. Gaideburov, 1863-65), vols. 1-2; for the German original, see Carl Vogt, *Vorlesungen über den Menschen, seine Stellung in der Schöpfung und in der Geschichte der Erde* (Giessen: Ricker, 1863).

7 For a voluminous biography of von Baer, see B. E. Raikov, *Karl Ber, ego zhizn' i trudy* (M.-L.: AN SSSR, 1961); on his anthropological works, see pp. 393-453.

8 See K. Ber [Karl von Baer], *Chelovek v estestvenno-istoricheskom otnoshenii* (SPb.: K. Vingeber, 1851).

9 The three sons of Noah: Shem (Asia), Ham (Africa), and Japheth (Europe); plus the native Americans, a group that was unknown in biblical times.

10 One of his principal sources for this section was an extensive treatise by Augustin Weisbach, "Beiträge zur Kenntniss der Schädelformen Österreicher Volker," *Wienese Medizinische Jahrbuch*, 1864, 1: 49-127.

11 Florinskii did not give an exact reference and thus it is impossible to ascertain to which particular work of Trémaux's he refers. As far as I was able to determine, none of Trémaux's works had been translated into Russian. Florinskii must have read some of his works in French. It could have been one of Trémaux's numerous publications on his travels and research in Africa. But it is also possible that it was his monumental treatise on "the origin and transformations of humans and other beings," that has been published in Paris at the beginning of 1865. See Pierre Trémaux, *Origine et transformations de l'homme et des autres êtres* (Paris: L. Hachette, 1865). Although neither the Russian State Library, nor the IMSA Library has the book, the Library of the Academy of Sciences (BAN) does have a copy.

12 Florinskii cites a Russian translation of Edwards's treatise by T. N. Granovskii, "O fiziologicheskikh priznakakh chelovecheskikh porod i ikh otnoshenii k istorii. Pis'ma V. F. Edvardsona i A. T'erri," in *idem, Sochineniia* (M.: V. Got'e, 1856), vol. 1, p. 33. For the original, see William Frédéric Edwards, *Des caractères physiologiques des races humaines considérés dans leurs rapports avec l'histoire* (Paris: Mondey-Dupre, 1841).

13 Florinskii here borrows data from and refers to a voluminous investigation by Shchapov on "the historical-geographical organization of the Russian population" serialized in *Russian Word*, see A. P. Shchapov, "Istoriko-geograficheskoe raspredelenie russkogo narodonaseleniia," *RS*, 1864, 8(I): 1-54; 9(I): 95-130; 10(I): 179-211; and *idem*, "Istoriko-geograficheskoe raspredelenie narodonaseleniia," *RS*, 1865, 6(I): 1-30; 7(I): 1-36; 8(I): 1-44; 9(I): 1-41.

14 As an illustration, see Google Ngram for the word *nasledstvennost'* in Russian publications from 1800 to 1900, at https://books.google.com/ngrams/graph?c

ontent=%D0%9D%D0%B0%D1%81%D0%BB%D0%B5%D0%B4%D1%81%D1%82%D0%B2%D0%B5%D0%BD%D0%BD%D0%BE%D1%81%D1%82%D1%8C&year_start=1800&year_end=1900&corpus=12&smoothing=3&share=&direct_url=t1%3B%2C%D0%9D%D0%B0%D1%81%D0%BB%D0%B5%D0%B4%D1%81%D1%82%D0%B2%D0%B5%D0%BD%D0%BD%D0%BE%D1%81%D1%82%D1%8C%3B%2Cc0

15 See the entry *"nasledit'"* in Dal's dictionary at http://dic.academic.ru/dic.nsf/enc2p/276362.

16 See F. Toll', *Nastol'nyi slovar' dlia spravok po vsem otrasliam znaniia*, 3 vols. (SPb.: F. Toll', 1863-66). See the entry for "hereditary bleeding" in vol. 2, p. 974.

17 For a detailed analysis of the emergence of the notions of "heredity" in French medicine, see C. López-Beltrán, "Forging Heredity: From Metaphor to Cause, a Reification Story," *Studies in History and Philosophy of Science*, 1994, 25(2): 211-35; and *idem*, "In the Cradle of Heredity: French Physicians and L'Hérédité Naturelle in the Early 19th Century," *JHB*, 2004, 37: 39–72; for its uses in British medicine, see John Waller, "Ideas of Heredity, Reproduction, and Eugenics in Britain, 1800-1875," *Studies in History and Philosophy of Biological and Biomedical Sciences*, 2001, 32(3): 457-89; on the development of the notions of heredity in Germany, see Frederick B. Churchill, "From Heredity to Vererbung: The Transmission Problem, 1850-1915," *Isis*, 1987, 78: 337-64; for general overviews, see S. Müller-Wille and H. J. Rheinberger, eds., *Heredity Produced: At the Crossroads of Biology, Politics, and Culture, 1500-1870* (Cambridge, MA: MIT Press, 2007); and S. Müller-Wille and Christina Brandt, eds., *Heredity Explored: Between Public Domain and Experimental Science* (Cambridge: MIT Press, 2016).

18 See Prosper Lucas, *Traite philosophique et physiologique de l'Hérédité naturelle* (Paris: J. B. Bàillière, 1847). For Darwin's references to Lucas's views, see *Origin*, pp. 12, 275.

19 Florinskii does not cite Virchow's work, but there is no doubt that he knew it well. Delivered in Berlin during the spring of 1858, Virchow's famous lectures were translated into Russian almost instantaneously. An abridged version appeared as an appendix to the November-December issue of the *Moscow Physicians' Journal* (see *Moskovskii vrachebnyi zhurnal*, 1858, 5-6). A full version (in a different translation) came out in book format a few months later, see R. Virkhov, *Patologiia, osnovannaia na teorii iacheek (tselularnaia patologiia)*, transl. Ia. Rozenblat and I. Chatskin (M.: Moskovskaia vrachebnaia gazeta, 1859). For the German original, see R. Virchow, *Die Cellularpathologie in ihrer Begründung auf physiologische und pathologische Gewebelehre: Zwanzig Vorlesungen* (Berlin: Hirschwald, 1858). For a historical analysis of the Russian reception of Virchow's ideas, see Larisa Shumeiko, *Die Rezeption Der Zellularpathologie*

*Rudolf Virchows (1821-1902) in Der Medizin Russlands und Der Sowjetunion* (Marburg: Tectum Verlag, 2002).

20 The story is found in *Genesis*, Leviticus 30: 25-43: While shepherding for his uncle Laban, Jacob noted that when sheep in heat looked at, and mated in front of, striped branches, they gave birth to speckled and spotted young. So he convinced his uncle to grant him all the speckled and spotted sheep born to Laban's flock as payment for his work. Jacob then placed striped branches in watering pools, so that sheep will always have to look at them when they come to drink. And as a result, more and more speckled sheep were born to Laban's flocks, which Jacob could claim for himself.

21 See J. M. Boudin, "De l'influence de l'âge relatif des parents sur le sexe des enfants," *Comptes rendus hebdomadaires des séances de l'Académie des Sciences*, 1863, 56: 353. A detailed account of Boudin's report appeared in the *Gazette Médicale de Paris*, 28 February 1863, 9: 137-39. In October, *Medical Herald* published a Russian translation of the report, see Buden [Boudin], "O vliianii sravnitel'nogo vozrasta roditelei na pol detei," *MV*, 1863 (12 October): 387-88.

22 Florinskii clearly avoided using the word "material" since "materialism" was a red flag for the censors, but that is exactly what he means here.

23 For instance, he cites examples of exceptional fecundity described in the memoirs of V. A. Nashchokin, *Zapiski* (SPb.: Akademiia nauk, 1842), p. 148; and in J. Lewis Brittain, "Repeated Twin Births," *Edinburgh Medical Journal*, 1862, 8(2): 468.

24 V. Florinskii, 1865, 10(I): 1-43. The exact citations from this essay are indicated in the text by specific page number in square brackets. The issue came out in early December. See *Golos*, 12 December 1865, p. 4.

25 Pisarev, "Razrushenie estetiki," p. 3.

26 See the discussion of "utilitarian aesthetics" in the previous chapter.

27 The book was reissued more than ten times from the second half of the eighteenth to the mid-nineteenth century. Florinskii cites its fifth edition, N. G. Kurganov, *Pis'movnik* (SPb.: Imperatorskaia akademiia nauk, 1793), Part 1, p. 283.

28 See Adolphe Quételet, *Sur l'homme et le développement de ses facultés, ou Essai de physique sociale*, 2 vols (Paris: Bachelier, 1835). St. Petersburg libraries have numerous copies and the book was well known in Russia (see, for instance, extensive citations from and references to this work in Shchapov's articles cited above). Furthermore, its first volume appeared in Russian translation exactly at the time Florinskii was working on his treatise in the spring of 1865,

see A. Ketle, *Chelovek i razvitie ego sposobnostei, ili Opyt obshchestvennoi fiziki* (SPb.: O. I. Bakst, 1865).

29 See A. Veidengammer, "Sel'sko-khoziaistvennoe skotovodstvo kak argument darvinovskoi teorii," *Zapiski Imperatorskogo russkogo obshchestva akklimatizatsii,* 1865: 143-81.

30 Much of the evidence Florinskii cited in this section comes from a popular manual on hygiene and medical police compiled from several German publications, which he had apparently used in preparation for his doctoral examination. See L. Pappenheim, *Rukovodstvo k gigiene i meditsinskoi politsii,* 2 vols. (SPb.: Biblioteka med. nauk d-ra M. Khana, 1860-61). For the German sources of this compilation, see Louis Pappenheim, *Handbuch der Sanitäts-Polizei: nach eigenen Untersuchungen* (Berlin: Hirschwald, 1858); and Friedrich Oesterlen, *Der Mensch und seine physische Erhaltung. Hygienische Briefe für weitere Leserkreise* (Leipzig: F. A. Brockhaus, 1859). The latter book also appeared in a separate Russian translation, see F. Esterlen, *Chelovek i sokhranenie ego zdorov'ia* (SPb.: O. I. Bakst, 1863).

31 Florinskii did not identify which particular work of Becquerel he quotes here. The quote actually comes from a highly popular "treatise on private and public hygiene," which since its first appearance in 1851 went through numerous editions and translations. See Louis Alfred Becquerel, *Traité élémentaire d'hygiène, privée et publique* (Paris: Labé, 1851), https://archive.org/details/traitlmentairedh00becq. Indeed, already in 1852, Iakov Chistovich, the IMSA professor of hygiene and medical police, published a Russian translation and used it as a textbook in his course on hygiene, which Florinskii had to take. See A. Bekkerel', *Elementarnoe nachertanie chastnoi i obshchestvennoi gigieny (nauki o sokhranenii chelovecheskogo zdorov'ia)* (SPb.: Med. Departament, 1852). The quote comes from p. 82 of the Russian edition.

32 For a detailed analysis of the notion of hereditary diseases in contemporary British medicine, see John C. Waller, "'The Illusion of an Explanation': The Concept of Hereditary Disease, 1770-1870," *Journal of the History of Medicine and Allied Sciences,* 2002, 57: 410-48; in the US, Rosenberg, "The Bitter Fruit."

33 Florinskii took this example from S. A. Usov, "Zubr," *Zapiski Imperatorskogo Russkogo obshchestva akklimatizatsii,* 1865: 1-64.

34 The November issue came out in late December, and the December issue only in late January 1866. See *Golos,* 24 December 1865, p. 4; and *ibid.,* 29 January 1866, p. 4.

35 V. Florinskii, 1865, 11(I): 1-25; 12(I): 27-43. The exact citations from this essay are indicated in square brackets in the text.

36 He took his numbers from Thomas Willis, *Facts Connected with the Social and Sanitary Condition of the Working Classes in the City of Dublin: With Tables of Sickness, Medical Attendance, Deaths, Expectation of Life, &c., &c; Together with Some Gleanings from the Census Returns of 1841* (Dublin: T. O'Gorman, 1845).

37 Florinskii borrowed much of his statistical data on the Germanic lands from Johann Ludwig Casper, *Ueber die wahrscheinliche Lebensdauer des Menschen* (Berlin: F. Dümmler, 1843).

38 The quotation came from Part I, Chapter 21, which in John Ormsby's 1885 English translation reads: "...there are two kinds of lineages in the world; some there be tracing and deriving their descent from kings and princes, whom time has reduced little by little until they end in a point like a pyramid upside down; and others who spring from the common herd and go on rising step by step until they come to be great lords; so that the difference is that the ones were what they no longer are, and the others are what they formerly were not." https://books.google.co.uk/books?id=dig6CwAAQBAJ&pg=PT133 &lpg=PT133&dq.

39 Florinskii used the adjectives "consanguineous" (*krovnye*) and "kin" (*rodstvennye*) synonymously.

40 For a historical assessment of the French debate, see Mauro Sebastián Vallejo, "El problema de la consanguinidad en la medicina francesa (1850-1880): cuando heredar demasiado era un riesgo y un deseo," *Asclepio. Revista de Historia de la Medicina y de la Ciencia*, 2012, 64(2): 517-40; for analyses of the debate in the English speaking countries, see A. H. Bittles, "The Bases of Western Attitudes to Consanguineous Marriage," *Developmental Medicine and Child Neurology*, 2003, 45: 135-38; and, especially, Diane B. Paul and Hamish G. Spencer, "Eugenics without Eugenists?: Anglo-American Critiques of Cousin Marriage in the Nineteenth and Early Twentieth Centuries," in Müller-Wille and Brandt, eds., *Heredity Explored*, pp. 49-79.

41 See K. T[olstoi], "O krovnykh brakakh," *MV*, 1865 (26 June): 233-35; (3 July): 249-53; (10 July): 257-59; (17 July): 269-72; (24 July): 277-80; (31 July): 285-88.

42 For an English language account of the long history of the studies of this condition, see H. Werner, *History of the Problem of Deaf-mutism until the 17th Century* (Jena: Gustav Fischer, 1932); for summaries of nineteenth-century studies, see Henry W. Hubbard, *Deaf-mutism: A Brief Account of the Deaf and Dumb Human Race, from the Earliest Ages to the Present Time* (London: Leisure Hour, 1894); and Holger P. T. Mygind, *Deaf Mutism* (London: F. J. Rebman, 1894).

43 See J.-Ch.-M. Boudin, *Dangers des unions consanguines et nécessité des*

*croisements dans l'espèce humaine et parmi les animaux* (Paris: J.-B. Baillière et Fils, 1862), originally published in *Annales d'hygiène publique et de médicine légale*, 1862, 18: 5-82; *idem*, "De la Nécessité des Croisements, et du Danger des Unions Consanguines dans l'Espèce Humaine et parmi les Animaux," *Recueil de mémoires de médecine, de chirurgie et de pharmacie militaires*, 1862 (3rd series), 8: 193-241; and *idem*, "Etudes statistiques sur les Dangers des Unions Consanguines dans l'Espèce Humaine et parmi les Animaux," *Journal de la Société de Statistique*, 1862, 3: 69-84; 103-20. Florinskii also cited a dissertation by L. T. Chazarain, *Du mariage entre consanguins considéré comme cause de dégénériscence organique, et plus particulièrement de surdi-mutité congéniale* (Collection des Thèses de l'Ecole de Médecine de Montpellier, No. 63, 1859). The dissertation is absent in the IMSA Library.

44 See Francis Devay, *Du Danger des Mariages consanguins au point de vue sanitaire* (Paris: Labé, 1857); and its much expanded second edition, *idem*, *Du Danger des Mariages consanguins sous le rapport sanitaire* (Paris: V. Masson et fils, 1862); the IMSA Library holds only the second volume, which Florinskii probably used. The anti-consanguinists argued that exogamous marriage replenished "hereditary blood" with fortifying traits. They cited the degeneration of the *Ancien Régime* to argue their case against consanguinity and thus called for the "regeneration of the French race" through cross-breeding.

45 P. Meniere, "Recherches sur l'origine de la surdi-mutité," *Gazette Médicale de Paris*, 1846, 3: 223-26; 243-46; *idem*, "Du mariage entre parents considéré comme cause de la surdi-mutité congénitale," *ibid.*, 1856: 303-06.

46 All of these periodicals were available in St. Petersburg's libraries.

47 In addition to the authors mentioned above, Florinskii cited a special report on idiocy by the director of the Perkins Institution and Massachusetts Asylum in Boston: S. G. (Samuel Gridley) Howe, *Report Made to the Legislature of Massachusetts, Upon Idiocy by Howe, S. G.* (Boston, MA: Coolidge and Wiley, 1848); and the monograph by Swiss physician Johann Jakob Guggenbühl, *Die Heilung und Verhütung des Cretinismus und ihre neuesten Fortschritte* (Bern: Huber, 1853); the last book had earned Guggenbühl an honorary membership in the IMSA, see RGVIA, f. 316, op. 28, d. 35, ll. 1-16.

48 R. Liebreich, "Abkunft aus Ehen unter Blutsverwandten als Grund von *Retinitis pigmentosa*," *Deutsche Klinik* (Berlin), 1861 (9 February), 13: 53-55.

49 In addition to the works by Devay and Boudin mentioned above, Florinskii also cited many observations of animals collected by French entomologist Charles Nicolas Aubé, "Notes sur les inconvénients qui peuvent résulter du défaut de croisement dans la propagation des espèces animales," *Bulletin de la Société impériale zoologique d'acclimatation*, 1857, 4: 509-18; and in humans

reported by American physician Samuel M. Bemiss, "On the Evil Effects of Marriages of Consanguinity," *North American Medico-Chirurgical Review*, 1857, 1: 97-108 (which also was republished as "On Marriages of Consanguinity" in the London-based *Journal of Psychological Medicine and Mental Pathology*, 1857, 10(6): 368-79); and *idem*, "Report on Influence of Marriages of Consanguinity upon Offspring," *Transactions of the American Medical Association*, 1858, 11: 319-425. Since none of these journals was available at the IMSA Library, most likely, Florinskii cited these studies from Boudin's and Devay's publications.

50 Here he referred to the second edition of Devay's book, *Du Danger des Mariages consanguins* (Paris: V. Masson et fils, 1862), which was available at the IMSA Library.

51 Here Florinskii referred to Devay's earlier monumental, two-volume study of "the physical and moral perfection of man," which was also available at the IMSA Library, see Francis Devay, *Hygiène des familles, ou Du perfectionnement physique et moral de l'homme: considéré particulièrement dans ses rapports avec l'éducation et les besoins de la civilisation moderne*, 2 vols. (Paris: Labé; Lyon: Dorier, 1846), http://gallica.bnf.fr/ark:/12148/bpt6k65137691.

52 V. Florinskii, 1865, 12(I): 27-43.

53 Florinskii here actually gave the English phrase "breeding in and in" without translation.

54 Here Florinskii summarized Sanson's presentation at the Society's meeting. See A. Sanson, "Unions Consanguines chez les Animaux," *Bulletins de la Société d'Anthropologie*, 1862, 1(3): 254-64.

55 He cited works by Jean Magne (1804-1885), professor at l'École impériale vétérinaire d'Alfort; Jean Gourdon (1824-1876), professor at l'École vétérinaire de Toulouse; and Antoine Richard "du Cantal" (1802-1891), French doctor, veterinarian, agronomist, and politician. All of these references apparently came from J.-Ch.-M. Boudin, *Du croisement des familles et des races et réponse a M. Dally* (Paris: Louis Guerin, 1863).

56 Robert Bakewell (1725–1795), British agriculturalist, now recognized as one of the most important figures in the British Agricultural Revolution, was one of the first to implement systematic selective breeding of livestock.

57 He cited articles by Alfred Bourgeois, "Sur les Résultats attribues aux Alliances Consanguines," *Comptes Rendus Hebdomadaires des Séances de l'Académie des Sciences*, 1863, 56: 177-81; J. A. N. Perier, "Essai sur les Croisements Ethniques," *Mémoires de la Société d'Anthropologie*, 1863, 1: 69-92; 2: 187-236; and 1865, 2: 261-374; August Voisin, "Contribution a l'Histoire des mariages entre consanguins," *Mémoires de la société d'anthropologie*, 1863, 2: 433-59, which

also came out in book format a few years later, see August Voisin, *Contribution a l'Histoire des mariages entre consanguins* (Paris: Bailliere et fils, 1866); and M. Seguin, "Sur les Mariages Consanguins," *Comptes Rendus Hebdomadaires des Séances de l'Académie des Sciences*, 1863, 57: 253-54.

58 Eugène Dally, "Recherches sur les Mariages consanguins et sur les races pures," *Bulletins de la Société d'Anthropologie de Paris*, 1863, 4: 515-75. Dally presented this long report to the Anthropological Society on 5 November 1863, i.e. after Florinskii had left Paris. But the journal that carried it was available at the IMSA Library. The next year the report was also issued as a booklet under the same title, see Eugène Dally, *Recherchés sur les Mariages consanguins et sur les races pures* (Paris: V. Masson, 1864). An English translation of this report by H. J. C. Beavan appeared in the London *Anthropological Review*, 1864 (May): 65-108.

59 J.-Ch.-M. Boudin, "Du Croisement des Familles et des Races, et Réponse a M. Dally," *Bulletins de la Société d'Anthropologie*, 1862, 6: 662-94.

60 In arguing this point, both the proponents and opponents of kin marriages mostly relied on data provided in Francisque-Michel, *Histoire des races maudites de la France et de l'Espagne* (Paris: A. Franck, 1847).

61 Meniere, "Recherches sur l'origine de la surdi-mutité," 1846; and *idem*, "Du mariage entre parents considéré comme cause de la surdi-mutité congénitale," 1856.

62 Boudin, 1862, p. 21. The original reads: "A notre sens, les mariages consanguins, loin de militer en faveur d'une hérédité toute imaginaire, constituent la protestation la plus flagrante contre les lois mêmes de l'hérédité. Comment, voilà des parents consanguins, pleins de force et de santé, exempts de toute infirmité appréciable, *incapables de donner à leurs enfants ce qu'ils ont, et leur donnant au contraire ce qu'ils n'ont pas, ce qu'ils n'ont jamais eu,* et c'est en présence de tels faits que l'on ose prononcer le mot hérédité (3)! Nous croyons inutile de prolonger cette discussion; citons quelques faits." Emphasis in original.

63 On Morel and his concept, see Daniel Pick, *The Faces of Degeneration: A European Disorder, c.1848-c.1918* (Cambridge: Cambridge University Press, 1989); and Kelly Hurley, "Hereditary Taint and Cultural Contagion: The Social Etiology of Fin-de-Siècle Degeneration Theory," *Nineteenth-Century Contexts*, 1990, 14(2): 193-214.

64 Florinskii probably read Morel's magnum opus: B. A Morel, *Traite des Dégénérescences physiques, intellectuelles, et morales de l'espèce humaine; et des causes qui produisent ces variétés maladives* (Paris: J. B. Baillières, 1857). It was

certainly available, along with Morel's other works, at the IMSA Library. With the exception of Perier, none of the participants in the French debate had cited Morel's work, though Tolstoi in his overview of the debate did recount very briefly some of Morel's ideas.

65 Although Florinskii did not cite this work in his treatise, his text strongly suggests that he had borrowed much of the analysis of the existing Russian laws on marriage from their extensive discussion in *The Contemporary*, see M. A. Filippov, "Vzgliad na russkie grazhdanskie zakony," *Sovremennik*, 1861, 2: 523-62; 3: 217-66.

66 For a contemporary overview of the Orthodox Church regulations of marriage, see "Obzor tserkovnykh postanovlenii o brake v pravoslavnoi tserkvi," *Pravoslavnyi sobesednik*, 1859, 2: 369-413; 3: 1-45; 119-52; 217-34; 325-51. For detailed instructions on determining the degree of kinship, see Sergei Grigorovskii, *O rodstve i svoistve* (SPb.: Trud, 1903).

67 See Grigorovskii, *O rodstve i svoistve*.

68 According to the Russian law, the children of such mixed marriages would be by law Orthodox.

69 Since the essays appeared in a "literary-political" journal addressed to the general reader and largely followed the then acceptable reference style in scientific/medical periodicals, Florinskii did not give the exact reference for *every* source he had consulted. Quite often he simply mentioned the last name of a scientist or a physician whose work he had used (giving foreign names in Russian transliteration or transcription). Equally often, he did not bother to give even a name. But the contents of his treatise allow one to identify even those publications (such as Lyell's and Huxley's books on the origins of man) that do not appear in his references.

70 For the intricacies of correct translations of Florinskii's vocabulary see the last chapter, "Apologia."

71 In *Origin*, Darwin used this expression in various forms more than 200 times.

72 On *raznochintsy*, see the classic work by Christopher Becker, "*Raznochintsy*: The Development of the Word and of the Concept," *American Slavic and East European Review*, 1959, 18(1): 63-74; for a more recent and much more detailed analysis, see Elise Kimerling Wirtschafter, *Social Identity in Imperial Russia* (DeKalb, IL: Northern Illinois University Press, 1997), pp. 66-99.

73 For an analysis of the early articulations of the *raznochintsy*'s ideology, see G. V. Zykova, *Zhurnal Moskovskogo universiteta "Vestnik Evropy" (1805-1830 gg.): Raznochintsy v epokhu dvorianskoi kul'tury* (M.: Dialog-MGU, 1998).

74 See, for instance, N. G. O. Pereira, "Challenging the Principle of Authority: The Polemic between *Sovremennik* and *Russkoe Slovo*, 1863-65," *Russian Review*, 1975, 34(2): 137-50.

75 On this motto and its role, see a classic study by Nicholas V. Riasanovsky, *Nicholas I and Official Nationality in Russia, 1825-1855* (Berkeley, CA: University of California Press, 1959).

76 For an excellent analysis of the formation of the "middle class" and its values in Britain, see Harold J. Perkin, *The Origins of Modern English Society 1780-1880* (London: Routledge, 1969) and *idem*, *The Rise of Professional Society: England Since 1880* (London: Routledge, 1989).

# Chapter 4

1 See "Ob"iavlenie ob izdanii literaturno-politicheskogo zhurnala 'Russkoe slovo' v 1866 godu," *RS*, 1865, 12: front matter. The same announcement appeared in the newspapers, see, for instance, *Golos*, 24 December 1865, p. 4.

2 See the announcement of this order in the official newspaper of the Ministry of Internal Affairs, *Severnaia pochta*, 17 February 1866, p. 1; also *Golos*, 24 February 1866, p. 4.

3 See, for instance, announcements of the publication in *Golos*, 11 March 1866, p. 4; and *Knizhnik*, 1866, 4: 211-12.

4 See *Luch: Ucheno-literaturnyi sbornik* (SPb.: Riumin i Komp., 1866), vol. 1. The first volume came out at the end of March, see *Golos*, 28 March 1866, p. 4.

5 See *Golos*, 28 March 1866, p. 4. Fortunately, several copies of the second volume survived. See *Luch: Ucheno-literaturnyi sbornik* (SPb.: Riumin i Komp., 1866), vol. 2. For the censorship materials regarding the collection, see RGIA, f. 777, op. 2 (1866), d. 53. For a detailed description of Blagosvetlov's battle with the censorship over the publication of *The Ray*, see Kuznetsov, *Nigilisty*, pp. 528-32.

6 For details of the Karakozov Affair, see Claudia Verhoeven, *The Odd Man Karakozov: Imperial Russia, Modernity and the Birth of Terrorism* (Ithaca, NY: Cornell University Press, 2009).

7 For details, see Kuznetsov, *Nigilisty*, p. 527.

8 Unfortunately, for some unknown reasons, though listed in the directory, a separate file with the documents related to the establishment of the IMSA

pediatrics clinic in the Russian State Military-Historical Archive has been destroyed, see RGVIA, f. 316, op. 34, d. 252. For the Academy Council's decision to put Florinskii in charge of the new clinics and his report on the first year of its operations, see RGVIA, f. 316, op. 69, d. 200, ll. 309 rev-311; d. 201, ll. 258 rev-259. For brief descriptions of Florinskii's involvement with pediatrics, see N. I. Bystrov, *Kratikii ocherk istorii kliniki detskikh boleznei Imperatorskoi VMA* (SPb.: Tip. MVD, 1899), pp. 14, 26-28, 34-37; and V. S. Vail', *Ocherki po istorii russkoi pediatrii* (Stalinabad: n. p., 1960), pp. 41-50.

9 See, V. Florinskii, "Obzor trudov [na nemetskom, frantsuzskom, angliiskom, latinskom iazykakh] po chasti akusherstva i zhenskikh boleznei za 1863 i 1864 gg.," *VMZh*, 1865, 94 (October): 23-58; (November): 59-102; (December): 103-39.

10 See, for instance, V. M. Florinskii, "Soderzhanie kormilitsy i rebenka," in Zh.-Zh. Russo [J.-J. Rousseau], *Sobranie sochinenii* (SPb.: N. Tiblen, 1866), vol. 1. *Teoriia vospitaniia*, pp. 625-36.

11 See [V. Florinskii], "Otchet o sostoianii i deiatel'nosti Obshchestva russkikh vrachei v S. Peterburge," *PZORV*, 1865-1866: 4-9. For the voting results and Florinskii's response to his re-election, see *ibid.*, p. 10.

12 For a rather favorable account of the newspaper's history written half a century later by one of its regular contributors, see N. A. Skorobotov, *"Peterburgskii listok" za tridtsat' piat' let. 1864-1899* (SPb.: "Gerol'd," 1914).

13 *Peterburgskii listok*, 11 January 1866, p. 3.

14 *Peterburgskii listok*, 15 January 1866, p. 4.

15 *Ibid.*, pp. 4-5.

16 See RGIA, f. 776, op. 3, d. 460, ll. 1-2.

17 Unfortunately, I was unable to find any 1866 documents of St. Petersburg Criminal Court in the Russian archives. Thus my reconstruction of the events surrounding the trial is based largely on newspaper accounts.

18 See *Peterburgskii listok*, 8 February 1866, p. 3; 10 February 1866, p. 3; 15 February 1866, p. 3.

19 When Mrs. Andreeva attempted to change her story, saying that it might have been not Monday 13 December, but perhaps Saturday 11 or Sunday 12 when she had come to Florinskii's apartment, the IMSA administration confirmed that on those days Florinskii too had been at the clinics from early morning till late afternoon. See the correspondence between the court and the academy in RGVIA, f. 316, op. 35, d. 142, ll. 1-4.

20 See announcements in "Sudebnaia khronika," *Golos*, 12 May 1866, p. 3; "Sudebnaia khronika," *ibid.*, 19 May 1866, p. 3; and reports on the court proceedings in *Peterburgskii listok*, 26 May 1866, pp. 1-3; 29 May 1866, pp. 1-2; and 5 June 1866, p. 1.

21 Reportedly, Florinskii asked the court to commute Mrs. Andreeva's sentence, since she had to take care of her child.

22 I have failed to find any information about Mr. Balabolkin.

23 Florinskii's suit was just one of several cases against *Petersburg Leaf*, which had been addressed by the court. See "Sudebnaia khronika," *Golos*, 19 May 1866, p. 3; and K. Arsen'ev, "Pis'mo v redaktsiiu," *Sankt Peterburgskie vedomosti*, 28 May 1866, p. 3; "Sudebnye prigovory," *ibid.*, 6 June 1866, p. 2.

24 See *MV*, 1866 (25 June): 308.

25 For the full text of the edict, see "Vysochaishii rescript ot 13 maia 1866 g.," *Sobranie uzakonenii i rasporiazhenii Pravitel'stva*, 1866, 44: 326. All the subsequent quotations are from this source.

26 See RGVIA, f. 316, op. 60, d. 585, ll. 1-10. All the following quotations are from this source.

27 His last publication in the newspaper was a dissertation review, which appeared in July 1866 and had likely been written a few months earlier, see *MV*, 1866 (30 July): 363-66.

28 See, for instance, V. Florinskii, "O rezul'tatakh nabliudenii nad upotrebleniem khloroforma," *PZORV*, 1867-1868: 109-11; 117-18; and *idem*, "O smertnosti rodil'nits v Sankt Peterburge," *ibid.*, 1872-73: 229-45.

29 Blagosvetlov to Shelgunov, 30 January 1866, cited in L. P. Shelgunova, "Iz dalekogo proshlogo," in Shelgunov, Shelgunova, Mikhailov, *Vospominaniia*, vol. 2, p. 197.

30 For detailed histories of the journal, see B. I. Esin, *Demokraticheskii zhurnal "Delo"* (M.: MGU, 1959); and M. A. Benina, "Zhurnal 'Delo' v 1860-70-e gg. ('Epokha Blagosvetlova')," in V. E. Kel'ner, ed., *Knizhnoe delo v Rossii vo vtoroi polovine XIX – nachale XX v.* (L.: GPB, 1988), vol. 3, pp. 7-20; on the censorship permission and other documents related to the journal publication, see RGIA, f. 777, op. 2 (1866), d. 76; and op. 3 (1866), dd. 398-99.

31 Cited in Kuznetsov, *Nigilisty*, p. 533.

32 See announcement in *Golos*, 28 September 1866, p. 4.

33 See Florinskii, 1866. The announcements of the publication appeared in newspapers, see, for instance, *Golos*, 4 September 1866, p. 4; and *Syn otechestva*, 14 September 1866, p. 1736.

34 Most books published during this period listed either a publisher (*izdatel'*), or a printer (*tipografiia*) on their front pages. For the archival documents that definitively identify the printing shop owned by Blagosvetlov that produced Florinskii's book, see RGIA, f. 777, op. 2 (1866), d. 7, ll. 12-15 rev.

35 In the newspaper announcements of the book's publication, however, the name of its author was spelled correctly, see *Golos*, 4 September 1866, p. 4; and *Syn otechestva*, 14 September 1866, p. 1736.

36 To give just one example, when a few years later Blagosvetlov issued as a separate volume a series of articles by Veniamin Portugalov, which had first been published in *Deed*, the volume carried a brief "foreword" by its author that provided the readers with the necessary information. See Avtor, "Predislovie," in V. O. Portugalov, *Voprosy obshchestvennoi gigieny* (SPb.: A. Morigerovskii, 1873), p. i.

37 There were no copyright laws in existence yet.

38 For the French original, see Auguste Debay, *Hygiène et physiologie du mariage: histoire naturelle et médicale de l'homme et de la femme mariés, dans ses plus curieux détails* (Paris: E. Dentu, 1862). This was the 29th edition of the book! For Russian translations, see O. Debe, *Gigiena i fiziologiia braka*, 2 vols. (M.: S. Orlov, 1862); and O. Debe, *Gigiena i fiziologiia braka*, 3 vols. (SPb.: D. F. Fedorov, 1862-63).

39 N. Shul'gin, "Ob"iavlenie ob izdanii ezhemesiachnogo zhurnala 'Delo'," *Golos*, 17 November 1866, p. 4. Emphasis in original. Although the announcement was signed by the "official" editor Shul'gin, there is no doubt that it had been written by the actual editor Blagosvetlov.

40 Iakobii entered Zurich University's Medical School to study psychiatry in 1864, and it was in Zurich that he met and married Zaitsev's sister, Varvara. For a biography of Iakobii, see I. I. Shchigolev, *Otechestvennyi psikhiatr P. I. Iakobii* (Briansk: Izd-vo BGU, 2001). This biography devotes much space to Iakobii's involvement with "revolutionary circles" and his psychiatric work after returning to Russia in 1890. But, unfortunately, it skips over nearly thirty years, from 1862 to 1890, which Iakobii spent studying, publishing, and working as a physician in Germany, Switzerland, and France.

41 See E. K-di [P. Iakobii], "Razvitie rabstva v Amerike," *RS*, 1865, 7(I): 91-115; 12(I): 1-30.

42 See P. Ia[kobii], "Dusha cheloveka i zhivotnykh. Lektsii professora geidel'bergskogo universiteta V. Vundta. Per. s nemetskovo E. K. Kemnitsa. Tom pervyi. Izdanie P. A. Gaideburova. SPb. 1865," *RS*, 1865, 10(II): 75-102. For the German original, see Wilhelm Wundt, *Vorlesungen über die Menschen- und Thierseele* (Leipzig: Leopold Voss, 1863).

43 See RGIA, f. 776, op. 3, d. 161, ll. 19-19 rev.

44 *Ibid.*, l. 23.

45 See Ern. Kalonn [P. Iakobii], "Psikhologicheskie etiudy. Organicheskie elementy mysli," *Luch. Ucheno-literaturnyi sbornik*, 1866, vol. 2, pp. 43-90.

46 See De-Kalonn [P. Iakobii], "Fizicheskie usloviia pervonachal'noi tsivilizatsii cheloveka," *Delo*, 1866, 1: 104-28.

47 After the Karakozov Affair, the secret police closely monitored Blagosvetlov's correspondence. This letter was copied and placed in Blagosvetlov's file. See GARF, f. 109, op. 1, d. 2045, l. 1.

48 De-Kalonn [P. Iakobii], "Khronika estestvenno-nauchnykh otkrytii," *Delo*, 1867, 5: 27-48; 6: 1-26.

49 De-Kalonn [P. Iakobii], "Fiziologiia mysli," *Delo*, 1867, 8: 1-48.

50 See L. E. Kalonn [P. Iakobii], "Psikhologicheskie etiudy," *Delo*, 1867, 11: 92-122.

51 See his dissertation, Paul Jacoby [P. Iakobii], *Considérations sur les monomanies impulsives* (Bernae: n. p., 1868). Iakobii would resume writing "natural science chronicle" for *Deed* almost three years later, see De Kalonn [P. Iakobii], "Estestvennye nauki v 1869 g (estestvenno-nauchnaia khronika)," *Delo*, 1870, 4: 39-78; 9: 35-75.

52 For a severely truncated biography of Portugalov that focuses predominantly on his work in Samara from the late 1870s through the 1890s, see P. S. Kabytov, S. I. Stegunin, and V. Iu. Kuz'min, *Zemskii vrach Veniamin Osipovich Portugalov (1835-1896 gg.)* (Samara: Samarskoe knizhnoe izd-vo, 2006). For Portugalov's much more informative autobiography, see the Manuscript Collection of the Institute of Russian Literature in St. Petersburg (hereafter RO IRLI), f. 377, op. 7, d. 2847; see also, "Portugalov, Veniamin Osipovich," in V. A. Mysliakov, ed., *Russkaia intelligentsia. Avtobiografii i bibliograficheskie materialy v sobranii S. A. Vengerova. Annotirovannyi ukazatel'* (SPb.: Nauka, 2010), vol. 2, "M-Ia," pp. 244-45.

53 His first publications appeared in the *Archive of Legal Medicine and Social Hygiene* in 1867, see V. O. Portugalov, "Shadrinsk i Cherdyn'," *Arkhiv sudebnoi meditsiny i obshchestvennoi gigieny* (hereafter *ASMiOG*), 1867, 4: 36-60.

54 See V. Portugalov, "Prichiny bolezni," *ASMiOG*, 1868, 2: 1-46; 3: 1-43; 4: 1-21; 1869, 1: 1-23; and *idem, Prichiny boleznei* (SPb.: Tip. Akademii Nauk, 1869).

55 V. Portugalov, "Istochniki boleznei," *Delo*, 1869, 3: 81-114.

56 The December 1868 issue of *Deed* carried an advertisement for the *Archive of Legal Medicine and Social Hygiene*, which perhaps attracted Blagosvetlov's attention to the journal and its contents.

57 See Charles Darwin, *The Variation of Animals and Plants Under Domestication*, 2 vols. (London: John Murray, 1868). The Russian translation appeared even before the English original, see Ch. Darvin, *Proiskhozhdenie vidov. Otd. 1. Izmeneniia zhivotnykh i rastenii vsledsvie prirucheniia*, 2 vols., ed. by I. M. Sechenov, transl. by V. Kovalevskii (SPb.: F. S. Sushchinskii, 1868). On the history of this translation, see Ia. M. Gall, "Vladimir Kovalevskii kak perevodchik i izdatel' truda Charl'za Darvina, 'The Variation of Animals and Plants under Domestication'," *Vestnik VOGiS*, 2007, 11(1): 40-44.

58 Portugalov, *Istochniki boleznei*, p. 114.

59 Portugalov's main source was Reich's recently published voluminous treatise, *Ueber die Entartung der Menschen, ihre Ursachen und Verhütung* (Erlangen: Ferdinand Enke, 1868). On Reich and his work on the treatise, see Karl-Heinz Karbe, "Eduard Reich (1836-1919) und sein Wirken für die 'gesamte Hygiene' in der Gothaer Schaffensperiode von 1861 bis 1869," *Beiträge zur Hochschul- und Wissenschaftsgeschichte Erfurts*, 1987-88, 21: 243-56. Portugalov also referred extensively to Virchow's famous series of articles on the typhus epidemic in Upper Silesia, which had been published in Virchow's journal, *Archiv für pathologische Anatomie und Physiologie und für klinische Medicin*, 1849, 2: 143-322. On Virchow's views on social hygiene, see Ian F. McNeely, *"Medicine on a Grand Scale": Rudolf Virchow, Liberalism, and the Public Health* (London: The Wellcome Trust, 2014).

60 V. Portugalov, "Bespredel'nost' gigieny," *Delo*, 1869, 8: 1-39.

61 This is how Russian specialists translated the journal's title. See Anon., "Zhurnal obshchestvennoi gigieny. Deutsche Vierteljahrsschrift für öffentliche Gesundheitspflege, Red. Von Prof. Reclam," *ASMiOG*, 1870, 1: 23-36. Surprisingly, despite his many contributions to legal medicine and social hygiene, my search for Reclam's biography or an historical assessment of his works yielded no results, aside from a brief entry in the German Biographical Lexicon, see Julius L. Pagel, "Reclam, Karl Heinrich," *Allgemeine Deutsche Biographie*, 1907, 53: 246, https://www.deutsche-biographie.de/gnd116374098.html#adbcontent.

62 On pangenesis, see Conway Zirkle, "The Inheritance of Acquired Characters

and the Provisional Hypothesis of Pangenesis," *American Naturalist*, 1935, 69: 417-45; and Kate Holterhoff, "The History and Reception of Charles Darwin's Hypothesis of Pangenesis," *JHB*, 2014, 47: 661-95.

63 See Blagosvetlov to Portugalov, [March], 1870, RGALI, f. 613, op. 1, d. 5661, ll. 174-76. Released from the Peter-Paul Fortress in November 1866, Pisarev drowned in July 1868 in a boating accident.

64 See K. Reklam, *Populiarnaia gigiena: nastol'naia kniga dlia sokhraneniia zdorov'ia i rabochei sily v narode* (SPb.: A. Morigerovskii, 1869). Altogether, Blagosvetlov issued five different editions of the book, in 1869, 1870, 1872, 1878, and 1882. Portugalov's essay was first included in the second edition. For the German original, see Carl H. Reclam, *Das Buch der vernünftigen Lebensweise. Für das Volk zur Erhaltung der Gesundheit und Arbeitsfähigkeit. Eine populäre Hygieine* (Leipzig: Winter, 1863).

65 See Blagosvetlov to Portugalov, 8 February 1870, RGALI, f. 613, op. 1, d. 5661, ll. 170-71.

66 V. Portugalov, "Poslednee slovo nauki," *Delo*, 1869, 11: 104-28; 12: 1-31; 1870, 2: 86-121; 4: 41-94; 6: 104-26; 7: 1-42.

67 See V. O. Portugalov, "Razvitie i porcha," *Delo*, 1870, 11: 114-71; 12: 102-36.

68 See V. O. Portugalov, "O vyrozhdenii," *Delo*, 1871, 1: 77-118; 2: 199-226; 3: 150-88.

69 Portugalov, "Razvitie i porcha," p. 136.

70 Reich, *Ueber die Entartung der Menschen*. Although Portugalov undoubtedly read Reich's work in the original German, it is worth noting that its Russian rendering appeared just a few months prior to the publication of Portugalov's essays. See Okt. Mil'chevskii, *Prichiny vyrozhdeniia cheloveka, nepolnota i nepravil'nost' ego telesnogo i dushevnogo razvitiia v nastoiashchee vremia* [compiled on the basis of a German work by Ed. Reich] (M.: A. I. Mamontov, 1870).

71 See Eduard Reich, *System der Hygiene*, 2 vols. (Leipzig: F. Fleischer, 1870-1871).

72 Morel, *Traite des Dégénérescences*.

73 V. O. Portugalov, *Voprosy obshchestvennoi gigieny* (SPb.: A. Morigerovskii, 1873). See an advertisement for the book in *Delo*, 1873, 5: back matter.

74 Charles Darwin, *The Descent of Man, and Selection in Relation to Sex*, 2 vols. (London: John Murray, 1871).

75 See *Znanie*, 1871, 4-9, supplements.

76 Ch. Darvin, *Proiskhozhdenie cheloveka i polovoi podbor* (SPb.: "Znanie," 1871).

77 He was also not pleased with a different translation, also published in the early fall under Ivan Sechenov's editorship. See Charl'z Darvin, *Proiskhozhdenie cheloveka i podbor po otnosheniiu k polu*, 2 vols., ed. and transl. by I. Sechenov (SPb.: Cherkesov, 1871-1872). On his critique of the numerous mistakes in this edition, see G. Blagosvetlov, "Ob"iasneniia s redaktsiei zhurnala 'Znanie' po povodu Darvina," *Delo*, 1871, 11: 31-46.

78 [G. Blagosvetlov], "Darvin. Proiskhozhdenie cheloveka i polovoi podbor. Sokrashchennyi perevod s angliiskogo. Izdanie redaktsii zhurnala 'Znanie'. SPb. 1871," *Delo*, 1871, 9: 19-23 (p. 19).

79 *Ibid.*, pp. 19-20.

80 Blagosvetlov was likely aware that during the same time another St. Petersburg publisher was preparing one more translation of Darwin's book edited by Ivan Sechenov, but he proceeded with his own translation anyway.

81 The Knowledge publishers responded in kind to Blagosvetlov's critique and, after the appearance of the first two volumes of his translation, published a lengthy critical review in *St. Petersburg News*, see [Redaktsiia zhurnala 'Znanie'], "Darvin i redaktsiia zhurnala 'Delo'," *Sankt Peterburgskie vedomosti*, 4 November 1871, 1(2): 1.

82 Charl's Darvin, *Proiskhozhdenie cheloveka i polovoi podbor*, 3 vols., ed. and transl by G. E. Blagosvetlov (SPb.: Tip. A. Morigerovskogo, 1871-1872). The delay in the appearance of the third volume was the result of the censorship interference, see RGIA, f. 776, op. 2, dd. 8-9; and op. 11, d. 142; and f. 777, op. 2, d. 73.

83 See also a similar critique of a Russian translation of A. Wallace's *Contributions to the Theory of Natural Selection* issued under the auspices of the Moscow popular-science journal *Priroda* (*Nature*) and its unfavorable comparison to Blagosvetlov's own translation of the same book: I. P., "Estestvennyi podbor," *Delo*, 1878, 6: 70-89.

84 For some observations on the reception of the book, see Ruth Schwartz Cowan, "Nature and Nurture: The Interplay of Biology and Politics in the Work of Francis Galton," in William Coleman and Camille Limoges, eds., *Studies in the History of Biology* (Baltimore, MD: Johns Hopkins University Press, 1977), vol. 1, pp. 133-208; and Emel Aileen Gökyigit, "The Reception of Francis Galton's 'Hereditary Genius' in the Victorian Periodical Press," *JHB*, 1994, 27(2): 215-40. Indeed, Russian appears to be the only language

into which the book was translated during the nineteenth century. The first German translation came out only in 1910, while a French one was never produced at all.

85 See F. Gal'ton, "Liudi nauki," *Znanie*, 1874, 5: 40-53. The translation was made not from the original English publication, Francis Galton, "On Men of Science, their Nature and their Nurture," *Proceedings of the Royal Institution*, 1874, 7: 227-36, but from its French translation, Francis Galton, "Les Hommes de Science leur Education et leur Régime," *Revue Scientifique*, 1874, 44 (2 May): 1035-40.

86 See F. Gal'ton, "Nasledstvennost' talanta, ee zakony i posledsviia," *Znanie*, 1874, 11-12 (supplement): 1-299. The editors' choice to translate this particular book seems strange. It would have been more logical to publish in Russian Galton's next book, *English Men of Science: Their Nature and Nurture* (London: Macmillan, 1874), which came out later in the same year and presented Galton's much more elaborate views on "hereditary genius" than those advanced in the eponymous volume, as well as his response to the criticism it evoked in various quarters, especially from the Swiss botanist Alphonse de Candolle. Yet, this choice actually supports the suggestion that it was Darwin's references to *Hereditary Genius* that incited its Russian translation.

87 F. Gal'ton, *Nasledstvennost' talanta, ee zakony i posledstviia* (SPb.: Znanie, 1875).

88 N. Ia. [P. Iakobii], "Sovremennaia bezdarnost' (Gal'ton, Nasledstvennost' talanta, ee zakony i posledstviia. Perevod s angliiskogo. SPb. 1875)," *Delo*, 1875, 5: 50-75.

89 See S. Sh., "Inostrannaia literatura. English Men of Science: Their Nature and Nurture. By Francis Galton. London, 1874. English Eccentrics and Eccentricities. By John Timbs. London, 1875," *Delo*, 1875, 10: 161-72. Although the reviews were signed "S. Sh.," which was the pen-name used by the journal's regular contributor, well-known historian Serafim Shashkov, it is likely that the reviews were actually written by Blagosvetlov himself, for their subjects were way beyond Shashkov's interests and expertise, while his knowledge of the English language was much more limited than that of Blagosvetlov's.

90 See A. R. Uolles, *Teoriia estestvennogo podbora* (SPb.: Tip. G. E. Blagosvetlova, 1878). For the English original of the "second edition with corrections and additions," see Alfred Russel Wallace, *Contributions to the Theory of Natural Selection* (New York: Macmillan, 1871).

91 V. Portugalov, "Gigienicheskie usloviia braka," *Delo*, 1876, 9: 1-39.

92 John Lubbock, *The Origin of Civilization and the Primitive Condition of Man* (London: Longmans, Green and Co., 1870). A Russian translation of Lubbock's book appeared as supplements to *Knowledge* in 1874-1875, see Dzh. Lebbok, "Nachalo tsivilizatsii," *Znanie*, 1874, 10; 1875, 1-6 (*prilozheniia*). The next year a different translation came out under the auspices of the Moscow-based popular-science magazine *Priroda*, see Dzh. Lebbok, *Doistoricheskie vremena* (M.: "Priroda", 1876).

93 Edward B. Taylor, *Primitive Culture*, 2 vols. (London: John Murray, 1871). A Russian translation of Taylor's monograph came out in 1872-1873 also under the auspices of *Knowledge*, see E. B. Teilor, *Pervobytnaia kul'tura*, 2 vols. (SPb.: "Znanie", 1872-73).

94 The first volume of his textbook, modestly titled *Introduction to Gynecology*, was published as two separate tomes of more than 650 pages in total in 1869 and 1870, respectively. See V. Florinskii, *Kurs akusherstva i zhenskikh boleznei (ginekologiia). T. 1. Vvedenie v ginekologiiu. Vyp. 1. Istoricheskii i anatomo-physiologicheskii otdely* (SPb.: Tip. Ia. Trei, 1869); idem, *Kurs akusherstva i zhenskikh boleznei (ginekologiia). T. 1. Vvedenie v ginekologiiu. Vyp. 2. Obshchaia diagnostika i terapiia zhenskikh boleznei* (SPb.: Tip. Ia. Trei, 1870).

95 See, for instance, V. Florinskii, "O rezul'tatakh nabliudenii nad upotrebleniem khloroforma," *PZORV*, 1867-1868: 109-11, 117-18.

96 For a voluminous general overview of the history of medicine in Russia from the tenth to the twentieth century, see M. B. Mirskii, *Meditsina Rossii X-XX vekov* (M.: ROSSPEN, 2005). For a detailed analysis of the creation of the system of state medicine in Russia in the early nineteenth century, see Elena Vishlenkova, "The State of Health: Balancing Power, Resources, and Expertise and the Birth of the Medical Profession in the Russian Empire," *Ab Imperio*, 2016, 3: 39-75.

97 Despite the fact that his name is mentioned in every textbook on the history of medicine and/or public health, there is still no detailed scholarly biography of this remarkable man, to say nothing of an examination of his time in Russia. For a largely hagiographic account, see Harald Breyer, *Johann Peter Frank: "Fürst unter den Ärzten Europas"* (Leipzig: Hirzel Teubner, 1983). For Frank's autobiography written before his appointment in St. Petersburg, see George Rosen, "Biography of Dr. Johann Peter Frank," *Journal of the History of Medicine*, 1948, 3(1): 11-46; (2): 279-314. For a detailed examination of his ideas about medical police and public health, see a recent dissertation by Rüdiger Haag, "Johann Peter Frank (1745-1821) und seine Bedeutung für die öffentliche Gesundheit," doctoral dissertation, Saarbrücken University, 2010.

98 To give but one example, it was Frank who insisted on adopting Latin as the

language of instruction and abolishing studies of the French language at the academy, even though at the time the majority of European medical schools were offering instruction in native languages, while French medicine was universally considered the world leader in nearly every medical specialty, thus making the knowledge of French absolutely critical to the profession.

99 Wylie was a personal surgeon to three consecutive Russian emperors: Paul I, Alexander I, and Nicholas I. As with Frank, there is still no fully-fledged scholarly biography of this remarkable man, either in English or in Russian. There are only some brief notices in various encyclopedia and "jubilee" articles. See, for instance, A. A. Novik, V. I. Mazurov, and P. D. A. Semple, "The Life and Times of Sir James Wylie Bt., Md., 1768–1854, Body Surgeon and Physician to the Czar and Chief of the Russian Military Medical Department," *Scottish Medical Journal*, 1996, 41(4): 116-20. Although there is a highly fictionalized account of his life (which used no Russian language sources at all), see Mary McGrigor, *The Tsar's Doctor: The Life and Times of Sir James Wylie* (Edinburgh: Birlinn, 2010). For a brief analysis of Wylie's role in the development of Russian medical services, see Vishlenkova, "The State of Health," 50-65.

100 Biographical details for many German physicians practicing in nineteenth-century Russia could be found in a biographical lexicon issued as part of a general project on "Scientific relations in the 19th century between Germany and Russia in the fields of chemistry, pharmacy and medicine" at the Saxon Academy of Sciences in Leipzig. See Marta Fischer, *Russische Karrieren. Leibärzte im 19. Jahrhundert* (Aachen: Shaker Verlag, 2010).

101 Florinskii, 1865, 10(I): 1-43 (p. 12).

102 Zinin retired in 1864.

103 The confrontation between the two factions became so vicious that it even found a way into the official history prepared for the academy's centennial, see N. P. Ivanovskii, ed., *Istoriia Imperatorskoi voenno-meditsinskoi (byvshei mediko-khirurgicheskoi) akademii za sto let, 1798-1898* (SPb.: Tip. MVD, 1898), pp. 602-05.

104 A good indication of belonging to one or the other faction is the membership of IMSA professors in either the St. Petersburg Society of Russian Physicians or the St. Petersburg Society of German Physicians, or both. For the membership lists for various years, see *PZORV* and *St. Petersburger medizinische Zeitschrift* (1861-1869).

105 Some relevant observations on the patriotic ethos of Russian physicians during the nineteenth century can be found in Nancy M. Frieden, *Russian*

*Physicians in an Era of Reform and Revolution, 1856-1905* (Princeton, NJ: Princeton University Press, 1981); and E. Vishlenkova, "'Vypolniaia vrachebnye obiazannosti, ia postig dukh narodnyi': Samosoznanie vracha kak prosvetitelia rossiiskogo gosudarstva," *Ab Imperio,* 2011, 2: 47-79; for more general analyses, see Elizabeth A. Hachten, "Science in the Service of Society: Bacteriology, Medicine and Hygiene in Russia, 1855-1907," doctoral dissertation, University of Wisconsin at Madison, 1991; and Lisa Kay Walker, "Public Health, Hygiene and the Rise of Preventive Medicine in Late Imperial Russia, 1874-1912," doctoral dissertation, University of California, Berkeley, 2003.

106 For a vivid depiction of the confrontation by one of its active participants, see Ia. A. Chistovich, *Dnevniki, 1855-80,* vol. 2 (1857-74), pp. 310-430. These unpublished diaries are held in the Manuscript collections of the Fundamental Library of the Military-Medical Academy (St. Petersburg), f. VIII, d. 16-18. Hereafter references to these diaries will be given as Chistovich, *Dnevniki.*

107 V. M. [Florinskii], "Mnenie russkogo inostrantsa o nashem meditsinskom byte," *MV,* 1862 (20 January): 17-21.

108 See V. M. [Florinskii], "Meditsinskii otchet Imperatorskogo S. Peterburgskogo vospitatel'nogo doma za 1857 god. SPb., 1860," *MV,* 1861 (24 June): 125-28; K. Raukhfus, "Otvet na stat'iu V. M. O meditsinskom otchete Imperatorskogo S-Peterburgskogo vospitatel'nogo doma za 1857g," *ibid.,* 1861 (5 August): 177-81; V. M. [Florinskii], "Otvet g. Raukhfusu, po povodu ego antikritiki," *ibid.,* 1861 (30 December): 385-89; K. Raukhfus, "Otvet g. V. M.," *ibid.,* 1862 (26 May): 203-06.

109 See Anon. [V. M. Florinskii], "Kurs prakticheskogo akusherstva A. Krasovskogo," *MV,* 1865 (21 August): 315-20. Although, the review was published anonymously, its authorship was not a heavily guarded secret.

110 See V. M. Florinskii, "Bibliografiia," *MV,* 1866 (30 July): 363-66. Bredov, in turn, published a highly critical review of the Russian translation of Gustav Braun's obstetrics manual edited by Florinskii, see R. K. Bredov, "Kurs operativnogo akusherstva prof. G. G. Brauna. Per. s nemetskogo pod red. adiunkt-professora V. Florinskogo," *VMZh,* 1866, 44 (January-April): 46-56.

111 It was the last act of unwavering support Dubovitskii extended to Florinskii over the years. Two months later, he died. For the documents related to Florinskii's promotion, see RGVIA, f. 316, op. 36, d. 315, ll. 1-20.

112 See the documents related to the failed promotion in RGVIA, f. 316, op. 69, d. 205. A detailed description of the intrigue surrounding the affair is provided

in Iastrebov, *Vasilii Markovich Florinskii v Peterburgskoi mediko-khirurgicheskoi akademii*, pp. 67-78.

113 Many years later, in his eulogy on the death of his colleague, Florinskii specifically emphasized Botkin's role as the leader of the IMSA Russian party: "In his patriotic aspirations he was, so to say, an extension of P. A. Dubovitskii and I. T. Glebov. ... His talent, inexhaustible energy, and, one could say, good fortune gave him an opportunity to lead the Russian scientific movement and to do much more than anyone else among his colleagues and comrades." See V. M. Florinskii, "Pamiati prof. S. P. Botkina," *Izvestiia Imperatorskogo Tomskogo universiteta*, 1890, 2: 64-70. For a biography of Botkin, see A. A. Budko and A. V. Shabunin, *Velikii Botkin: serdtse, otdannoe liudiam* (SPb.: VMM, 2006). For an English language account of Botkin's work at the academy, see Kichigina, *The Imperial Laboratory*, pp. 97-130, 201-24.

114 In 1863, when the IMSA Council voted on Florinskii's appointment as an adjunct professor, the votes split roughly two to one, with thirteen members voting for and six against, which was quite typical. See RGVIA, f. 316, op. 32, d. 29, ll. 1-2.

115 G. G. Skorichenko, *Imperatorskaia Voenno-meditsinskaia (mediko-khirurgicheskaia) akademiia: istoricheskii ocherk* (SPb.: M. Vol'f, 1910), vol. 2, pp. 81-82. The author had mistaken Roman K. Bredov for his father Karl von Bredow, a German physician who had come to Russia during the Napoleonic wars.

116 At that time, the most common causes of children mortality under the age of five were intestinal (during the summer) and respiratory (during the winter) infections. Since Sergei died in October, it is likely that he had contracted a respiratory infection, perhaps, scarlet fever that was nearly endemic in St. Petersburg.

117 See Bystrov, *Kratkii ocherk*, pp. 26-28, 34-37.

118 In 1875, Florinskii's students published a conspectus of his lectures, see [V. M. Florinskii], *Zapiski po akusherstvu po lektsiiam ekstraordinarnogo professora Mediko-khirurgicheskoi akademii V. M. Florinskogo* (SPb.: Arnol'd, 1875).

119 See PZORV, 1871-1872: 3-9.

120 Florinskii recounted certain details of this appointment in his memoirs. See [V. M. Florinskii], "Zametki i vospominaniia V. M. Florinskogo, 1875-1880," *Russkaia starina*, 1906, 126 (1): 78-95. For the documents related to his appointment, see RGIA, f. 733, op. 121, d. 19, ll. 1-68; f. 734, op. 5, d. 2; and RGVIA, f. 316, op. 60, d. 382, ll. 43-44, 50-52 rev.

121 For a partial overview of the Committee's activities, see a treatise by its chairman from 1873 to 1898, Alexander I. Georgievskii, *K istorii Uchenogo komiteta Ministerstva narodnogo prosveshcheniia* (SPb.: Senat. Tip., 1902). For the ministry's detailed official history prepared for its centennial, see S. V. Rozhdestvenskii, *Istoricheskii obzor deiatel'nosti Ministerstva narodnogo prosveshcheniia: 1802-1902* (SPb.: MNP, 1902).

122 For an overview of Tolstoi's life and career, see V. L. Stepanov, "Dmitrii Andreevich Tolstoi," in A. N. Bokhanov, ed., *Rossiiskie conservatory* (M.: "Russkii mir," 1997), pp. 233-87. For a detailed analysis of Tolstoi's work as a minister of people's enlightenment, see Allen Sinel, *The Classroom and the Chancellery: State Educational Reform in Russia under Count Dmitry Tolstoi* (Cambridge, MA: Harvard University Press, 1973).

123 See A. I. Georgievskii, *Kratkii istoricheskii ocherk pravitel'stvennykh mer i prednachertanii protiv studencheskikh besporiadkov* (SPb.: V. S. Balashev, 1890).

124 The academy was subordinate to the Ministry of People's Enlightenment from 1810 to 1822.

125 The diaries of the war minister Dmitrii A. Miliutin for the years of 1874-1875 contain numerous references to and details of this conflict, see *Dnevnik D. A. Miliutina, 1873-75* (M.: n. p., 1947), vol. 1. Similarly, the diaries of Iakov Chistovich contain numerous entries on the conflict, but mention Florinskii's involvement only obliquely. See Chistovich, *Dnevniki, 1855-80*, vol. 3.

126 For a forceful contemporary expression of this concern, see Vl. Snegirev, "Mediko-khirurgicheskaia akademiia ili meditsinskii fakul'tet," *ASMiOG*, 1870, 1: 1-18. For historical analyses, see, for instance, Peter F. Krug, "The Debate Over the Delivery of Health Care in Rural Russia: The Moscow Zemstvo, 1864-1878," *Bulletin of the History of Medicine*, 1976, 50: 226-41; Samuel Ramer, "The Zemstvo and Public Health," in Terence Emmons and Wayne S. Vucinich, eds., *The Zemstvo in Russia: An Experiment in Local Self-Government* (Cambridge: Cambridge University Press, 1982), pp. 279-314; and Frieden, *Russian Physicians*.

127 Despite the enormous influence the grand duke had exerted on the course of the Great Reforms and the history of Russia writ large, there is still no fully-fledged scholarly biography of the man. The most complete biographical account remains N. P. Pavlov-Sil'vanskii, "Velikii kniaz' Konstantin Nikolaevich. Biograficheskii ocherk," in *idem*, *Sochineniia: Ocherki po russkoi istorii XVIII-XIX vv.* (SPb.: M. M. Stasiulevich, 1910), vol. 2, pp. 304-73. For a detailed analysis of the early part of the duke's life and work, see V. E. Voronin, *Velikii kniaz' Konstantin Nikolaevich. Stanovlenie gosudarstvennogo deiatelia* (M.: Russkii mir, 2002).

128 There is very little documentary evidence related to this event. True to his "Hippocratic oath," Florinskii left not a word on the subject in his personal papers. The grand duke's personal journals for 1874-1878 (which are preserved in GARF, f. 722, op. 1, dd. 103-12) might have relevant information, but, unfortunately, they are written in code and have not yet been deciphered.

129 On imperial marriages, see Greg King and Penny Wilson, *Gilded Prism: The Konstantinovichi Grand Dukes and the Last Years of the Romanov Dynasty* (East Richmond Heights, CA: Eurohistory, 2006).

130 See the grand duke's recently published diaries of 1846 that recorded his feelings about Alexandra, in I. N. Zasypkina and I. S. Chirkov, "Konstantin i Aleksandra. Pervaia liubov' velikogo kniazia Konstantina Nikolaevicha po ego dnevnikam 1846 goda," *Vestnik arkhivista*, 2009, 1: 220-44; 2: 222-40.

131 For a detailed description of the grand duke's marriage and family, see L. V. Zav'ialova and K. V. Orlov, *Velikii kniaz' Konstantin Nikolaevich i velikie kniaz'ia Konstantinovichi: istoriia sem'i* (SPb.: Vita Nova, 2009).

132 The reference is reported in the diaries of a State Council member, A. A. Polovtsov, *Dnevnik gosudarstvennogo sekretaria* (M.: Tsentrpoligraf, 2005), vol. 2, p. 238.

133 See S. A. Sapozhnikov, "Potomstvo velikogo kniazia Konstantina Nikolaevicha (1827-1892) ot Anny Vasil'evny Kuznetsovoi," *Istoricheskaia genealogiia*, 1993, 2: 22-27; and M. M. Medvedkova, "Dopolnenie i ispravleniia k stat'e Sapozhnikova," *ibid.*, 1994, 3: 4.

134 [D. A. Miliutin], *Dnevnik D. A. Miliutina, 1873-75* (M.: n. p., 1947), vol. 1, pp. 198-201.

135 On the university reform and the 1863 Statute, see the recently published memoirs of their main architect, A. Golovin, *Zapiski dlia nemnogikh* (SPb.: Nestor-Istoriia, 2004); for a historical analysis, see R. G. Eimontova, *Russkie universitety na putiakh reformy: shestidesiatye gody XIX veka* (M.: Nauka, 1993).

136 Delianov served from 1861 to 1882 as director of the St. Petersburg Public Library and from 1867 to 1874 as a deputy-minister. For some details of his life and career, see Paul W. Johnson, "Taming Student Radicalism: The Educational Policy of I. D. Delianov," *Russian Review*, 1974, 33(3): 259-68; and Iu. P. Gospodarik, "Ministr iz komandy Aleksandra III: Graf Ivan Davydovich Delianov," in V. M. Filippov, ed., *Ocherki istorii rossiiskogo obrazovaniia* (M.: MGUP, 2002), vol. 2, pp. 105-35.

137 V. M. Florinskii, *Svedeniia o sostoianii i potrebnostiakh russkikh meditsinskikh*

*fakul'tetov, predstavlennye na rassmotrenie v vysochaishe utverzhdennuiu komissiiu dlia peresmotra nyne deistvuiushchego universitetskogo ustava* (SPb.: V. S. Balyshev, 1876).

138 Now remembered mostly as a friend and publisher of Anton Chekhov, Suvorin was not only a very successful publisher, but also left a noticeable imprint on Russian journalism. See Effie Ambler, *The Career of Aleksei S. Suvorin, Russian Journalism and Politics, 1861-1881* (Detroit, MI: Wayne State University Press, 1972); for a detailed biography of Suvorin's, see E. A. Dinershtein, *A. S. Suvorin. Chelovek, sdelavshii kar'eru* (M.: Rosspen, 1998).

139 Florinskii himself described in detail his involvement with this project in his memoirs, see V. Florinskii, "Zametki i vospominaniia," *Russkaia starina*, 1906, 125(1): 75-109; 125(2): 288-311; 125(3): 564-96; 126(1): 109-56; 126(2): 280-323; 126(3): 596-621. For a detailed description of Florinskii's work on the establishment of Tomsk University, see Iastrebov, *Vasilii Markovich Florinskii*, pp. 46-128.

140 See, for instance, a detailed description of its history by the idea's active proponent Nikolai Iadrinskii, which appeared in Blagosvetlov's *Deed* in October 1875, N. Iadrinskii, "Potrebnost' znaniia na Vostoke (po povodu uchrezhdeniia Sibirskogo universiteta)," *Delo*, 1875, 10: 33-69.

141 See V. Florinskii, "Sibirskii universitet," *Novoe vremia*, 6 March 1876, p. 1; and *idem*, "Prigoden li Omsk dlia universiteta," *ibid.*, 2 September 1876, p. 3.

142 See *Trudy komissii, uchrezhdennoi po vysochaishemu poveleniiu dlia izucheniia voprosa ob izbranii goroda dlia Sibirskogo universiteta* (SPb.: Balashov, 1878).

143 Florinskii, "Zametki i vospominaniia," *Russkaia starina*, 1906, 125(3), p. 564.

144 See, for instance, V. M. Florinskii, *Materialy dlia izucheniia chumy* (Kazan: Tip. Universiteta, 1879). In early 1878, a plague epidemic hit the southern regions of the empire and threatened to spread all over the country, which explains a particular attention paid by Russian physicians to the disease in the late 1870s.

145 See V. M. Florinskii, *Kurs akusherstva: Lektsii, chitannye v Imperatorskom Kazanskom universitete* (Kazan: Tip. Universiteta, 1883).

146 See V. M. Florinskii, *Domashniaia meditsina: Lechebnik dlia narodnogo upotrebleniia* (Kazan: Tip. Universiteta, 1880). The manual went through nine editions and stayed in print long after its author's death. Reportedly, even the writer Leo Tolstoy used it to treat some liver problems.

147 See, V. M. Florinskii, "Proekt publichnogo istoriko-etnograficheskogo

museia," *Izvestiia Obshchestva arkheologii, istorii i etnografii pri Kazanskom universitete*, 1878, 1: 125-40.

148 See V. M. Florinskii, *Russkie prostonarodnye travniki i lechebniki* (Kazan: Tip. Universiteta, 1880). He also published (with extensive commentaries) a 500-page collection of manuscripts pertaining to the diplomatic relations between Russia and China during the seventeenth and eighteenth centuries, which had been compiled by Nikolai Bantysh-Kamenskii and which Florinskii acquired after the latter's death. See V. M. Florinskii, ed., *Diplomaticheskoe sobranie del mezhdu Rossiiskim i Kitaiskim gosudarstvami s 1619 po 1792 god* (Kazan: Tip. Universiteta, 1882).

149 For a detailed two-volume history of Tomsk University, see S. A. Nekrylov, *Tomskii universitet — pervyi nauchnyi tsentr v aziatskoi chasti Rossii (seredina 1870-kh gg. – 1919)*, 2 vols. (Tomsk: Izd-vo Tomskogo universiteta, 2010-2011).

150 See [V. M. Florinskii,] *Rech' Professora V. M. Florinskogo, proiznesennaia pri zakladke Sibirskogo universiteta 26 avgusta 1880* (Tomsk: Tip. Mikhailova i Makushina, 1880).

151 For detailed, though contradictory, historical assessments of the new statute, see G. I. Shchetinina, *Universitety Rossii i ustav 1884 g.* (M.: Nauka, 1976); Samuel D. Kassow, *Students, Professors, and the State in Tsarist Russia* (Berkeley, CA: University of California Press, 1989); E. V. Oleseiuk, ed., *Otechestvennye universitety v dinamike zolotogo veka russkoi kul'tury* (SPb.: Izd-vo "Soiuz", 2005); and many others.

152 For an analysis of the changing attitudes to universities see, for instance, E. S. Liakhovich and A. S. Revushkin, *Universitety v istorii i kul'ture dorevoliutsionnoi Rossii* (Tomsk: TGASU, 1998).

153 His official title was *popechitel' Zapadno-Sibirskogo uchebnogo okruga*. On the establishment of the new "educational region" and the responsibilities of its supervisor, see A. V. Blinov, "Organizatsiia i razvitie Zapadno-Sibirskogo uchebnogo okruga," doctoral dissertation, Kemerovo, 2000.

154 See annual reports and registries issued by Florinskii's office, *Pamiatnaia knizhka Zapadnno-Sibirskogo uchebnogo okruga* (Tomsk, 1887-1900).

155 For descriptions of Florinskii's work at his new post, in addition to the literature cited above, see A. A. Sechenova, "Vasilii Markovich Florinskii — pervyi popechitel' Zapadno-Sibirskogo uchebnogo okruga," *Vestnik Tomskogo gosudarstvennogo pedagogicheskogo instituta*, 2009, 12: 139-41; and A. A. Bubnov, "Vasilii Markovich Florinskii — organizator i sozdatel' nauki v Sibiri," *Vestnik Kurganskogo gosudarstvennogo universiteta*, 2014, 1(6): 3-15.

156 Which he certainly did. Unlike in all other Russian universities, there were no students' strikes, unrest, or "revolutionary circles" in Tomsk during Florinskii's tenure, which earned him the label of a "reactionary" in the Soviet-era histories of the institution. For a recent, much more balanced analysis of Florinskii's "anti-revolutionary" activities and measures, see E. A. Degal'tseva, "V. M. Florinskii v Sibiri: satrap ili patriot," in V. A. Skubnevskii and Iu. M. Goncharov, eds., *Aktual'nye voprosy istorii Sibiri* (Barnaul: AGU, 2007), vol. 1, pp. 221-27.

157 See V. M. Florinskii, *Rech' popechitelia Zapadno-Sibirskogo uchebnogo okruga, proiznesennaia pri otkrytii Imperatorskogo Tomskogo universiteta 22 iiulia 1888 g.* (Tomsk: n. p., 1888); and a special volume published for the opening of the university, which contained a detailed description of the new school, including the biographies of its first professors, *Pervyi universitet v Sibiri* (Tomsk: Sibirskii vestnik, 1889), http://access.bl.uk/item/pdf/lsidyv36c696d1.

158 Its law school would be opened only a decade later, in 1898. See M. F. Popov, *Kratkii istoricheskii ocherk Imperatorskogo Tomskogo universiteta za pervye 25 let ego suchchestvovavniia (1888-1913)* (Tomsk: n. p., 1913); and V. F. Volovich, ed., *Iuridicheskoe obrazovanie v Tomskom universitete: Ocherk istorii (1898-1998)* (Tomsk: Izd-vo Tomskogo universiteta, 1998).

159 [M. V. Florinskii], *Opisanie prazdnovaniia zakladki Tomskogo Tekhnologicheskogo instituta 6 iiulia 1896 goda* (Tomsk: P. I. Makushin, 1896).

160 See "Trudy Tomskogo obshchestva estestvoispytatelei," *Izvestiia Imperatorskogo Tomskogo Universiteta* (hereafter *ITU*), 1889, 1: 1-32; and V. P. Puzyrev, "V. M. Florinskii i Tomskoe obshchestvo estestvoispytatelei," *Sibirskii meditsinskii zhurnal (Tomsk)*, 1999, 14(1-2): 87-91.

161 See V. M. Florinskii, "Zametka ob influentse," *ITU*, 1890, 2: 98-106; idem, "Materialy dlia izucheniia chumy," ibid., 1897, 12: 1-25.

162 See Florinskii, "Proekt publichnogo istoriko-etnograficheskogo museia."

163 See V. M. [Florinskii], *Arkheologicheskii muzei Tomskogo universiteta [Katalog]* (Tomsk: 1888). In the subsequent years, Florinskii regularly updated the catalogue. See, for instance, V. M. Florinskii, *Vtoroe pribavlenie k katalogu arkheologicheskogo muzeia Tomskogo universiteta* (Tomsk: Makushin, 1898).

164 See, for instance, V. M. Florinskii, "Topograficheskie svedeniia o kurganakh Semipalatinskoi i Semirechenskoi oblasti," *ITU*, 1889, 1: 15-31; idem, "Nekotorye svedeniia o kurganakh iugo-zapadnoi chasti Semirechenskoi oblasti," ibid., 32-49; idem, "Kurgany Tomskoi oblasti," ibid., 58-86; idem, *Topograficheskie svedeniia o kurganakh Zapadnoi Sibiri* (Tomsk: Mikhailov i

Makushin, 1889); *idem*, "Dvadtsat' tri chelovecheskikh cherepa Tomskogo arkheologicheskogo muzeia," *ITU*, 1890, 2: 16-46.

165 V. M. Florinskii, *Pervobytnye slaviane po pamiatnikam ikh doistoricheskoi zhizni: Opyt slavianskoi arkheologii* (Tomsk: Makushin, 1894-1898).

166 For an assessment of Florinskii's archeological works, see A. V. Zhuk, "Vasilii Markovich Florinskii kak arkheolog," *AB ORIGINE: Problemy genezisa kul'tur Sibiri* (Tiumen': TiumGU, 2013), 5: 5-22; and *idem*, "Vasilii Markovich Florinskii, ego mesto i znachenie v otechestvennoi arkheologii," *Vestnik Omskogo universiteta. Seriia "Istoricheskie nauki"*, 2015, 1(5): 100-15.

167 On the history of this railway, see Christian Wolmar, *To the Edge of the World: The Story of the Trans-Siberian Railway* (London: Atlantic, 2013).

168 For a brief overview of the literary type, see Ellen Chances, "The Superfluous Man in Russian Literature," in Neil Cornwell, ed., *The Routledge Companion to Russian Literature* (London: Routledge, 2001), pp. 111-22.

169 D. Pisarev, "Shkola i zhizn'," *RS*, 1865, 7(I), p. 132.

170 See Anon., "Zhurnalistika," *Golos*, 29 October 1865, pp. 1-3; 30 October 1865, p. 1.

171 See Anon., "Zhurnalistika," *Golos*, 14 December 1865, pp. 1-2.

172 Anon., "Zhurnalistika," *Golos*, 17 February 1866, p. 1.

173 See RGIA, f. 777, op. 2, d. 7, ll. 12-13 rev. All the following quotations are from this source.

174 See *Golos*, 31 August 1866, p. 4. The same advertisement was repeated a few days later, see *ibid.*, 4 September 1866, p. 4.

175 See, for instance, *Syn Otechestva*, 14 September 1866, p. 1736; and *Sovremennaia meditsina*, 1866, 36 (November 18): 583.

176 See *Delo*, 1866, 1: back matter; *ibid.*, 1867, 2: back matter.

177 See *Knizhnyi vestnik*, 1866, 18-19 (October): 362.

178 See "Kritika i bibliografiia," *Zhenskii vestnik*, 1867, 3: 55-56. For the original of the reviewed book, see S. I. Baranovskii, *Gigiena, rukovodstvo k sokhraneniiu zdorov'ia* (SPb.: V. Prokhorov, 1860).

179 See "Kritika i bibliografiia," *Zhenskii vestnik*, 1867, 4: 13-20. For the original of the reviewed book, see A. T. Ronchevskii, *Populiarnye lektsii o kholere, chitannye v Tiflise v 1866 g.* (Tiflis: Tip. Gl. Upr. Namestnika Kavkaza, 1866).

180 In addition to the journals and newspapers mentioned above, I have examined the 1866-67 issues of *Literaturnaia biblioteka, Nedelia, Otechestvennye zapiski, Russkii vestnik, Sankt Peterburgskie vedomosti, Sovremennaia letopis', Sovremennoe obozrenie, Syn Otechestva, Vestnik Evropy, Vest'*, and *Zapiski dlia chteniia*.

181 For a general overview of the history of marriage regulations in Russia, see N. S. Nizhnik, *Pravovoe regulirovanie semeino-brachnykh otnoshenii v russkoi istorii* (SPb.: Iuridicheskii tsentr Press, 2006). For some aspects of marriage in late nineteenth-century Russia, see Laura Engelstein, *The Keys to Happiness: Sex and the Search for Modernity in Fin-de-Siècle Russia* (Ithaca, NY: Cornell University Press, 1992).

182 For a concise analysis of the situation, see Freeze, "Bringing Order to the Russian Family."

183 For a general overview of the changes in the Russian Church in the era of the Great Reforms, see a voluminous study by S. V. Rimskii, *Rossiiskaia tserkov' v epokhu velikikh reform: Tserkovnye reformy v Rossii 1860-1870-kh godov* (M.: Krutits, 1999).

184 This sentiment appears in the diary entry of 10 April 1858, written by the daughter of an eminent St. Petersburg architect and hostess of a popular literary salon, Elena A. Shtakenshneider, *Dnevnik i zapiski* (M.-L.: Academia, 1934), p. 199.

185 See, for instance, a detailed critique of the existing civic laws on marriage and family, published on the eve of the emancipation of the serfs by M. A. Filippov, "Vzgliad na russkie grazhdanskie zakony," *Sovremennik*, 1861, 2: 523-62; 3: 217-66; which Florinskii used extensively in his own essays. See also N. Leskov, "Svodnye braki v Rossii," *Otechestvennye zapiski*, 1861, 3: 37-47; "Dva mneniia po voprosu o brakakh," *Biblioteka dlia chteniia*, 1863, 11(II): 64-68; "O brake pravoslavnykh s nepravoslavnymi," *Pravoslavnyi sobesednik*, 1863, 1: 57-86; and many others.

186 Freeze, "Bringing Order to the Russian Family," p. 711.

187 That it was not just the "nihilist" or "revolutionary" circles with their open endorsement of "fictitious marriages," but a much broader segment of Russian society that was calling for changes in the existing marital order is demonstrated by the close involvement of Petr Valuev, the minister of internal affairs, in reforming the Orthodox Church's regulations on marriage. For details, see Paul W. Werth, "Empire, Religious Freedom, and the Legal Regulation of 'Mixed' Marriages in Russia," *Journal of Modern History*, 2008, 80: 296-331. For a discussion of "fictitious marriages" largely within the framework of women's emancipation in Russia, see Richard

Stites, *The Women's Liberation Movement in Russia: Feminism, Nihilsm, and Bolshevism, 1860-1930* (Princeton, NJ: Princeton University Press, 1978); and Barbara A. Engel, *Mothers and Daughters: Women of the Intelligentsia in Nineteenth-Century Russia* (Cambridge: Cambridge University Press, 1983).

188 See N. I. Cherniavskii, *Grazhdanskii brak: Komediia v 5 deistviiakh* (SPb.: A. P. Cherviakov, 1867). On the author and the play, see Abram Reitblat, "P"esa 'Grazhdanskii brak' i ee avtor glazami agenta III otdeleniia," in idem, *Pisat' poperek: Stat'i po biografike, sotsiologii i istorii literatury* (M.: NLO, 2014), pp. 339-55.

189 For critical reviews of the play, see, for instance, M. Stebnitskii, "Russkii dramaticheskii teatr v Peterburge," *Otechestvennye zapiski*, 1866, 169, 11-12: 258-87; and M. R. [M. P. Rozengeim], "Teatral'naia letopis'," *Syn Otechestva*, 1866, 48: 1-3. For Cherniavskii's "defense," see N. I. Cherniavskii, "Ot avtora," in idem, *Grazhdanskii brak: Komediia v 5 deistviiakh* (SPb.: E. I. Ekshurskii, 1868), pp. i-xii.

190 Blagosvetlov to Iakobii, GARF, f. 109, op. 1, d. 2045, l. 1.

191 I have examined the 1866-67 issues of *Arkhiv sudebnoi meditsiny i obshchestvennoi gigieny*, *Drug zdraviia*, *Meditsinskii vestnik*, *Moskovskaia meditsinskaia gazeta*, *Sovremennaia meditsina*, and *Voenno-meditsinskii zhurnal*. Only *Modern Medicine*, a "newspaper for physicians" published in Kiev, carried an announcement of the publication of Florinskii's book in its regular list of new books, but not a review. See *Sovremennaia meditsina*, 1866, 6(36): 583.

192 See *Naturalist*, SPb., 1864-1867, vols. 1-4.

193 During 1868, the book was advertised in the third, seventh, ninth, tenth, and twelfth issues; in 1869, in the second, forth, and seventh issues. It is possible that I have missed some of the advertisements — during the rebinding of the journal for library storage, some of the ads could have been either lost or deliberately excised. This is certainly true of the available electronic copies of the journal, many of which do not include back and front matters at all.

194 See an advertisement in *Deed*, 1869, 7: back matter. The price was dropped from 75 to fifty kopeks in store and from one ruble to 75 kopeks with mail order.

195 I found two "versions" of the book in various Russian libraries. One is smaller in size (176 mm x 115 mm) and carries the name of the Russian Word printing shop (Riumin and Co.) on its titlepage, indicating that it is this version that had been originally issued in August 1866. Another is larger (186 mm x 124 mm) and its titlepage has no name of a printing shop

at all. The two versions are identical in all other features. The frequency of advertisements for the book printed in *Deed* indicates that the second (larger) version of the book probably came out in the late summer of 1871. In the entire 1870 run, the journal carried an advertisement of Florinskii's book in only one (the eleventh) issue. The advertisement reappeared in September 1871 and was then run regularly in almost every issue.

196 See Ludwik Fleck, *Genesis and Development of a Scientific Fact* (Chicago, IL: University of Chicago Press, 1979).

197 A concise introduction to the comparative history of anthropology could be found in Henrika Kuklick, ed., *A New History of Anthropology* (Malden, MA: Blackwell, 2008).

198 See *Izvestiia Antropologicheskogo otdeleniia Obshchestva liubitelei estestvoznaniia, sostoiashchego pri Moskovskom universitete*, M., 1865, vol. 1; and *Izvestiia Obshchestva liubitelei estestvoznaniia, antropologii i etnografii, sostoiashchego pri Moskovskom universitete*, vol. 4(1). *Antropologicheskie materialy*, Part 1. M., 1867. For a contemporary account of the establishment and development of the Anthropological Section, see V. N. Benzengr, *Istoricheskii ocherk deiatel'nosti Antropologicheskogo otdela* (M.: M. N. Lavrov, 1878); on the general history of Russian physical anthropology in English, see Marina Mogilner, *Homo Imperii: A History of Physical Anthropology in Russia* (Lincoln, NE: University of Nebraska Press, 2013); on the general history of Russian ethnography in English, see Nathaniel Knight, "Constructing the Science of Nationality: Ethnography in Mid-Nineteenth Century Russia," doctoral dissertation, Columbia University, 1995.

199 On the exhibition, see Nathaniel Knight, *The Empire on Display: Ethnographic Exhibition and the Conceptualization of Human Diversity in Post-Emancipation Russia* (Washington, DC: NCEEER, 2001), https://www.ucis.pitt.edu/nceeer/2001-814-11g-Knight.pdf; for a Russian version, see *idem*, "Imperiia napokaz: vserossiiskaia etnograficheskaia vystavka 1867 g.," *Novoe literaturnoe obozrenie*, 2001, 51: 111-31. On a more general issue of the relationship between ethnographic exhibits and the formation of the discipline of anthropology, see Sadia Qureshi, *Peoples on Parade: Exhibitions, Empire, and Anthropology in Nineteenth Century Britain* (Chicago, IL: University of Chicago Press, 2011).

200 See *Zapiski Russkogo geograficheskogo obshchestva po otdeleniiu etnografii* (SPb., 1867). The regularly published *Memoirs* replaced an occasional publication, titled *Ethnographic Collection* (*Etnograficheskii sbornik*), which from 1853 to 1864 had been issued only six times.

201 See A. P. Bogdanov, "Antropologiia i etnografiia," *Naturalist*, 1866, 20-21: 309-14. On the role of Bogdanov in the development of Russian anthropology

and ethnography, in addition to the literature cited above, see M. G. Levin, *Ocherki istorii antropologii v Rossii* (M.: Izd-vo AN SSSR, 1960); N. G. Zalkind, *Moskovskaia shkola antropologov v razvitii otechestvennoi nauki o cheloveke* (M.: Nauka, 1974); and Galina Krivosheina, "Long Way to the Anthropological Exhibition: The Institutionalization of Physical Anthropology in Russia," *Centaurus*, 2014, 56(4): 275-304.

202 See A. I. Moiseev, *Meditsinskii sovet Ministerstva vnutrennikh del. Kratkii istoricheskii ocherk* (SPb.: Tip. MVD, 1913).

203 For a detailed overview of the development of legal and forensic medicine during this period, see S. V. Shershavkin, *Istoriia otechestvennoi sudebno-meditsinskoi sluzhby* (M.: Meditsina, 1968); and Elisa M. Becker, *Medicine, Law, and the State in Imperial Russia* (Budapest: CEU Press, 2011).

204 The available literature on the role of zemstvos in providing medical services and articulating public health needs is massive. For general overviews of the development of "social medicine" in post-Crimean Russia, see P. E. Zabludovskii, *Meditsina v Rossii v period kapitalizma: Razvitie gigieny. Voprosy obshchestvennoi meditsiny* (M.: Medgiz, 1956); M. M. Levit, *Stanovlenie obshchestvennoi meditsiny v Rossii* (M.: Meditsina, 1974); Frieden, *Russian Physicians*; and V. Iu. Kuz'min, *Vlast', obshchestvo i zemskaia meditsina (1864-1917 gg.)* (Samara: Samarskii un-t, 2003).

205 See, "Konkurs na zameshchenie novootkrytoi kafedry obshchestvennoi gigieny v MKhA," *MV*, 1866, 25: 295-96. The difficulties in the organization of "social hygiene" were clearly reflected in the inability of the IMSA Council to find a suitable candidate for the post of the department chair in nearly four years. The position was filled only in 1870. For detailed histories of the two departments, see A. K. Evropin, *Istoricheskii ocherk kafedry sodebnoi meditsiny v Voenno-meditsinskoi akademii, 1798-1898* (SPb.: V. P. Meshcherskii, 1898); and Z. G. Surovtsov, *Materialy dlia istorii kafedry gigieny v Voenno-meditsinskoi akademii* (SPb.: Vladimirskaia tipo-litografiia, 1898).

206 Characteristically, in 1870, the *Archive of Legal Medicine and Social Hygiene* explicitly excluded epidemiology from its pages and began to issue a special supplement, *Epidemiological Leaf* (*Epidemiologicheskii listok*), that became the country's first specialized epidemiological journal. On the agendas of Russian social hygiene and social medicine, see A. P. Zhuk, *Razvitie obshchestvenno-meditsinskoi mysli v Rossii v 60-70-e gg. XIX veka* (M.: Meditsina, 1963).

207 For the society's journal, see *Trudy Russkogo obshchestva okhraneniia narodnogo zdraviia* (SPb., 1884-1890); for a detailed history of the society, see E. I. Lotova, *Russkaia intelligentsiia i voprosy obshchestvennoi gigieny* (M.: GIZM, 1962).

208 On the exhibit, see A. P. Bogdanov, ed., *Antropologicheskaia vystavka Obshchestva liubitelei estestvoznaniia, antropologii i etnografii*, 4 vols. (M.: Komitet vystavki, 1878-1886). For a clear articulation of the then current agendas of Russian anthropology, see A. P. Bogdanov, *Antropologiia i universitet* (M.: Imp. Mosk. Un-t, 1876); for the concurrent understanding of the relationship between anthropology and medicine, see V. G. Emme, *Antropologiia i meditsina* (Poltava: N. Pigurenko, 1882). Needless to say, none of these publications mentioned Florinskii or *Human Perfection and Degeneration*.

209 For a detailed exposition of the then current agendas of Russian social hygiene, see a textbook by I. I. Arkhangel'skii, *Lektsii obshchestvennoi gigieny i dietetiki* (Minsk: Gub. Tipografiia, 1872), or a treatise "on the fundamental issues of public hygiene" by Ir. P. Skvortsov, *Osnovnye voprosy narodnoi gigieny* (SPb.: M. M. Stasiulevich, 1877). Neither publication mentioned Florinskii, or *Human Perfection and Degeneration*.

210 For a brief account of the 1866 epidemic of cholera in Russia, see the first chapter of Charlotte E. Henze, *Disease, Health Care and Government in Late Imperial Russia: Life and Death on the Volga, 1823-1914* (New York: Routledge, 2011).

211 See, for instance, A. Z-n [A. Zabelin], "O prichinakh izmel'chaniia i boleznennosti naroda," *Zhurnal zemlevladel'tsev*, 1858, 5(VI): 19-29; and D*** [V. V. Deriker], *O vyrozhdenii chelovecheskogo roda i sredstvakh prepiatstvovat' etomu vyrozhdeniiu fizicheskim i nravstvennym usovershenstvovaniem cheloveka* (M.: L. Stepanova, 1860). The latter book was written by Vasilii V. Deriker, a one-time IMSA student, who would become one of the founders of Russian homeopathy. Much like Florinskii's, Deriker's treatise went completely unnoticed by the contemporary medical and scientific communities. Florinskii himself was apparently unaware of its existence. For a brief account of Deriker's life and work, which, tellingly, does not mention this book, see K. K. Boianus, *Gomeopatiia v Rossii: Istoricheskii ocherk* (M.: V. V. Davydov, 1882).

212 The statement appears in Chernyshevskii's discussion of Thomas Malthus's theory of overpopulation in the commentaries to his 1860 translation of J. S. Mill's *Principles of Political Economy*. See, N. G. Chernyshevskii, "Zamechaniia na poslednie chetyre glavy pervoi knigi Millia," in *idem, Polnoe sobranie sochinenii v 15 t.* (M.: GIKhL, 1949), vol. 9, pp. 251-336 (p. 307).

213 Unfortunately, there are no detailed historical studies of nineteenth-century Russian demography. For some general observations, largely on the late nineteenth-century developments, see the first three chapters of Juliette Cadiot, *Le laboratoire imperial. Russie-URSS, 1890-1940* (Paris: CNRS Editions,

2007); the Russian translation, Zhul'et Kadio, *Laboratoriia imperii: Rossiia/ SSSR, 1890-1940* (M.: Novoe literaturnoe obozrenie, 2010); and Martine Mespoulet, *Statistique et révolution en Russie. Un compromis impossible (1880-1930)* (Rennes: Presses Universitaires de Rennes, 2001); idem, "Statisticiens des zemstva: formation d'une nouvelle profession intellectuelle en Russie dans la période prérévolutionnaire (1880-1917): Le cas de Saratov," *Cahiers du monde russe*, 1999, 40(4): 573-624.

214 On the role of demographic and vital statistics in the rise of concerns with degeneration in various countries, see, for instance, Paul Weindling, *Health, Race and German Politics between National Unification and Nazism, 1870-1945* (Cambridge: Cambridge University Press, 1989); Richard A. Soloway, "Counting the Degenerates: The Statistics of Race Deterioration in Edwardian England," *Journal of Contemporary History*, 1982, 17(1): 137-64; idem, *Demography and Degeneration: Eugenics and the Declining Birth Rate in Twentieth-Century Britain* (Chapel Hill, NC: The University of North Carolina, 1990); Kelly Hurley, "Hereditary Taint and Cultural Contagion: The Social Etiology of Fin-de-Siècle Degeneration Theory," *Nineteenth-Century Contexts*, 1990, 14(2): 193-214; Edmund Ramsden, "Social Demography and Eugenics in the Interwar United States," *Population and Development Review*, 2003, 29: 547-99; and Cassata, *Building the New Man*.

215 For some observations on Russian "medico-topographical surveys," see B. S. Sigal, "Pervye mediko-topograficheskie opisaniia v Rossii," *Voprosy gigieny*, 1949, 1: 175-208; E. A. Vishlenkova, "Mediko-biologicheskie ob"iasneniia sotsial'nykh problem Rossii (vtoraia tret' XIX veka)," *Istoriia i istoricheskaia pamiat'*, 2011, 4: 37-65. For an overview of the history of medical geography in Russia, see A. P. Markovin, *Razvitie meditsinskoi geografii v Rossii* (SPb.: Nauka, 1993).

216 For a comparison to the contemporary European studies, see, for instance, an analysis of the sex distributions in the Russian population by the "founding father" of Russian statistics Konstantin I. Arsen'ev, "Issledovaniia o chislennom otnoshenii polov v narodonaselenii Rossii," *Zhurnal MVD*, 1844, 1: 5-47.

217 See V. Ia. Buniakovskii, *Opyt o zakonakh smertnosti v Rossii i o raspredelenii pravoslavnogo naseleniia po vozrastam* (SPb.: Imp. Ak. Nauk, 1865); idem, *Issledovaniia o vozrastnom sostave zhenskogo pravoslavnogo naseleniia Rossii* (SPb.: Imp. Ak. Nauk, 1866); idem, *Upotreblenie tablits smertnosti i narodonaseleniia* (SPb.: Imp. Ak. Nauk, 1867); idem, *Antoropobiologicheskie issledovaniia i ikh prilozhenie k muzhskomy naseleniiu Rossii* (SPb.: Imp. Ak. Nauk, 1874); and many others. For a brief biography of Buniakovskii, see V. E. Prudnikov, *V. Ia. Buniakovskii — uchenyi i pedagog* (M.: Uchpedgiz, 1954); more specifically on his work on the probability theory, see O. B. Sheinin, *O rabotakh V.*

Ia. Buniakovskogo po teorii veroiatnostei (M.: n. p., 1988). Buniakovskii's demographic works, however, remain almost completely forgotten.

218 See Ia. A. Chistovich, "O sobiranii materialov dlia meditsinskoi geografii i statistiki Rossii," *PZORV*, 1865-1866: 31-47.

219 Florinskii himself published his first work on differential mortality only in 1873, see V. Florinskii, "O smertnosti rodil'nits v g. S.-Peterburge i po uezdam S.-Peterburgskoi gubernii," *PZORV*, 1872-1873: 229-45.

220 For a contemporary argument for the importance of medical statistics in Russia, see P. A. Peskov, *Meditsinskaia statistika i geografiia kak otdel'nye otrasli obshchestvennykh nauk i metody statisticheskogo issledovaniia v meditsine* (Kazan: Univ. tip., 1874). For historical overviews of Russian vital statistics, see A. M. Merkov, ed., *Ocherki istorii otechestvennoi sanitarnoi statistiki* (M.: Meditsina, 1966); and P. E. Zabludovskii, *Razvitie meditsinskoi statistiki. Istoricheskii obzor: Meditsinskaia statistika v Rossii v XVIII-XIX vekakh* (M.: n. p., 1974). For an analysis of the role of zemstvos in advancing social hygiene agendas, see L. N. Karpov, *Zemskaia sanitarnaia organizatsiia v Rossii* (L.: Meditsina, 1964).

221 The increased availability of such statistical data in the later decades of the nineteenth and the first decades of the twentieth century certainly was an important factor in the emerging popularity of degeneration concepts in Russia. Alas, the existing literature on the subject completely ignores this factor. See Daniel Beer, *Renovating Russia: The Human Sciences and the Fate of Liberal Modernity, 1880-1930* (Ithaca, NY: Cornell University Press, 2008).

222 Florinskii, *Mysli i vospominaniia*, NMRT, #205, l. 6.

223 See, for instance, "Florinskii, Vasilii Markovich," in L. F. Zmeev, *Russkie vrachi-pisateli* (SPb.: V. Demakov, 1886), vol. 2, pp. 139-40; V. S. Gruzdev, *Istoricheskii ocherk kafedry akusherstva i zhenskikh boleznei Imperatorskoi voenno-meditsinskoi akademii i soedinennoi s neiu akademicheskoi akushersko-ginekologicheskoi kliniki* (SPb.: Imperatorskaia akademiia nauk, 1898), pp. 214-41; and idem, "Florinskii, Vasilii Markovich," in N. P. Zagoskin, *Biograficheskii slovar' professorov i prepodavatelei Imperatorskogo Kazanskogo universiteta* (Kazan': Tip. Imperatorskogo universiteta, 1904), vol. 2, pp. 353-62.

224 An outline of this project is preserved in NMRT, #260.

225 M. N. Pargamin, *O vyrozhdenii* (Voronezh: V. V. Iurkevich, 1891). Originally, the essay had appeared in *Medical Conversation*, a popular medical journal published in Voronezh, see idem, "O vyrozhdenii," *Meditsinskaia beseda*, 1891, 10 (May): 253-63.

226 M. N. Pargamin, *Nasledstvennost' i gigiena braka* (Voronezh: V. V. Iurkevich,

1896). Originally, this more than 100-page survey had appeared in installments in *Medical Conversation* in the previous year, see *idem*, "Nasledstvennost' i gigiena braka," *Meditsinskaia beseda*, 1895, 6: 171-80; 7-8: 200-13; 9:251-58; 11: 313-22; 13-14: 392-402; 15: 432-38; 16: 457-64; 17: 498-500; 19: 553-63; 20: 618-22; 23: 713-31.

227 P. Jacoby to V. Florinskii, [no date], NMRT, #813, ll. 1-2. All the following quotations are from this source. The letter is undated, but its contents indicate that it was written not earlier than 1883 and not later than 1889.

228 In fact it was the Royal Madrid Academy of Medicine that ran the competition and granted the prize. See Paul Jacoby [P. Iakobii], "Préface," in *idem, Etudes sur la sélection dans les rapports avec l'hérédité chez l'homme* (Paris: Bailliere, 1881), p. V.

229 See, for instance, a review of Iakobii's book in *Mysl'*, 1881, 10-11: 228-29.

230 Here Iakobii apparently refers to Galton's 1883 *Inquiries into Human Faculty and Its Development*, not to his 1869 *Hereditary Genius*, or the 1865 article.

231 Here Iakobii refers to the Swiss botanist Alphonse de Candolle, who in his 1873 book *Histoire des Sciences et des Savants depuis deux Siècles*, criticized Galton's *Hereditary Genius*. For details on the polemics between the two scientists and its influence on Galton's views, see Raymond E. Fancher, "Alphonse de Candolle, Francis Galton, and the Early History of the Nature-Nurture Controversy," *Journal of the History of the Behavioral Sciences*, 1983, 19(4): 341-52.

232 See Paul Jacoby, *Études sur la sélection chez l›homme* (Paris: F. Alcan, 1904). Tarde's preface along with Iakobii's was also published in *Archives de l'Anthropologie Criminelle*, 1904, 19: 937-42.

233 See [V. M. Florinskii], *Stat'i i rechi Vasiliia Markovicha Florinskogo* (Kazan: Imperatorskii universitet, 1903).

# Chapter 5

1 For details, see Michael Bulmer, "The Development of Francis Galton's Ideas on the Mechanism of Heredity," *JHB*, 1999, 32: 263-92; and *idem, Francis Galton: Pioneer of Heredity and Biometry* (Baltimore, MD: Johns Hopkins University Press, 2003).

2 Francis Galton, "The Possible Improvement of the Human Breed under the Existing Conditions of Law and Sentiment," *Nature*, 1901, 64 (31 October): 659-65 (p. 659).

3 Karl Pearson, *The Life, Letters and Labours of Francis Galton* (London: Oxford University Press, 1930), vol. 3, part 1, p. 412. Hereafter references to this publication will be given as Pearson, 3(1): 412.

4 For a brief history of the journal during its first years, see John Aldrich, "Karl Pearson's *Biometrika*: 1901–36," *Biometrika*, 2013, 100(1): 3-15.

5 In its early years, *Biometrika* carried a variety of items on eugenics, ranging from short reviews to lengthy research reports, see W. P. E., "Probability, the Foundation of Eugenics, by Francis Galton," *Biometrika*, 1907, 5(4): 477; idem, "The Scope and Importance to the State of the Science of National Eugenics: The Fourteenth Boyle Lecture by Karl Pearson," *ibid.*, 1908, 6(1): 124; Edgar Schuster, "Hereditary Deafness: A Discussion of the Data Collected by Dr. E. A. Fay in America," *ibid.*, 1906, 4(4): 465-82; Edgar Schuster and E. M. Elderton, "The Inheritance of Psychical Characters," *ibid.*, 1907, 5(4): 460-69; K. P[earson], "Note on Inheritance in Man," *ibid.*, 1908, 6(2/3): 327-28; Ethel M. Elderton, "On the Association of Drawing with Other Capacities in School Children," *ibid.*, 1909, 7(1/2): 222-26; and David Heron, "Note on Reproductive Selection," *ibid.*, 1914, 10(2/3): 419-20.

6 Galton, "The Possible Improvement of the Human Breed"; the same article also appeared in *Annual Report of the Board of Regents of the Smithsonian Institution*, 1901: 523-38; *Man*, 1901, 1(132): 161-64; and *Popular Science Monthly*, 1901-02, 60 (January 1902): 218-33.

7 See, for instance, F. Galton, "Our National Physique — Prospects of the British Race — Are We Degenerating?" *Daily Chronicle*, 29 July 1903.

8 For further details on the society and its role in Galton's campaign, see Chris Renwich, *British Sociology's Lost Biological Roots: A History of Futures Past* (Basingstoke: Palgrave Macmillan, 2012).

9 F. Galton, "Eugenics: Its Definition, Scope and Aims," *Sociological Papers*, 1905, 1: 45-51 (p. 45). The same article in an abridged form appeared in *Nature*, 1904, 70 (26 May): 82; while its full text, along with its discussion, was also published by *The American Journal of Sociology*, 1904, 10(1): 1-25.

10 See Galton, "Eugenics: Its Definition, Scope and Aims." The paper was supplemented by the texts of Pearson's opening remarks, the subsequent discussion, and written comments, as well as excerpts from the press coverage, see *Sociological Papers*, 1905, 1: 52-84.

11 Pearson, 3(1): 261.

12 Francis Galton, *Memories of My Life* (London: Methuen, 1908), p. 320.

13 Pearson, 3(1): 222.

14 This definition was publicly introduced for the first time in Galton's Herbert Spencer Lecture delivered at Oxford University on 5 June 1907, see F. Galton, "Probability, The Foundation of Eugenics," in *idem, Essays on Eugenics*, pp. 73-99.

15 See Pearson, 3(1): 222 and passim.

16 See *Sociological Papers*, 1905, 2: 1-53. ("Restrictions in Marriage," pp. 3–13; "Studies in National Eugenics," pp. 14-17; "Reply to the Speakers," pp. 49-51; "Eugenics as a Factor in Religion," pp. 52-53).

17 For a detailed history of the society, see Pauline M. H. Mazumdar, *Eugenics, Human Genetics and Human Failings: The Eugenics Society, its Sources and Critics in Britain* (London: Routledge, 1992).

18 Francis Galton, *Essays on Eugenics* (London: The Eugenics Education Society, 1909).

19 A clear allusion to Samuel Butler's *Erewhon*, an 1872 novel about the future evolution of humanity, whose sequel, *Erewhon Revisited Twenty Years Later*, appeared in 1901.

20 Sybil Gotto, "Preface," in *Problems in Eugenics* (London: The Eugenics Education Society, 1912), vol. 1, p. i.

21 The literature on the early history of eugenics in these countries is vast. Useful introductions could be found in Bashford and Levine, eds., *Oxford Handbook*.

22 For detailed analyses of the formation of the "transnational eugenics movement" in the early decades of the twentieth century, see Deborah Barrett and Charles Kurzman, "Globalizing Social Movement Theory: The Case of Eugenics," *Theory and Society*, 2004, 33: 487-527; Alison Bashford, "Internationalism, Cosmopolitanism and Eugenics," in Bashford and Levine, eds., *Oxford Handbook*, pp. 254-86; Stefan Kühl, *For the Betterment of the Race: The Rise and Fall of the International Movement for Eugenics and Racial Hygiene* (New York: Palgrave Macmillan, 2013); and Nikolai Krementsov, "The Strength of a Loosely Defined Movement: Eugenics and Medicine in Imperial Russia," *Medical History*, 2015, 59(1): 6-31.

23 See "1-i Mezhdunarodnyi s"ezd po evgenike (rasovoi gigiene)," *Vrachebnaia gazeta*, 1911, 40: 1260.

24 Gotto, "Preface," p. i.

25 See A. S. Sholomovich, "Novoe techenie v uchenii o nasledstvennoti (po povodu Gissenskogo kongressa o nasledstvennosti)," *Sovremennaia psikhiatriia* (hereafter *SP*), 1912, 6: 392-401; and *idem*, "Pervyi kongress po genealogii,"

*Nevrologicheskii vestnik*, 1912, 19(3): 582-602. On Somner's involvement with eugenics, see Volker Roelcke, "'Prävention' in Hygiene und Psychiatrie zu Beginn des 20 Jahrhunderts. Krankheit, Gesellschaft, Vererbung und Eugenik bei Robert Sommer und Emil Gotschlich," in Ulrike Enke and Volker Roelcke, eds., *Die Medizinische Fakultät der Universität Giesen. Institutionen, Akteure und Ereignisse von der Gründung 1607 bis ins 20. Jahrhundert* (Stuttgart: Steiner, 2007), pp. 395-416.

26 For the history of Russian eugenics during this period, see Adams, "Eugenics in Russia"; Krementsov, "From 'Beastly Philosophy' to Medical Genetics"; Felder, "Rasovaia gigiena v Rossii"; and Krementsov, "The Strength of a Loosely Defined Movement."

27 See, for example, Iogannes Rutgers [Johannes Rutgers], *Uluchshenie chelovecheskoi porody* (SPb.: A. S. Suvorin, 1909); for the Dutch original, see J. Rutgers, *Rasverbetering en bewuste aantalsbeperking: kritiek van het Malthusianisme en van het Nieuw-Malthusianisme* (Rotterdam: W. J. van Hengel, 1905); however, the Russian translation was probably made from a German-language edition, see J. Rutgers, *Rassenverbesserung: Malthusianismus und Neumalthusianismus* (Dresden; Leipzig: Minden, 1908); see also an excerpt from Davenport's book, *Heredity in Relation to Eugenics* (New York: H. Holt, 1911) published as Charl'z Davenport, *Evgenika kak nauka ob uluchshenii prirody cheloveka* (M.: V. Kariakin, 1913); and many others.

28 See Liudvik Krzhivitskii [Ludwik Krzywicki], *Psikhicheskie rasy* (SPb.: XX vek, 1902), pp. 54-73, 212-23. Tellingly, when six years earlier, Krzhivitskii published a lengthy (350-page long) overview of contemporary anthropology (see L. Krzhivitskii, *Antropologiia* (SPb.: F. Pavlenkov, 1896), he mentioned Galton only in passing and did not use the term "eugenics" at all. But he did discuss various ideas and projects of "human betterment" propounded by other authors. It was in the latter volume that Krzhivitskii first used the term "anthropotechnique" to describe such ideas and projects. Nowadays Krzywicki is remembered mostly as a foremost Polish Marxist, sociologist, and economist. See Ludwik Krzywicki, *Wspomnienia*, 3 vols. (Warsaw: Czytelnik, 1957-59); Tadeusz Kowalik and Henryka Hołda-Róziewicz, *Ludwik Krzywicki* (Interpress, 1976); Henryk Holland, *Ludwik Krzywicki – nieznany* (Warsaw: Książka i Prasa, 2007); and Wojciech Olszewski, "Ludwik Krzywicki: An Inconveniently Labelled Marxist," *Journal of Classical Sociology*, 2006, 6(3): 359-80. His role in the development of eugenics either in Russia or in Poland, however, remains almost completely forgotten. See, for instance, Grzegorz Radomski, "Eugenika i przejawy jej recepcji w polskiej mysli politycznej do 1939 roku," *Historia i Polityka*, 2010, 4(11): 85-100; and Magdalena Gawin, "Early Twentieth-Century Eugenics in Europe's Peripheries: The Polish Perspective," *East Central Europe*, 2011, 38(1): 1-15.

29 See K. Timiriazev, "Galton," *Entsiklopedicheskii slovar' Br. A. i I. Granat i Ko.* (M.: "Granat," 1911), vol. 12, pp. 469-73. It is worth noting that in 1901, the same encyclopedia carried only a very brief note on Galton, which did not mention eugenics at all. See "Galton, Francis," *Nastol'nyi entsiklopedicheskii slovar'* (M.: "Granat" i Ko., 1901), vol. 2, p. 1090.

30 K. A. Timiriazev, "Evgenika," *Entsiklopedicheskii slovar' Br. A. i I. Granat i Ko.* (M.: "Granat," 1913), vol. 19, pp. 391-95.

31 See Liudvik Krzhivitskii [*Krzywicki*], "Antropotekhnika," in *Entsiklopedicheskii slovar' Br. A. i I. Granat i Ko.* (M.: "Granat," 1911), vol. 3, pp. 249-50; K. A. Timiriazev, "Evgenika," *ibid.* (M.: "Granat," 1913), vol. 19, pp. 391-95; L. Krzhivitskii, "Antropotekhnika," in *Novyi entsiklopedicheskii slovar'* (SPb.: Brokgauz-Efron: 1911), vol. 3, pp. 99-101; Anon., "Evgenika," *ibid.* (SPb.: Brokgauz i Efron: 1914), vol. 17, p. 173.

32 Kropotkin's speech was published in the congress's proceedings, see *Problems in Eugenics* (London: The Eugenics Education Society, 1913), vol. 2, pp. 50-51. All the subsequent quotations are from this source.

33 Dioneo [I. Shklovskii], "Iz Anglii. Zverinaia filosofiia," *Russkoe bogatstvo*, 1912, 10: 296-323 (p. 302). The essay was also reprinted in the two-volume collection of Shklovskii's writings issued two years later under the general title *Changing England*. See Dioneo [I. Shklovskii], "Glava Piatnadtsataia. Zverinaia filosofiia," in *idem, Meniaiushchaiasia Angliia* (M.: Tovarishchestvo pisatelei v Moskve, 1914-1915), vol. 2, pp. 217-50.

34 See, for instance, V. Chizh, *Kriminal'naia antropologiia* (Odessa: G. Beilenson i I. Iurovskii, 1895); P. I. Kovalevskii, *Vyrozhdenie i vozrozhdenie. Prestupnik i bor'ba s prestupnost'iu* (SPb.: M. I. Akinfeev i I. V. Leont'ev, 1903); and I. A. Sikorskii, *Chto takoe natsiia i drugie formy etnicheskoi zhizni* (Kiev: S. V. Kul'zhenko, 1915).

35 For a general history of physical anthropology and the concept of race in Imperial Russia, see Mogilner, *Homo Imperii: A History of Physical Anthropology in Russia*. See also Nathaniel Knight, "Vocabularies of Difference: Ethnicity and Race in Late Imperial and Early Soviet Russia," *Kritika: Explorations in Russian and Eurasian History*, 2012, 13(3): 667-83.

36 See, for instance, E. Chepurkovskii, "Biologicheskii i statisticheskii metody v izuchenii nasledstvennosti u cheloveka," *Russkii antropologicheskii zhurnal* (hereafter *RAZh*), 1916, 1-2: 15-32; 3-4: 17-43.

37 See Krzhivitskii, "Antropotekhnika," 1911, vol. 3, pp. 249-50; *idem*, "Antropotekhnika," 1914, vol. 3, pp. 99-101.

38 Krzhivitskii, "Antropotekhnika," 1914, p. 100.

39 See Vlad. Nabokov, "'Poslednee slovo' kriminalistiki," *Pravo*, 1908, 14: 808-12.

40 See, for instance, A. A. Zhizhilenko, "Mery sotsial'noi bor'by s opasnymi prestupnikami," *Pravo*, 1910, 35: 2078-91; 36: 2136-43; 37: 2167-77; N. N. Lebedev, "Bor'ba s prestupnost'iu v Amerike: operativnyi sposob uluchsheniia roda chelovecheskogo (sterilizatsiia)," *Vestnik obshchestvennoi gigieny, sudebnoi i prakticheskoi meditsiny*, 1911, 1: 1-11; and many others.

41 P. I. Liublinskii, "Novaia mera bor'by s vyrozhdeniem i prestupnost'iu," *Russkaia mysl'*, 1912, 3: 31-56.

42 See, for instance, S. Ukshe, "Vyrozhdenie, ego rol' v prestupnosti i mery bor'by s nim," *Vestnik obshchestvennoi gigieny, sudebnoi i prakticheskoi meditsiny*, 1915, 6: 798-816.

43 See P. I. Kovalevskii, *Otstalye deti (idioty, otstalye i prestupnye deti), ikh lechenie i vospitanie* (SPb.: Vestnik dushevnykh boleznei, 1906).

44 See "Khronika," *Gigiena i sanitarnoe delo*, 1914, 1: 118.

45 See I. G. Orshanskii, "Rol' nasledstvennosti v peredache boleznei," *Prakticheskaia meditsina*, 1897, 8-9: 1-120; and T. Iudin, "Psikhozy u bliznetsov," *Zhurnal nevropatologii i psikhiatrii*, 1907, 7: 68-83.

46 V. M. Bekhterev, "Vorporsy vyrozhdenia i bor'ba s nim," *Obozrenie psikhiatrii i nevrologii*, 1908, 9: 518-21; and T. Iudin, "O kharaktere nasledstvennykh vzaimootnoshenii pri dushevnykh bolezniakh," *SP*, 1913, 8: 568-78.

47 I. G. Orshanskii, "Izuchenie nasledstvennosti talanta," *Vestnik vospitaniia*, 1911, 1: 1-41, 2: 95-127. See also, Vl. Chizh, "Nasldstvennost' talanta u nashikh izvestnykh deiatelei," *Nauka i zhizn*, 1906, 2-3: 267-90.

48 See, for instance, N. Kabanov, *Rol' nasledstvennosti v etiologii boleznei vnutrennikh organov* (M.: G. I. Prostakov, 1899); and P. P. Tutyshkin, *Rol' otritsiatel'nogo otbora v protsesse semeinogo vyrozhdeniia* (Khar'kov: M. Zil'berberg, 1902);

49 A. Sholomovich, *Nasledstvennost' i fizicheskie priznaki vyrozhdeniia u dushevno-bol'nykh i zdorovykh* (Kazan: Tip. Imp. Un-ta, 1913); and idem, *Nasledstvennost' i fizicheskoe vyrozhdenie* (Kazan: Tip. Imp. Un-ta, 1915).

50 See, for example, S. A. Preobrazhenskii, "Mendelizm i eigenika," *Vrachebnaia gazeta*, 1913, 11: 409; M. G. Zaidner, "Braki mezhdu krovnymi rodstvennikami s tochki zreniia rasovoi gigieny," *ibid.*, 1914, 16: 656; and L. G. Lichkus, "Nalsedstvennost' i eigenika," *ibid.*, 1914, 22: 893.

51 [N. Gamaleia], "[Programma zhurnala]," *Gigiena i sanitariia* (hereafter *GIS*), 1910, 1: 1-5 (p. 5).

52 See, for instance, I. V. Sazhin, *Nasledstvennost' i spirtnye napitki* (SPb.: Soikin, 1908).

53 See K. Kuchuk, "Kratkii ocherk sovremennykh vzgliadov na nasledstvennost'," *GIS*, 1912, 21-22: 437-41.

54 See E. A. Shepilevskii, "Osnovy i sredstva rasovoi gigieny (gigiena razmnozheniia)," *Trudy i protokoly zasedanii Meditsinskogo obshchestva im. N. I. Pirogova pri Imperatorskom Iur'evskom Universitete*, 1913-14, 6: 61-137. On Shepilevskii and his involvement with "racial hygiene," see Felder, "Rasovaia gigiena v Rossii."

55 See L. P. Kravets, "Nasledstvennost' u cheloveka," *Priroda*, 1914, 6: 722-43; and Kr. L., "Evgenetika," *ibid.*, 10: 1229.

56 See N. Kol'tsov, "Alkogolizm i nasledstvennost'," *Priroda*, 1916, 4: 502-05; *idem*, "K voprosu o nasledovanii posledstvii alkogolizma," *ibid.*, 1916, 10: 1189; and Iu. Filipchenko, "O vidovykh gibridakh," in V. A. Vagner, ed., *Novye idei v biologii. Nasledstvennost'* (SPb.: Obrazovanie, 1914), pp. 124-49.

57 Iu. Filipchenko, "Evgenika," *Russkaia mysl'*, 1918, 3-6: 69-96.

58 For a detailed analysis of this situation, see Krementsov, "The Strength of a Loosely Defined Movement."

59 For a discussion of the concept of *ozdorovlenie* (healthification) and its role in the ideology and activities of Russian physicians, see John Hutchinson, *Politics and Public Health in Revolutionary Russia, 1890-1918* (Baltimore, MD: Johns Hopkins University Press, 1990), pp. xv-xx.

60 See [N. Gamaleia], "[Programma zhurnala]," *GIS*, 1910, 1: 5.

61 See K. V. Karaffa-Korbutt, "Ocherki po evgenike," *GIS*, 1910, 1: 41-48; 2: 138-45; 3: 276-81. Judging from the contents of these three articles, Karaffa-Korbutt had originally planned a much longer series, with separate articles on biometrics, Galton, and German Rassenhygiene. But although the last published article promised "to be continued," no further articles appeared. I was unable to discover any reasons for this abrupt end of the series.

62 See, for instance, Kazimir Karaffa-Korbut, "I. Rutgers. Uluchshenie chelovecheskoi porody. Agnessa Blium. Etika i evgenika. SPb 1909," *GIS*, 1910, 1: 75-76; N. G[amaleia], "Bertillon," *ibid.*, 1910, 4: 292-93; and N. Avgustovskii, "N. Norre, O zachatii v sostoianii op'ianeniia," *ibid.*, 1910, 9: 670-71.

63 See "Istoriia evgeniki. I. A. Fields, The progress of eugenics," *GIS*, 1913, 17-20: 286-90; [N. Gamaleia], "Pervyi mezhdunarodnyi evgenicheskii kongress

v Londone 24-30 iiulia 1912," *ibid.*, 1912, 15-16: 175-82; and Kuchuk, "Kratkii ocherk sovremennykh vzgliadov na nasledstvennost'."

64 See, for instance, N. G[amaleia], "Bertillion," *GIS*, 1910, 4: 292-93; and *idem*, "1 iiulia 1910 goda," *ibid.*, 1910, 13: 1-5.

65 N. F. Gamaleia, "Ob usloviiakh, blagopriiatstvuiushchikh uluchsheniiu prirodnykh svoistv liudei," *GIS*, 1912, 19-20: 340-61.

66 *Ibid.*, p. 361.

67 Gamaleia did not limit this activity to his journal. For instance, while teaching bacteriology and hygiene in Iur'ev (now Tartu, Estonia), he delivered a series of public lectures on eugenics, which were reported in the city's major Estonian-language daily, see N. F. Gamaleja [Gamaleia], "Toutervendluse-opetuse, eugeenika pohjus, motteist ja ulesannetest," *Postimees*, 5–9 November 1912, http: //dea.nlib.ee/fullview.php?frameset=3&showset=1&wholepage=keskmine&pid=s474228&nid=7093. I am grateful to Julia Laius for her assistance in finding this source.

68 For instance, contrary to the opinion of Bjorn M. Felder, it was Gamaleia who incited Evgenii A. Shepilevskii, a professor at Iur'ev University (where Gamaleia was teaching in 1911-13), to take up the discussion of, and research in, racial hygiene. See Felder, "Rasovaia gigiena v Rossii."

69 T. I. Iudin, "Ob evgenike i evgenicheskom dvizhenii," *SP*, 1914, 4: 319-36. All the subsequent quotations are from this source.

70 Iudin had conducted extensive research and published a series of articles on the subject, see T. Iudin, "Psikhozy u bliznetsov," *Zhurnal nevropatologii i psikhiatrii*, 1907, 7(1): 68-83; *idem*, "O skhodstve psikhozov u brat'ev i sester," *SP*, 1907, 10: 337-42; 11: 401-9; 12: 451-59; *idem*, "O forme dushevnykh zabolevanii, vstrechaiushchikhsia v sem'e progressivnykh paralitikov," *ibid.*, 1911, 1-2: 126-43; and *idem*, "O kharaktere nasledstvennykh vzaimootnoshenii pri dushevnykh bolezniakh," *ibid.*, 1913 (August): 568-79.

71 Just a few months earlier, a Moscow publisher had issued a Russian translation of the third, revised and expanded, edition of Punnett's classic textbook, *Mendelism* (London: Macmillan, 1911), which Iudin used in his survey, see R. Pennet, *Mendelizm* (M.: Bios, 1913).

72 Iudin used a Russian translation of Correns's book, *Die neuen Vererbungsgesetze* (Berlin: Verlag von Gebrüder Borntraeger, 1912), which had appeared just a few months earlier, see K. Korrens, *Novye zakony nasledstvennosti* (M.: Bios, 1913).

73 My reconstructions of Volotskoi's life and career are largely based on his personnel files, especially CVs, preserved at various institutions where he had studied and worked. See the Central State Archive of the City of Moscow (hereafter TsGAM), f. 418, op. 325, d. 310, ll. 1-12; the Archive of the Russian Academy of Sciences (hereafter ARAN), f. 356, op. 3, d. 60, ll. 298-303; ARAN, f. 669, op. 2, d. 30, ll. 1-43; and RGALI, f. 117, op. 1, d. 77, ll. 5-5 rev.

74 On Anuchin and his role in Russian anthropology, see G. V. Karpov, *Put' uchenogo: Ocherki zhizni, nauchnoi i obshchestvennoi deiatel'nosti D. N. Anuchina* (M.: Geografgiz, 1958); and S. S. Alymov, "Dmitrii Nikolaevich Anuchin: 'estestvennaia istoriia cheloveka v obshirnom smysle etogo slova'," in V. A. Tishkov and D. D. Tumarkin, eds., *Vydaiushchiesia otechestvennye etnologi i antropologi XX veka* (M.: Nauka, 2004), pp. 7-48.

75 For illuminating memoirs (in English) of the February Revolution, see Semion Lyandres, *The Fall of Tsarism: Untold Stories of the February 1917 Revolution* (Oxford: Oxford University Press, 2013).

76 See Michael David-Fox, *Revolution of the Mind: Higher Learning among the Bolsheviks, 1918–1929* (Ithaca, NY: Cornell University Press, 1997).

77 For details, see Nikolai Krementsov, *Stalinist Science* (Princeton, NJ: Princeton University Press, 1997); and *idem*, "Big Revolution, Little Revolution: Science and Politics in Bolshevik Russia," *Social Research*, 2006, 73(4): 1173-204.

78 For a general discussion of the Bolsheviks' social, economic, and cultural policies during this period, see D. P. Koenker, W. G. Rosenberg, and R. G. Suny, eds., *Party, State, and Society in the Russian Civil War* (Bloomington, IN: Indiana University Press, 1989).

79 For details of the conflict, see the memoirs of the university rector, zoologist Mikhail M. Novikov, *Ot Moskvy do N'iu Iorka. Moia zhizn' v politike i nauke* (New York: Izdatel'stvo imeni Chekhova, 1952). For a general, though dated, assessment of Narkompros's activities in the first years of the Bolshevik regime, see Sheila Fitzpatrick, *The Commissariat of Enlightenment: Soviet Organization of Education and the Arts under Lunacharsky, October 1917-1921* (Cambridge: Cambridge University Press, 1970).

80 Numerous dairies vividly depict the difficulties of survival in Moscow during the civil war years. See, for example, N. P. Okunev, *Dnevnik moskvicha, 1914-1924*, 2 vols. (SPb.: Voenizdat, 1998). For a thorough historical analysis of urban life during this period, see A. A. Il'iukhov, *Zhizn' v epokhu peremen* (M.: ROSSPEN, 2007).

81 On Bunak, see M. F. Nesturkh, "Viktor Valerianovich Bunak," in *idem*, ed., *Sovremennaia antropologiia* (M.: MGU, 1964), pp. 9-18; and S. V. Vasil'ev and M.

I. Urynson, "Viktor Valerianovich Bunak: patriarch antropologii," in Tishkov and Tumarkin, eds., *Vydaiushchiesia otechestvennye etnologi i antropologi*, pp. 233-60. The former article appeared in a jubilee volume celebrating Bunak's seventieth birthday and did not mention his involvement with eugenics at all. The list of Bunak's publications appended to the article did not include a single reference to his numerous works on the subject. This omission is particularly telling, since Bunak was the only Soviet eugenicist whose paper appeared in the proceedings of an international eugenics congress, see V. Bunak, "Sex-Ratio of New-Born Infants, as an Index of Vitality," in *A Decade of Progress in Eugenics. Scientific Papers of the Third International Congress of Eugenics* (Baltimore, MD: Williams and Wilkins, 1934), pp. 431-35.

82 On the establishment of the State Museum of Social Hygiene and its first director, see M. P. Kuzybaeva, "Gigienicheskii muzei professora A. V. Mol'kova," *Gigiena i sanitariia*, 2013, 4: 94-97. On the museum's consultants, see GARF, f. A1571, op. 1, dd. 1-2.

83 On the creation of Narkomzdrav and its activities during the first decade of operation, see Neil B. Weissman, "Origins of Soviet Health Administration, Narkomzdrav, 1918-28," in S. G. Solomon and J. F. Hutchinson, eds., *Health and Society in Revolutionary Russia* (Bloomington, IN: Indiana University Press, 1990), pp. 97-120.

84 T. Ia. Tkachev, *Sotsial'naia gigiena* (Voronezh: Gubzdravotdel, 1924), pp. 11, 153.

85 On the history of Soviet social hygiene, see Susan G. Solomon, "Social Hygiene and Soviet Public Health, 1921-1930," in Solomon and Hutchinson, eds., *Health and Society in Revolutionary Russia*, pp. 175-99; and idem, "The Limits of Government Patronage of Sciences: Social Hygiene and the Soviet State, 1920–1930," *Social History of Medicine*, 1990, 3(3): 405-35.

86 V. Mol'kov, "Piat' let raboty Gosudarstvennogo Instituta Sotsial'noi Gigieny," *Sanitarnoe prosveshchenie*, 1924, 2: 31.

87 On Kol'tsov's life and activities, see B. L. Astaurov and P. F. Rokitskii, *Nikolai Konstantinovich Kol'tsov* (M.: Nauka, 1975).

88 On the history of the institute, see Mark B. Adams, "Science, Ideology, and Structure: The Kol'tsov Institute, 1900-1970," in Linda L. Lubrano and Susan G. Solomon, eds., *The Social Context of Soviet Science* (Boulder, CO: Westview, 1980), pp. 173-204.

89 For a focused analysis of Kol'tsov's early efforts to build working relations with the Bolshevik regime, see Nikolai Krementsov, *Revolutionary Experiments: The Quest for Immortality in Bolshevik Science and Fiction* (New York: Oxford University Press, 2014), pp. 183-85.

90 One of the first projects Kol'tsov offered to Semashko in 1919 was breeding rabbits, mice, guinea-pigs, and chickens for Narkomzdrav's research institutions, which, given the total absence of a system of supply of laboratory animals in the country, was a very timely and appealing offer. See GARF, f. A-482, op. 1, d. 34, ll. 346-47.

91 For the early history of this institution, see L. A. Tarasevich and V. A. Liubarskii, eds., *Gosudarstvennyi Institut Narodnogo zdravookhraneniia imeni Pastera ("GINZ"), 1919-1924* (M.: GINZ, 1924).

92 See ARAN, f. 450, op. 4, d. 7, ll. 1-4; d. 8, ll. 22-23.

93 See the descriptions of the meeting in V. Bunak, "O deiatel'nosti Russkogo evgenicheskogo obshchestva za 1921 god," *Russkii evgenicheskii zhurnal* (hereafter *REZh*), 1922, 1(1): 99-101; and "Khronika," *RAZh*, 1922, 12(1-2): 215-16.

94 See, for instance, Volotskoi's reports to Narkompros on the RES activities in 1922 and 1923, in GARF, f.A2307, op.8, d. 278, ll.81-83; 85-92.

95 Kol'tsov invited Bunak to head the department and reserved for himself its "general scientific direction," see ARAN, f. 570, op. 1, d. 1, ll. 27, 58.

96 On the "rejuvenation craze" in early 1920s Russia, see Krementsov, *Revolutionary Experiments*, pp. 127-58.

97 For detailed analyses of the US "eugenic" sterilization, see Philip R. Reilly, *The Surgical Solution: A History of Involuntary Sterilization in the United States* (Baltimore, MD: Johns Hopkins University Press, 1991); and Mark A. Largent, *Breeding Contempt: The History of Coerced Sterilization in the United States* (New Brunswick, NJ: Rutgers University Press, 2008).

98 See Dr. Sharp, "Sterilizatsiia v shtate Indiana," in N. K. Kol'tsov, ed., *Omolozhenie* (M.-Pg.: GIZ, 1923), pp. 121-23. For the original, see "Discussion," *Eugenics Review*, 1912, 4(2): 204-05.

99 For a highly readable account of the civil war, see W. Bruce Lincoln, *Red Victory: A History of Russian Civil War, 1918-1921* (New York: Da Capo, 1989).

100 For a general discussion of the social, economic, cultural, and political aspects of NEP, see Sheila Fitzpatrick, Alexander Rabinowitch, and Richard Stites, eds., *Russia in the Era of NEP* (Bloomington, IN: Indiana University Press, 1991); and Lewis H. Siegelbaum, *Soviet State and Society between Revolutions, 1918-1929* (Cambridge: Cambridge University Press, 1992). For more recent assessments based on the newly available archival documents, see materials of a series of conferences held at the Institute of Russian History in Moscow, A. K. Sokolov, ed., *NEP v kontekste*

*istoricheskogo razvitiia Rossii XX veka* (M.: IRI, 2001); A. S. Siniavskii, ed., *NEP: ekonomicheskie, politicheskie i sotsiokul'turnye aspekty* (M.: ROSSPEN, 2006); and many others.

101 For details, see Krementsov, *Stalinist Science*, and *idem*, "Big Revolution, Little Revolution."

102 For details on the post-revolutionary literacy and science-popularization campaigns, see James T. Andrews, *Science for the Masses: The Bolshevik State, Public Science and the Popular Imagination in Soviet Russia, 1917-1934* (College Station, TX: Texas A&M University Press, 2003).

103 For instance, in order to get more information on the Indiana sterilization law, Volotskoi initiated correspondence with one of its instigators, Indiana State Health Officer John N. Hurty. See Sharp, "Sterilizatsiia v shtate Indiana," pp. 121-23.

104 N. K. Kol'tsov, "Uluchshenie chelovecheskoi porody," *REZh*, 1922, 1: 3-27. All the following citations are from this source. This and several other articles that had been published by Soviet eugenicists were reprinted in V. V. Babkov, *Zaria genetiki cheloveka* (M.: Progress-Traditsiia, 2008). These publications are also available in an English translation; see V. V. Babkov, *The Dawn of Human Genetics*, transl. by Victor Fet, ed. by James Schwarz (Cold Spring Harbor, NY: Cold Spring Harbor Press, 2013). Although in my own work I have always used the original Russian texts, for the convenience of my English-reading audience, I include references to their English translations available in the latter volume. Hereafter such references will be given as [Babkov, pp. 66-86].

105 Kol'tsov used the word "religion" in the same sense as we use the word "ideology" today.

106 A year later, when Kol'tsov published this speech as a brochure, he excised its last two paragraphs that had advanced the view of eugenics as religion, see N. K. Kol'tsov, *Uluchshenie chelovecheskoi porody* (Pg.: Vremia, 1923).

107 For a short biography of Filipchenko in English, see Mark B. Adams, "Filipchenko, Iurii Aleksandrovich," in Frederic Holmes, ed., *Dictionary of Scientific Biography* (New York: Scribner, 1990), vol. 17, suppl. 2, pp. 297-303; for a much more detailed Russian-language biography, see N. N. Medvedev, *Iurii Aleksandrovich Filipchenko* (M.: Nauka, 2006). From the very beginning of his organizational efforts, Kol'tsov sought Filipchenko's support. He even invited Filipchenko to head the IEB eugenics department and managed to approve his candidacy by Narkomzdrav (see ARAN, f. 570, op. 1, d. 1, ll. 29, 34, 58). But, busy with building his own institutional base in Petrograd, Filipchenko declined to move to Moscow. In September 1920, Kol'tsov also invited Filipchenko to join him in founding the Russian Eugenics Society.

On the latter's visit to Moscow in November, Filipchenko and Kol'tsov discussed the strategy and agreed that Filipchenko would act in Petrograd independently of whatever Kol'tsov would do in Moscow. For a description of the meeting, see Filipchenko's diaries held in the Manuscript Department of the Russian National Library in St. Petersburg (hereafter RO RNB), f. 813, op. 1, d. 1283, l. 3.

108 See the St. Petersburg branch of the ARAN, f. 132, op. 1, d. 217, ll. 2-6; and Iu. Filipchenko, "Biuro po evgenike," *Izvestiia Biuro po Evgenike*, 1922, 1: 1-4; for a brief history of the bureau, see M. B. Konashev, "Biuro po evgenike, 1922-1930," *Issledovaniia po genetike*, 1994, 11: 22-28.

109 In 1924, this society became a branch of the RES and Filipchenko joined Kol'tsov as a co-editor of its oracle, the *Russian Eugenics Journal*.

110 For a voluminous, though far from complete bibliography of eugenics publications up to 1928, see K. Gurvich, "Ukazatel' literatury po voprosam evgeniki, nasledstvennosti i selektsii i sopredel'nykh oblastei, opublikovannoi na russkom iazyke do 01.01.1928 g.," *REZh*, 1928, 6(2-3): 121-43.

111 See "Evgenika i biologicheskie voprosy," *Ginekologiia i akusherstvo*, 1924, 4: 409-13.

112 Many Russian gynecologists preferred the term *evgenetika* (eugénnetique) introduced by their prominent French colleague, obstetrician Adolphe Pinard. See, V. Wallich, "L'eugénnetique," in E. Brissaud, A. Pinard, P. Reclus, eds., *Nouvelle pratique médico-chirurgicale illustrée. Premier supplément* (Paris: Masson, 1911-1912), pp. 547-50, http://gallica.bnf.fr/ark:/12148/bpt6k54465064. On Pinard's involvement with eugenics, see Alain Drouard, "Eugenics in France and in Scandinavia: Two Case Studies," in Robert A. Peel, ed., *Essays in the History of Eugenics* (London: The Galton Institute, 1998), pp. 173-207; and idem, *L'eugénisme en questions: L'exemple de l'eugénisme "français"* (Paris: Elilipses, 1999).

113 N. M. Kakushkin, "Evgenetika i ginekologiia," in *Trudy VI s"ezda vsesoiuznogo obshchestva ginekologov i akusherov* (M.: T. Dortman, 1925), p. 415.

114 See *Trudy VI s"ezda vsesoiuznogo obshchestva ginekologov i akusherov*, pp. 448-55.

115 For more details on Russian geneticists' and eugenicists' international activities, see Nikolai Krementsov, *International Science between the World Wars: The Case of Genetics* (London: Routledge, 2005).

116 Vavilov to Davenport, 21 September 1921. This letter is preserved among Davenport's papers held in the Manuscript Division of the American

Philosophical Society (hereafter APS), Mss. B. D27; hereafter references to this collection will be given as "Davenport Papers."

117 See "Foreign Notes," *Eugenical News*, 1921, 6: 72-73.

118 See Koltzoff [Kol'tsov] to Davenport, 25 June 1921, Davenport Papers.

119 See Philiptschenko [Filipchenko] to Davenport, 28 October 1921, Davenport Papers. Davenport also printed Filipchenko's plea in the ERO newsletter, see "Russian Eugenics Bureau," *Eugenical News*, 1922, 7:13.

120 Richard Stites, *Revolutionary Dreams: Utopian Vision and Experimental Life in the Russian Revolution* (Oxford: Oxford University Press, 1989).

121 L. Trotskii, *Literatura i revoliutsiia* (M.: Krasnaia nov', 1923), pp. 195-97.

122 See "Russkii evgenicheskii zhurnal," *Izvestiia*, 7 October 1922, p. 8.

123 See T. I. Iudin, "Nasledstvennost' dushevnykh boleznei," *REZh*, 1922, 1(1): 28-39; and V. V. Bunak, "Evgenicheskie opytnye stantsii, ikh zadachi i plan ikh rabot," *ibid.*, 83-99.

124 See A. S. Serebrovskii, "O zadachakh i putiakh antropogenetiki," *REZh*, 1923, 1(2): 107-16. For a brief biography of Serebrovskii in English, see Mark B. Adams, "Serebrovskii, Aleksandr Sergeevich," in Holmes, ed., *Dictionary of Scientific Biography*, vol. 18, Suppl. II, pp. 803-11; for a much more detailed description of Serebrovskii's life and works in Russian, see N. N. Vorontsov, ed., *Aleksandr Sergeevich Serebrovskii* (M.: Nauka, 1993).

125 V. V. Bunak, "Metody izucheniia nasledstvennosti u cheloveka," *REZh*, 1923, 1(2): 137-200.

126 M. V. Volotskoi, "O polovoi sterilizatsii nasledstvenno-defektivnykh," *REZh*, 1923, 1(2): 201-22.

127 M. V. Volotskoi, *Podniatie zhiznennykh sil rasy. Novyi put'* (M.: Zhizn' i znanie, 1923), p. 5.

128 See, for instance, T. Ia. Tkachev, "Polovaia sterilizatsiia, kak problema sotsial'noi gigieny," *Voronezhskoe zdravookhranenie*, 1923, 2: 51-55; and *idem*, *Sotsial'naia gigiena*.

129 See, for instance, Russian discussions of the Norwegian and Swedish eugenic legislation, Anon., "Sovremennoe sostoianie voprosa o sterilizatsii v Shvetsii," *REZh*, 1925, 3(1): 78-81; and Iu. Filipchenko, "Obsuzhdenie norvezhskoi evgenicheskoi programmy na zasedaniiakh Leningradskogo Otdeleniia R. E. O.," *ibid.*, 1925, 3(2): 139-43; the latter article was reprinted

in *Eugenics Review*, see Ju. A. Philiptschenko, "The Norwegian Eugenic Programme: Discussed at Meetings of the Eugenic Society of Leningrad," *Eugenics Review*, 1928, 19(4): 294-98.

130 See Iu. A. Filipchenko, "M. V. Volotskoi, Podniatie zhiznennykh sil rasy. Novyi put'. Iz-vo 'Zhizn' i znanie'. M. 1923 g. Str. 96," *Pechat' i revoliutsiia*, 1924, 6: 248-50.

131 See, for instance, N. Lialin, "Khirurgicheskaia sterilizatsiia zhenshchin i ee sotsial'noe znachenie," *Zdravookhranenie*, 1929, 11-12: 162-68.

132 V. M. Volotskoi, *Podniatie zhiznennykh sil rasy. Novyi put'* (M.: Zhizn' i znanie, 1926).

133 *Ibid.*, p. 24.

134 Volotskoi published a "historical inquiry" on Peter the Great's "anthropotechnical projects" in *Russian Eugenics Journal*, see M. V. Volotskoi, "Antropotekhnicheskie porekty Petra I-go (istoricheskaia spravka)," *REZh*, 1923, 1(2): 235-36 (p. 336), [Babkov, pp. 20-22].

135 He did notice in the treatise's text several references to it being "a journal article" and tried to find where it had originally been published. He took as his guide a reference to the journal *Deed* as the publisher of Florinskii's book that appeared in an article on Florinskii in a popular encyclopedia (see A. [M. T. Alekseev], "Florinskii, Vasilii Markovich," *Entsiklopedicheskii slovar' Brokgauza i Efrona* (SPb.: Brokgauz i Efron, 1902), vol. 36, p. 169). But, of course, he did not find any of Florinskii's publications in *Deed*. See the first "editor's commentary" in Florinskii, 1926, p. 157.

136 Anuchin knew and corresponded with Florinskii during the 1890s regarding the latter's archeological research in Siberia (see a sample of this correspondence in Iastrebov, *Sto neizvestnykh pisem*, pp. 21-25), and it seems likely that he owned a copy of Florinskii's treatise. However, Anuchin's book collection currently held at the Rare Book Section of the Scientific Library of Moscow University does *not* have a copy of Florinskii's 1866 book (though it does have a copy of the 1926 edition issued by Volotskoi). According to the head of the Rare Book Section, Alexander Livshits (personal communication, 10 November 2016), it is possible that the 1866 copy has been lost, as a result of either the numerous "purges" the library had suffered during the Soviet era, or its evacuation from Moscow during World War II. It is also possible that Volotskoi had "borrowed" the copy to prepare and typeset his 1926 edition of the book and never returned it to the library. However, none of these possibilities could be verified by the available materials.

137 Volotskoi, *Podniatie zhiznennykh sil rasy*, p. 91.

138 V. M. Volotskoi, "K istorii evgenicheskogo dvizheniia," *REZh*, 1924, 2(1): 50-55, [Babkov, pp. 23-28]; all the subsequent quotations are from this source.

139 *Ibid.*, pp. 53-54.

140 *Ibid.*, p. 54.

141 At the time of writing his book, Volotskoi was clearly unaware of Galton's 1865 article and apparently thought that Galton's first "eugenic" work was *Hereditary Genius* published four years later.

142 See M. V. Volotskoi, "O dvukh formakh chelovecheskoi kisti preimushchestvenno v sviazi s polovymi i rasovymi otlichiiami," *RAZh*, 1924, 13(3-4): 70-82.

143 Only two years older than Volotskoi, Bunak apparently saw the younger man as a formidable threat to his own administrative ambitions at both the anthropology department and the IEB eugenics department. Judging by available materials, the relations between the two men were less than cordial and might well have also contributed to Volotskoi's decision to leave the university. Tellingly, in 1930 when Bunak was fired, Volotskoi returned to his alma mater.

144 On the establishment of this institute and its work during the first five years, see *Piatiletie Gosudarstvennogo tsentral'nogo instituta fizicheskoi kul'tury, 1918-1923* (M.: TsIT, 1923); for a Soviet-era history of the institute, which of course does not even mention eugenics, see I. G. Chudinov, *Gosudarstvennyi Tsentral'nyi ordena Lenina institut fizicheskoi kul'tury: Istoricheskii ocherk* (M.: Fizkul'tura i sport, 1966).

145 *Podgotovka rabotnikov po fizicheskoi kul'ture* (M.: Izd-vo Vysshego i Moskovskogo soveta fiz.kul'tury, 1924), p. 8.

146 ARAN, f. 669, op. 2, d. 30, l.1.

147 For details on the Bolsheviks' efforts to create "Communist" and "proletarian" science, see Krementsov, *Stalinist Science*; and *idem*, "Big Revolution, Little Revolution."

148 For details on Bogdanov's concept of "proletarian science," see Nikolai Krementsov, *A Martian Stranded on Earth: Alexander Bogdanov, Blood Transfusions, and Proletarian Science* (Chicago, IL: University of Chicago Press, 2011).

149 For details, see David-Fox, *Revolution of the Mind*; and Krementsov, *Stalinist Science*.

150 For detailed analyses of the role of Marxism in 1920s Russian, especially biomedical, science, as well as an overview of voluminous historical literature on the subject, see Nikolai Krementsov, "Marxism, Darwinism, and Genetics in the Soviet Union," in Denis Alexander and Ron Numbers, eds., *Biology and Ideology: From Descartes to Dawkins* (Chicago, IL: University of Chicago Press, 2010), pp. 215-46; and Daniel P. Todes and Nikolai Krementsov, "Dialectical Materialism and Soviet Science in the 1920s and 1930s," in William Leatherbarrow and Derek Oxford, eds., *A History of Russian Thought* (Cambridge: Cambridge University Press, 2010), pp. 340-67.

151 In his 1922 article, titled "On the Significance of Militant Materialism," Lenin proclaimed that every "scientist must be an up-to-date materialist, a deliberate follower of the materialism presented by Marx, that is, he must be a dialectical materialist." See V. I. Lenin, "O znachenii voinstvuiushchego materializma," *Pod znamenem marksizma* (hereafter *PZM*), 1922, 3: 29. The rules of appointment are preserved among the Timiriazev Institute's materials, see ARAN, f. 669, op. 2, d. 30, l. 8.

152 Aside from an occasional mention of his name in various publications on the history of Soviet pediatrics and physical culture, I was able to find only very brief biographies of Radin. See "Radin Evgenii Petrovich," *Bol'shaia meditsinskaia entsiklopediia* (M.: Gosmedizdat, 1962), vol. 27, p. 739; the same entry was reprinted in B. D. Petrov, ed., *Vrachi-soratniki V. I. Lenina, uchastniki revoliutsionnogo dvizheniia. Biobliograficheskii spravochnik* (M.: n. p., 1970), 95-96. The following reconstruction is based on Radin's personnel file preserved in the Narkomzdrav archive, GARF, f. A482, op. 41, d. 2820, ll. 1-10. Alas, the file covers only the 1923-29 period.

153 The anthem was "*Smelo, tovarishchi, v nogu*" (Bravely, Comrades, in Step). See "Radin, Leonid Petrovich," in P. A. Nikolaev, ed., *Russkie pisateli. Biobibliograficheskii slovar'* (M.: Prosveshchenie, 1990), vol. 2, "M-Ia," pp. 185-86.

154 See E. Radin, *Okhranenie nervnogo i dushevnogo zdorov'ia uchashikhsia* (SPb.: B. M. Vol'f, 1910). He also published several interesting studies on what at the time was termed "social psychiatry," investigating manifestations of mental illness in contemporary literature and social attitudes. See, for example, E. P. Radin, "Vyrozhdaiushchiesia vysshego poriadka," *SP*, 1908, October: 433-44; November: 483-93; idem, *Problema pola v sovremennoi literature i bol'nye nervy* (SPb.: Montvida, 1910); idem, *Dushevnoe nastroenie sovremennoi uchasheisia molodezhi po dannym Peterburgskoi obshchestudencheskoi ankety 1912 goda* (SPb.: N. P, Karbasnikov, 1913); and idem, "*Futurizm i bezumie*". *Paralleli tvorchestva i analogii novogo iazyka kubo-futuristov* (SPb.: N. P. Karbasnikov, 1914).

155 For a general examination of the Bolsheviks' concern with children, see Alan

M. Ball, *And Now My Soul Is Hardened: Abandoned Children in Soviet Russia, 1918-1930* (Berkeley, CA: University of California Press, 1994); and Loraine de la Fe, "Empire's Children: Soviet Childhood in the Age of Revolution," doctoral dissertation, Florida International University, 2013.

156 See, for instance, E. P. Radin, *Chto delaet sovetskaia vlast' dlia okhrany zdorov'ia detei* (M.: Komissiia pamiati V. M. Bonch-Bruevich [Velichkinoi], 1919). He fought fiercely to wrestle control over school physicians from Narkompros and instigated several decrees by the SNK to subordinate all the issues of children's health to Narkomzdrav's authority. See E. P. Radin, *Okhrana zdorov'ia detei i podrostkov i sotsial'naia evgenika* (Orel: GIZ, 1923), pp. 23-58. For a brief contemporary English-language description of the OZDP department, see Anon., "Children's Health Protection in the Soviet Union," *Russian Review*, 1925, 3(22): 461-62. In 1927 Radin became the director of the first research institute for the protection of children's and adolescents' health, see E. P. Radin, *Gosudarstvennyi nauchnyi institut okhrany zdorov'ia detei i podrostokov Narkomzdrava* (Orel: Orlovskoe pedologicheskoe obscshestvo, 1929).

157 For general histories of early Soviet physical culture, which, alas, do not mention Radin, see Susan Grant, "The Politics and Organization of Physical Culture in the USSR during the 1920s," *Slavonic and East European Review*, 2011, 89(3): 494-515; and *idem*, *Physical Culture and Sport in Soviet Society: Propaganda, Acculturation, and Transformation in the 1920s and 1930s* (New York: Routledge, 2013).

158 For accounts of the history of Soviet pedology, see N. Kurek, *Istoriia likvidatsii pedologii i psikhotekhniki* (SPb.: Aleteia, 2004); and E. M. Balashov, *Pedologiia v Rossii v pervoi treti XX veka* (SPb.: Nestor-Istoriia, 2012); on the history of Russian pedology in English, see a series of publications by Andy Byford, "Professional Cross-Dressing: Doctors in Education in Late Imperial Russia (1881-1917)," *Russian Review*, 2006, 65(4): 586-616; *idem*, "The Mental Test as a Boundary Object in Early-20th-Century Russian Child Science," *History of the Human Sciences*, 2014, 27(4): 22-58; and *idem*, "Imperial Normativities and the Sciences of the Child: The Politics of Development in the USSR, 1920s-1930s," *Ab Imperio*, 2016, 2: 71-124. Alas, none of the above publications mention Radin's contributions to the field.

159 See, for instance, his book (co-written with his wife) on "New games for new children" that went through five editions during the 1920s, M. A. Kornil'eva-Radina and E. P. Radin, *Novym detiam — novye igry. Podvizhnye igry shkol'nogo i vneshkol'nogo vozrastov (ot 7 do 18 let) v refleksologicheskom i pedologicheskom osveshchenii*, 5th edn. (M.: Medgiz, 1929).

160 Radin, *Okhrana zdorov'ia detei i podrostkov*.

161 N. Semashko, "Predislovie," in E. P. Radin et al., eds., *Fizicheskaia kul'tura v nauchnom osveshchenii* (M.: Izdanie Vysshego i Moskovskogo sovetov fizicheskoi kul'tury, 1924), pp. 3-4 (p. 3).

162 M. V. Volotskoi, "Fizicheskaia kul'tura s tochki zreniia evgeniki," in Radin, *Fizicheskia kul'tura v nauchnom osveshchenii*, pp. 62-75; idem, "O nekotorykh techeniiakh v sovremennoi evgenike," in *ibid.*, pp. 76-85. The first article was actually the text of a report he had delivered in November 1923 to an all-Russia conference on the protection of children's and adolescents' health. He also published an updated version of this article three years later in a new journal, *Theory and Practice of Physical Culture*, established by Narkomzdrav, see V. M. Volotskoi, "Fizicheskaia kul'tura i evgenika," *Teoriia i praktika fizicheskoi kul'tury*, 1927, 1: 19-26.

163 V. M. Volotskoi, *Klassovye interesy i sovremennaia evgenika* (M.: Zhizn' i znanie, 1925).

164 Volotskoi used Siemens's article, "Die Proletarisierung unseres Nachwuchses, eine Gefahr unrassenhygienischer Bevölkerungspolitik," *Archiv für Rassen- und Gesellschafts-Biologie* (hereafter *ARGB*), 1916-17, 12(1): 43-55; and his two-volume treatise on *Grundzüge der Rassenhygiene* (Munich: J. F. Lehmann, 1923); along with the second volume of the infamous collection by Erwin Baur, Eugen Fischer, and Fritz Lenz, *Grundriss der menschlichen Erblichkeitslehre und Rassenhygiene* (Munich: J. F. Lehmann, 1921), written and published by Lenz under the title, *Menschliche Auslese und Rassenhygiene*.

165 Volotskoi, *Klassovye interesy*, p. 45.

166 See the tables of contents of the two oracles of Russian eugenics, *Russian Eugenics Journal* and *Herald of the Eugenics Bureau*, in Babkov, pp. 196-203; and pp. 287-89.

167 See N. K. Kol'tsov, "Genealogiia Ch. Darvina i F. Gal'tona," *REZh*, 1922, 1(1): 64-73; and A. S. Serebrovskii, "Genealogiia roda Aksakovykh," *ibid.*, 82-97.

168 See, for instance, D. M. D'iakonov and Ia. Ia. Lus, "Raspredelenie i nasledonvanie spetsial'nykh sposobnostei," *Izvestiia biuro po evgenike*, 1922, 1: 72-112; and G. G. Shefter, *Vyrozhdenie i evgenika* (M.-L.: GIZ, 1927).

169 Kol'tsov, "Uluchshenie chelovecheskoi porody," p. 10.

170 He described this "experiment" in detail in his article, Volotskoi, "O nekotorykh techeniiakh v sovremennoi evgenike." All the following quotations are from this source.

171 Volotskoi took this citation from Florinskii, 1866, p. 73.

172 He repeated the same talk in April 1926 at the Institute of Physical Culture. Its text, however, appeared in print only two years later, see M. V. Volotskoi, *Sistema evgeniki kak biosotsial'noi distsipliny* (M.: Izd. Timiriazevskogo instituta, 1928). All the subsequent quotations are from this source.

173 See V. M. Volotskoi, "Alkogolizm i sifilis kak factory, vliiaiushchie na potomstvo," *Fizicheskaia kul'tura v nauchno-prakticheskom osveshchenii*, 1928, 1-2: 134-47.

174 See V. M. Volotskoi, *Professional'nye vrednosti i potomstvo* (Vologda: Severnyi pechatnik, 1929). This 250-page book had actually been finished in 1926, but appeared in print only three years later.

175 He gave a talk on the subject at the Timiriazev Institute, but never published its text. See ARAN, f. 356, op. 1, d. 38, ll. 72-73 rev. Unfortunately, the archive does not contain a stenographic record of the talk.

176 Florinskii, 1926.

177 As did all other Soviet publications, on the back of its front page, Florinskii's volume bore a specific inscription — "*Gublit № 1114 (Vologda)*" — that "identified" the concrete censor who had reviewed the book and approved it for publishing. The inscription indicates that the volume was submitted to the censor in the city of Vologda where it was printed. Alas, I was unable to find any archival records illuminating the censorship process.

178 Florinskii, 1926, pp. 163-64.

179 *Ibid.*, pp. 159-60. He took his picture from *Vestnik mody*, 1922, 5.

180 M. V. Volotskoi, "K istorii i sovremennomu sostoianiiu evgenicheskogo dvizheniia, v sviazi s knigoi V. M. Florinskogo," in Florinskii, 1926, pp. VII-XIX. All the subsequent citations are from this source.

181 Volotskoi cited an English translation of Niceforo's report to the 1912 London congress, see A. Niceforo, "The Causes of the Inferiority of Physical and Moral Characters in the Lower Classes," *Problems in Eugenics*, vol. 1, pp. 189-94. On Niceforo and his place in Italian eugenics, see Cassata, *Building a New Man*.

# Chapter 6

1 N. K. Koltzoff [Kol'tsov], "Die rassenhygienische Bewegung in Russland," *ARGB*, 1925/26, 17: 96-99; all the subsequent quotations are from this source. It is worth noting that in his Russian-language texts on the subject, Kol'tsov almost never used the term "racial hygiene," and it seems likely that the title of this particular article, as well as some of its language, was supplied by the journal's editor.

2 The only source of information on Vaisenberg's biography I was able to find is his obituary published in the *Russian Anthropological Journal*, see V. Bunak, "S. A. Vaisenberg (1867-1928). Nekrolog," *RAZh*, 1929, 18(1-2): 71-72.

3 S. Weisenberg [S. Vaisenberg], "Theoretische und praktische Eugenik in Sowjetrussland," *ARGB*, 1926, 18: 69-83 (pp. 71-72).

4 S. Weisenberg [S. Vaisenberg], "Florinsky, W. M. Die Verbesserung und Entartung des menschlichen Geschlechts. (Russisch.) XIX und 165 Seiten. Wologda 1926," *ARGB*, 1927, 19: 105-07.

5 See "Florinsky, W. M., 1926, Die Verbesserung und Entartung des menschlichen Geschlechts. (Russisch.) Wologda, XIX u. 165 S. Eine Neuauflage eines 1866 erschienenen Buches, das in mancher Beziehung als Vorläufer der gegenwärtigen eugenischen Ideen zu betrachten," in Anon., "Bibliographie," *Anthropologischer Anzeiger*, 1927, 4(1): 1-19 (p. 6).

6 Vas. Slepkov, "Prof. V. M. Florinskii. Usovershenstvovanie i vyrozhdenie chelovecheskogo roda. 'Severnyi pechatnik'. Vologda. 1926. 164 str.," *Pravda*, 17 March 1926, p. 3; all the following quotations are from this source. On Slepkov and his involvement with genetics and eugenics, see A. I. Ermolaev, *Istoriia geneticheskikh issledovanii v Kazanskom universitete* (Kazan': Izd-vo Kazanskogo universiteta, 2004), pp. 29-76.

7 V. S[lepkov], "Gosudarstvennyi Timiriazevskii nauchno-issledovatel'skii institut," *Izvestiia*, 18 April 1926, p. 5.

8 B. Vishnevskii, "Zabytyi russkii evgenik," *Priroda*, 1926, 3-4: 100-01. There are almost no biographical materials for this interesting man, only brief obituaries in various journals. See, for instance, M. S. Spirov and N. G. Zalkind, "B. N. Vishnevskii (1892-1965)," *Voprosy antropologii*, 1967, 26: 182.

9 N. K[ol'tsov], "Ot redaktsii," *REZh*, 1924, 2(1): 50.

10 See Nik. Kol'tsov, "V. M. Florinskii. Usovershenstvovanie i vyrozhdenie chelovecheskogo roda. Izdanie 2-e. Pod red. i so vstup. stat'ei V. M.

Volotskogo. Gos. Timiriazevskii nauchno-issledovatel'skii institut. Seriia V. Bib-ka materialista. Vyp. 1. 'Severnyi pechatnik'. Vologda. 1926. Str. 164. Tir. 3000 ekz. Ts. 1 r. 75 kop.," *Pechat' i revoliutsiia*, 1926, 4: 191-93; all the subsequent quotations are from this source.

11 Here Kol'tsov is somewhat exaggerating, for Darwin had introduced the concept of sexual selection already in *Origin*. But he indeed discussed its application to humans only in the *Descent of Man*, six years after the first publication of Florinskii's book.

12 N. Kol'tsov, "R. Gets. Nasledstvennost' i evgenika. Avtorizovannyi perevod s angl. pod red. Iu. A. Filipchenko. 'Seiatel'. L. 1926. Str. 267," *Pechat' i revoliutsiia*, 1926, 2: 206-07 (p. 207). For the English original of the reviewed book, see R. Ruggles Gates, *Heredity and Eugenics* (London: Constable; New York: Macmillan, 1923).

13 M. Beliaev, "Obzor literatury po evoliutsionnomu ucheniiu," *Narodnyi uchitel'*, 1927, 9: 112-18 (p. 116).

14 Volotskoi, "K istorii i sovremennomu sostoianiiu evgenicheskogo dvizheniia," p. xix.

15 See, for instance, T. Iudin, *Evgenika. Uchenie ob uluchshenii prirodnykh svoistv cheloveka* (M.: M. i S. Sabashnikovy, 1925), p. 40; idem, *Evgenika. Uchenie ob uluchshenii prirodnykh svoistv cheloveka* (M.: M. i S. Sabashnikovy, 1928), pp. 53-54; and V. Slepkov, *Evgenika. Uluchshenie chelovecheskoi prirody* (M.: GIZ, 1927), pp. 18-19.

16 T. Iudin, "Evgenika," *Bol'shaia meditsinskaia entsiklopediia* (M.: Sovetskaia entsiklopediia, 1929), vol. 9, pp. 663-70 (p. 670); hereafter references to this encyclopedia will be given as *BME*, 1929, 9: 663-70.

17 For a brief overview, see Becky L. Glass and Margaret K. Stolee, "Family Law in Soviet Russia, 1917-1945," *Journal of Marriage and the Family*, 1987, 49(4): 893-902. For more detailed analyses written largely from a feminist perspective, see Dorothy Atkinson, Alexander Dallin, and Gail Lapidus, eds., *Women in Russia* (Stanford, CA: Stanford University Press, 1977); Wendy Z. Goldman, *Women, the State and Revolution: Soviet Family Policy and Social Life, 1917-1936* (Cambridge: Cambridge University Press, 1993); and Elizabeth A. Wood, *The Baba and the Comrade: Gender and Politics in Revolutionary Russia* (Bloomington, IN: Indiana University Press, 1997).

18 See "O grazhdanskom brake, o detiakh i o vedenii knig aktov sostoianiia," *Sobranie uzakonenii RSFSR*, 1917, 11: 160. For an English-language collection of the Soviet legal documents on family, see Rudolf Schlesinger, ed., *The Family in the USSR* (London: Routledge and Kegan Paul, [1949] 1998).

19 See "Kodeks zakonov ob aktakh grazhdanskogo sostoianiia, brachnom, semeinom i opekunskom prave," *Sobranie uzakonenii RSFSR*, 1918, 76-77: 818.

20 See Anon., "Okhrana zdorov'ia brachuiushchikhsia," *Izvestiia*, 11 August 1923, p. 4; and L. Vasilevskii, "Zdorov'e i brak. K proektu Narkomzdrava," *Pravda*, 9 September 1923, p. 1.

21 A. Sysin, "Brak, zdorov'e i potomstvo," *Izvestiia*, 18 September 1923, p. 3.

22 See an announcement of the report in *Pravda*, 23 November 1923, p. 4.

23 See "Kafedra sotsialnoi gigieny," *Sotsial'naia gigiena*, 1923/1924, 2: 170-72.

24 See A. N. Sysin, "Pervye shagi evgenicheskogo zakonodatel'stva v Rossii," *Sotsial'naia gigiena*, 1924, 3-4: 11-20. All the subsequent quotations are from this source.

25 See, for instance, P. I. Liublinskii, "Evgenicheskaia sterilizatsiia," *Vestnik znaniia*, 1925, 6: 443-50; and idem, "Evgenicheskie tendentsii i noveishee zakonodatel'stvo o detiakh," *REZh*, 1925, 3(1): 3-29.

26 For an analysis of certain aspects of this campaign, especially in relation to sex education, see Frances Lee Bernstein, "'What Everyone Should Know about Sex': Gender, Sexual Enlightenment, and the Politics of Health in Revolutionary Russia, 1918-1931." doctoral dissertation, Columbia University, 1998; and idem, *The Dictatorship of Sex: Lifestyle Advice for the Soviet Masses* (DeKalb, IL: Northern Illinois University Press, 2007).

27 See D. I. Kurskii, ed., *Sbornik statei i materialov po brachnomu i semeinomu pravu* (M.: Iuridicheskoe izdanie, 1926). Many documents from this collection appeared in English translations in Schlesinger, ed., *The Family in the USSR*.

28 N. I. Mikulina-Ivanova, "Oplodotvorenie i nasledstvennost'," *Zhenskii zhurnal*, 1927, 1: 14-15.

29 See Il. Poltavskii, "Institut sotsial'noi gigieny," *Vecherniaia Moskva*, 13 December 1926, p. 3. The State Institute of Social Hygiene established the consultation jointly with the Moscow Venereal Dispensary and the Moscow Psycho-Neurological Dispensary.

30 S. N. Davidenkov, "Geneticheskoe biuro pri M. O. N. i P.," *REZh*, 1928, 6: 55-56.

31 See S. N. Davidenkov, "Geneticheskoe napravlenie v nervno-psikhiatricheskoi profilaktike," *Klinicheskaia meditsina*, 1929, 23-24: 1478-92.

32 See T. I. Iudin, *Zdorov'e, brak i sem'ia* (M.: Narkomzdrav, 1928); and P. Rokitskii,

"Evgenika i brak." *Zhenskii zhurnal,* 1930, 6: 18. See also Tat'iana Pletneva, "Obshchestvenno-nauchnyi podkhod k materinstvu," *Zhenskii zhurnal,* 1927, 7: 13.

33 See also a series of popular articles by obstetrician N. I. Mikulina-Ivanova, "Oplodotvorenie i nasledstvennost'," *Zhenskii zhurnal,* 1927, 1: 14-15; and *idem,* "Brak i evgenika: O kontrole nad zdorov'em lits, vstupaiushchikh v brak," *ibid.,* 1929, 1: 22; as well as her voluminous manual on *Woman's Health, idem, Zdorov'e zhenshchiny* (M.: OMM NKZ, 1928).

34 See, GARF, f. A2307, op. 8, d. 278, ll. 81-83.

35 See Iu. A. Filipchenko, *Frensis Gal'ton i Gregor Mendel'* (M.: GIZ, 1924). The second printing came out in a slightly larger format and under a slightly different title, see *idem, Gal'ton i Mendel': Ikh zhizn' i trudy* (M.: GIZ, 1924).

36 Kol'tsov was a member of the editorial board of the series and profoundly influenced the choice of books that were to comprise it. See GARF, f. R395, op. 9, dd. 315-318.

37 See GARF, f. R395, op. 9, d. 313, ll. 14-14 rev; d. 315, ll. 180, 184-85.

38 See Gregor Mendel', *Opyty nad rastitel'nymi gibridami* (M.-Pg.: GIZ, 1923).

39 Volotskoi recounted the story in a lengthy footnote in his 1925 pamphlet on *Class Interests and Modern Eugenics,* see Volotskoi, *Klassovye interesy,* p. 13.

40 One could only speculate that, if Volotskoi did complete his translation in 1922, the translation of Galton's book might have perhaps been published by the GIZ in the next year, along with Mendel's.

41 Apparently, the manuscript never went into production, see GARF, f. R395, op. 9, d. 77, l. 51; d. 92, l. 40; d. 313, l. 14.

42 GARF, f. R395, op. 9, d. 313, l. 30 rev.

43 Iu. A. Filipchenko, *Puti uluchsheniia chelovecheskogo roda (Evgenika)* (L.: GIZ, 1924).

44 Anon., "Predislovie redaktsii," in Filipchenko, *Puti uluchsheniia chelovecheskogo roda,* pp. 3-7.

45 See GARF, f. R395, op. 9, d. 77, l. 46; d. 92, l. 40.

46 On Solov'ev's life and works, see V. A. Solov'eva, ed., *Zhizn' i deiatel'nost' Z. P. Solov'eva po vospominaniiam sovremennikov* (M.: Meditsina, 1980).

47 Z. P. Solov'ev, "Neskol'ko slov o 'razvedenii porody cheloveka'," in *idem,*

*Stroitel'stvo sovetskogo zdravookhraneniia* (M.: Medgiz, 1932), pp. 320-34. All the subsequent quotations are from this source.

48 See T. Iudin, *Evgenika. Uchenie ob uluchshenii prirodnykh svoistv cheloveka* (M.: M. i S. Sabashnikovy, 1925). The second updated and expanded edition was released by the same private press three years later.

49 I failed to find any materials that could explain why Kol'tsov did not publish the translation of Galton's book with a different press.

50 I was unable to find out who had written the "editorial foreword." The GIZ Petrograd office had several regular reviewers for biological literature, including such well-known specialists as expert in biology education Boris Raikov, zoologist Petr Shmidt, and parasitologist Evgenii Pavlovskii. None of them had ever published anything even remotely similar to the "Marxist" assessment of eugenics presented in the foreword. To the contrary, Raikov, for instance, published several favorable reviews of Filipchenko's various publications on eugenics in the journal *Natural Sciences in School*, see *Estestvoznanie v shkole*, 1922, 1-2: 78; 1926, 2: 90; 1926, 3: 95-96.

51 See, for instance, E. I. Berman, "Kak uluchshit' prirodnye svoistva cheloveka? (Evgenika)," *Molodaia gvardiia*, 1926, 2: 108-16. The article, written by young Moscow pediatrician Efim Berman (1894-1963), was basically a recapitulation of Volotskoi's arguments advanced in his pamphlet on *Class Interests and Modern Eugenics*. For a similar reiteration of Volotskoi's work, see also B. A. Ivanovskii, *Sovetskaia meditsina i fizkul'tura* (M.: Gosmedgiz, 1928), pp. 25-29.

52 It is worth noting that, a year earlier Filipchenko published in the same journal his review of Volotskoi's book, *Elevating the Vital Forces of the Race*, that first introduced Florinskii's treatise to its Soviet audiences. See Iu. A. Filipchenko, "M. V. Volotskoi, Podniatie zhiznennykh sil rasy. Novyi put'. Iz-vo 'Zhizn' i znanie'. M. 1923 g. Str. 96," *Pechat' i revoliutsiia*, 1924, 6: 248-50.

53 N. A. Semashko, "Ikh evgenika i nasha," *Vestnik sovremennoi meditsiny*, 1927, 10: 639-49. The article was an excerpt from a public lecture Semashko has delivered. Alas, I was unable to find either the date or the venue of this lecture.

54 See Iu. A. Filipchenko, "Intelligentsiia i talanty," *Izvestiia biuro po evgenike*, 1925, 3: 83-96, [Babkov, pp. 279-87].

55 G. Shmidt, "Ne iz verkhnikh desiati tysiach, a iz nizhnikh millionov," *PZM*, 1925, 7: 128-33.

56 *Ibid.*, p. 128.

57 See, for instance, Vas. Slepkov, "Nasledstvennost' i otbor u cheloveka," *PZM*,

1925, 4: 102-22; *idem*, "Biologiia cheloveka," *ibid.*, 1925, 10-11: 115-42; *idem*, *Evgenika* (M.-L.: GIZ, 1927); and B. M. Zavadovskii, "Darvinizm i marksizm," *Vestnik Kommunisticheskoi akademii* (hereafter *VKA*), 1926, 14: 226-74.

58 For a brief sketch of Batkis's life and works, see Nora Karlsen, "Grigorii Abramovich Batkis (k 110-letiiu so dnia rozhdeniia)," *Demoskop Weekly*, 2005, 225, http://demoscope.ru/weekly/2005/0225/nauka03.php.

59 See G. A. Batkis, "Sovremennye evgenicheskie techeniia v svete sotsial'noi gigieny," *Sotsial'naia gigiena*, 1927, 1(9): 97-98; and, especially, *idem*, "Sotsial'nye osnovy evgeniki," *ibid.*, 1927, 2(10): 7-25. All the subsequent quotations are from the latter article.

60 For a detailed historical analysis of the interrelations among Marxism, Darwinism, Lamarckism, and genetics in Russia, see Krementsov, "Marxism, Darwinism, and Genetics," pp. 215-46. As any "ism," the label "Lamarckism" meant many different things to its various users, see Snait B. Gissis and Eva Jablonka, eds., *Transformations of Lamarckism: From Subtle Fluids to Molecular Biology* (Cambridge, MA: MIT Press, 2011). In the 1920s, this label was largely used (often in the form of "neo-Lamarckism") to signify a particular concept of biological evolution that considered the inheritance of acquired characteristics to be its major mechanism. Neo-Lamarckism, thus, contradicted both the contemporary notions of heredity developed by geneticists on the basis of Mendel's and Weismann's works and the contemporary notions of evolution developed by Darwin's followers, which considered natural selection to be the major mechanism of biological evolution.

61 See, for example, *Preformizm ili epigenezis?* (Volodga: Severnyi pechatnik, 1926); P. A. Novikov, *Teoriia epigeneza v biologii: Istoriko-sistematicheskii ocherk* (M.: Izdatel'stvo komakademii, 1927); E. S. Smirnov, *Problema nasledovaniia priobretennykh priznakov: Kriticheskii obzor literatury* (M.: Izdatel'stvo komakademii, 1927). For a historical assessment of the debate between geneticists and Lamarckists, see A. E. Gaissinovitch, "The Origins of Soviet Genetics and the Struggle with Lamarckism, 1922-1929," *JHB*, 1980, 13: 1-51; and Krementsov, "Marxism, Darwinism, and Genetics."

62 On Kammerer's work, see Sander Gliboff, "The Case of Paul Kammerer: Evolution and Experimentation in the Early 20th Century," *JHB*, 2006, 39: 525-63.

63 P. Kammerer, *Obshchaia biologiia* (M.-L.: GIZ, 1925). The translation was made from the second updated edition of the original, Paul Kammerer, *Allgemeine biologie* (Stuttgart: Deutsche Verlags-Anstalt, 1920).

64 On Kammerer's life story, see Arthur Koestler, *The Case of the Midwife-Toad* (New York: Random House, 1971).

65 See P. Kammerer, *Zagadka nasledstvennosti* (L.: Priboi, 1927) and (M.: GIZ, 1927). For the original, see *idem, Das Rätsel der Vererbung: Grundlagen der allgemeinen Vererbungslehre* (Berlin: Ullstein, 1925). See also, P. Kammerer, *Pol, razmnozhenie i plodovitost': Biologiia vosproizvedenia* (L.: Priboi, 1927). Furthermore, also in 1927, the head of Narkompros Anatolii Lunacharskii — the official patron of all "pure" science in the country and an active member of the Communist Academy — wrote a movie script loosely based on Kammerer's life-story. Directed by a young talented director, Grigorii Roshal', and featuring Lunacharskii's wife in the lead female part, the movie, entitled *Salamander*, hit screens across the country in late 1928. The movie portrayed Kammerer as a progressive scientist, whose research on the inheritance of acquired characteristics in salamanders earned him the hatred of the clergy and the bourgeoisie. His opponents (who apparently subscribed to the views of modern genetics on the impossibility of Lamarckian inheritance) plot to destroy his reputation by tampering with his research. Driven out of his lab and haunted by lack of money, the scientist is saved from poverty and disgrace (and his salamanders from inevitable death) by the invitation of the Soviet government to come to Moscow to continue his research. The official press greeted the movie with much enthusiasm. See, for example, Kh. Khersonskii, "Salamandra," *Pravda*, 30 December 1928, p. 5.

66 See, for instance, M. I. Lifshits, *Uchenie o konstitutsiiakh cheloveka s kratkim ocherkom sovremennogo polozheniia voprosa o nasledstvennosti* (Kiev: GIZ Ukrainy, 1924); A. A. Krontovskii, *Nasledstvennost' i konstitutsiia* (Kiev: GIZ Ukrainy, 1925); S. Davidenkov, *Nasledstvennye bolezni nervnoi sistemy* (Kiev: GIZ Ukrainy, 1925); T. I. Iudin, *Psikhopaticheskie konstitutsii* (M.: Sabashnikov, 1926); and many others.

67 See, for instance, S. Levit, "Evoliutsionnye teorii v biologii i marksizm," *Vestnik sovremennoi meditsiny*, 1925, 9: 15-24; *idem*, "Evoliutsionnye teorii v biologii i marksizm," *Meditsina i dialekticheskii materialism*, 1926, 1: 15-32; *idem*, "Dialekticheskii materializm v meditsine," *Vestnik sovremennoi meditsiny*, 1927, 23: 1481-90. For a brief biography of Levit, see Mark B. Adams, "Levit, Solomon Grigorevich," in Holmes, ed., *Dictionary of Scientific Biography*, vol. 18, suppl. II, pp. 546-49.

68 See, *Vestnik sovremennoi meditsiny*, 1925, 9: 27; and *ibid.*, 1927, 23: 1506-07.

69 S. Levit, "Problema konstitutsii v meditsine i dialekticheskii materialism," in *Meditsina i dialekticheskii materializm*, 1927, 2: 7-34 (pp. 20-21).

70 See, for instance, I. I. Rozenblium, "Popytka marksistkogo podkhoda k nekotorym problemam konstitutsii i nasledstvennosti," *Leningradskii meditsinskii zhurnal*, 1926, 4: 48-63; L. Syrkin, "Uluchshenie chelovecheskogo roda ili evgenika," *V pomoshch' sanitarke*, 1928, 8: 1-2, and Berman, "Kak uluchshit' prirodnye svoistva cheloveka? (Evgenika)."

71 See, for instance, "Preniia po dokladu M. V. Volotskogo," *VKA*, 1927, 20: 232-54 (p. 237).

72 *Ibid.*, p. 246.

73 See, for instance, B. M. Zavadovskii, "Darvinizm i lamarkizm i problema nasledovaniia priobretennykh priznakov," *PZM*, 1925, 10-11: 79-114; and *idem, Darvinizm i marksizm* (M.: GIZ, 1926).

74 This quote comes from a commentary by Mikhail Mestergazi, a geneticist and Bolshevik party member, to Volotskoi's report on "issues in modern eugenics" delivered to a meeting of the Society of Materialist-Biologists in December 1926, see "Preniia po dokladu M. V. Volotskogo," *VKA*, 1927, 20: 232-54 (p. 234).

75 See, for instance, his favorable review of the book published by a well-known supporter of Lamarckism under the title "Are Acquired Characteristics Transmitted by Heredity?", M. Volotskoi, "A. P. Vladimirskii. Peredaiutsia li po nasledstvu priobretennye priznaki? Darvinovskaia bib-ka pod red. prof. M. M. Zavadovskogo. GIZ. M.-L. 1927. Str. 184. Tirazh 4000. Ts. 1 r. 25 k.," *Pechat' i revoliutsiia*, 1927, 6: 179-80.

76 For a sample of publications by members of the Lamarck Circle, see E. S. Smirnov, Iu. M. Vermel', and B. S. Kuzin, *Ocherki po teorii evolutsii* (M.: Krasnaiia nov', 1924); and E. S. Smirnov, *Problema nasledovaniia priobretennykh priznakov: Kriticheskii obzor literatury* (M.: Izdatel'stvo Komakademii, 1927).

77 See ARAN, f. 356, op. 1, d. 38, ll. 53-54. Unfortunately, the file does not contain the text of the report.

78 See, for instance, lengthy analyses of the concept of the inheritance of acquired characteristics produced by Filipchenko's student Theodosius Dobzhansky in F. G. Dobrzhanskii, "K voprosu o nasledovanii priobretennykh priznakov," in *Preformizm ili epigenezis?*, pp. 27-47; and *idem.*, *Chto i kak nasleduetsia u zhivykh sushchestv?* (L.: GIZ, 1926).

79 See, for instance, N. K. Kol'tsov, "Noveishie popytki dokazat' nasledstvennost' blagopriobretennykh priznakov," *REZh*, 1924, 3(2-3): 159-67.

80 T. H. Morgan and Iu. A. Filipchenko, *Nasledstvenny li priobretennye priznaki* (Leningrad: Seiatel', 1925). For Morgan's original article, see T. H. Morgan, "Are Acquired Characteristics Inherited?" *Yale Review*, 1924, 13(4): 712-29.

81 Iu. A. Filipchenko, "Nasledstvennost' priobretennykh priznakov," in Morgan and Filipchenko, *Nasledstvenny li priobretennye priznaki*, p. 57. Here Filipchenko used the same "translation" of Florinskii's "hereditary potentials" as "genes," which had been used by Volotskoi.

82 See ARAN, f. 1595, op. 1, d. 389, l. 1.

83 See a stenographic record of the report in ARAN, f. 350, op. 2, d. 112. A full text of the report soon appeared on the pages of *Under the Banner of Marxism*, see A. S. Serebrovskii, "Teoriia nasledstvennosti Morgana i Mendelia i marksisty," *PZM*, 1926, 3: 98-117. All the following quotations are from this source. Unless noted otherwise, all the emphasis is Serebrovskii's.

84 Volotskoi, "O nekotorykh techeniiakh v sovremennoi evgenike," p. 85. For the original citation see Florinskii, 1866, p. 73.

85 See A. S. Serebrovskii, "Genogeografiia i genofond sel'skokhoziaistvennykh zhivotnykh SSSR," *Nauchnoe slovo*, 1928, 9: 3-22.

86 On the further development of this concept within the framework of population genetics and evolutionary synthesis, see Mark B. Adams, "From 'Gene Fund' to 'Gene Pool': On the Evolution of Evolutionary Language," in William Coleman and Camille Limoges, eds., *Studies in the History of Biology* (Baltimore, MD: Johns Hopkins University Press, 1979), vol. 3, pp. 241–85.

87 See, for instance, Anon., "U nas ogromnyi genofond," *30 dnei*, 1926, 12: 84-85.

88 N. K. Kol'tsov, "Rodoslovnye nashikh vydvizhentsev," *REZh*, 1926, 4(3-4): 103-43, [Babkov, pp. 152-95].

89 For a stenographic record of the report and its discussion, see ARAN, f. 350, op. 2, d. 68, ll. 1-71. The report and its discussion soon appeared in print in the *Herald of the Communist Academy*, see M. V. Volotskoi, "Spornye voprosy evgeniki," *VKA*, 1927, 20: 212-32; and "Preniia po dokladu M. V. Volotskogo," *ibid.*, 232-54.

90 "Preniia po dokladu M. V. Volotskogo," p. 236.

91 "Preniia po dokladu M. V. Volotskogo," p. 243.

92 M. Mestergazi, "Epigenezis i genetika," *VKA*, 1927, 19: 197-233 (p. 232).

93 The debates over "bourgeois" and "proletarian" eugenics also found vivid expression in a variety of literary, theatrical, and cinematographic productions. For a brief introduction to this theme, see my analysis of the play *I Want a Baby*, written in 1926 by Sergei Tret'iakov, a popular poet and playwright; Krementsov, "From 'Beastly Philosophy' to Medical Genetics." I am currently preparing a manuscript that explores these productions in their scientific and social contexts.

94 N. Kol'tsov, "Evfenika," *BME*, 1929, 9: 689-92. All the following quotations are from this source. Although there were slight differences in the Russian

spelling of the term — Kol'tsov spelled it *evfenika*, Andreev *eufenika*, Mestergazi *eutenika* — their explanations and use in the respective texts leave no doubt that all three meant exactly the same thing, representing merely different Russian transliterations of the term "eu-phenics" modeled after "eu-genics."

95 See ARAN, f. 1595, op. 1, d. 323, ll. 1-2.

96 See S. L[evit], "Otchet o rabote kabineta nasledstvennosti i konstitutsii cheloveka pri Mediko-biologicheskom institute za 1928-29 akad. g.," *Mediko-biologicheskii zhurnal*, 1929, 5: 115-16.

97 See N. Kol'tsov, "Zadachi i metody izucheniia rasovoi patologii," *REZh*, 1929, 7(2-3): 69-87, [Babkov, pp. 479-92]; and Anon., "Kratkii otchet o deiatel'nosti obshchestva po izucheniiu rasovoi patologii i geograficheskogo rasprostraneniia boleznei," *ibid.*: 113.

98 A. Serebrovskii, "Galton," *Bol'shaia sovetskaia entsiklopediia* (hereafter *BSE*) (M.: Sovetskaia entsiklopediia, 1929), vol. 14, pp. 443-44; idem, "Galton," *BME*, 1929, 6: 253-54.

99 T. Iudin, "Evgenika," *BME*, 1929, 9: 663-70; N. Kol'tsov, "Evfenika," *ibid.*, 9: 689-92.

100 For details, see Robert C. Tucker, *Stalin in Power: The Revolution from Above, 1928-1941* (New York: Norton, 1990); also Sheila Fitzpatrick, ed., *Cultural Revolution in Russia, 1928-1931* (Bloomington, IN: Indiana University Press, 1978).

101 On the Shakhty trial, see Kendal Bailes, *Science and Russian Culture in an Age of Revolutions* (Bloomington, IN: Indiana University Press, 1990); Loren R. Graham, *The Ghost of the Executed Engineer* (Cambridge, MA: Harvard University Press, 1993); for much more extensive and informative Russian-language publications, see S. A. Kislitsyn, *Shakhtinskoe delo: nachalo stalinskikh repressii protiv nauchno-tekhnicheskoi intelligentsii v SSSR* (M.: Logos, 1992) and, especially, S. A. Krasil'nikov, ed., *Shakhtinskii protsess 1928 g.: podgotovka, provedenie, itogi*, 2 vols. (M.: ROSSPEN, 2011).

102 For a general assessment of the relations between specialists and the state in the field of health protection in Russia, see Susan G. Solomon, "The Expert and the State in Russian Public Health: Continuities and Changes Across the Revolutionary Divide," in D. Parker, ed., *The History of Public Health and the Modern State* (Amsterdam: Rodopi, 1994), pp. 183-223.

103 For a detailed analysis of the impact of the "revolution from above" on the Soviet science system, see Krementsov, *Stalinist Science*.

104 See P. Rokitskii, *Mozhno li uluchshit' chelovecheskii rod* (M.-L.: GIZ, 1928).

105 See ARAN, f. 356, op. 3, d. 60, l. 301.

106 V. M. Volotskoi, *Professional'nye vrednosti i potomstvo* (M.: Timiriazevskii Institut, 1929).

107 See *Trudy Vsesoiuznogo s"ezda po genetike i selektsii*, 6 vols. (L.: Redkollegiia, 1930). See also a brief report on the congress by Erwin Baur, one of its German guests, E. B., "Der Allrussische Kongres für Genetik, Tier- und Pflanzenzüchtung in Leningrad, Januar 1929," *Der Züchter*, 1929, 1(1): 24-25.

108 A. Sh. Shorokhova, "Novye puti v selektsii cheloveka i mlekopitaiushchikh," *Vrachebnaia gazeta*, 1929, 3-4: 179-84.

109 See RO RNB, f. 813, op. 1, d. 363 and d. 736.

110 A. S. Serebrovskii, "Antropogenetika i evgenika v sotsialisticheskom obshchestve," *Trudy Kabineta nasledstvennosti i konstitutsii cheloveka pri Mediko-biologicheskom institute*, published as a special issue of *Mediko-biologicheskii zhurnal*, 1929, 5: 3-19, [Babkov, pp. 505-16]; S. Levit, "Genetika i patologiia (v sviazi s sovremennym krizisom v meditsine)," *ibid.*, 20-39, [Babkov, pp. 552-65].

111 For a detailed, though dated analysis of the Bolshevization of the Academy of Sciences, see Loren R. Graham, *The Soviet Academy of Sciences and the Communist Party, 1927-1932* (Princeton, NJ: Princeton University Press, 1967); for a more recent assessment based on newly available archival materials, see F. F. Perchenok, "'Delo Akademii nauk' i 'velikii perelom' v sovetskoi nauke," in *Tragicheskie sud'by: repressirovannye uchenye Akademii nauk SSSR* (M.: Nauka, 1995), pp. 201-35.

112 For detailed, but also dated analyses of these campaigns in English, see David Joravsky, *Soviet Marxism and Natural Science, 1917–1932* (New York: Columbia University Press, 1961); and Loren R. Graham, *Science and Philosophy in the Soviet Union* (New York: Knopf, 1974); for a more recent analysis, focusing specifically on the "ideological" impact of the campaigns in biological sciences, see E. I. Kolchinskii, *V poiskakh sovetskogo 'soiuza' filosofii i biologii* (SPb.: D. Bulanin, 1999).

113 D. Bednyi, "Evgenika," *Izvestiia*, 4 June 1930, p. 4. On Bednyi and his role in Soviet literature, see Robert Horvath, "The Poet of Terror: Dem'ian Bednyi and Stalinist Culture," *Russian Review*, 2006, 65(1): 53-71.

114 The texts of Bednyi's poem and Serebrovskii's response have been reprinted in R. A. Fando, "Polemika o sud'be evgeniki (v poeticheskom zhanre)," *VIET*,

2002, 3: 604-17. However, in this publication, Bednyi's poem is misdated.

115 Anon., "Po povodu proizvodstvennogo plana 'sotsialisticheskoi evgeniki'," *Moskovskii meditsinskii zhurnal*, 1930, 9: 77-87.

116 A. S. Serebrovskii, "Pis'mo v redaktsiiu," *Mediko-biologicheskii zhurnal*, 1930, 4-5: 447-48, [Babkov, pp. 517-18].

117 A. I. Abrikosov, "Trudy kabineta nasledstvennosti i konstitutsii cheloveka. Vyp.1," *Russkaia klinika*, 1930, 13(72): 522-23 (p. 522); see also V. Bunak, "Medio-biologicheskii zhurnal, vyp. 5, 1929," *RAZh*, 1930, 19(3-4): 209; N. P. Dubinin, "Trudy kabineta nasledstvennosti i konstitutsii cheloveka. Vyp.1. 1929," *Nauchnoe slovo*, 1930, 4: 122-26; *idem*, "Uspekhi sovremennoi genetiki i meditsina," *Tsentral'nyi referativnyi meditsinskii zhurnal*, 1930, 6: 330-50; and Ia. I. Cherniak, *Genetika i meditsina* (Kharkov: Nauchnaia mysl', 1930).

118 For the standard instructions to a workers' brigade conducting an inspection, see GARF, f. A2307, op. 17, d. 151, l. 3. On the general campaign to reform "medical-scientific societies," see S. Subotnik, "Meditsinskie nauchnye obshchestva nuzhdaiutsia v ozdorovlenii," *Izvestiia*, 11 January 1930, p. 4; and Anon., "Nauchno-meditsinskie obshchestva otorvany ot shirokoi nauchnoi obshchestvennosti. Obsledovanie nauchnykh meditsinskikh obshchestv rabochimi brigadami," *Izvestiia*, 12 February 1930, p. 5.

119 See G. Sobolev, "Russkoe evgenicheskoe obshchestvo," *VARNITSO*, 1930, 5: 49-50. On the campaign directed personally against Kol'tsov, see B. Zavadovskii, "Prof. N. K. Kol'tsov," *Izvestiia*, 18 January 1930, p. 2; and against his IEB, M. Rokhlina, "Obshchestvennyi smotr In-ta eksperimental'noi biologii," *VARNITSO*, 1930, 5: 44-48.

120 The plan is preserved among Kol'tsov's personal papers in ARAN, f. 450, op. 4, d. 7, ll. 3-7.

121 A manuscript of this article is preserved among Kol'tsov's personal papers in ARAN, f. 450, op. 5, d. 29, ll. 1-25 rev. All the subsequent quotations are from this source. For an English translation, see Babkov, pp. 48-56.

122 Tkachev, *Sotsial'naia gigiena*, pp. 11, 153.

123 G. Batkis, "Evgenika," *BSE*, 1931, 23: 812-19, [Babkov, pp. 519-24].

124 Solov'ev, "Neskol'ko slov o 'razvedenii porody cheloveka'," in *idem*, *Stroitel'stvo sovetskogo zdravookhraneniia* (M.: Medgiz, 1932), pp. 320-34. After its first 1932 publication the article was regularly reprinted in various collections of Solov'ev's works, see Z. P. Solov'ev, *Voprosy zdravookhraneniia* (M.-L.: Medgiz, 1940), pp. 300-11; *idem*, *Izbrannye proizvedeniia* (M.: Izd.

Meditsinskoi literatury, 1965), pp. 83-96; *idem, Izbrannye proizvedeniia* (M.: Meditsina, 1970), pp.189-99.

125 On "public discussion" as an instrument of Bolshevik policies, see Krementsov, *Stalinist Science.*

126 See, for instance, P. I. Valeskaln and B. P. Tokin, eds., *Uchenie Darvina i marksizm-leninzm* (M.: Partizdat, 1932).

127 Engels's brochure appeared in Russia in numerous editions and translations, see F. Engels, *Ot obez'iany k cheloveku* (Gomel': Gomel'skii rabochii, 1922); and *idem, Rol' truda v protsesse ochelovechivaniia obez'iany* (M.: Partizdat, 1932).

128 See L. Vygotskii, "Sotsialisticheskaia peredelka cheloveka," *VARNITSO,* 1930, 9-10: 36-44. For a dated, but still useful historical account of this newfound emphasis on nurture versus nature in the specific area of "the conception of personality," see Raymond A. Bauer, *The New Man in Soviet Psychology* (Cambridge, MA: Harvard University Press, 1952). On *VARNITSO,* see I. A. Tugarinov, "VARNITSO i AN SSSR," *VIET,* 1989, 4: 46-55.

129 V. I. Kremianskii, "Perekhod ot vedushchei roli estestvennogo otbora k vedushchei roli truda," *Uspekhi sovremennoi biologii,* 1941, 14(2): 356-71.

130 See, for instance, materials of the First All-Union Conference of Teachers of Deaf-Mute Children held in Moscow in May 1929, which appeared under the title "Socio-Political Upbringing of Mentally-Retarded and Physically-Defective Children," M. M. Pistrak, ed., *Obshchestvenno-politicheskoe vospitanie umstvenno-otstalykh i fizicheski-defektivnkh detei* (M.-L.: GIZ, 1930).

131 See F. Galton, "Experiments in Pangenesis," *Proceedings of the Royal Society,* 1871, 19: 393-410.

132 The main apostles of this value system — Nikolai Chernyshevskii, Nikolai Dobroliubov, and Dmitrii Pisarev — became the "cultural heroes" of the foundational myths generated by both the Russian intelligentsia and the Bolsheviks. Lenin's numerous pronouncements on the subject, endlessly reproduced during the 1920s and beyond, placed Chernyshevskii, Dobroliubov, and Pisarev in the pantheon of the Russian Revolution as the rightful successors to the Decembrists and Alexander Herzen, and the predecessors (via *narodniki*) of the Bolsheviks themselves. See, for instance, Lenin's articles "What Legacy Do We Reject?" (1897); "What is To Be Done?" (1902); "On the Occasion of a Jubilee" (1911); "To Herzen's Memory" (1912); and many others.

133 Kol'tsov, "Uluchshenie chelovecheskoi porody," p. 20.

134 See P. I. Liublinskii, "Evgenicheskie tendentsii i noveishee zakonodatel'stvo o detiakh," *REZh*, 1925, 3(1): 3-29; *idem*, "Brak i evgenika (O kontrole nad zdorov'em lits, vstupaiushchikh v brak)," *REZh*, 1927, 5(2): 49-89; R. Bravaia, "Mal'tuziantsy XX veka," *Zhurnal po izucheniiu rannego detskogo vozrasta*, 1928, 7(1): 63-69; Z. O. Michnik, "Soznatel'noe materinstvo i regulirovanie detorozhdeniia," *ibid.*, 70-76; N. I. Mikulina-Ivanova, "Brak i evgenika: O kontrole nad zdorov'em lits, vstupaiushchikh v brak," *Zhenskii zhurnal*, 1929, 1: 22; and P. Rokitskii, "Evgenika i brak." *ibid.*, 1930, 6: 18.

135 Galton, "Eugenics: Its Definition, Scope and Aims," p. 45.

136 Galton, "Probability, The Foundation of Eugenics," p. 81.

137 M. Gremiatskii, "Evgenika," *Malaia sovetskaia entsiklopediia* (M.: Sovetskaia entsiklopediia, 1936), vol. 3, pp. 150-52.

138 See G. A. Batkis, *Sotsial'naia gigiena. Sanitarnoe sostoianie naseleniia i sanitarnaia statistika* (M.-L.: Biomedgiz, 1936), pp. 273-89 (pp. 277-78).

139 The history of human genetics in Russia has attracted considerable attention and generated substantial literature. See, for instance, Mark B. Adams, "The Politics of Human Heredity in the USSR, 1920-1940," *Genome*, 1989, 31: 879-84; *idem*, "The Soviet Nature-Nurture Debate," in Loren R. Graham, ed., *Science and the Soviet Social Order* (Cambridge, MA: Harvard University Press, 1990), pp. 94-138; and *idem*, Garland E. Allen, and Sheila F. Weiss, "Human Heredity and Politics: A Comparative Institutional Study of the Eugenics Record Office at Cold Spring Harbor (United States), the Kaiser Wilhelm Institute for Anthropology, Human Heredity, and Eugenics (Germany), and the Maxim Gorky Medical Genetics Institute (USSR)," *Osiris*, 2005, 20: 232-62. See also numerous Russian language studies, such as V. V. Babkov, "Meditsinskaia genetika v SSSR," *Vestnik RAN*, 2001, 71(10): 928-37; *idem*, "Moskva, 1934: Rozhdenie meditsinskoi genetiki," *Vestnik VOGIS*, 2006, 10(3): 455-78; *idem*, *Zaria genetiki cheloveka*; M. B. Konashev, "Ot evgeniki k meditsinskoi genetike," *Rossiiskii biomeditsinskii zhurnal*, 2002, 3, http://www.medline.ru/public/art/tom3/eumedgen.phtml; M. D. Golubovskii, "Stanovlenie genetiki cheloveka," *Priroda*, 2012, 10: 53-63; R. A. Fando, *Stanovlenie othechestvennoi genetiki cheloveka. Na perekrestke nauki i politiki* (M.: Maks-Press, 2013); and many others. Alas, with the exception of the last book, all other historical works focus almost exclusively on the history of medical genetics research and institutions in Moscow and, to a lesser extent, Leningrad, passing over in silence important developments in other centers, such as Kazan and Kharkov. Furthermore, despite the existence of this voluminous literature, the history of Russian human genetics is either misrepresented or altogether ignored in the *general* histories of human and medical genetics available in English. See Diane B. Paul, *Controlling*

*Human Heredity: 1865 to the Present* (Atlantic Highlands, NJ: Humanities Press, 1995); Robert G. Resta and Diane B. Paul, eds., *Historical Aspects of Medical Genetics*, a special issue of the *American Journal of Medical Genetics*, 2002, 115(2): 73-110; Peter S. Harper, *A Short History of Medical Genetics* (New York: Oxford University Press, 2008); Bernd Gausemeier, Saffan Muller-Wille, and Edmund Ramsden, eds., *Human Heredity in the Twentieth Century* (London: Pickering and Chatto, 2013); and Petermann, Harper, and Doetz, eds., *History of Human Genetics*.

140 The same year, Serebrovskii became chairman of the first genetics department at Moscow University and chairman of the genetics department at the Narkomzem Institute of Animal Breeding, which perhaps also contributed to his move away from research in human genetics.

141 S. Levit, "Chelovek kak geneticheskii ob"ekt i izuchenie bliznetsov kak metod antropogenetiki," *Trudy geneticheskogo otdeleniia pri Mediko-biologicheskom institute*, 2, published as a special issue of *Mediko-biologicheskii zhurnal*, 1930, 4-5: 273-87, [Babkov, pp. 566-76]. All the subsequent quotations are from this source. For the tables of contents of the four volumes of the MBI's *Proceedings*, see Babkov, pp. 621-24.

142 The term "Phänogenetik" was coined in 1918 by the one-time student of August Weismann, Halle University zoology professor Valentin Haecker, *Entwicklungsgeschichtliche Eigenschaftsanalyse: (Phänogenetik): gemeinsame Aufgaben der Entwicklungsgeschichte, Vererbungs- und Rassenlehre* (Jena: Fisher, 1918). On Haecker, see Uwe Hosfeld, Elizabeth Watts, and Georgy S. Levit, "Valentin Haecker (1864–1927) as a Pioneer of Phenogenetics: Building the Bridge Between Genotype and Phenotype," *Epigenetics*, 2017, 12(4): 247-53. For a detailed historical analysis of the development of phenogenetics, see K. B. Sokolova, *Razvitie fenogenetiki v pervoi polovine XX veka* (M.: Nauka, 1998); for a brief English-language overview of Russian contributions to the field, see Leonid I. Korochkin, Boris V. Konyukhov, and Alexander T. Mikhailov, "From Genes to Development: Phenogenetic Contributions to Developmental Biology in Soviet Russia from 1917 to 1967," *International Journal of Developmental Biology*, 1997, 4: 763-70.

143 On the Rockefeller fellowships for Russian scientists and physicians, see Susan G. Solomon and Nikolai Krementsov, "Giving and Taking Across Borders: The Rockefeller Foundation and Soviet Russia, 1919-1928," *Minerva*, 2001, 3: 265-98.

144 Muller apprised his Russian colleagues of the remarkable progress the Morgan group had made during the last ten years and presented them a large collection of Drosophila stocks he brought from the United States. See G. G. Meller [H. J. Muller], "Rezul'taty desiatiletnikh geneticheskikh issledovanii

s Drosophila," *Uspekhi eksperimental'noi biologii*, 1923, 1: 292-21; for Muller's description of his trip, see H. J. Muller, "Observations of Biological Science in Russia," *Scientific Monthly*, 1923, 16: 539-52. For a voluminous biography of Muller, see Elof Axel Carson, *Genes, Radiation and Society: The Life and Work of H. J. Muller* (Ithaca, NY: Cornell University Press, 1981). Alas, this biography glosses over both Muller's involvement with eugenics and his relations to Soviet Russia and socialist ideas.

145 See a letter from the RF officer Daniel O'Brian to Serebrovskii, of 11 February 1931, in ARAN, f. 1595, op. 1, d. 377, l. 101; and Levit's report on his fellowship in GARF, f. A2307, op. 19, d. 241, ll. 2-5. Isaak Agol, another active "materialist-biologist" who had converted to genetics under Serebrovskii's tutelage also received an RF fellowship, which he took together with Levit at Muller's lab in Texas, see his report in GARF, f. A2307, op. 19, d. 232, ll. 4-7.

146 See A. Serebrovskii, "Chetyre stranitsy, kotorye vzvolnovali uchenyi mir," *Pravda*, 11 September 1927, p. 5. The title of the article, "Four Pages that Shook the Scientific World," alluded to the title of the famous book about the Bolshevik Revolution by John Reed, *Ten Days that Shook the World* (New York: Boni and Liveright, 1919).

147 See, H. J. Muller, "Mutation," in *Eugenics, Genetics and the Family: Scientific Papers of the Second International Congress of Eugenics* (Baltimore, MD: Williams and Wilkins, 1923), vol. 1, pp. 106-12. On Muller's involvement with eugenics, see Garland E. Allen, "Biology and Culture: Science and Society in the Eugenic Thought of H. J. Muller," *BioScience*, 1970, 20(6): 346-53; Diane B. Paul, "Eugenics and the Left," *Journal of the History of Ideas*, 1984, 45(4): 567-90; and Elof Axel Carlson, "Eugenics and Basic Genetics in H. J. Muller's Approach to Human Genetics," *History and Philosophy of the Life Sciences*, 1987, 9(1): 57-78.

148 On Davenport and his genetics department, see Jan A. Witkowski, *The Road to Discovery: A Short History of Cold Spring Harbor Laboratory* (Cold Spring Harbor, NY: Cold Spring Harbor Laboratory Press, 2016).

149 [S. Levit], "Ot redaktora," *Trudy mediko-biologicheskogo instituta* (M.-L.: Biomedgiz, 1934), vol. 3, p. i.

150 See S. Levit, "Darvinizm, rasovyi shovinizm i sotsial-fashizm," in Valeskaln and Tokin, eds., *Uchenie Darvina i marksizm-leninizm*, pp. 107-25.

151 G. I. Meller [H. J. Muller], "Evgenika v usloviiakh kapitalisticheskogo obshchestva," *Uspekhi sovremennoi biologii*, 1933, 2(3): 3-11; for the English language original, see H. J. Muller, "The Dominance of Economics over

Eugenics," *Scientific Monthly*, 1933, 37: 44-47.

152 G. Meller [H. J. Muller], "Evgenika na sluzhbe natsional-sotsializma," *Priroda*, 1934, 1: 100-6.

153 See also H. J. Muller, "Lenin's Doctrines in Relation to Genetics," in *Pamiati V. I. Lenina* (M.: Partizdat, 1934), pp. 565-92. The article has been reprinted in Graham, *Science and Philosophy in the Soviet Union*, pp. 453-69.

154 See *Konferentsiia po meditsinskoi genetike. Doklady i preniia*, issued as a supplement to the journal *Sovetskaia klinika*, 1934, 20(7-8); all the subsequent quotations are from this source.

155 See Todes, *Ivan Pavlov*.

156 See, for instance, S. G. Levit and S. N. Ardashnikov, eds., *Trudy Mediko-geneticheskogo nauchno- issledovatel'skogo instituta imeni Maksima Gorkogo* (M.-L.: Biomedgiz, 1936), vol. 4; N. N. Anichkov, ed. *Nevrologiia i genetika* (M.-L.: VIEM, 1936), vol. 1; and S. N. Davidenkov, ed., *Nevrologiia i genetika* (M.-L.: VIEM, 1936), vol. 2.

157 See, for instance, a report on a 1935 visit to Moscow and Levit's IMG by the secretary of *Eugenics Review* E. A. Palmer, in "Notes of the Quarter," *Eugenics Review*, 1935, 27(3): 187-88.

158 See, for instance, S. G. Levit and N. N. Malkova, "A New Mutation in Man: Haemophilia-a," *Journal of Heredity*, 1930, 21(2): 73-78; S. Levit, "On the Heredity of Atheroma," *ibid.*, 1931, 22(3): 3-5; *idem*, "Twin investigations in the USSR," *Character and Personality*, 1935, 3: 188-93; S. G. Levit, S. G. Ginsburg, V. S. Kalinin, and R. G. Feinberg, "Immunological Detection of the Y-chromosome in *Drosophilia Melanogaster*," *Nature*, 1936, 139: 78-79; V. V. Bunak, "Changes in the Mean Values of Characters in Mixed Populations," *Annals of Eugenics*, 1936, 7(3): 195-206; and many others.

159 For a detailed history of the Seventh International Genetics Congress, see Nikolai Krementsov, *International Science between the World Wars: The Case of Genetics* (London: Routledge, 2005).

160 H. J. Muller, *Out of the Night. A Biologist's View of the Future* (New York: Gollancz, 1935). See a contemporary review of the book by an eminent US geneticist, P. W. Whiting, "Communist Eugenics: Review of a Preview of a Possible Tomorrow for Human Society," *Journal of Heredity*, 1936, 27(3): 132-35.

161 Mark B. Adams found a draft of Muller's letter to Stalin among Muller's papers, see Adams, "Eugenics in Russia," in *idem*, ed., *The Wellborn Science*,

p. 195. A Russian translation of the letter has been found in Stalin's personal archive, see "Pis'mo Germana Miollera — I. V. Stalinu," *VIET*, 1997, 1: 65-78. The English text was published in John Glad, "Hermann J. Muller's 1936 letter to Stalin," *Mankind Quarterly*, 2003, 43(3): 305-20. The following quotations from Muller's letter are given from the English original located in the Lilly Library of the University of Indiana (Bloomington), Muller MSS, Writings, Box 3, Folder 1936.

162 For details on Muller's further development of these ideas, see Chapter 8.

163 "Notes of the Quarter," *Eugenics Review*, 1935, 27(3): 188.

164 Charles Davenport to the Secretary of State, 17 December 1936. APS, Davenport Papers.

165 For details, see Adams, Allen, and Weiss, "Human Heredity and Politics," pp. 232-62; and Krementsov, "Eugenics, *Rassenhygiene*, and Human Genetics in the late 1930s," in Solomon, ed., *Doing Medicine Together: Germany and Russia between the Wars*, pp. 369-404.

166 See, for instance, G. I. Petrov, *Rasovaia teoriia na sluzhbe u fashizma* (M.-L.: Sotsekizdat, 1934); E. A. Finkel'shtein, "Evgenika i fashizm," in *Rassovaia teoriia na sluzhbe fashizma* (Kiev: Medizdat, 1935), pp. 37-88; Z. A. Gurevich, "Fashizm, 'rasovaia gigiena' i meditsina," in *ibid.*, pp. 89-125.

167 G. Frizen, "Genetika i fashizm," *PZM*, 1935, 3: 86-95.

168 E. Kol'man, "Chernosotennyi bred fashizma i nasha mediko-biologicheskaia nauka," *PZM*, 1936, 11: 64-72.

169 L. Karlik, "Trudy Mediko-geneticheskogo instituta im. M. Gor'kogo," *PZM*, 1936, 12: 169-86.

170 See, Anon., "Po lozhnomu puti," *Pravda*, 26 December 1936, p. 4.

171 For the meeting's transcripts, see GARF, f. R8009, op. 1, d. 113.

172 For details, see Krementsov, *International Science*; and *idem*, "Eugenics, *Rassenhygiene*, and Human Genetics".

173 See "Men and Mice at Edinburgh," *Journal of Heredity*, 1939, 30: 371-74. All the subsequent quotations are from this source. The text of the "manifesto" also appeared in *Nature*, see "Social Biology and Population Improvement," *Nature*, 1939, 144: 521-22. For a detailed analysis of Muller's role in preparing the manifesto, see Krementsov, *International Science*, pp. 121-22.

174 The original signatories were F. A. Crew, J. B. S. Haldane, S. C. Harland, L.

Hogben, J. S. Huxley, H. J. Muller, and J. Needham. Additional signatures were those of G. P. Child, P. R. David, G. Dahlberg, Th. Dobzhansky, R. A. Emerson, John Hammond, C. L. Huskins, W. Landauer, H. H. Plough, E. Price, J. Schultz, A. G. Steinberg, and C. H. Waddington.

175 For details, see Krementsov, "Printsip konkurentnogo iskliucheniia," pp. 107-64; and *idem*, *Stalinist Science*.

176 For published materials on the conference, see M. Mitin, "Za peredovuiu sovetskuiu geneticheskuiu nauku," *PZM*, 1939, 10: 147-76; and V. Kolbanovskii, "Spornye voprosy genetiki i selektsii (Obshchii obzor soveshchaniia)," *PZM*, 1939, 11: 86-126. For the English translations of the three main speeches at the conference by Vavilov, Lysenko, and the PZM editor-in-chief Mitin, see "Genetics in the Soviet Union: Three Speeches from the 1939 Conference on Genetics and Selection," *Science and Society*, 1940, 4(3): 183-233.

177 Kolbanovskii, "Spornye voprosy genetiki i selektsii," p. 116.

178 See, the Russian State Archive of Socio-Political History (hereafter RGASPI), f. 71, op. 3, d. 109, ll. 291-80, (l. 287).

179 See Peter Pringle, *The Murder of Nikolai Vavilov* (New York: Simon and Schuster, 2011).

180 For a detailed description of the establishment of the new institution and its operations in the immediate post-World-War-II years, see Nikolai Krementsov, *The Cure: A Story of Cancer and Politics from the Annals of the Cold War* (Chicago, IL: University of Chicago Press, 2002), pp. 56-62.

181 S. N. Davidenkov, *Evoliutsionno-geneticheskie problemy v nevropatologii* (L.: GIDUV, 1947).

182 *Ibid.*, p. 94. Here Davidenkov cited the 1875 Russian edition of *Hereditary Genius*.

183 The literature on the so-called Lysenko affair is vast. For recent overviews, see Krementsov, *Stalinist Science*; Nils Roll-Hansen, *The Lysenko Effect: The Politics of Science* (Amherst, NY: Humanity Books, 2005); William deJong-Lambert, *The Cold War Politics of Genetic Research: An Introduction to the Lysenko Affair* (Dordrecht: Springer, 2012); for an overview of available historical research on the subject, see Nikolai Krementsov and William deJong-Lambert, "Lysenkoism Redux," in William deJong-Lambert and Nikolai Krementsov, eds., *The Lysenko Controversy as a Global Phenomenon* (New York: Palgrave Macmillan, 2017), vol. 1, pp. 1-34.

184 A. N. Studitskii, "Mukholiuby-chelovekonenavistniki," *Ogonek*, 1949, 11: 14-16.

185 See, Anon, "Gal'ton, Fransis," *BSE*, 1952, 10: 179; Anon, "Genetika," *ibid.*, 10: 430-38; and Anon., "Evgenika," *ibid.*, 15: 372-73.

186 See, for instance, M. V. Volotskoi, "Sluchai nasledstvennoi anomalii pigmentatsii zubov v sviazi s problemoi tsertatsii u cheloveka," in Davidenkov, ed., *Nevrologiia i genetika*, vol. 2, pp. 277-86; and *idem*, "K voprosu o genetike papiliarnykh uzorov pal'tsev," in Levit and Ardashnikov, eds., *Trudy mediko-geneticheskogo instituta*, pp. 404-39.

187 See *Zhurnal'naia letopis' gosudarstvennoi tsentral'noi knizhnoi palaty*, 1926-39.

# Chapter 7

1 See, for instance, Conway Zirkle, ed., *Death of a Science in Russia: The Fate of Genetics as Described in "Pravda" and Elsewhere* (Philadelphia, PA: University of Pennsylvania Press, 1949); and Michael I. Lerner, *Genetics in the USSR: An Obituary* (Vancouver: University of British Columbia, 1950).

2 For details, see Mark B. Adams, "Genetics and the Soviet Scientific Community, 1948-1965," doctoral dissertation, Harvard University, 1972; *idem.*, "Biology After Stalin: A Case Study," *Survey: A Journal of East and West Studies*, 1977-78, 23: 53-80; and *idem.*, *Networks in Action* (Trondheim: Trondheim Studies on East European Cultures and Societies, 2000).

3 For the full text of the report, see N. S. Khrushchev, "O kul'te lichnosti i ego posledstviiakh," *Izvestiia TsK KPSS*, 1989, 3: 128-70; an English translation appeared in *The New York Times* on 5 June 1956, p. 13; its text is also available online at numerous sites, see https://www.marxists.org/archive/khrushchev/1956/02/24.htm.

4 See S. V. Shalimov, "Razvitie genetiki v Novosibirskom nauchnom tsentre vo vtoroi polovine 1960-kh gg.: Sotsial'no-istoricheskii aspect," *Istoriko-biologicheskie issledovaniia*, 2013, 5(1): 16-32. For the history of Akademgorodok in English, see Paul R. Josephson, *New Atlantis Revisited: Akademgorodok, the Siberian City of Science* (Princeton, NJ: Princeton University Press, 1997).

5 On Beliaev's life and works, see a volume of memoirs by his colleagues and family, V. K. Shumnyi, ed., *Dmitrii Konstantinovich Beliaev: Kniga vospominanii* (Novosibirsk: "Geo," 2002); for a brief history of the institute, see N. A. Kupershtokh, "Institut tsitologii i genetiki Sibirskogo otdeleniia RAN," *Vestnik*

*RAN*, 2009, 79 (6): 546-55; for a lively popular account of Beliaev's personal research interests, see Lee Alan Dugatkin and Lyudmila Trut, *How to Tame a Fox (and Build a Dog): Visionary Scientists and a Siberian Tale of Jump-Started Evolution* (Chicago, IL: University of Chicago Press, 2017).

6 See, for instance, a detailed but rather self-aggrandizing account of the controversy in N. P. Dubinin, *Vechnoe dvizhenie*, 2nd ed. (M.: Politizdat, 1975). The book, titled *Perpetual Motion*, went through several successive editions and was dubbed by Dubinin's opponents, "Perpetual Self-Promotion." For a detailed historical account of the controversy, see Adams, "Soviet Nature-Nurture Debate," pp. 94-138.

7 For a detailed examination of how exactly the AMN dealt with Lysenko's takeover of biological research in 1948, see Krementsov, *Stalinist Science*, pp. 191-254.

8 S. N. Davidenkov, "Genetika meditsinskaia," *BME*, 1958, 6: 841-55.

9 See V. P. Efroimson, "Osnovnye dostizheniia meditsinskoi genetiki i ee neotlozhnye zadachi," *Vestnik AMN SSSR*, 1962, 17 (7): 74-82.

10 See, for instance, P. B. Gofman-Kadochnikov, *Dve lektsii po meditsinskoi genetike* (M.: n. p., 1961); and V. P. Efroimson, *Vvedenie v meditsinskuiu genetiku* (M.: Meditsina, 1964).

11 See, for instance, Dzh. Nil' and U. Shell, *Nasledstvennost' u cheloveka* (M.: Izdatel'stvo inostrannoi literatury, 1958); for the English original, see James V. Neel and William Shull, *Human Heredity* (Chicago, IL: University of Chicago Press, 1954); R. Vagner i G. Mitchell, *Genetika i obmen veshchestv* (M.: Izdatel'stvo inostrannoi literatury, 1958); for the English original, see Robert P. Wagner and Herschel K. Mitchell, *Genetics and Metabolism* (New York: Willey; London: Chapman, 1955); Sh. Auerbakh, *Genetika v atomnom veke* (M.: Atomizdat, 1959); for the English original, see Charlotte Auerbach, *Genetics in the Atomic Age* (Edinburgh: Oliver and Boyd, 1956).

12 S. N. Davidenkov and V. P. Efroimson, "Nasledstvennost' cheloveka," *BME*, 1961, 19: 1009-65.

13 See *Vestnik AMN SSSR*, 1962, 17(12).

14 See Michael D. Gordin, "Lysenko Unemployed: Soviet Genetics after the Aftermath," *Isis*, 2018, 109(1): 56-78.

15 S. I. Alikhanian, ed., *Aktual'nye voprosy sovremennoi genetiki* (M.: MGU, 1966). Given the paucity of textbooks on genetics, this collection of essays was often used as a "textbook" by young biologists in the USSR in the 1960s and 1970s.

16 See G. Mendel', *Opyty nad rastitel'nymi gibridami* (M.: Nauka, 1965). Furthermore, three years later, yet another edition of Mendel's *Selected Works* appeared under the auspices of the AMN publishing house, while a fictionalized biography of Mendel's came out in the popular series "The Lives of Remarkable People," see G. Mendel', *Izbrannye raboty* (M.: Meditsina, 1968); and B. G. Volodin, *Mendel': (Vita alterna)* (M.: Molodaia gvardiia, 1968).

17 See M. Popovskii, "Tysiacha dnei akademika Vavilova," *Prostor (Alma-Ata)*, 1966, 7-8; and S. Reznik, *Nikolai Vavilov* (M.: Molodaia gvardiia, 1968).

18 By that time the institute had been renamed the Institute of Developmental Biology. On Astaurov's life and career see, O. G. Stroeva, ed., *Boris L'vovich Astaurov. Ocherki, vospominaniia, materialy* (M.: Nauka, 2004).

19 See *Konferentsiia po meditsinskoi genetike. Doklady i preniia, Sovetskaia klinika*, 1934, 20(7-8): Supplement.

20 Bochkov described his path to genetics in a popular book, titled *Genes and Fates*, see Bochkov, *Geny i sud'by* (M.: Molodaia gvardiia, 1990). For telling details on Bochkov's appointment and the creation of the institute, see the transcript of Peter Harper's interview with Bochkov conducted on 27 May 2005, https://genmedhist.eshg.org/fileadmin/content/.../interviewees.../Bochkov%20N.pdf.

21 See V. P. Efroimson, "K istorii izucheniia genetiki cheloveka v SSSR," *Genetika*, 1967, 10: 114-27.

22 See V. P. Efroimson, *Vvedenie v meditsinskuiu genetiku* (M.: Meditsina, 1964).

23 See, for instance, E. N. Pavlovskii and P. S. Pervomaiskii, "Ob eksperimental'nom izmenenii nasledovaniia okraski u krolika," *Izvestiia AN SSSR, ser. Biologiia*, 1949, 6: 702-08. For a Soviet era, laudatory biography of Pavlovskii, see N. P. Prokhorova, *Akademik E. N. Pavlovskii* (M.: Meditsina, 1972).

24 See E. Pavlovskii, "Evgenika," *BME*, 1959, 9: 961-67.

25 See V. Polynin, *Mama, papa i ia* (M.: Sovetskaia Rossiia, 1967). For a brief sketch of Blanter's work in the journal, see O. O. Astakhova and O. I. Shutova, "O prirode 'Prirody'," *Priroda*, 2012, 1: 3-10.

26 Polynin, *Mama, papa i ia*, pp. 182, 293-300.

27 He would slightly expand this very impoverished portrayal of early Soviet eugenics in a fictionalized biography of Kol'tsov published two years later, see V. Polynin, *Prorok v svoem otechestve* (M.: Sovetskaia Rossiia, 1969), pp. 10-13.

28 B. Astaurov, "Predislovie," in Polynin, *Mama, papa i ia*, pp. 5-11. All the subsequent quotations are from this source.

29 See I. I. Kanaev, *Frensis Gal'ton, 1822-1911* (L.: Nauka, 1972). In 2000 the book was reprinted in a volume of Kanaev's *Selected Works on the History of Biology*. See I. I. Kanaev, "Frensis Gal'ton (1822-1911)," in *idem, Izbrannye trudy po istorii nauki* (SPb.: Aleteiia, 2000), pp. 356-475.

30 See, for instance, N. A. Pan'kov, "Gete v Saranske. Pis'ma M. M. Bakhtina k I. I. Kanaevu," *Dialog. Karnaval. Khronotop*, 1999, 3: 79-97; Ben Taylor, "Kanaev, Vitalism and the Bakhtin Circle," in David Shepherd, Craig Brandist, and Galin Tihanov, eds., *The Bakhtin Circle: In the Master's Absence* (Manchester: Manchester University Press, 2004), pp. 150-66; Iu. P. Medvedev and D. A. Medvedeva, "Trudy i dni kruga M. M. Bakhtina," *Zvezda*, 2008, 7, http://magazines.russ.ru/zvezda/2008/7/me14.html; and many others.

31 For a brief biography of Kanaev, see K. V. Manoilenko, "I. I. Kanaev i ego rol' v razvitii evoliutsionnoi biologii," in Kanaev, *Izbrannye trudy*, pp. 7-16; and *idem*, "Biolog-evoliutsionist I. I. Kanaev kak istorik nauki," in *Russkaia nauka v biograficheskikh ocherkakh* (SPb.: Dmitrii Bulanin, 2003), pp. 129-36.

32 The following biographical information comes from Kanaev's "autobiography" and other materials preserved in his personnel file held at the St. Petersburg Branch of the ARAN, f. 806, op. 2, d. 259.

33 See I. I. Kanaev, "Iz istorii biologii v Leningrade v nachale 20-kh godov," *Nauka i tekhnika (voprosy istorii i teorii)*, 1972, 7(2): 44-46; see also L. Z. Kaidanov, "Formirovanie kafedry genetiki i eksperimental'noi zoologii v Petrogradskom Universitete (1913-1920)," *Issledovaniia po genetike*, 1994, 11: 6-12.

34 See, for instance, I. I. Kanaev, *Nasledstvennost': Nauchno-populiarnyi ocherk* (L.: Priboi, 1925).

35 Kanaev's magnum opus on the subject appeared in 1952 and was translated into English in 1969, see I. I. Kanaev, *Hydra: Essays on the Biology of Fresh Water Polyps* (Irving, CA.: Howard M. Lenhoff, 1969).

36 See, for instance, I. Kanajew [Kanaev], "Über den Porus aboralis bei Pelmatohydra oligactis Pall.," *Zoologischer Anzeiger*, 1928, 76: 37-44.

37 In 1930, as part of the sweeping reorganizations of the Great Break, all medical schools were separated from universities and re-constituted as independent teaching facilities.

38 For details, see Todes, *Ivan Pavlov*.

39 See I. P. Pavlov, *Dvadtsatiletnii opyt ob"ektivnogo izucheniia vysshei nervnoi deiatel'nosti (povedeniia) zhivotnykh* (L.: GIZ, 1923).

40 On Pavlov's flirtation with Lamarckism and its consequences, see Krementsov, "Marxism, Darwinism, and Genetics," pp. 225-28.

41 See N. K. Kol'tsov, "I. P. Pavlov: Trud zhizni velikogo biologa," *Biologicheskii zhurnal*, 1936, 5(3): 387-402.

42 Some of Krasnogorskii's experiments with conditional reflexes in children had been recorded in the famous 1926 film, *Mechanics of the Brain*. An abridged copy of the film with English subtitles is available at http://vimeo.com/20583313.

43 See I. I. Kanaev, "Izuchenie bliznetsov, kak geneticheskii metod," *Priroda*, 1934, 12: 37-45; *idem*, "Opyt izucheniia uslovnykh refleksov u odnoiaitsevykh bliznetsov," *Arkhiv biologicheskikh nauk*, 1934, 34(5-6): 569-77; and *idem*, "Dal'neishee izuchenie fiziologicheskoi deiatel'nosti mozga u odnoiaitsevykh bliznetsov," *ibid.*, 1936, 44(1): 12-42.

44 See I. I. Kanaev, "Bezuslovnye sliunnye refleksy u chelovecheskikh bliznetsov," *Doklady Akademii nauk SSSR*, 1939, 25(3): 255-58; *idem*, "Spontannaia sliunnaia sekretsiia u chelovecheskikh bliznetsov," *ibid.*: 252-54; *idem*, "Uslovnye sliunnye refleksy u chelovecheskikh bliznetsov," *ibid.*: 259-60; *idem*, "K voprosu o podvizhnosti uslovnykh refleksov u bliznetsov," *ibid.*, 1941, 30(9): 851-53; and *idem*, "Sliunnye refleksy u bliznetsov," *Priroda*, 1940, 11: 81-85. See also I. I. Kanaev, "Genetika i embriologiia papiliarnykh risunkov chelovecheskikh pal'tsev," *Priroda*, 1935, 4: 37-47; and *idem*, "Physiology of the Brain in Twins," *Character and Personality*, 1938, 6(3): 177-87.

45 I. I. Kanaev, "Eksperimental'naia genetika vysshei nervnoi deiatel'nosti," *Uspekhi sovremennoi biologii*, 1948, 25(1): 149-55.

46 For details, see N. L. Krementsov, "Ot sel'skogo khoziaistva do ... meditsiny," in *Repressirovannaia nauka* (L.: Nauka, 1991), vol. 1, pp. 91-116.

47 See, for instance, I. I. Kanaev, "K izucheniiu nervnykh protsessov pri dvigatel'nykh reaktsiiakh ruk u detei," *Fiziologicheskii zhurnal SSSR*, 1954, 40(1): 9-17; and *idem*, "Materialy k fiziologii otscheta vremeni det'mi," *ibid.*, 1956, 42(4): 341-47.

48 For instance, he was one of the nearly 300 signatories to the famous letter detailing the detrimental effects of Lysenko's monopoly on the development of Soviet science sent in 1955 by leading biologists, physicists, chemists, and mathematicians to the Central Committee of the Communist Party. See, L. G. Dubinina and I. F. Zhimulev, "K 50-letiiu 'Pis'ma trekhsot'," *Vestnik VOGiS*, 2005, 9(1): 12-33.

49 On the history of this institution and Kanaev's work there, see E. I. Kolchinskii, *Istoriko-nauchnoe soobshchestvo v Leningrade-Sankt-Peterburge v 1950-2010 gody: liudi, traditsii, sversheniia* (SPb.: Nestor-Istoriia, 2013).

50 See, I. I. Kanaev, "Gete-naturalist," in I. V. Gete, *Izbrannye sochineniia po estestvoznaniiu* (M.: Izd. AN SSSR, 1957), pp. 418-87.

51 I. I. Kanaev, *Bliznetsy. Ocherki po voprosam mnogoplodiia* (M.-L.: Izd. AN SSSR, 1959).

52 Many, but not all of these articles were reprinted in the volume of his "selected works," see Kanaev, *Izbrannye trudy*, pp. 17-338. A bibliography of Kanaev's publications on the history of biology is appended to the same volume, see "Trudy I. I. Kanaeva," in Kanaev, *Izbrannye trudy*, pp. 488-91.

53 For an interesting comparison of these visions, see John Griffiths, *Three Tomorrows: American, British and Soviet Science Fiction* (London: Macmillan, 1980).

54 See Gordon Wolstenholme, ed., *Man and his Future* (Boston, MA: Little, Brown and Co, 1963), https://archive.org/details/manhisfutureciba00wols. For a detailed analysis of the symposium and its role in the formation of new eugenics, see the next chapter.

55 On "new" eugenics, see Kevles, *In the Name of Eugenics*; R. A. Soloway, "From Mainline to Reform Eugenics – Leonard Darwin and C. P. Blacker," in Peel, ed., *Essays in the History of Eugenics*, pp. 52-80; Paul, *The Politics of Heredity*; Pauline Mazumdar, "'Reform' Eugenics and the Decline of Mendelism," *Trends in Genetics*, 2002, 18(1): 48-52; and Comfort, *The Science of Human Perfection*. For a more detailed discussion of "new" eugenics, see the next chapter.

56 See, for instance, Anon., "Glubzhe razrabatyvat' metodologicheskie problemy biologii," *Voprosy filosofii*, 1964, 12: 25-32. This anonymous editorial, titled "To Deepen the Elaboration of the Methodological Problems of Biology," inaugurated biologists' direct attack on philosophers' superficial "ideological" and "methodological" critiques of science, especially, genetics. The campaign was part of a much broader movement by Soviet scientists to get out from under the strict "ideological" control of the party-state apparatus and to limit interference by Marxist philosophers, who in the 1930s had become spokesmen of the apparatus in the scientific community, in the development of Soviet science. See Krementsov, *Stalinist Science*; and Adams, *Networks*.

57 See, for instance, Dzh. B. S. Kholdein, "O vozmozhnosti sotsial'nykh prilozhenii antropogenetiki," in V. N. Stoletov, ed., *Nauka o nauke* (M.:

Progress, 1966), pp. 179-87. This was a Russian translation of Haldane's essay "The Proper Social Application of the Knowledge of Human Genetics" in the famous 1964 volume *The Science of Science*, edited by Maurice Goldsmith and Alan Mackay.

58 See, for instance, Ia. S. Iorish, "Budushchee biologii i obshchestvo," *Voprosy filosofii*, 1966, 9: 169-77. For details on the symposium and the discussion, see the next chapter.

59 See an overview of the discussion in I. K. Liseev and A. Ia. Sharov, "Genetika cheloveka, ee filosofskie i sotsial'no-eticheskie problemy. Kruglyi stol 'Voprosov filosofii'," *Voprosy filosofii*, 1970, 7: 106-15; 8: 125-34.

60 On Neifakh and his contributions, see M. E. Aspiz, "Ob A. A. Neifakhe kak ob uchenom," in A. A. Neifakh, *Vzgliady, idei, razdum'ia* (M.: Nauka, 2001), pp. 114-18.

61 Partially this discussion has been examined in Loren R. Graham, "Reasons to Studying Soviet Science: The Example of Genetic Engineering," in Lubrano and Solomon, eds., *Social Context of Soviet Science*, pp. 205-40.

62 See, for instance, B. A. Nikitiuk, ed., *Sootnoshenie biologicheskogo i sotsial'nogo v razvitii cheloveka* (M.: n. p., 1974); V. M. Banshchikov, ed., *Sootnoshenie biologicheskogo i sotsial'nogo v cheloveke* (M.: n. p., 1975); B. F. Lomov, ed., *Biologicheskoe i sotsial'noe v razvitii cheloveka* (M.: Nauka, 1977); A. F. Polis, ed., *Biologicheskoe i sotsial'noe v formirovanii tselostnoi lichnosti* (Riga: LGU, 1977); V. V. Orlov, ed., *Sotsial'noe i biologicheskoe* (Perm: n. p., 1975); idem, ed., *Sootnoshenie biologicheskogo i sotsial'nogo* (Perm: PGU, 1981); and many others. The 1975 publication of Edward O. Wilson's *Sociobiology* and extensive discussions it provoked in the West fueled these debates further. For an analysis of these later debates, see Yvonne Howell, "The Liberal Gene: Sociobiology as Emancipatory Discourse in the Late Soviet Union," *Slavic Review*, 2010, 69(2): 356-76.

63 Compare, for instance, special brochures on "Genetics and the Future of Humanity" published in the aftermath of the discussion by Nikolai Dubinin, *Genetika i budushchee chelovechestva* (M.: Znanie, 1971) and Nikolai Bochkov, *Progress obshchestva i genetika cheloveka* (M.: Znanie, 1971). The brochures were issued in huge print runs of 100,000 copies each by the "All-Union Society Knowledge," as preparatory material for the society's cadres of lecturers and were distributed to the society's local offices, as well as public libraries, all over the country. On the early history of the society, see Michael Froggatt, "Science in Propaganda and Popular Culture in the USSR under Khrushchev, 1953-1964," doctoral dissertation, Oxford University, 2005; and James T. Andrews, "Inculcating Materialist Minds: Scientific Propaganda and Anti-

Religion in the USSR during the Cold War," in P. Betts and S. Smith, eds., *Science, Religion and Communism in Cold War Europe* (London: Palgrave Macmillan, 2016), pp. 105-25.

64 The campaign also spread to literary and popular journals. For instance, in 1971, the leading journal *New World* carried a controversial article by Efroimson on the "Genealogy of Altruism: Ethics from the Viewpoint of Human Evolutionary Genetics," see V. Evroimson, "Rodoslovnaia al'truizma (Etika s pozitsii evoliutsionnoi genetiki cheloveka), *Novyi mir*, 1971, 10: 193-213. The article sparked a prolonged debate in various periodicals; see Howell, "The Liberal Gene."

65 For instance, in 1965 a special collection, under the revealing title "The Novelty in Science and the Problems of Preparing the Term-list for the Third Edition of the Great Soviet Encyclopedia," was issued for discussions over the contents of the new edition. See *Novoe v nauke i problemy podgotovki slovnika 3-go izdaniia BSE* (M.: Sovetskaia entsiklopediia, 1965).

66 With a short time lag, each volume of the entire encyclopedia was translated into English and published by Macmillan, see A. M. Prokhorov, ed., *Great Soviet Encyclopedia*, 30 vols. (New York: Macmillan, 1973-1982).

67 M. G. Iaroshevskii, "Gal'ton, Frensis," *BSE*, 1971, 6: 229-30.

68 Anon, "Gal'ton, Fransis," *BSE*, 1952, 10: 179.

69 M. E. Lobashov and Iu. E. Vel'tishchev, "Evgenika," *BSE*, 1971, 8: 584-85. All the subsequent quotations are from this source.

70 See E. F. Davidenkova, "Genetika meditsinskaia," *BSE*, 1971, 6: 692-94.

71 See D. K. Beliaev, "Genetika," *BSE*, 1971, 6: 677-91, and A. A. Prokof'eva-Bel'govskaia, K. N. Grinberg, "Genetika cheloveka," *ibid.*, 1971, 6: 700-03.

72 N. V. Glotov, A. A. Liapunov, and N. V. Timofeev-Resovskii, "Biometriia," *BSE*, 1970, 3: 1061-62.

73 I. I. Kanaev and V. A. Pokrovskii, "Bliznetsy," *BSE*, 1970, 3: 1249.

74 See a reference to Galton's research in Kanaev's first publication on the subject, I. Kanaev, "Izuchenie bliznetsov, kak geneticheskii metod," *Priroda*, 1934, 12: 37-45.

75 Kanaev, *Bliznetsy*, p. 29.

76 See I. I. Kanaev, *Bliznetsy i genetika* (L.: Nauka, 1968), p. 49.

77 See I. I. Kanaev to Th. Dobzhansky, 10 February 1969. Theodosius Dobzhansky Papers, Mss. B. D65. Series I. Correspondence. Folder "Kanaev," in APS (hereafter Dobzhansky Papers). Alas, the folder contains only six letters by Kanaev to Dobzhansky, but no copies of Dobzhansky's letters to Kanaev, which so far remain unfound.

78 On Dobzhansky, see Mark B. Adams, ed., *The Evolution of Theodosius Dobzhansky: His Life and Thought in Russia and America* (Princeton, NJ: Princeton University Press, 1994); and M. B. Konashev, *Stanovlenie evoliutsionnoi kontseptsii F. G. Dobrzhanskogo* (SPb.: Nestor-Istoriia, 2011). Kanaev's letters to Dobzhansky have recently been published in M. B. Konashev and E. I. Kolchinksii, "'... v SShA vsei genetikoi rukovodit russkii,' (perepiska othechestvennykh biologov s F. G. Dobrzhanskim)," *Istoriko-biologicheskie issledovaniia*, 2010, 2(3): 116-41; Kanaev's letters appear on pp. 132-35.

79 See I. I. Kanaev, *Gete kak estestvoispytatel'* (L.: Nauka, 1970); idem, *Ocherki iz istorii problem fiziologii tsvetogogo zreniia ot antichnosti do XX veka* (L.: Nauka, 1971); and idem, *Abraam Tramble* (L.: Nauka, 1972).

80 Kanaev, *Frensis Gal'ton*, p. 5.

81 This rehabilitation was continued, albeit in a limited way, in the biographies of the two leaders of Soviet eugenics, Kol'tsov and Filipchenko, published in the 1970s by their students in the same series of "Scientific Biographies." See B. L. Astaurov and P. F. Rokitskii, *Nikolai Konstantinovich Kol'tsov* (M.: Nauka, 1975); and N. N. Medvedev, *Iurii Aleksandrovich Filipchenko* (M.: Nauka, 1978). Each book had a short chapter on "human genetics" that actually discussed the involvement of their protagonists with eugenics.

82 P. F. Rokitskii, "Osnovopolozhnik genetiki cheloveka," *Priroda*, 1973, 5: 116-18; all the subsequent quotations are from this source. See also a similarly praising review by M. Reidiboim, "U istokov genetiki cheloveka," *Defektologiia*, 1973, 5: 87-90.

83 Kanaev, *Frensis Gal'ton*, p. 117.

84 He certainly knew of Volotskoi's active involvement with eugenics and human genetics. In the book he referred to an article on twins that appeared in the *Russian Eugenics Journal*, as well as to Volotskoi's contribution to the fourth volume of the *IMG Proceedings*.

85 Kanaev to Dobzhansky, 14 March 1969, Dobzhansky Papers.

86 I. I. Kanaev, "Na puti k meditsinskoi genetike," *Priroda*, 1973, 1: 52-68. All the subsequent quotations are from this source and indicated in the text by page number in square brackets.

87 Kanaev's main sources for Florinskii's biography were the 1898 history of the IMSA gynecology department and Florinskii's eulogy of Sergei Botkin, see Gruzdev, *Istoricheskii ocherk kafedry akusherstva i zhenskikh boleznei*, pp. 214-41; and Florinskii, "Pamiati prof. S. P. Botkina," 64-70. In his references Kanaev "renamed" Gruzdev into Gur'ev.

88 A copy of Florinskii's book held in this library has numerous pencil marks on the margins indicating most of the cuts made by Volotskoi in the original text, which might have been left by Kanaev.

89 N. P. Bochkov, *Genetika cheloveka* (M.: Meditsina, 1978), pp. 11-12; see also *idem*, Geny i sud'by, p. 61.

90 N. P. Bochkov, "Meditsinskaia genetika," *BME*, 1980, 14: 1092-98.

91 Iu. E. Vel'tishchev and B. V. Koniukhov, "Nasledstvennye bolezni," *BME*, 1981, 16: 183-88.

92 See E. I. Gusev, "Genealogicheskii metod," *BME*, 1977, 5: 720-31; N. P. Bochkov, E. T. Lil'in, and R. P. Martynova, "Bliznetsovyi metod," *ibid.*, 1976, 3: 718-27; and V. A. Alpatov and V. M. Akhutin, "Biometriia," *ibid.*, 1976, 3: 509-16.

93 N. E. Granat, "Florinskii, Vasilii Markovich," *BME*, 1985, 26: 1034-35. The entry was a slightly expanded version of an unsigned article published in the previous edition, see Anon., "Florinskii Vasilii Markovich," *BME*, 1963, 33: 854.

94 V. P. Bisiarina and M. S. Maslov, "Pediatriia," *BME*, 1982, 18: 1301-19 (p. 1307).

95 In the early 1970s, Kanaev published brief accounts of research conducted at Levit's Institute of Medical Genetics, which was clearly part of this project. See, I. I. Kanaev, "Antropogenetika i praktika," *Nauka i tekhnika: Vorposy istorii i teorii*, 1971, 6: 169-72; and *idem*, "O rabotakh MGI im. M. Gor'kogo," *ibid.*, 1973, 8(2): 155-58.

96 Alas, I was unable to obtain a copy of the manuscript that reportedly is still kept in the family.

97 For concise depictions of the Brezhnev era, see William Tompson, *The Soviet Union under Brezhnev* (Harlow: Pearson/Longman, 2003); and Dina Fainberg and Artemy M. Kalinovsky, eds., *Reconsidering Stagnation in the Brezhnev Era: Ideology and Exchange* (London: Lexington, 2016).

98 Around the same time, Nikolai Medvedev, another student of Filipchenko, experienced similar difficulties in publishing a biography of his teacher. But he

decided to compromise and agreed to rewrite and shorten the chapters dealing with eugenics, thus assuring that the biography would come out. For a new edition of Medvedev's book, which contains the full text of the original 1976 manuscript, along with a vivid depiction of his negotiations with the publisher, see N. N. Medvedev, *Iurii Aleksandrovich Filipchenko* (M.: Nauka, 2006).

99 For a succinct overview of the war, see Mark Galeotti, *Russia's Wars in Chechnya 1994-2009* (Oxford: Osprey, 2014).

100 For informative introductions to political and economic developments in Russia, see Michael McFaul, *Russia's Unfinished Revolution: Political Change from Gorbachev to Putin* (Ithaca, NY: Cornell University Press, 2001); David E. Hoffman, *The Oligarchs: Wealth and Power in the New Russia* (New York: Public Affairs, 2002); Neil Robinson, ed., *The Political Economy of Russia* (Toronto: Rowman and Littlefield, 2013); and Svetlana Stephenson, *Gangs of Russia: From the Streets to the Corridors of Power* (Ithaca, NY: Cornell University Press, 2015).

101 See, for instance, Donald J. Raleigh, ed., *Soviet Historians and Perestroika: The First Phase* (London: Routledge, 1989).

102 On the demise of Soviet censorship in the late 1980s and early 1990s, see the published materials of seven conferences on "Censorship in Russia: History and the Present" held in St. Petersburg in the 1990s and 2000s, *Tsenzura v Rossii: Istoriia i sovremennost'* (SPb.: n. p., 2001-2015), vols. 1-7.

103 See, for instance, four volumes of historical materials published during 1992-93 under the general title "Unknown Russia. XX century," *Neizvestnaia Rossiia. XX Vek: Arkhivy, Pis'ma, Memuary*, 4 vols. (M.: Istoricheskoe nasledie, 1992-1993).

104 See M. B. Konashev, "Biuro po evgenike, 1922-1930," *Issledovaniia po genetike*, 1994, 11: 22-28; D. A. Aleksandrov, "Osobennosti Petrograda-Leningrada kak tsentra razvitiia evgeniki," *Nauka i tekhnika: voprosy istorii i teorii*, 1996, 10: 113-19; and V. V. Babkov, "Biologicheskie i sotsial'nye ierarkhii. Konteksty pis'ma G. G. Mellera I. V. Stalinu," *VIET*, 1997, 1: 76-94.

105 For a brief account of Motkov's life and work, see Iakov Pravdin, "Posviatil zhizn' probleme uluchsheniia genofonda," *Chestnoe slovo (Kazan)*, 28 December 2016, http://chskaz.wixsite.com/index/170-09.

106 S. E. Motkov, "Vvedenie," *Sovetskaia evgenika*, 1991, 1: 1-3 (p. 1).

107 See, for instance, S. Gershenzon and T. Buzhievskaia, "Evgenika: 100 let spustia," *Chelovek*, 1996, 1: 23-29; and I. Frolov, "Nachalo puti (kratkie zametki o neoevgenike)," *ibid.*, 1997, 1: 34-37.

108 For the transcript of the discussion, see "Ne khotim byt' klonami (Problemy sovremennoi evgeniki)," *Chelovek*, 1996, 4: 22-33; 5: 21-37.

109 "Nauka, kotoraia sebia izzhila. Beseda s N. P. Bochkovym," *Chelovek*, 1997, 2: 20-30. Both the materials of the round table and Bochkov's assessment were included in a special volume put together by the journal's editor-in-chief under the title *Bioethics: Principles, Rules, Problems*, see B. G. Iudin, ed., *Bioetika: printsipy, pravila, problemy* (M.: Editorial URSS, 1998).

110 See F. Gal'ton, *Nasledstvennost' talanta: Zakony i posledstviia* (M.: Mysl',1996).

111 I. I. Kanaev, "Frensis Gal'ton (1822-1911)," in idem, *Izbrannye trudy po istorii nauki* (SPb.: Aleteiia, 2000), pp. 356-475.

112 See V. V. Babkov, "Meditsinskaia genetika v SSSR," *Vestnik RAN*, 2001, 71(10): 928-37; idem, "Moskva, 1934: Rozhdenie meditsinskoi genetiki," *Vestnik VOGIS*, 2006, 10(3): 455-78; M. B. Konashev, "Ot evgeniki k meditsinskoi genetike," *Rossiiskii biomeditsinskii zhurnal*, 2002, 3, http://www.medline.ru/public/art/tom3/eumedgen.phtml; E. B. Muzrukova, "Raboty Iu. A. Filipchenko i ego shkoly po izucheniiu nauchnogo soobshchestva Petrograda," *Sotsiokul'turnye problem nauki i tekhniki*, 2006, 4: 32-42; E. B. Muzrukova and R. A. Fando, "Evgenicheskie raboty Iu. A. Filipchenko i A. S. Serebrovskogo," *Nauka i tekhnika v pervye desiatiletiia sovetskoi vlasti: sotsiokul'turnoe izmerenie (1917-1940)* (M.: Academia, 2007), pp. 257-78; Iu. V. Khen, "Evgenika: osnovateli i prodolzhateli," *Chelovek*, 2006, 3: 80-88; L. I. Korochkin, L. G. Romanova, "Genetika povedeniia cheloveka i evgenika," ibid., 2007, 2: 32-43; A. M. Polishchuk, "Meditsinskaia genetika v Rossii," *Khimiia i zhizn'*, 2010, 2: 44-51; and many others.

113 See, for instance, R. A. Fando, "Polemika o sud'be evgeniki," *VIET*, 2002, 3: 604-17.

114 See V. Babkov, *Zaria genetiki cheloveka* (M.: Progress-Traditsiia, 2008).

115 E. V. Pchelov, ed., *Rodoslovnaia genial'nosti: iz istorii otechestvennoi nauki 1920-kh gg.* (M.: Staraia Basmannaia, 2008).

116 Fando, *Proshloe nauki budushchego*.

117 See R. A. Fando, "Tragicheskaia sud'ba othechestvennoi evgeniki," in *Nauka i tekhnika v pervye desiatiletiia sovetskoi vlasti: sotsiokul'turnoe izmerenie (1917-1940)* (M.: Academia, 2007), pp. 279-305.

118 E. V. Iastrebov, *Bibliografiia opublikovannykh trudov Vasiliia Markovicha Florinskogo* (M.: n. p., 1992); idem, *Vasilii Markovich Florinskii* (Tomsk: Izd-vo Tomskogo Universiteta, 1994); idem, *Sto neizvestnykh pisem russkikh uchenykh*

*i gosudarstvennykh deiatelei Vasiliiu Markovichu Florinskomu* (Tomsk: Izd-vo Tomskogo Universiteta, 1995); and *idem, Vasilii Markovich Florinskii v Peterburgskoi mediko-khirurgicheskoi akademii* (M.: MPU "Signal", 1999).

119 See, for instance, a brief biography of his maternal grandfather in N. A. Medvedeva, "K biografii sviashchenika Rezhevskoi Ioanno-Predtecheskoi tserkvi Petra Ladyzhnikova," in A. M. Britvin, ed., *Pravoslavie na Urale* (Ekaterinburg: Ural'skoe tserkovno-istoricheskoe obshchestvo, 2017), pp. 67-76.

120 Before the war, the Urals University did not have a geography department. On the early history of this university, see M. E. Glavatskii, *Istoriia rozhdeniia Ural'skogo universiteta* (Ekaterinburg: UrGU, 2000).

121 See his memoirs on the subject in E. V. Iastrebov, "O Borise Pavloviche Kolesnikove i pervykh shagakh deiatel'nosti komissii po okhrane prirody pri Ural'skom filiale Akademii nauk SSSR," *Izvestiia UrGU*, 2002, 23: 158-68.

122 See, for instance, his popular publications, E. V. Iastrebov, *Po reke Chusovoi*, 2nd ed. (Sverdlovsk: Knizhnoe izdatel'stvo, 1963); and N. P. Arkhipova and E. V. Iastrebov, *Kak byli otkryty Ural'skie gory* (Perm': Knizhnoe izdatel'stvo, 1971); the latter book has been reprinted three times.

123 A collection of Iastrebov's personal papers deposited in the state archive of the Sverdlovsk region contains several unpublished manuscripts on his grandparents and other members of the extended family. See *Gosudarstvennyi arkhiv sverdlovskoi oblasti* (hereafter GASO), f. R-2787, op. 1 (1939-1999), 263 dd.

124 Iastrebov, *Vasilii Markovich Florinskii*, p. 6.

125 See GASO, f. R-2787, op. 1, d. 119, ll. 1-226.

126 See, for instance, David M. Woodruff, *Money Unmade: Barter and the Fate of Russian Capitalism* (Ithaca, NY: Cornell University Press, 2000).

127 See the data on the series compiled in Z. K. Sokolovskaia and V. I. Sokolovskii, *550 knig ob uchenykh, inzhenerakh i izobretateliakh* (M.: Nauka, 1999).

128 See V. A. Makarov, *Ivan Timofeevich Glebov* (M.: Nauka, 1995).

129 See V. P. Puzyrev, "Predislovie," in Iastrebov, *Sto neizvestnykh pisem*, pp. 3-7.

130 Although his name as the actual publisher of the new edition does not figure in the book's bibliographic record, several oblique statements in "preliminary materials" definitively point to his key role in its publication.

131 Iastrebov, *Vasilii Markovich Florinskii*, p. 130.

132 See a brief biography published on the occasion of his sixtieth birthday, "Puzyrev Valerii Pavlovish (k 60-letiui so dnia rozhdeniia)," *Sibirskii meditsinskii zhurnal*, 2007, 2: 141-42.

133 The following story is based to a certain degree on my correspondence and conversations with Valerii Puzyrev (in St. Petersburg in October 2016), as well as his recollections written at my request.

134 V. P. Puzyrev, "Kliniko-genealogicheskoe i biokhimicheskoe issledovanie nasledstvennoi predraspolozhennosti k aterosklerozu i ishemicheskoi bolezni serdtsa," doctoral dissertation, Novosibirsk, 1977; some materials from this dissertation were included in a book on *Heredity and Atherosclerosis* that Puzyrev co-authored with a member of the Institute of Cytology and Genetics, see A. A. Dzizinskii and V. P. Puzyrev, *Nasledstvennost' i ateroskleroz* (Novosibirsk: Nauka, 1977).

135 For a brief history of the institute, see its official website at http://www.medgenetics.ru/about/history.

136 Certain elements of this gigantic project, largely those carried out by the Moscow IMG, are described in Susanne Bauer, "Mutations in Soviet Public Health Science: Post-Lysenko Medical Genetics, 1969-1991," *Studies in History and Philosophy of Biological and Biomedical Sciences*, 2014, 30: 1-10; and *idem*, "Virtual Geographies of Belonging: The Case of Soviet and Post-Soviet Human Genetic Diversity Research," *Science, Technology, and Human Values*, 2014, 39(4): 511-37.

137 See, for instance, V. P. Puzyrev, *Mediko-geneticheskoe islledovanie naseleniia pripoliarnykh regionov* (Tomsk: Izd. Tomskogo Universiteta, 1991); *idem*, ed., *Genetika cheloveka i patologiia* (Tomsk: Izd. Tomskogo universiteta, 1992); and many others.

138 See Puzyrev's correspondence with Iastrebov in GASO, f. R-2787, op. 1, d. 213, ll. 1-12.

139 See V. P. Puzyrev, "Granitsy chelovecheskoi zhizni (V. M. Florinskii: sovremennoe zvuchanie myslei uchenogo)," *Krasnoe znamia*, 21 September 1993, # 209; *idem*, "Podvizhnik nauki. 160 let so dnia rozhdeniia professor V. M. Florinskogo," *ibid.*, 30 March 1994, # 78; *idem*, "Professor V. M. Florinskii (k 160-letiiu so dnia rozhdeniia)," *Nauka v Sibiri*, 16 April 1994; *idem*, "Tomskii period zhizni i tvorchestva V. M. Florinskogo," *Vrach*, 1994, 11: 47; *idem*, "V. M. Florinskii i ego evgenicheskie vzgliady na uluchshenie i vyrozhdenie chelovecheskogo roda," *Genetika*, 1994, 30, Prilozhenie: 129.

140 Iastrebov misdated the photograph as being taken in 1861.

141 See V. P. Puzyrev, "Evgenicheskie vzgliady V. M. Florinskogo na 'usovershenstvovanie i vyrozhdenie chelovecheskogo roda'," in Florinskii, 1995, pp. 120-26; the same text also appeared in the *Bulletin of the Tomsk Scientific Center of the Academy of Medical Sciences, Biulleten' Tomskogo nauchnogo tsentra,* 1995, 6: 23-33; while its abstract was published in the Russian journal of genetics, *Genetika,* 1994, 30, Prilozhenie: 129.

142 K. Garver and B. Garver, "Proekt 'Genom cheloveka' i evgenicheskie problemy," in Florinskii, 1995, pp. 127-47; for the original, see Kenneth L. Garver and Bettylee Garver, "The Human Genome Project and Eugenic Concerns," *American Journal of Human Genetics,* 1994, 54: 148-58.

143 V. P. Puzyrev, "About V. M. Florinskii and his Book 'Improvement and Degeneration of the Human Race' (1865)," in Florinskii, 1995, pp. 148-51.

144 On the debate, see Kenneth L. Garver and Bettylee Garver, "Eugenics: Past, Present, and the Future," *American Journal of Human Genetics,* 1991, 49: 1109-18; Daniel J. Kevles and Leroy Hood, eds., *The Code of Codes: Scientific and Social Issues in the Human Genome Project* (Cambridge, MA: Harvard University Press, 1993); Diana Paul, *Controlling Human Heredity: 1865 to the Present* (Atlantic Highlands, NJ: Humanities Press International, 1995); *idem, The Politics of Heredity: Essays on Eugenics, Biomedicine, and the Nature-Nurture Debate* (Albany, NY: SUNY Press, 1998); and many others.

145 See V. P. Puzyrev, "'Physiological laws of heredity' by V. M. Florinsky," *Brazilian Journal of Genetics,* 1996, 19(2): 211 (Abstracts of the 9th International Congress of Human Genetics, Rio de Janeiro, 1996).

146 V. P. Puzyrev, *About V. M. Florinsky and his Book "Improvement and Degeneration of the Human Race" (1865)* (Tomsk: Tomsk Scientific Center of the RAMS, 1996).

147 V. P. Puzyrev, "'Hygiene of Marriage' Concept (to the 130th Anniversary of Publishing 'Improvement and Degeneration of the Human Race' by Vassily Florinsky)," *European Journal of Human Genetics,* 1996, 4, Suppl. 1: 121.

148 V. P. Puzyrev, "Vol'nosti genoma i meditsinskaia patogenetika," *Biulleten' sibirskoi meditsiny,* 2002, 2: 16-29. After the dissolution of the Soviet Union, Tomsk Medical Institute was renamed "Tomsk Medical University."

149 The only brief notice I found has appeared in a footnote on Florinskii in the introduction to Pchelov's compilation, see E. V. Pchelov, "Evgenika i genealogiia v othechestvennoi nauke 1920-kh godov," in *idem,* ed. *Rodoslovnaia genial'nosti: Iz istorii othechestvennoi nauki 1920-kh gg.,* pp. 7-60, the reference is in fn. 11 (p.13).

150 See E. V. Iastrebov, *Vasilii Markovich Florinskii v Peterburgskoi mediko-khirurgicheskoi akademii* (M.: MPU "Signal", 1999), pp. 91-92.

151 See, for instance, A. A. Sechenova, "Vasilii Markovich Florinskii — pervyi popechitel' Zapadno-Sibirskogo uchebnogo okruga," *Vestnik Tomskogo gosudarstvennogo pedagogicheskogo instituta*, 2009, 12: 139-41; I. A. Dunbinskii, "Rol' V. M. Florinskogo v formirovanii uchebno-vspomogatel'nykh uchrezhdenii Tomskogo imperatorskogo universiteta," *Kul'tura, Dukhovnost', Obshchestvo (Novosibirsk)*, 2013, 5: 46-49; A. I. Teriukov, "V. M. Florinskii i M. K. Sidorov," *Vestnik Tomskogo gosudarstvennogo universiteta, Istoriia*, 2013, 2(22): 199-202; N. A. Kachin, "V. M. Florinskii: V poiskakh sibirskoi modeli 'klassicheskogo universiteta'," *ibid.*, 2015, 6(38): 11-18; *idem*, "Sobiraia 'Khranilishche dukhovnoi pishchi': iz istorii sozdaniia V. M. Florinskim knizhnogo fonda nauchnoi biblioteki Tomrskogo universiteta," *ibid.*, 2016, 406: 90-97; V. A. Bubnov, "Vasilii Markovich Florinskii — organizator i sozdatel' nauki v Sibiri," *Vestnik Kurganskogo gosudarstvennogo universiteta. Fiziologiia*, 2014, 1(32): 3-16; and many others.

152 V. P. Puzyrev, "V. M. Florinskii i Tomskoe obshchestvo estestvoispytatelei," *Sibirskii meditsinskii zhurnal* (Tomsk), 1999, 14(1-2): 87-91.

153 See, Iu. A. Ozheredov, "Nauchnaia konferentsiia, posviashchennaia 130-letiiu muzeia arkheologii i etnografii Sibiri im. V. M. Florinskogo TGU," *Vestnik Tomskogo gosudarstvennogo universiteta, Istoriia*, 2013, 2(22): 7-9.

154 Florinskii, 2012, pp. 44-134.

155 Terry Martin, *The Affirmative Action Empire: Nations and Nationalism in the Soviet Union, 1923-1939* (Ithaca, NY: Cornell University Press, 2001). See also Francine Hirsch, *Empire of Nations: Ethnographic Knowledge and the Making of the Soviet Union* (Ithaca, NY: Cornell University Press, 2005).

156 See, for instance, Marlène Laruelle, *Russian Eurasianism: An Ideology of Empire* (Baltimore, MD: Johns Hopkins University Press, 2008); Stephen Shenfield, *Russian Fascism: Traditions, Tendencies and Movements* (London: Routledge, 2001); Wendy Helleman, ed., *The Russian Idea: In Search of a New Identity* (Bloomington, IN: Slavica, 2004); Hilary Pilkington, Elena Omel'chenko, and Al'bina Garifzianova, *Russia's Skinheads: Exploring and Rethinking Subcultural Lives* (London: Routledge, 2010); John Garrard and Carol Garrard, *Russian Orthodoxy Resurgent: Faith and Power in the New Russia* (Princeton, NJ: Princeton University Press, 2014); Pal Kolstø and Helge Blakkisrud, eds., *The New Russian Nationalism* (Edinburgh: Edinburgh University Press, 2016); and many others.

157 On the place of raciology among other contemporary ideological constructs, see Mark Bassin, "'What is More Important: Blood or Soil?': Rasologiia

contra Eurasianism," in Mark Bassin and Gonzaldo Pozo, eds., *The Politics of Eurasianism: Identity, Popular Culture and Russia's Foreign Policy* (London: Rowman and Littlefield, 2017), pp. 39-58.

158 For Avdeev's "literary and scientific biography" up to 2012, published on the occasion of his fiftieth birthday in the notorious nationalist newspaper *Za russkoe delo*, see "Literaturnaia nauchnaia biografiia V. B. Avdeeva," http://www.zrd.spb.ru/news/2012-01/news-0236.htm; Avdeev's personal page on the Russian social media site "In-Contact" contains a wealth of relevant materials: https://vk.com/racology. Avdeev also maintains his own website for "raciology" at http://racology.ru/avdeyev-vladimir-borisovich or http://racology.ru. However, several websites maintained by Russian nationalists since 2001 that carried nearly all of Avdeev's publications, as well as numerous commentaries, are no longer available, for instance, http://www.xpomo.com/ruskolan/start.htm; or have changed their address, for instance, http://www.velesova-sloboda.org became https://www.velesova-sloboda.info/start/index.html.

159 See, V. B. Avdeev, *Strasti po Gabrieliu* (M.: Stolitsa, 1990); and *idem*, *Protezist* (Kharkov: n. p., 1992).

160 After the disintegration of the Soviet Union in 1991, the Union of Soviet Writers (established at Stalin's direction in 1934) split into two separate, antagonistic organizations, a "democratic" Russia's Union of Writers (Rossiiskii soiuz pisatelei) and a "patriotic" Writers Union of Russia (Soiuz pisatelei Rossii). For a general overview of these unions, see N. N. Shneidman, *Russian Literature, 1995-2002: On the Threshold of the New Millennium* (Toronto: University of Toronto Press, 2004).

161 V. B. Avdeev, *Preodolenie khristianstva* (M.: Kap', 1994). The book was reprinted in 2006 and 2011 and became the subject of scholarly works and dissertations. See, for instance, O. V. Aseev, "Iazychestvo v sovremennoi Rossii: Sotsial'nyi i etnopoliticheskii aspekty," doctoral dissertation, Moscow, 1999; A. V. Gaidukov, "Ideologiia i praktika slavianskogo neoiazychestva," doctoral dissertation, St. Petersburg, 2000; A. V. Puchkov, "Neoiazychestvo v sovremennoi evropeiskoi kul'ture na primere rassovykh teorii," doctoral dissertation, Rostov-na-Donu, 2005; and S. M. Petkova, A. V. Puchkov, *Neoiazychestvo v sovremennoi evropeiskoi kul'ture* (Rostov-na-Donu: RGUPS, 2009).

162 Avdeev later collected many of these early articles into a single volume published under the ambiguous title "Metaphysical Anthropology," see V. B. Avdeev, *Metafizicheskaia antropologiia* (M.: Belye al'vy, 2002); the volume was hailed as a symbol of neo-paganism, see, for instance, A. B. Iartsev,

*Kniga V. B. Avdeeva "Metafizicheskaia antropologiia" kak kharakternyi logotip sovremennogo iazychestva* (M.: MAKS, 2009).

163 For a detailed general analysis of racism in contemporary Russia and the role of Avdeev in its popularization, see Nikolay Zakharov, *Race and Racism in Russia* (Basingstoke: Palgrave Macmillan, 2015).

164 For a general analysis of the resurgence of racism in post-communist regimes, see Ian Law, *Red Racisms: Racism in Communist and Post-Communist Contexts* (Basingstoke: Palgrave Macmillan, 2012).

165 V. B. Avdeev and A. N. Savel'ev, eds., *Rasovyi smysl russkoi idei*, 2 vols. (M.: Belye al'vy, 2002). See the table of contents of the two volumes https://www.velesova-sloboda.info/antrop/rasovyy-smysl-russkoy-idei.html.

166 V. B. Avdeev, ed., *Russkaia rasovaia teorii do 1917 goda*, 2 vols. (M.: FERI-V, 2002-04). This anthology, along with many other publications by Avdeev, is freely available online at various sites, for instance, at Librusec, a major Russian internet library, http://lib.rus.ec/a/26150.

167 See Liudvig Vol'tman, *Politicheskaia antropologiia* (M.: Belye Al'vy, 2000); for the German original see Ludwig Woltmann, *Politische Anthropologie* (Jena: Diederichs, 1903). See also Karl Shtrats, *Rasovaia zhenskaia krasota* (M.: Belye Al'vy, 2004); for the German original see Carl H. Stratz, *Die Rassenschönheit des Weibes* (Stuttgart: Enke, 1901). For various works by Ernst Krieck, the leading Nazi theorist in the field of pedagogy, see Ernst Krik, *Preodolenie idealizma. Osnovy rasovoi pedagogiki* (M.: Belye Al'vy, 2004). For a number of articles by the leading Nazi racial hygienist Hans Friedrich Karl Günther published during the 1930s and 1940s, see Gans F. K. Giunter, *Izbrannye rabory po rasologii* (M.: Belye Al'vy, 2004).

168 V. B. Avdeed, *Rasologiia: nauka o nasledstvennykh kachestvakh liudei* (M.: Belye Al'vy, 2005).

169 See, T. I. Alekseeva et. al., "Retsidivy shovinizma i rasovoi neterpimosti," *Priroda*, 2003, 6: 80-81.

170 See V. A. Shnirel'man, "Rasizm vchera i segodnia," *Pro et Contra*, 2005, 5: 41-65, http://www.demoscope.ru/weekly/2006/0233/analit02.php; *idem*, "Rasologiia v deistvii: mechty deputata Savel'eva," in A. M. Verkhovskii, ed., *Verkhi i nizy russkogo natsionalizma* (M.: Tsentr "Sova", 2007), pp. 162-87; and *idem*, "'Tsepnoi pes rasy': divannaia rasologiia kak zashchitnitsa 'belogo cheloveka'," in *ibid*., pp. 188-208; an expanded version of the latter article is available at https://scepsis.net/library/id_1597.html; see also a lengthy review of Avdeev's *Raciology* by the leading specialist of the St. Petersburg

Museum of Anthropology and Ethnography Alexander Kozintsev, "Rasolog Vladimir Avdeev izuchaet izviliny v mozge vraga," in *Kritika rasizma v sovremennoi Rossii i nauchnyi vzgliad na problemu etnokul'turnogo raznoobraziia* (M.: Academia, 2008), pp. 19-40.

171 See V. A. Shnirel'man, *"Porog tolerantnosti": Ideologiia i praktika novogo racizma*, 2 vols. (M.: NLO, 2011-14).

172 Shnirel'man, "'Tsepnoi pes rasy'," pp. 192-93.

173 See, for instance, his response to *Priroda*'s article in the foreword to the second volume of his anthology, V. B. Avdeev, "Russkaia rasovaia teoriia do 1917 goda (prodolzhenie nachatoi temy)," in *idem*, ed., *Russkaia rasovaia teoriia do 1917 goda* (M.: FERI-V, 2004), vol. 2, pp. 5-67.

174 See V. B. Avdeev and A. N. Sevast'ianov, *Rasa i etnos* (M.: Knizhnyi mir, 2007); and V. B. Avdeev, *Istoriia angliiskoi rasologii* (M.: Belye al'vy, 2010). For the full list of the publications on raciology under the White Elves trademark, see http://shop.influx.ru/-c-30_34.html?page=2&sort=products_sort_order.

175 Vladimir Avdeyev, *Raciology: The Science of the Hereditary Traits of Peoples*, transl. by Patrick Cloutier (Morrisville, NC: Lulu, 2011).

176 On Draper and his "Pioneer Fund," see William H. Tucker, *The Funding of Scientific Racism: Wickliffe Draper and the Pioneer Fund* (Urbana, IL: University of Illinois Press, 2002).

177 See http://www.zrd.spb.ru/news/2012-01/news-0236.htm.

178 See Dzh. F. Rashton, *Rasa, evoliutsiia i povedenie* (M.: Profit-Stail, 2011). For Avdeev's own description of his involvement with the publication of this book, see http://концептуал.рф/vladimir-avdeev-obzor-rasologicheskih-novinok.

179 See https://majorityrights.com/weblog/comments/vladimir_avdeyevs_raciology. One of the latest additions to Avdeev's "Library of Racial Thought" was a Russian translation of Taylor's 2014 book *Face to Face with Race*, see Dzhared Teilor, *Litsom k litsu s rasoi* (M.: Belye al'vy, 2016).

180 See the text of the court decision at http://www.zrd.spb.ru/news/2011-02/news-0756.htm. See also a response by the publisher at http://www.zrd.spb.ru/news/2010-02/news-0830.htm.

181 The journal's title was probably a veiled reference to "Ahnenerbe," an infamous Nazi society for the "study of the German ancestral heritage" founded by Heinrich Himmler in 1935.

182 V. Avdeev, "Svoboda lichnosti i rasovaia gigiena," *Nasledie predkov*, 1997, 3, http://cultoboz.ru/np3/292-cboo----.

183 Vladimir Avdeev, "Geneticheskii socialism," *Nasledie predkov*, 1997, 4: 9-15; all the subsequent quotations are from this source.

184 N. K. Kol'tsov, "Rassovo-gigienicheskoe dvizhenie v Rossii," *Nasledie predkov*, 1998, 5: 41-42; for the German original, see N. Koltzoff [Kol'tsov], "Die rassenhygienische Bewegung in Russland," *ARGB*, 1925, 17: 96-103.

185 V. Avdeev, "Nikolai Konstantinovich Kol'tsov," *Nasledie predkov*, 1998, 5: 41.

186 Thus, he included his articles on eugenics published during the late 1990s into his collections on *Metaphysical Anthropology* and *The Racial Essence of the Russian Idea*.

187 V. M. Avdeev, ed., *Russkaia evgenika. Sbornik original'nykh rabot russkikh uchenykh (khrestomatiia)* (M.: Belye al'vy, 2012).

188 It also included publications by several much less known proponents of eugenics in Russia, such as the hygienist Kazimir Karafa-Korbut, the biochemist Boris Slovtsov, the psychiatrist Viktor Osipov, and the forensic expert Aleksander Kriukov.

189 On the "patriotic" campaigns, see Krementsov, *Stalinist Science*.

190 See Florinskii, 2012, pp. 44-134.

191 Vladimir Avdeev, "Ideologiia russkoi evgeniki," in *idem*, *Russkaia evgenika*, pp. 3-43.

192 See https://www.youtube.com/watch?v=OXAhjTM6Ta8; there are more than seventy videos featuring Avdeev on YouTube.

193 See, for instance, A. Ia. Ivaniushkin, Iu. E. Lapin, V. I. Smirnov, "Evgenika: ot utopii k nauke i ...ot nauki k utopii," *Rossiiskii pediatricheskii zhurnal*, 2013, 2: 55-59; K. A. Barsht, "Medved'-kuznets iz povesti A. Platonova 'Kotlovan' i opyty I. I. Ivanova po sozdaniiu gibrida cheloveka i obez'iany," *Novoe literaturnoe obozrenie*, 2015, 6, http://magazines.russ.ru/nlo/2015/6/medved-kuznec-iz-povesti-platonova-kotlovan-i-opyty-ii-ivanova-.html#_ftnref3; N. V. Mikhalenko, "Simvolika Vavilonskoi bashni v 'Puteshestvii moego brata Alekseia v stranu krest'ianskoi utopii' A. V. Chaianova," *Problemy istoricheskoi poetiki*, 2016, 14: 428-40; and R. N. Gaishun, "Teoreticheskie i institutsional'nye predposylki stanovleniia evgeniki v Rossii i SSSR," in *Chelovek v Mire. Mir v Cheloveke* (Perm: PGNIU, 2016), pp. 38-46.

194 See Roman Raskol'nikov at http://zavtra.ru/blogs/apostrof-42; and http://lurkmore.to/%D0%95%D0%B2%D0%B3%D0%B5%D0%BD%D0%B8%D0%BA%D0%B0; and Igor Sokolov at https://vk.com/topic-27125505_26811584; and http://russian7.ru/post/russkaya-evgenika-kak-sozdat-idealn.

195 Adams, *Networks in Action*.

196 See R. Vagner i G. Mitchell, *Genetika i obmen veshchestv* (M.: Izdatel'stvo inostrannoi literatury, 1958); Dzh. Nil' and U. Shell, *Nasledstvennost' u cheloveka* (M.: Inostrannaia literatura, 1958); Sh. Auerbakh, *Genetika v atomnom veke* (M.: Atomizdat, 1959); K. Shtern, *Osnovy genetiki cheloveka* (M.: Meditsina, 1965); V. A. Makk'iusik, *Genetika cheloveka* (M.: Mir, 1967); A. Stivenson and B. Davison, *Mediko-geneticheskoe konsul'tirovanie* (M.: Mir, 1972).

197 See, S. N. Ardashnikov, "Predislovie k russkomu izdaniiu," in Nil' and Shell, *Nasledstvennost' u cheloveka*, pp. 5-6.

198 See the last chapter, "Eugenika," in *ibid.*, pp. 374-88.

199 See, for instance, a foreword to the translation of Curt Stern's *Principles of Human Genetics*, Redaktory, "Predislovie k russkomu izdaniiu," in Kurt Shtern, *Osnovy genetiki cheloveka* (M.: Meditsina, 1965), p. 5.

200 See N. P. Bochkov, ed., *Genetika i meditsina: Itogi XIV Mezhdunarodnogo geneticheskogo kongressa* (M.: Meditsina, 1979); and M. E. Vartanian, ed., *Genetika i blagosostoianie chelovechestva: Trudy XIV mezhdunarodnogo geneticheskogo kongressa, Moskva, 21–30 avgusta 1978* (M.: Nauka, 1981).

201 For a general analysis of the situation, see Loren R. Graham and Irina Dezhina, *Science in the New Russia: Crisis, Aid, Reform* (Bloomington, IN: Indiana University Press, 2008).

202 For a detailed analysis of the International Science Foundation's work in 1990s Russia, see Irina Dezhina, *The International Science Foundation: The Preservation of Basic Science in the Former Soviet Union* (New York: OSI, 2000).

203 See, for instance, Mark G. Field, "The Health Crisis in the Former Soviet Union: A Report from the 'Post-War' Zone," *Social Science and Medicine*, 1995, 41(11): 1469-78; and Mark G. Field and Judith L. Twigg, eds., *Russia's Torn Safety Nets: Health and Social Welfare during the Transition* (New York: St. Martin's, 2000).

204 See, for instance, John M. Kramer, "Drug Abuse in Russia: An Emerging Threat," *Problems of Post-Communism*, 2011, 58(1): 31-43;

and Eugene Raikhel, *Governing Habits: Treating Alcoholism in the Post-Soviet Clinic* (Ithaca, NY: Cornell University Press, 2016).

205 See, for instance, Michele Rivkin-Fish, *Women's Health in Post-Soviet Russia: The Politics of Intervention* (Bloomington, IN: Indiana University Press, 2005).

206 See, for instance, K. K. Borisenko, L. I. Tichonova, and A. M. Renton, "Syphilis and other sexually transmitted infections in the Russian Federation," *International Journal of STDs and AIDS*, 1999, 10(10): 665-8; and Michael Z. David, "Social Welfare or Wasteful Excess?: The Legacy of Soviet Tuberculosis Programs in Post-Soviet Russia," in T. Lahusen and P. H. Solomon, eds., *What is Soviet Now?: Identities, Legacies, Memories* (Berlin: Lit Verlag, 2008), pp. 214-33.

207 At the administrative level, they managed to accomplish this task quite successfully. At the very end of 1993, the Russian Ministry of Health Protection issued a special order (no. 316) that outlined an extensive programme of creating "medico-genetic services" throughout the country. But the order was not backed by funding and, for almost a decade, its implementation depended on the ability of medical geneticists to marshal necessary resources elsewhere. For the complete text of the order and its subsequent revisions in 2001 and 2003, see http://lawrussia.ru/texts/legal_382/doc382a880x203.htm.

208 This line of research became an important part of the Tomsk IMG agendas in the late 1990s and early 2000s. See, for instance, *Genetiko-ekologicheskaia otsenka sostoianiia zdorov'ia zhitelei Iakutii* (Iakutsk: n. p., 2001); and *Iaderno-khimicheskoe proizvodstvo i geneticheskoe zdorov'e* (Tomsk: Pechatnaia manufaktura, 2004).

209 See V. Avdeev, "Restavrator russkogo mirovozzreniia V. M. Florinskii," in V. M. Florinskii, *Pervobytnye slaviane po pamiatnikam ikh doistoricheskoi zhizni: opyt slavianskoi arkheologii* (M.: Belye Al'vy, 2015), pp. 3-19.

## Chapter 8

1 N. P. Bochkov, *Klinicheskaia genetika* (M.: GEOTAR-Media, 2002), p. 9. In the last ten years the book appeared in five "updated and revised" editions. But its "historical introduction" remains basically the same, see, for instance, N. P. Bochkov, V. P. Puzyrev, and S. A. Smirnikhina, *Klinicheskaia genetika*, 4th edn. (M.: GEOTAR-Media, 2011), pp. 16-17.

2 Terentianus Maurus, *De litteris, syllabis, pedibus et metris* (London: H. Bohn, 1825), verse 1286, p. 57.

3 The story is available in numerous editions and translations. I use here the translation by Anthony Bonner, see Jorge Luis Borges, "Pier Menard, Author of the Quixote," in *Ficciones* (New York: Grove, 1962), pp. 45-55. All the subsequent quotations are from this edition and all the emphasis is in the original.

4 The phrase first appeared in the "Preface to the 1888 English Edition of the Manifesto of the Communist Party." See K. Marx and F. Engels, *Selected Works* (M.: Progress, 1969), vol. 1, p. 8.

5 See, for instance, V. V. Bunak, "O smeshenii chelovecheskikh ras," *REZh*, 1925, 3(2): 121-38.

6 On the western studies of the "unfit," see Nicole H. Rafter, *White Trash: The Eugenic Family Studies, 1877-1919* (Boston, MA: Northeastern University Press, 1988); idem, *Creating Born Criminals* (Chicago, IL: University of Illinois Press, 1998); and Richard F. Wetzell, *Inventing the Criminal: A History of German Criminology, 1880-1948* (Chapel Hill, NC: University of North Carolina Press, 2000).

7 E. K. Krasnushkin, "Chto takoe prestupnik?" in *Prestupnik i prestupnost'*, 1926, 1: 6-33 (p. 32). For a general overview of the early Soviet studies of the criminal, see Kenneth M. Pinnow, "From All Sides: Interdisciplinary Knowledge, Scientific Collaboration, and the Soviet Criminological Laboratories of the 1920s," *Slavic Review*, 2017, 76(1): 122-46.

8 My analysis here is based on numerous historical studies of eugenics in various countries, and especially on several explicitly comparative collections and compilations, including Adams, ed., *The Wellborn Science*; Broberg and Roll-Hansen, eds., *Eugenics and the Welfare State*; Stepan, *"The Hour of Eugenics"*; Raphael Falk, Diane B. Paul, and Garland Allen, eds., *Science in Context* (a special issue on eugenics), 1998, 11(3-4): 329-627; Pauline M. H. Mazumdar, ed., *The Eugenics Movement: An International Perspective*, 6 vols. (London: Routledge, 2007); Turda and Weindling, eds., *"Blood and Homeland"*; Bashford

and Levine, eds., *Oxford Handbook*; Promitzer, Trubeta, and Turda, eds., *Health, Hygiene and Eugenics*; Felder and Weindling, eds, *Baltic Eugenics*; Kühl, *For the Betterment of the Race*; Turda, ed., *The History of East-Central European Eugenics*; Turda and Gillette, *Latin Eugenics*; and Paul, Stenhouse, Spencer, eds., *Eugenics at the Edges of Empire*.

9 For a fine-grained analysis of the "local" and the "general" in the history of genetics, see Jonathan Harwood, "National Styles in Science: Genetics in Germany and the United States between the World Wars," *Isis*, 1987, 78: 390-414; and *idem*, *Styles of Scientific Thought: The German Genetics Community, 1900-1933* (Chicago, IL: University of Chicago Press, 1993); for a concise application of similar arguments to the history of eugenics, see Adams, "Eugenics," in V. Ravitsky, A. Fiester, and A. Caplan, eds., *The Penn Center Guide to Bioethics* (Cham: Springer, 2008), pp. 371-82.

10 The glossaries attached to the chapters dealing with separate countries in Marius Turda's collection of materials on the development of eugenics in eastern and central Europe provide perfect examples of such local vocabularies, as well as differences and similarities among them. Alas, the collection's authors have neglected to utilize these rich sources for a comparative analysis of the "domestication" of various "imported" versions of eugenics in these countries, see Turda, ed., *History of East-Central European Eugenics*.

11 Leonard Darwin, "Presidential Address," in *Problems in Eugenics*, vol. I, p. 2.

12 Barrett and Kurzman, "Globalizing Social Movement Theory."

13 For a general analysis of the attitudes of the Russian educated elites to the state's authority, see Joseph Bradley, "Subjects into Citizens: Societies, Civil Society, and Autocracy in Tsarist Russia," *American Historical Review*, 2002, 107(4): 1094-123.

14 For instance, an article published in *Russian Thought* in 1893 under the characteristic title "Biologists on the Women Question" denied the "inherent" "biological inferiority" of women. See. L. E. Obolenskii, "Biologi o zhenskom voprose," *Russkaia mysl'*, 1893, 2: 61-78. On the Russian intelligentsia's general attitudes towards the "women question," see Richard Stites, *The Women's Liberation Movement in Russia: Feminism, Nihilism, and Bolshevism, 1860-1930* (Princeton, NJ: Princeton University Press, 1978). For a more general analysis of eugenics and gender, see Alexandra Minna Stern, "Eugenics, Gender and Sexuality: A Global Tour and Compass," in Bashford and Levine, eds., *Oxford Handbook*, pp. 173-91; and Susan Klausen and Alison Bashford, "Fertility Control: Eugenics, Neo-Malthusianism, and Feminism," in *ibid.*, pp. 98-115.

15 See, for instance, Richard Cleminson, *Anarchism, Science and Sex: Eugenics in Eastern Spain, 1900-1937* (Bern: Peter Lang, 2000); *idem*, "Eugenics without the State: Anarchism in Catalonia, 1900-1937," *Studies in History and Philosophy of Biological and Biomedical Sciences*, 2008, 39(2): 232-39; and Richard Sonn, "'Your Body Is Yours': Anarchism, Birth Control, and Eugenics in Interwar France," *Journal of the History of Sexuality*, 2005, 14(4): 415-32.

16 For a useful comparative overview of the place of eugenics in a "welfare state," see Véronique Mottier, "Eugenics and the State: Policy-Making in Comparative Perspective," in Bashford and Levine, eds., *Oxford Handbook*, pp. 134-53.

17 I cannot discuss here the actual implementation of this *decreed* gender equality in Russia, which is the subject of numerous historical studies. For some general observations, see Wendy Z. Goldman, *Women, the State and Revolution: Soviet Family Policy and Social Life, 1917-1936* (Cambridge: Cambridge University Press, 1993); and Elizabeth A. Wood, *The Baba and the Comrade: Gender and Politics in Revolutionary Russia* (Bloomington, IN: Indiana University Press, 1997).

18 The issues surrounding the actual realization of the egalitarian ethnic policies in the Soviet Union generated an extensive body of historical literature. For some general observations, see Hirsch, *Empire of Nations*; and Martin, *Affirmative Action Empire*.

19 See Birte Kohtz, "Gute Gene, schlechte Gene. Eugenik in der Sowjetunion zwischen Begabungsforschung und genetischer Familienberatung," *Jahrbücher für Geschichte Osteuropas*, 2013, 61(4): 591-610.

20 The involvement of several early proponents of eugenics, such as, for instance, eminent American zoologist David Starr Jordan, in the nascent peace movement also suggests that perhaps a more general ideology of "international dialogue" also played a role in the eugenics movement's success on the international scene before, and especially after, World War I.

21 See, for instance, Paul Weindling, "The 'Sonderweg' of German Eugenics: Nationalism and Scientific Internationalism," *British Journal of the History of Science*, 1989, 22: 321-33; *idem*, "International Eugenics: Swedish Sterilization in Context," *Scandinavian Journal of History*, 1999, 24: 179-97; and Alison Bashford, "Internationalism, Cosmopolitanism and Eugenics," in Bashford and Levine, eds., *Oxford Handbook*, 254-86.

22 Garland E. Allen, "The Eugenics Record Office at Cold Spring Harbor, 1910-1940: An Essay in Institutional History," *Osiris*, 1986, 2: 225-64.

23 See, for instance, Jean-Jacques Salomon, "The 'Internationale' of Science,"

*Science Studies*, 1971, 1: 24-42; Paul Weindling, ed., *International Health Organisations and Movements, 1918–1939* (Cambridge: Cambridge University Press, 1995); Krementsov, *International Science*; a special issue on "Transnational History of Science," *British Journal for the History of Science*, 2012, 45(3); and many others.

24 The subsequent meetings were held in London (1919), New York City (1921), Brussels (1922), Lund (1923), Milan (1924), London (1925), Paris (1926), Amsterdam (1927), Munich (1928), Rome (1929), Farnham, England (1930), New York City (1932), Zurich (1934), and Schweningen, the Netherlands (1936). Reports on the meetings appeared regularly on the pages of *Eugenics Review*.

25 For detailed examinations of the role of such networks in the history of science in general and of Soviet science in particular, see Adams, *Networks in Action*; and Krementsov, *International Science*.

26 For an examination of the role of international networks in the history of genetics and eugenics in Russia, see Krementsov, "Eugenics, *Rassenhygiene*, and Human Genetics."

27 See, for instance, Michael Burleigh, "Eugenic Utopias and the Genetic Present," *Totalitarian Movements and Political Religions*, 2000, 1(1): 56-77.

28 See, for instance, Anne-Laure Simmonot, *Hygiénisme et eugénisme au XXe siècle à travers la psychiatrie française* (Paris: S. Arslan, 1999).

29 On the co-evolution of the notions of social hygiene and public health, see the next chapter "Apologia."

30 Edwin Chadwick, *Report to Her Majesty's Principal Secretary of State for the Home Department, from the Poor Law Commissioners, on an inquiry into the sanitary condition of the labouring population of Great Britain; with appendices* (London: Clowes, 1842).

31 For a historical analysis of this process in France, see William Coleman, *Death is a Social Disease: Public Health and Political Economy in Early Industrial France* (Madison, WI: University of Wisconsin Press, 1982).

32 See J. P. Frank, "The Civil Administrator, Most Successful Physician," (1784), transl. by Jean Captain Sabine, *Bulletin of the History of Medicine*, 1944, 16: 289-318.

33 As we saw in Chapter 5, in his 1912 assessment of the First International Eugenics Congress, Nikolai Gamaleia pinpointed the close connection between the growth of social hygiene and the rise of eugenics in Britain.

34 David Starr Jordan, *The Blood of the Nation: A Study in the Decay of Races through the Survival of the Unfit* (Boston, MA: American Unitarian Association, 1902).

35 See, for instance, Paul Crook, "War as Genetic Disaster?: The First World War Debate over the Eugenics of Warfare," *War & Society*, 1990, 8(1): 47-70; and *idem*, *Darwinism, War and History: The Debate over the Biology of War from the "Origin of Species" to the First World War* (Cambridge: Cambridge University Press, 1994).

36 M. M. Zavadovskii, *Pol i razvitie ego priznakov* (M.: GIZ, 1922), p. 235.

37 For a general assessment of the influence of Haldane's essay, see Krishna R. Dronamraju, ed., *Haldane's "Daedalus" Revisited* (Oxford: Oxford University Press, 1995).

38 See his groundbreaking analysis in Mark B. Adams, "Last Judgment: The Visionary Biology of J. B. S. Haldane," *JHB*, 2000, 33: 457-91; and *idem*, "The Quest for Immortality: Visions and Presentiments in Science and Literature," in Stephen G. Post and Robert H. Binstock, eds., *The Fountain of Youth: Cultural, Scientific, and Ethical Perspectives on a Biomedical Goal* (New York: Oxford University Press, 2004), pp. 38-71.

39 On polemics between Haldane and Russell, see Adams, "Last Judgment"; and Charles T. Rubin, "Daedalus and Icarus Revisited," *The New Atlantis: The Journal of Technology and Society*, 2005, 8: 73-91. Both Haldane's *Daedalus* and Russell's *Icarus* appeared in Russian as a single volume, see D. B. S. Holden and B. Rassel, *Dedal i Ikar (Budushchee nauki)* (M.-L.: Petrograd, 1926).

40 See, for instance, Jon Turney, *Frankenstein's Footsteps: Science, Genetics and Popular Culture* (New Haven, CT: Yale University Press, 1998); and the first chapter in Jon Towlson, *Subversive Horror Cinema: Countercultural Messages of Films from Frankenstein to the Present* (Jefferson, NC: McFarland, 2014). The Google Ngram for the word "Frankenstein" shows a doubling of its use in English books from 1910 to 1940, see https://books.google.com/ngrams/graph?content=Frankenstein&year_start=1910&year_end=1940&corpus=15&smoothing=5&share=&direct_url=t1%3B%2CFrankenstein%3B%2Cc0

41 Philip J. Pauly, *Controlling Life: Jacques Loeb and the Engineering Ideal in Biology* (New York: Oxford University Press, 1987).

42 On the early history of the notion of "social engineering" and its popularity during this very period, see David Östlund, "A Knower and Friend of Human Beings, Not Machines: The Business Career of the Terminology of Social Engineering, 1894-1910," *Ideas in History: Journal of the Nordic Society for the History of Ideas*, 2007, 2(2): 43-82.

43 A. E. Hamilton, "Putting Over Eugenics," *Journal of Heredity*, 1915, 6(6): 281-88 (p. 281).

44 For an overview of eugenics' impact on the public imagination in the United States during the 1920s, see Betsy Lee Nies, *Eugenic Fantasies: Racial Ideology in the Literature and Popular Culture of the 1920s* (London: Routledge, 2002).

45 See Leslie C. Dunn, ed., *Genetics in the 20th Century: Essays on the Progress of Genetics during Its First 50 Years* (New York: Macmillan, 1951); especially the report on "Old and New Pathways in Human Genetics" by leading US human geneticist Lawrence H. Snyder, pp. 369-92. For an analysis of the Golden Jubilee's goals and means, see Audra J. Wolfe, "The Cold War Context of the Golden Jubilee, Or, Why We Think of Mendel as the Father of Genetics," *JHB*, 2012, 45(3): 389-414.

46 See Leslie C. Dunn, *A Short History of Genetics: The Development of Some of the Main Lines of Thought, 1864-1939* (New York: McGraw-Hill, 1965); and Alfred H. Sturtevant, *A History of Genetics* (New York: Harper and Row, 1965). A similar process of distancing genetics from eugenics occurred in many other countries. See, for instance, a detailed analysis of this process in Italy in Cassata, *Building the New Man*.

47 The first issue of the renamed *Annals* carried a brief "editorial note" that "explained" the name change: "In the foreword to the first volume of the *Annals of Eugenics*, published in 1925, Karl Pearson stated that the time was ripe for a journal which should devote its pages wholly to the scientific treatment of racial problems in man but that contributions dealing with heredity in man from any scientific standpoint would be acceptable. From the outset, the journal contained many papers dealing with heredity and, in recent years, has consisted almost exclusively of contributions to the science of human genetics. It seems logical to recognize this trend by the alteration of the title from *Annals of Eugenics* to *Annals of Human Genetics* (beginning with Vol. 19, Part 1). The numbering of the volumes will follow on without any change." Anon., "Editorial Note," *Annals of Human Genetics*, 1954, 19(1): 79-80.

48 For a penetrating analysis of the movement and its connections to eugenics, see Matthew Connelly, *Fatal Misconception: The Struggle to Control World Population* (Cambridge, MA: Harvard University Press, 2008); for a more focused analysis, see Randall Hansen and Desmond King, *Sterilized by the State: Eugenics, Race, and the Population Scare in Twentieth Century North America* (Cambridge: Cambridge University Press, 2013).

49 To list only the most prominent examples, such alliances included the North Atlantic Treaty Organization (NATO, 1949), the Australia, New Zealand,

United States Security Treaty (ANZUS, 1951), the Southeast Asia Treaty Organization (SEATO, 1954), the World Bank (1944), and the Common Market (1958) in the "capitalist West"; the Warsaw Pact (1955) and the Council for Mutual Economic Assistance (COMECON, 1949) in the "socialist East"; and the Association of Southeast Asian Nations (ASEAN, 1967) formed by several of the nonaligned countries.

50 On "big science" and its development in western contexts, see Derek J. de Solla Price, *Little Science, Big Science* (New York: Columbia University Press, 1963); Peter Galison and Bruce Hevly, eds., *Big Science: The Growth of Large-scale Research* (Stanford, CA: Stanford University Press, 1992); and Olof Hallonsten, *Big Science Transformed: Science, Politics and Organization in Europe and the United States* (New York: Palgrave Macmillan, 2016).

51 Julian S. Huxley, *Evolution. The Modern Synthesis* (London: Allen and Unwin, 1942). For broad historical assessments of the synthesis, see Ernst Mayr and William B. Provine, eds., *The Evolutionary Synthesis: Perspectives on the Unification of Biology* (Cambridge, MA: Harvard University Press, 1980); and Vassiliki B. Smocovitis, *Unifying Biology: The Evolutionary Synthesis and Evolutionary Biology* (Princeton, NJ: Princeton University Press, 1996).

52 See Gisela Nass, *Moleküle des Lebens* (Munich: Kindler, 1970); *idem, Las moléculas de la vida* (Madrid: Ediciones Guadarrama, 1970); and *idem, The Molecules of Life* (London: Weidenfeld and Nickolson, 1970).

53 Lily E. Kay, *The Molecular Vision of Life: Caltech, the Rockefeller Foundation, and the Rise of the New Biology* (New York: Oxford University Press, 1996).

54 See V. I. Vernadskii, *Biosfera* (L.: VSNKh, 1926), for the first modern articulation of this notion.

55 See Nancy M. Jessop, *Biosphere: A Study of Life* (Englewood Cliffs, NJ: Prentice-Hall, 1970).

56 This new understanding was clearly conveyed in an extensive "science program with experiments and observations for the student" published in the popular textbook series "Today's Basic Science." See John Gabriel Navarra, Joseph Zafforoni, John E. Garone, *Today's Basic Science: The Molecule and the Biosphere* (New York: Harper and Row, 1965).

57 To list just a few, almost random titles, see Pierre Teilhard de Chardin, *L'avenir de l'homme* (Paris: Editions du Seuil, 1955), published in English as P. Teilhard de Chardin, *Phenomenon of Man* (London: Collins, 1959); Jean Rostand, *Peut-on modifier l'homme?* (Paris: Gallimard, 1956), published in English as J. Rostand, *Can Man be Modified?* (New York: Basic, 1959); Lewis Mumford, *The Transformations of Man* (London: Allen and Unwin, 1957); D. C. Rife, *Heredity*

*and Human Nature* (New York: Vantage, 1959); P. B. Medawar, *The Future of Man* (New York: Basic, 1960); C. H. Waddington, *The Ethical Animal* (London: Allen and Unwin, 1960); Garrett Hardin, *Nature and Man's Fate* (New York: New American Library, 1961); H. Hoagland and R. W. Burhoe, eds., *Evolution and Man's Progress* (New York: Columbia University Press, 1962), a collection of articles originally published in the summer of 1961 as a special issue of the journal *Daedalus*; Theodosius Dobzhansky, *Mankind Evolving: The Evolution of the Human Species* (New Haven, CT: Yale University Press, 1962); H. L. Carson, *Heredity and Human Life* (New York: Columbia University Press, 1963); R. Dubos, *Man Adapting* (New Haven, CT: Yale University Press, 1965); and many others.

58 See, for instance, the development of the notion of "environmental engineering" in B. MacKaye, *From Geography to Geotechnics* (Urbana, IL: University of Illinois Press, 1968); and B. H. Jennings, *Environmental Engineering: Analysis and Practice* (New York: International Textbook Company, 1970); for the notion of "biomedical engineering," see J. H. U. Brown, John E. Jacobs, and Lawrence Stark, eds., *Biomedical Engineering* (Philadelphia, PA: F. A. Davis, 1970); and Heinz Siegfried Wolff, *Biomedical Engineering* (New York: McGraw-Hill, 1970).

59 See, Wolstenholme, ed., *Man and his Future*; T. M. Sonneborn, ed., *The Control of Human Heredity and Evolution* (New York: Macmillan, 1965); and John D. Roslansky, ed., *Genetics and the Future of Man* (Amsterdam: North-Holland, 1966).

60 *Life Sciences: Recent Progress and Application to Human Affairs* (Washington, DC: National Academy of Sciences, 1970).

61 Philip Handler, ed., *Biology and the Future of Man* (New York: Oxford University Press, 1970).

62 Some elements of these debates have been examined by Kevles, *In the Name of Eugenics*; Paul, *The Politics of Heredity*; Comfort, *The Science of Human Perfection*; Diane B. Paul, "Genetic Engineering and Eugenics: The Uses of History," in Harold W. Baillie and Timothy K. Casey, ed., *Is Human Nature Obsolete?: Genetics, Bioengineering and the Future of Human Condition* (Cambridge, MA: MIT Press, 2005), pp. 134-63; and many others. For insightful but necessarily brief overviews, see Alison Bashford, "Epilogue: Where Did Eugenics Go?" in Bashford and Levine, eds., *Oxford Handbook*, pp. 539-58; and Carolyn Burdett, "Introduction: Eugenics Old and New," *New Formations*, 2007, 60: 7-12.

63 See Anon., "The Ciba Foundation," *Lancet*, 1949, 254 (6566): 25-26.

64 For an insider's history of the Ciba Foundation, see F. Peter Woodford, *The*

*Ciba Foundation: An Analytic History, 1949-1974* (Amsterdam: Elsevier, 1974); see also a favorable review of this history by one of the participants of the 1962 symposium, Alan S. Parkes, "The Ciba Foundation, 1949-1974: An Appreciation," *Journal of Biosocial Science*, 1976, 8(1): 69-73.

65 Gordon Wolstenholme, "Preface," in *idem*, ed., *Man and His Future*, p. v.

66 The following story of the symposium is based on its published proceedings and all of the subsequent quotations are taken from this volume, Wolstenholme, ed., *Man and his Future*.

67 Artur Glikson came from Tel Aviv, Israel, and Marc Klein from Strasbourg, France. Since Muller could not come to London due to illness, only 26 invited participants actually attended the symposium.

68 See H. J. Muller, "The Guidance of Human Evolution," *Perspectives in Biology and Medicine*, 1959, 3: 1-43; and *idem*, "Human Evolution by Voluntary Choice of Germ Plasm," *Science*, 1961, 154: 643-49.

69 On the development of artificial insemination in the twentieth-century United States, see Kara W. Swanson, "The Birth of the Sperm Bank," *The Annals of Iowa*, 2012, 71: 241-76.

70 See, for instance, Anon., "Man and His Future," *Lancet*, 6 July 1963, 33-34; Marjorie C. Meehan, "Man and His Future," *JAMA*, 1963, 187(2): 159; N. J. Berrill, "Man and His Future," *Perspectives in Biology and Medicine*, 1964, 7(3): 368-69; Charles D. Aring, "Man and His Future," *Archives of Internal Medicine*, 1964, 113: 458-59.

71 Berrill, "Man and His Future," pp. 368-69.

72 J. Lederberg, "Molecular Biology, Eugenics and Euphenics," *Nature*, 1963, 198: 428-29.

73 See J. F. Crow, "Modifying Man: Muller's Eugenics and Lederberg's Euphenics," *Science*, 1965, 148: 1579-80.

74 See, for instance, Rollin D. Hotchkiss, "Portents for a Genetic Engineering," *Journal of Heredity*, 1965, 56: 197-202.

75 Predictably, geneticists were the most vocal group in this debate. See, for instance, Theodosius Dobzhansky, *Heredity and the Nature of Man* (New York: Harcourt, Brace and World, 1964); *idem*, *The Biology of Ultimate Concern* (New York: New American Library, 1967); I. Michael Lerner, *Heredity, Evolution and Society* (San Francisco, CA: Freeman, 1968); Arne Muntzing, *Biological Points of View on Some Humanistic Problems* (Lund: Gleerup, 1968); and C.

H. Waddington, *Biology, Purpose and Ethics* (Barre, MA: Barre Publishing Company, 1971).

76 Sonneborn, ed., *The Control of Human Heredity*; see also the records of a similar discussion at a Nobel conference held in 1965 at Gustavus Adolphus College in John D. Roslansky, ed., *Genetics and the Future of Man* (Amsterdam: North-Holland, 1966).

77 Hotchkiss, "Portents for a Genetic Engineering."

78 Theodosius Dobzhansky, "Changing Man," *Science*, 1967, 155: 409-15.

79 K. Hirschhorn, "On Re-Doing Man," *Commonweal*, 1968, 88: 257-61; the article was reprinted in the *Annals of the New York Academy of Sciences*, 1971, 184: 103-12, as part of a special issue on "Environment and Society in Transition."

80 The address was published posthumously as Jack Schultz, "Human Values and Human Genetics," *American Naturalist*, 1973, 107: 585-97.

81 Joshua Lederberg, "Experimental Genetics and Human Evolution," *Bulletin of the Atomic Scientists*, 1966, 22(8): 4-11; also reprinted in *American Naturalist*, 1966, 100 (915): 519-31; and H. J. Muller, "Means and Aims in Human Genetic Betterment," in Sonneborn, ed., *The Control of Human Heredity*, pp. 100-22; and *idem*, "What Genetic Course Will Man Steer?" in James F. Crow and James V. Neel, eds., *Proceedings of the Third International Congress of Human Genetics* (Baltimore, MD: Johns Hopkins University Press, 1967), pp. 521-43.

82 James F. Crow, "The Quality of People: Human Evolutionary Changes," *Bioscience*, 1966, 16: 863-67 (p. 867).

83 See John J. Pauson, *Beyond Morality and the Law* (Pittsburgh, PA: Philosophical Press, 1966); Theodosius Dobzhansky, "On Genetics, Sociology and Politics," *Perspectives in Biology and Medicine*, 1968, 11: 544-54; Paul Ramsey, *Fabricated Man: The Ethics of Genetic Control* (New Haven, CT: Yale University Press, 1970); Joseph Fletcher, "Ethical Aspects of Genetic Controls: Designed Genetic Changes in Man," *New England Journal of Medicine*, 1971, 285: 776-83; B. Glass, "Prometheus and Pandora: 1971," *Bulletin of the New York Academy of Medicine*, 1971, 47: 1045-58; M. S. Fox, et al., "Reservations Concerning Gene Therapy," *Science*, 173: 195; Anon., "The Biologists' Dilemmas," *Nature*, 1970, 228: 900-01; P. Handler, "Can Man Shape His Future?" *Perspectives in Biology and Medicine*, 1971, 14: 207-27; P. R. Abelson, "Anxiety About Genetic Engineering," *Science*, 1971, 173: 285; James J. Nagle, "The Dilemma of Genetic Engineering," *Journal of Religion and Health*, 1972, 11(4): 370-76; James R. Sorenson, *Social and Psychological Aspects of Applied Human Genetics: A Bibliography* (Bethesda, MD: Fogarty International Center, 1973); and many others.

84 See, for instance, Karl H. Hertz, "What Man Can Make of Man: Genetic Programming," *Christian Century*, 1967, 84(25): 807-10; Michael Hamilton, "New Life for Old: Genetic Decisions," *ibid.*, 1969, 86(22): 741-44; H. B. Kuhn, "Prospect of Carbon-Copy Humans," *Christianity Today*, 1971, 15: 11-12; Charles T. Epstein, "Medical Genetics: Recent Advances with Legal Implications," *The Hastings Law Journal*, 1969, 21: 35-49; Bernard D. Davis, "Ethical and Technical Aspects of Genetic Intervention," *New England Journal of Medicine*, 1971, 285: 799-801; Michael P. Hamilton, ed., *The New Genetics and the Future of Man* (Grand Rapids, MI: W. B. Eerdmans, 1972); and many others.

85 See, for instance, Louis Lasogna, "Heredity Control: Dream or Nightmare?" *New York Times Magazine*, 5 August 1962, 7: 58-61; Lucy Eisenberg, "Genetics and the Survival of the Unfit," *Harper's Magazine*, 1966, 232: 53-58; "Man Into Superman: The Promise and Peril of the New Genetics," *Time*, 19 April 1971: 33-52; "Playing God," *Newsweek*, 23 November 1970, 76: 120; Albert Rosenfeld, "Science, Sex and Tomorrow's Morality," *Life*, 13 June 1969, 66: 37-50.

86 Fred Warshofsky, *The Control of Life* (New York: Viking, 1969).

87 Anon., "New Words in Biology and Related Subjects," *Nature*, 1964, 204: 628. See also, Panos D. Bardis, "Eudemics, Eugenics, Euphenics, Euthenics," *Phi Kappa Phi Journal*, 1972, 52(3): 37.

88 See J. Lederberg, "Letter to the editor," *The New York Times*, 26 September 1970, p. 20.

89 See Blacker, *Eugenics*; Haller, *Eugenics*; Pickens, *Eugenics*; and Ludmerer, *Genetics and American Society*.

90 The protestations of Muller's widow eventually forced Graham to drop Muller's name from his sperm bank. For details, see Cynthia R. Daniels and Janet Golden, "Procreative Compounds: Popular Eugenics, Artificial Insemination and the Rise of the American Sperm Banking Industry," *Journal of Social History*, 2004, 38(1): 5-27; David Plotz, *The Genius Factory: The Curious History of the Nobel Prize Sperm Bank* (New York: Random House, 2005); Martin Richards, "Artificial Insemination and Eugenics: Celibate Motherhood, Eutelegenesis and Germinal Choice," *Studies in History and Philosophy of Biological and Biomedical Sciences*, 2008, 39(2): 211-21; and Kara W. Swanson, *Banking on the Body: The Market in Blood, Milk, and Sperm in Modern America* (Cambridge, MA: Harvard University Press, 2014).

91 See a contemporary overview of the available techniques in E. Freese, ed., *The Prospects of Gene Therapy* (Bethesda, MD: Fogarty International Center, 1972); for a concise summary, see James J. Nagle, "Genetic Manipulations

in Humans I: The Potentials for Gene Therapy," *Bios*, 1976, 47(1): 3-13; for a brief historical account co-written by Lederberg himself, see J. A. Wolff and J. Lederberg, "A History of Gene Transfer and Therapy," in Jon A. Wolff, ed., *Gene Therapeutics: Methods and Applications of Direct Gene Transfer* (Boston, MA: Birkhäuser, 1994), pp. 3-25.

92 See, for instance, Robert G. McKinnell and Marie A. Di Berardino, "The Biology of Cloning: History and Rationale," *BioScience*, 1999, 49(11): 875-85; and a popular account in Gina Kolata, *Clone: The Road to Dolly, and the Path Ahead* (New York: William Morrow, 1998).

93 For an analysis of how western medical geneticists themselves dealt with the origins of their discipline in eugenics, see Diane B. Paul, "From Eugenics to Medical Genetics," *Journal of Policy History*, 1997, 9: 96-116; and Comfort, *The Science of Human Perfection*.

94 See, for instance, a long chapter on "Eugenics, Euphenics, and Human Welfare," in L. L. Cavalli-Sforza and W. F. Bodmer, *The Genetics of Human Populations* (San Francisco, CA: Freeman, 1971), pp. 753-804.

95 In contrast to the previous name change from *Eugenics News* to *Eugenics Quarterly* in 1954, which had been justified at length in a special editorial (see, Anon., "Editorial comment," *Eugenics Quarterly*, 1954, 1(1): 1-3), the 1969 renaming was not explained or even mentioned in any of the materials appearing in the journal, see *Social Biology*, 1969, 1-4.

96 Compare, for instance, Frederick H. Osborn, *The Future of Human Heredity: An Introduction to Eugenics in Modern Society* (New York: Weybright and Talley, 1968) and John K. Brierley, *Biology and the Social Crisis: A Social Biology for Everyman* (Madison, NJ: Fairleigh Dickinson University Press, 1970).

97 See Van Rensselaer Potter, *Bioethics: Bridge to the Future* (Englewood Cliffs, NJ: Prentice-Hall, 1971). For historical accounts of the new field, see Albert R. Jonsen, *A Short History of Medical Ethics* (New York: Oxford University Press, 1999); and John H. Evans, *The History and Future of Bioethics: A Sociological View* (New York: Oxford University Press, 2011).

98 See numerous references to publications in the British press in Turney, *Frankenstein's Footsteps*, such as, for instance, Anon., "New Hopes for the Childless," *The Guardian*, 14 February 1969, p. 1; and Anon, "Test-tube Fertility Hope for Women," *The Times*, 15 February 1969, p. 1. Similar publications appeared in US magazines, see Edward Grossman, "The Obsolescent Mother: Is the Artificial Womb Inevitable?," *The Atlantic*, 1971, 227 (May): 16-32; James D. Watson, "Moving toward the Clonal Man: Is This What We Want?," *ibid.*, 1971, 227 (May): 50-63; C. Stinson, "Theology and the Baron Frankenstein:

Cloning and Beyond," *Christian Century*, 1972, 89(3): 60-63; and many others.

99 Nancy M. Freedman, *Joshua, Son of None* (New York: Delacorte, 1973).

100 See, for instance, Judith Daar, *The New Eugenics: Selective Breeding in an Era of Reproductive Technologies* (New Haven, CT: Yale University Press, 2017).

101 To list just three publications that reflected these new developments, see Jürgen Habermas, *The Future of Human Nature* (Cambridge: Polity, 2003), originally published as idem, *Die Zukunft der menschlichen Natur. Auf dem Weh su einer liberalen Eugenik?* (Frankfurt am Main: Suhrkamp, 2001); Nicholas Agar, *Liberal Eugenics: In Defense of Human Enhancement* (Boston, MA: Blackwell, 2004); and John Glad, *Future Human Evolution: Eugenics in the Twenty-first Century* (Schuylkill Haven, PA: Hermitage, 2006).

102 The latest developments, from the 1990s through the 2010s, generated a huge body of literature both pro and contra the newest incarnations of eugenics, which is beyond the scope of this project. See, for instance, Harold W. Baillie and Timothy K. Casey, eds., *Is Human Nature Obsolete?: Genetics, Bioengineering, and the Future of the Human Condition* (Cambridge, MA: MIT Press, 2004); Jean Gayon and Daniel Jacobi, eds., *L'éternal retour de l'eugénisme* (Paris: Presses Universitaires de France, 2006); Calum MacKellar and Christopher Bechtel, eds., *The Ethics of the New Eugenics* (New York: Berghahn, 2014); Henry T. Greely, *The End of Sex and the Future of Human Reproduction* (Cambridge, MA: Harvard University Press, 2016); Daar, *The New Eugenics*; Jennifer A. Doudna and Samuel H. Sternberg, *A Crack in Creation: Gene Editing and the Unthinkable Power to Control Evolution* (Boston, MA: Houghton, Mifflin, Harcourt, 2017); and many others.

103 See http://www.tsu.ru/news/pamyatnik-osnovatelyam-tgu-florinskomu-i-mendeleev

# Apologia: The Historian's Craft

1 Cited from the complete text of the first 1605 edition of Miguel de Cervantes Saavedra, *El ingenioso hidalgo don Quijote de la Mancha*, available at http://cvc.cervantes.es/literatura/clasicos/quijote/edicion/parte1/cap09/cap09_02.htm; the full quote cited in the epigraph in English would read as: "... it is the job and duty of historians to be exact, truthful, and dispassionate, and neither interest nor fear, hatred nor love, should make them swerve from the path of truth, whose mother is history, the rival of time, storehouse of deeds, witness for the past, example and counsel for the present, and warning for the future." The translation is mine.

2 Josephine Tey, *The Daughter of Time* (London: Peter Davis, 1951). All the subsequent quotations are from this source.

3 For a recent analysis of the relations between More's account and Shakespeare's drama, see Douglas Bruster, "Thomas More's *Richard III* and Shakespeare," *Moreana*, 2005, 42(163): 79-92.

4 See, for instance, Robin W. Winks, ed., *The Historian as Detective: Essays on Evidence* (New York: Harper and Row, 1969).

5 I am profoundly grateful to its publisher Valerii Puzyrev for supplying me with a copy from his personal stash.

6 On the remarkable story of the accidental discovery of this collection, see I. Efremova, "Tsennaia nakhodka," *Sovetskaia Tatariia*, 10 August 1938, p. 4; *idem*, "Eshche raz o tsennoi nakhodke," *ibid.*, 15 August 1938, p. 4; and G. Zemlianitskii, "Tsennye rukopisi," *ibid.*, 30 November 1938, p. 4.

7 See NMRT, V. M. Florinskii's collection, №117959.

8 See, for instance, "Florinskii, Vasilii Markovich," in L. F. Zmeev, *Russkie vrachi-pisateli* (SPb.: V. Demakov, 1886), vol. 2, pp. 139-40.

9 See NMRT, #49, O. Florinskaia.

10 For a sample of this correspondence, see Iastrebov, *Sto neizvestnykh pisem*. All of the 100 letters published in this volume date from after 1879.

11 See NMRT, ##99, 204, 205.

12 See, V. Florinskii, "Tri pory zhizni," NMRT, #203.

13 See NMRT, #260.

14 See V. Florinskii, "Zametki i vospominaniia," *Russkaia starina*, 1906, 125(1): 75-109; 125(2): 288-311; 125(3): 564-96; 126(1): 109-56; 126(2): 280-323; 126(3): 596-621.

15 See Chistovich, *Dnevniki, 1855-1880*.

16 See, for instance, A. [M. T. Alekseev], "Florinskii, Vasilii Markovich," *Entsiklopedicheskii slovar' Brokgauza i Efrona* (SPb.: Brokgauz i Efron, 1902), vol. 36, p. 169; A. A. Dmitriev, "Florinskii, Vasilii Markovich," in *idem, Materialy dlia biografii pamiatnykh deiatelei iz permskikh urozhentsev* (*Trudy permskoi uchenoi arkhivnoi komissii*), 1902, 5: 70-71; and Gruzdev, "Florinskii, Vasilii Markovich," in Zagoskin, *Biograficheskii slovar'*, pp. 353-62.

17 See Gruzdev, *Istoricheskii ocherk kafedry akusherstva i zhenskikh boleznei*, pp. 214-41; and Vail', *Ocherki po istorii russkoi pediatrii*, pp. 41-54.

18 See, for instance, a brief assessment in K. A. Bogdanov, *Vrachi, patsienty, chitateli: patograficheskie teksty russkoi kul'tury XVIII-XIX vekov* (M.: OGI, 2005), pp. 258-59.

19 See L. C. Dunn, "Cross Currents in the History of Human Genetics," *American Journal of Human Genetics*, 1962, 14: 1-13 (pp. 9 and 12). In Dunn's bibliography it was listed as "Florinsky, W. M. 1866. Über die Vervollkommung und Entartung der Menschheit. Petersburg."

20 See N. V. Shelgunov, "Predislovie," in G. E. Blagosvetlov, *Sochineniia* (SPb.: E. A. Blagosvetlova, 1882), pp. iii-xxviii.

21 See RGALI, f. 613, op. 1, dd. 5660-63; and f. 629, op. 1, d. 422.

22 See, for instance, Kuznetsov, *Nigilisty*; Varustin, *Zhurnal "Russkoe slovo"*; T. I. Grazhdanova, "G. E. Blagosvetlov v russkom osvoboditel'nom dvizhenii 40-60-kh gg. XIX veka," doctoral dissertation, Leningrad, 1985; E. K. Murenina, "Literaturno-kriticheskaia deiatel'nost' G. E. Blagosvetlova," doctoral dissertation, Sverdlovsk, 1989; Feliks Kuznetsov, *Krug D. I. Pisareva* (M.: Khudozhestvennaia literatura, 1990); and many others.

23 See RGALI, f. 117. For the contents of this fond that includes only 47 files, see http://www.rgali.ru/object/215034575.

24 For the most detailed account of Volotskoi's life, focusing almost exclusively on his work on Dostoevsky's genealogy, see Nikolai N. Bogdanov, "Mikhail Volotskoi i ego "khronika roda Dostoevskogo'," in N. D. Shmeleva, ed., *Dostoevskii i sovremennost'. Materialy XXI Mezhdunarodnykh starorusskikh chtenii 2006 goda* (Velikii Novgorod: Novgorodskii gosudarstvennyi muzei zapovednik, 2007), pp. 407-34. A revised version of this article also appeared in *Voprosy literatury*, 2009, 4: 410-33, http://www.hrono.ru/text/2007/bogd0108.php

25 T. V. Tomashevich, "Pamiati M. V. Volotskogo — vydaiushchegosia antropologa, osnovopolozhnika i lidera rossiiskoi dermatoglifiki," *Moskovskii universitet*, October 2005, 34 (4139), http://www.getmedia.msu.ru/newspaper/newspaper/4139/all/mirnauki.htm

26 See A. M. Reshetov, *Materialy k biobibliograficheskomu slovariu rossiiskikh ethnografov i antropologov. XX vek* (SPb.: Nauka, 2012).

27 See I. G. Chudinov, *Bibliograficheskii ukazatel' nauchno-issledovatel'skikh i nauchno-metodicheskikh rabot sotrudnikov instituta [fizicheskoi kul'tury], 1920-1957* (M.: n. p., 1958).

28 See http://museum.sportedu.ru/category/muzei/v-binokl-vremeni/ vospominaniya, especially the memoirs by I. M. Sarkisov-Sarazini, who worked in the IFK from its very beginnings.

29 INION had "inherited" the Library of the Communist Academy and thus preserved many books and journals issued under the auspices of this bastion of "Marxist" science during the 1920s and early 1930s, including those published by the Timiriazev Institute. Tragically, in recent years, the collections of both BAN and INION have been damaged severely by fires, and many of the rare publications in their holdings have been lost.

30 On *spetskhran* and its uses, see M. B. Konashev, "Lysenkoizm pod okhranoi spetskhrana," in *Repressirovannaia nauka* (SPb.: Nauka, 1994), vol. 2, pp. 97-112; S. F. Varlamova, "K istorii sozdaniia i razvitiia spetsfondov Gosudarstvennoi publichnoi biblioteki im. M. E. Saltykova-Shchedrina," in *Tsenzura v tsarskoi Rossii i Sovetskom soiuze* (M.: Rudomino, 1995), pp. 161-67; and K. V. Liutova, *Spetskhran Biblioteki Akademii nauk: Iz istorii sekretnykh fondov* (SPb.: BAN, 1999).

31 The only but very brief account of Volotskoi's involvement with eugenics can be found in Fando, *Proshloe nauki budushchego*, pp. 179-86.

32 See, for instance, a review of my 2002 book *The Cure* by a practicing clinician who could not accept such reconstructions and accused me of writing "like a novelist," which I take as the highest compliment a historian could receive, Ross Camidge, "Nikolai Krementsov, The Cure: A Story of Cancer and Politics from the Annals of the Cold War," *British Medical Journal*, 2002, 324: 1589.

33 See the new edition of David Lowenthal's classic, *The Past is a Foreign Country: Revisited* (Cambridge: Cambridge University Press, 2015).

34 I am not concerned here with the philosophical and historiographical debates about the nature of this translation process, and even less with the possibilities or impossibilities of an "exact" translation of certain "past" meanings into "present" ones, see, for instance, W. A. DeVries, "Meaning and Interpretation in History," *History and Theory*, 1983, 22: 253-63; or Gary L. Hardcastle, "Presentism and the Indeterminacy of Translation," *Studies in History and Philosophy of Science*, 1991, 22(2): 321-45.

35 See, for instance, Vivian Nutton, "The Changing Language of Medicine, 1450-1550," in Olga Weijers, ed., *Vocabulary of Teaching and Research between Middle Ages and Renaissance* (Turnhout: Brepols, 1995), pp. 184-98; and the recent "Focus" section in the oracle of the History of Science Society, "Focus: Linguistic Hegemony and the History of Science," *Isis*, 2017, 108 (3): 606-50.

36 For a pioneering detailed examination of the importance of multi-language translations and interactions in the development of science, see Michael D. Gordin, *Scientific Babel: How Science Was Done Before and After Global English* (Chicago, IL: University of Chicago Press, 2015).

37 Pisarev, "Progress v mire zhivotnykh i rastenii," p. 1.

38 V. Dal', *Tolkovyi slovar' zhivogo velikorusskogo iazyka*, 4 vols. (M.: Ob-vo liubitelei rossiiskoi slovesnosti, 1863-1866). The entire dictionary is now available online at http://dic.academic.ru/contents.nsf/enc2p

39 See Toll', *Nastol'nyi slovar'*.

40 See L. P. Grinberg, *Vseobshchii terminologichesko-meditsinskii slovar' na latinskom, nemetskom i russkom iazykakh*, 4 vols. (Berlin, Leipzig: G. Reimer, 1840-42); and idem, *Terminologicheskii meditsinskii slovar' na latinskom, nemetskom, frantsuzskom i russkom iazykakh*, 2nd ed. (SPb.: Ia. I. Isakov, 1862-64).

41 See [Ia. Banks,] *Anglo-russkii slovar', sostavlennyi Iakovom Banksom*, 2 vols. (M.: A. Semen, 1838); and idem, *Russko-angliiskii slovar', sostavlennyi Iakovom Banksom* (M.: A. Semen, 1840); Ch. Ph. Reiff, *Dictionnaires parallèles des langues russe, française, allemande et anglaise: Dictionnaire français* (St.Pétersbourg; Carlsruhe, 1852), 2-e éd.; idem, *Parallel-Wörterbücher der russischen, französischen, deutschen und englischen Sprache für die russische Jugend* (SPb.; Karlsruhe, 1861), 4-te Aufl.; and idem, *New parallel dictionaries of the Russian, French, German and English languages: English dictionary* (SPb.; Karlsruhe, 1862).

42 See A. D. Mikhel'son, *Ob"iasnenie 7000 intostrannykh slov, voshedshikh v upotreblenie v russkii iazyk* (M.: Tip. Lazarevskogo in-ta vostochnykh iazykov, 1861); and I. F. Burdon and A. D. Mikhel'son, *Slovotolkovatel' 30,000 inostrannykh slov, voshedshikh v sostav russkogo iazyka* (M.: Universitetsakia tipografiia, 1866).

43 Sydney Ross, "'Scientist': The Story of a Word," *Annals of Science*, 1962, 18: 65-85.

44 For details, see Soboleva, *Organizatsiia nauki v poreformennoi Rossii*, pp. 35-43.

45 See Charles Rosenberg, "Toward an Ecology of Knowledge: On Discipline, Context, and History," in idem, *No Other Gods: On Science and American Social Thought* (Baltimore, MD: Johns Hopkins University Press, 1997), pp. 225-39.

46 See, for instance, a "Glossarial note" in the English translation of an excerpt from Paul Broca's treatise on hybridity, Paul Broca, *On the Phenomena of Hybridity in the Genus Homo*, transl. by C. Carter Blake (London: Longman, 1864), p. x; for the French original, which, of course, did not have such a

"glossary," see M. Paul Broca, *Recherches sur l'hybridité animale en général et sur l'hybridité humaine en particulier* (Paris: J. Claye, 1860).

47 See, for instance, Robley Dunglison's famous one-thousand-page "medical lexicon," which first appeared in 1833 and over the years went through more than twenty editions. It included not only English terms, with detailed explanations of their etymology and meaning, but also their French, Latin, and occasionally German synonyms, Robley Dunglison, *Medical Lexicon: A Dictionary of Medical Science* (Philadelphia: Blanchard and Lea, 1838), 2nd ed. On the dictionary and its author, see Chalmers L. Gemmill, "Robley Dunglison's Dictionary of Medical Science, 1833," *Bulletin of the New York Academy of Medicine*, 1972, 48(5): 791-98.

48 See, for instance, *Drug zdraviia* (Health Companion), one of the oldest medical weeklies, published in St. Petersburg from 1833 to 1866.

49 For a brief overview of this growing business, see A. G. Smoliakova, "Perevodnaia spetsial'naia kniga v Rossii XVI-XIX vv.," doctoral dissertation, Moscow, 1999.

50 See Karl Fogt [Carl Vogt, trans.], [Robert Chambers], *Estestvennaia istoriia mirozdaniia* (M.: A. Cherenin i A. Ushakov, 1863). For the German edition that had been used to prepare the Russian one, see Anon. [Robert Chambers], *Natürliche Geschichte der Schöpfung des Weltalls, der Erde und der auf ihr befindlichen Organismen, begründet auf die durch die Wissenschaft errungenen Thatsachen*, transl. by Carl Vogt (Braunschweig: Friedrich Vieweg, 1858), 2nd ed.

51 See T. G. Geksli, *Mesto cheloveka v tsarstve zhivotnom*. Per. Iu. Gol'dendakh s nem. Ed. V. Karusa (M.: Universitetskaia tipografiia, 1864); and T. G. Geksli, *O polozhenii cheloveka v riadu organicheskikh sushchestv*. Per. pod red. A. Beketova (SPb.: N. Tiblen, 1864).

52 N. L. Tiblen, "[Predislovie]," in G. Spenser, *Klassifikatsiia nauk* (SPb.: N. L. Tiblen, 1866), pp. i-iii (p. ii). Unfortunately, the process of creating scientific terminology in the biomedical sciences in Russia has attracted no attention from historians. For an analysis of this process in Russian physics, see L. L. Kutina, *Formirovanie terminologii fiziki v Rossii* (M.-L.: Nauka, 1966).

53 Charles Darwin, *On the Origin of Species by Means of Natural Selection, Or the Preservation of Favoured Races in the Struggle for Life* (London: John Murray, 1859). In this section I focus mostly on the actual translations of Darwin's vocabulary and its key terms and expressions into the Russian language, not the perception and reinterpretation of their meanings by Russian readers. For a detailed analysis of the latter processes as regards a key metaphor of

Darwin's *Origin*, "the struggle for life," see Todes, *Darwin without Malthus*.

54 See, for instance, the catalogues of foreign language books available in St. Petersburg's major bookstores in the second half of the nineteenth century, *Katalog knig, prodaiushchikhsia v magazine russkikh i inostrannykh knig D. E. Kozhanchikova v S. Peterburge* (SPb.: n. p., 1863); *Katalog udeshevlennykh meditsinskikh knig na russkom i inostrannykh iazykakh, prodaiushchikhsia po porucheniiu izdatelei v mediko-khirurgicheskom knizhnom magazine N. P. Petrova v S.-Peterburge* (SPb.: n. p. 1887); *Katalog knig po vsem otrasliam znaniia na russkom i inostrannykh iazykakh knizhnogo magazina V. L. Lebedeva* (SPb.: n. p. 1906); and many others.

55 See Charl's Darvin, *O proiskhozhdenii vidov v tsarstvakh zhivotnom i rastitel'nom putem estestvennogo podbora rodichei, ili sokhranenie usovershenstvovannykh porod v bor'be za sushchestvovanie.* Per. S. A. Rachinskogo (SPb.: A. I. Glazunov, 1864). In the second edition, Rachinskii excised these words, thus restoring Darwin's original title. See Charl's Darvin, *O proiskhozhdenii vidov putem estestvennogo podbora, ili sokhranenie usovershenstvovannykh porod v bor'be za sushchestvovanie.* Per. S. A. Rachinskogo (SPb.: A. I. Glazunov, 1865).

56 N. Strakhov, "Kritika i bibliografiia," *Grazhdanin*, 16 July 1873, http://smalt.karelia.ru/~filolog/grazh/1873/16jyN29

57 See, for instance, Diane B. Paul, "The Selection of the 'Survival of the Fittest'," *JHB*, 1988, 21(3): 411-24.

58 See Charles Darwin, *Über die Entstehung der Arten durch natürliche Zuchtwahl, oder die Erhaltung der begünstigten Rassen im Kampfe um's Dasein. Aus dem Englischen übersetzt von H. G. Bronn. Nach der vierten englischen sehr vermehrten Ausgabe durchgesehen und berichtigt von J. Victor Carus* (Stuttgart: E. Schweizerbart, 1867). Tellingly, Friedrich Rolle in his own rendering of Darwin's *Origin* excised the expression "natural selection" from his title.

59 Claparède, "M. Darwin et sa théorie de la formation des espèces," p. 531. The original reads: "*Je regrette d'employer une expression aussi paradoxale. Il est difficile, en effet, d'admettre qu'une élection puisse être inconsciente. L'expression employée par le naturaliste anglais a l'avantage de ne renfermer aucune contradiction dans les termes. Malheureusement, notre langue ne renferme aucun mot qui rende exactement le terme selection. J'ai choisi celui d'élection, malgré son insuffisance, plutôt que d'employer un néologisme de couleur trop étrangère.*" Needless to say, this passage is absent in the Russian translation of Claparède's review. For a similar discussion of the advantages and disadvantages of using the words *sélection* and *élection* as the correct renderings of Darwin's term, see Rey, *Dégénération de l'espèce humaine et sa régénération*, pp. 4, 12, 29.

60 See Charles Darwin, *De l'origine des espèces par sélection naturelle: ou, Des lois de transformation des êtres organisés*, traduit en français avec l'autorisation de l'auteur par Clémence Royer, avec une préface et des notes du traducteur, 2nd edn. (Paris: Guillaumin et Cie, 1866).

61 See, for instance, Charles Darwin, *L'origine des espèces au moyen de la sélection naturelle, ou, La lutte pour l'existence dans la nature*. Trad. de J.-J. Moulinié (Paris: C. Reinwald, 1873).

62 See, for instance, Ia. Sliaskii, *Iz selektsionnykh zametok* (Kiev: G. T. Korchak-Novitskii, 1893); *Ocherk deiatel'nosti Smelianskoi selektsionnoi stantsii grafov Bobrinskikh za piat' let: 1890-96* (Kiev: n. p., 1896); *Selektsionnaia stantsiia pri Vol'finskom imenii tainogo sovetnika N. A. Tereshchenko. 1890-97* (Kiev: I. N. Kushnerev, 1897); and many others.

63 See Ch. Darvin, "Proiskhozhdenie vidov putem estestvennogo otbora, ili sokhranenie izbrannykh porod v bor'be za zhizn'. Per. prof. K. A. Timiriazeva," in Ch. Darvin, *Sobranie sochinenii v 4-kh tomakh* (SPb.: O. N. Popova, 1896-98), vol. 1.

64 See Toll', *Nastol'nyi slovar'*. For specific entries, see *"plemia,"* vol. 3, p. 117; *"poroda,"* vol. 3, p. 172; *"rasa,"* vol. 3, p. 269.

65 Only in the 1870s, there appeared major English-language anthropological works, such as John Lubbock's *The Origin of Civilization and the Primitive Condition of Man* (1870) and Edward B. Taylor's *Primitive Culture* (1871), which within a few years were translated into Russian. See Dzh. Lebbok, "Nachalo tsivilizatsii," *Znanie*, 1874, 10; 1875, 1-6 (prilozheniia); Dzh. Lebbok, *Doistoricheskie vremena* (M.: "Priroda", 1876); and E. B. Teilor, *Pervobytnaia kul'tura*, 2 vols. (SPb.: "Znanie", 1872-73).

66 See, for instance, Louis Pappenheim, *Handbuch der Sanitäts-Polizei: nach eigenen Untersuchungen* (Berlin: Hirschwald, 1858), https://books.google.co.uk/books/about/Handbuch_der_Sanit%C3%A4ts_Polizei.html?id=3EVLAAAAcAAJ&redir_esc=y; and Friedrich Oesterlen, *Der Mensch und seine physische Erhaltung. Hygienische Briefe für weitere Leserkreise* (Leipzig: F. A. Brockhaus, 1859). Florinskii used both books in his own work.

67 See, for instance, J.-Ch.-M. Boudin, *Études d'hygiène publique sur l'état sanitaire, les maladies et la mortalité des armées de terre et de mer* (Paris: J. Corréard et J. Dumaine, 1846), http://gallica.bnf.fr/ark:/12148/bpt6k64732014; and Louis Alfred Becquerel, *Traité élémentaire d'hygiène*. Florinskii used both books in his own work.

68 See, for instance, Michael Ryan, *A Manual of Medical Jurisprudence and State Medicine* (London: Sherwood, Gilbert, and Piper, 1836).

69 See, for instance, a survey of the English "social hygiene legislation" translated from German (!), *Istoriia progressa angliiskogo zakonodatel'stva po chasti obshchestvennoi gigieny*. Per. s nem. (Kiev: Universitetskaia tipografiia, 1861). The German source for this survey was an article by Fr. Oesterlen, "Die neuere Sanitäts-Gesetzgebung und Sanitätsreform in England. Deren Geschichte und Resultate," *Zeitschrift für Hygiene, medizinische Statistik und Sanitätspolizei*, 1860, 1: 131-65.

70 In 1851 and 1859 the first two international "sanitary" conferences were held in Paris in an attempt, largely unsuccessful, to hammer out "sanitary conventions" to prevent the spread of infectious diseases across national borders. But the conferences did not even raise the issues of terminology and methodology of the nascent field of "public health." See Norman Howard-Jones, *The Scientific Background of the International Sanitary Conferences, 1851-1938* (Geneva: WHO, 1975).

71 There are numerous histories of the multiple "national" variants of the field of public health. For a concise overview, see Susan G. Solomon, Lion Murard, Patrick Zylberman, eds., *Shifting Boundaries of Public Health: Europe in the Twentieth Century* (Rochester, NY: Rochester University Press, 2008). Alas, only German historians have made an effort to trace its terminological development, see Rudolf Thissen, "Die Entwicklung der Terminologie auf dem Gebiet der Sozialhygiene und Sozialmedizin im deutschsprachigen Gebiet bis etwa zum Jahre 1930," doctoral dissertation, Düsseldorf, 1968.

72 See L. C. Dunn, "Cross Currents in the History of Human Genetics," *American Journal of Human Genetics*, 1962, 14: 1-13 (p. 12). Dunn visited Russia in 1927, and it seems likely that Kol'tsov gave him the reprint of the article he had published in *ARGB* the previous year. For Dunn's report on his visit, see L. C. Dunn, "Genetics at the Anikowo Station," *Journal of Heredity*, 1928, 19(6): 281-86; Dunn also wrote a report for the Rockefeller Foundation on his Russian tour, see Joe Cain and Iona Layland, "The Situation in Genetics I: Dunn's 1927 Russian Tour," *Mendel Newsletter*, 2003, 12: 10-15.

73 Adams, "Eugenics in Russia," p. 170.

74 See, for instance, Beer, *Renovating Russia*, p. 39; Krementsov, "From 'Beastly Philosophy' to Medical Genetics," p. 66; Babkov, *The Dawn of Human Genetics*, pp. 23-29.

75 This is the first meaning attributed to the word in all major English dictionaries, such as the Miriam-Webster and the Oxford dictionaries, see www.merriam-webster.com/dictionary/improvement, and https://en.oxforddictionaries.com/definition/improvement

76 See, for instance, Kol'tsov, "Uluchshenie chelovecheskoi porody"; B. I. Slovtsov, *Uluchshenie rasy (evgenika)* (Pg.: Akademicheskoe izd-vo, 1923); Filipchenko, *Puti uluchsheniia chelovecheskogo roda (evgenika)*; Slepkov, *Evgenika. Uluchshenie chelovecheskoi prirody*; and many others.

77 All the subsequent quotations are from the entry *sovershat'* at http://dic.academic.ru/dic.nsf/enc2p/356274

78 See, for instance, I. Gergard, *Piat'desiat i odno sviashchennykh razmyshlenii, sluzhashchikh k vozbuzhdeniiu istinnogo blagochestiia i k usovershenstvovaniiu vnutrennego cheloveka* (M.: Reshetnikov, 1802); and E. Iv-niso, *Blagost' bozhiia v priorode ili Sovershenstvo estestvennogo mira* (M.: Universitetskaia tipografiia, 1845).

79 See, for instance, V. Kh. Fribe, *Rukovodstvo k usovershenstvovaniiu v Rossii ovtsevodstva* (SPb.: Meditsinskaia tipografiia, 1807); N. I Abashev, *Prakticheskoe rukovodstvo k usovershenstvovaniiu sel'skogo khoziaistva v nechernozemnoi polose Rossii* (SPb.: Ia. Ionson, 1855); and *Trudy Komiteta dlia rassmotreniia razlichnykh sistem ventiliatsii, dlia priiskaniia sredstv k ikh usovershenstvovaniiu* (SPb.: N. Tiblen, 1864).

80 See, for instance, P. D. Markelov, *Sovershenstvovanie cheloveka v istine* (SPb.: A. Pliushar, 1839).

81 See, for instance, A. Dreier, *Kosmosomatika ili ob iskusstve proizvodit', usovershenstvovat', podderzhivat' krasotu tela i istrebliat' nedostatki ee* (M.: Universitetskaia tipografiia, 1840).

82 See, for instance, Jean-Alexis Borrelly, *Introduction à la connaissance et au perfectionnement de l'homme physique et moral* (Marseille: J. Mossy, 1796); Antoine Desmoulins, *Exposition des motifs d'un nouveau système d'hygiène, déduit des lois de la physiologie et appliqué au perfectionnement physique et moral de l'homme* (Paris: Didot, 1818), https://books.google.co.uk/books?id=csebxAc4ma0C; Léopold Deslandes, *Manuel d'hygiène publique et privée, ou Précis élémentaire des connaissances relatives à la conservation de la santé et au perfectionnement physique et moral des hommes* (Paris: Gabon et Cie, 1827); Victor Maquel, *Perfectionnement ou dégénération physique et morale de l'espèce humaine* (Paris: Desloges, 1860); and many others. For the further development of these perfectionist ideas in French ethnology, see François Souffret, *De la disparité physique & mentale des races humaines & de ses principes* (Paris: Felix Alcan, 1892), https://archive.org/details/deladisparitphy00soufgoog. Alas, I did not find a historical analysis of the interrelations between the perfectionist ideas of French biologists and early French public health. For some observations, see Sean M. Quinlan, *The Great Nation in Decline: Sex, Modernity and Health Crises in Revolutionary France c. 1750-1850* (Aldershot: Ashgate, 2007); and *idem*, "Heredity, Reproduction,

and Perfectibility in Revolutionary and Napoleonic France, 1789-1815," *Endeavour*, 2010, 34(4): 142-50.

83 See Francis Devay, *Hygiène des familles*; and idem, *Traité spécial d'hygiène des familles particulièrement dans ses rapports avec le mariage au physique et au moral et les maladies héréditaires* (Paris: Labé, 1858), http://gallica.bnf.fr/ark:/12148/bpt6k9734739v. Both editions were available at the IMSA Library.

84 For the French original, see Auguste Debay, *Hygiène et physiologie du mariage, histoire naturelle et médicale de l'homme et de la femme mariés, perfectionnement de l'espèce, hygiène du nouveau-né*, 4th edn. (Paris: l'auteur, 1853); the 1857 edition is available at https://archive.org/details/hygineetphysiol00debagoog. For Russian translations, see O. Debe, *Gigiena i fiziologiia braka*, 3 vols. (SPb.: D. F. Fedorov, 1862-63); and O. Debe, *Gigiena i fiziologiia braka*, 2 vols. (M.: S. Orlov, 1862).

85 See Auguste Debay, *Hygiène et perfectionnement de la beauté humaine dans ses lignes, ses formes et sa couleur: théorie nouvelle des aliments et boissons, digestion, nutrition: art de développer les formes en moins et de diminuer les formes en trop, orthopédie, gymnastique, éducation physique, hygiène des sens*, etc., 4th edn. (Paris: E. Dentu, 1864), http://gallica.bnf.fr/ark:/12148/bpt6k107746x. For a Russian translation, see O. Deve, *Gigiena i usovershenstvovanie chelovecheskoi krasoty v ee liniiakh, formakh i tsvete* (M.: O. Nazarova, 1866).

86 Recall, for instance, Royer's "translation" of the subtitle of Darwin's *Origin* as "*des lois du progrès chez les êtres organisés*" (the laws of progress in organized beings) and Pisarev's interpretation of its contents as "progress in the world of animals and plants."

87 See, for instance, Louis Agassiz, *The Diversity of Origin of the Human Races* (Boston, MA: n. p., 1850); Thomas Smyth, *The Unity of the Human Races Proved to be the Doctrine of Scripture, Reason, and Science* (New York: G. P. Putnam, 1850); Joseph Arthur de Gobineau, *Essai sur l'inégalité des races humaines*, 4 vols. (Paris: Firmin Didot Frères, 1853-1855); and John P. Jeffries, *The Natural History of the Human Races* (New York: E. O. Jenkins, 1869).

88 For the UNESCO declarations on race, see Michelle Brattain, "Race, Racism, and Antiracism: UNESCO and the Politics of Presenting Science to the Postwar Public," *American Historical Review*, 2007, 112(5): 1386-413; and Marcos C. Maio and Ricardo V. Santos, "Antiracism and the Uses of Science in the Post-World War II: An Analysis of UNESCO's First Statements on Race (1950 and 1951)," *Vibrant: Virtual Brazilian Anthropology*, 2015, 12(2): 1-26. See also the Google Ngram for the phrase "the human race" that illustrates its usage in print during the period of 1800-2008, https://books.google.com/ngrams/graph?content=%22the+human+race%22&year_start=1800&year_en

d=2008&corpus=15&smoothing=7&share=&direct_url=t1%3B%2C%22%20 the%20human%20race%20%22%3B%2Cc0

89 Toll', *Nastol'nyi slovar'*, vol. 3, p. 269.

90 The literature on the concept of degeneration is vast. Useful overviews can be found in Richard Walter, "What Became of the Degenerate?: A Brief History of a Concept," *Journal of the History of Medicine and Allied Sciences*, 1956, 11: 422-29; Robert A. Nye, *Crime, Madness and Politics in Modern France: The Medical Concept of National Decline* (Princeton, NJ: Princeton University Press, 1984); J. Edward Chamberlain and Sander Gilman, eds., *Degeneration: The Dark Side of Progress* (New York: Columbia University Press, 1985); Daniel Pick, *Faces of Degeneration: A European Disorder, c.1848-c.1914* (Cambridge: Cambridge University Press, 1989); Ian Dowbiggin, *Inheriting Madness: Professionalisation and Psychiatric Knowledge in Nineteenth-century France* (Berkeley, CA: University of California Press, 1991); and many others. For a rather superficial account of the popularity of Morel's concept in late-nineteenth and early-twentieth century Russia, see Beer, *Renovating Russia*.

91 The original reads: "*Le mot dégénérescence lui-même est un mot nouveau.*" See P. Buchez, "Rapport fait à la société médico-psychologiques sur le traité des dégénérescences physiques, intellectuelles et morales de l'espèce humaine et des causés qui les produisent," *Annales médico-psychologiques*, 1857, 3: 455-67 (p. 455).

92 See the Google Ngram for the word "degeneration" that illustrates its usage in print in 1800-2008, https://books.google.com/ngrams/graph?year_start=1800&year_end=2008&corpus=15&smoothing=7&case_insensitive=on &content=degeneration&direct_url=t4%3B%2Cdegeneration%3B%2Cc0%3 B%2Cs0%3B%3Bdegeneration%3B%2Cc0%3B%3BDegeneration%3B%2Cc0

93 All the subsequent quotations are from the entry *vyrozhdat'* in Dal's dictionary, http://dic.academic.ru/dic.nsf/enc2p/222446. See also the meaning of one of its derivative, "*vyrodok*," in Toll', *Nastol'nyi slovar'*, vol. 1, p. 545.

94 See the classic analysis by Herbert Butterfield, *The Whig Interpretation of History* (London: Bell, 1931). For an application of this critique to the history of science, see, for instance, Loren R. Graham, "Why Can't History Dance Contemporary Ballet?: Or Whig History and the Evils of Contemporary Dance," *Science, Technology, & Human Values*, 1981, 6(34): 3-6; and Nick Jardine, "Whigs and Stories: Herbert Butterfield and the Historiography of Science," *History of Science*, 2003, 41: 125-40.

# Index

abolitionism 108, 110, 112
abortion 4, 14, 269, 333, 405, 432
Academy of Medical Sciences (AMN) 345, 353–356, 386, 423, 609–610, 622
Academy of Sciences (AN) xxi, 137, 233, 235, 262, 267, 276, 296, 322, 337, 341, 344–345, 352–354, 358, 361, 366, 378, 381, 393–394, 446, 471, 478, 513, 518, 533–534, 553, 564–565, 577, 599, 601, 610, 613, 637, 639
   Nauka (publishers) 362, 374, 384
Adams, Mark B. xi, 13, 402, 441, 468, 488, 491, 502, 605–606
*Advances of Modern Biology* 326, 337
Alexander I 37, 159, 213
Alexander II 10, 48, 61, 63, 75, 81, 185, 188, 208, 210–211, 218, 227, 375, 414, 463, 469
Alexander III 218, 220
Alexandra of Saxe-Altenburg, Grand Duchess 210, 557
All-Union Society of Geneticists and Breeders. *See* VOGiS
*American Journal of Human Genetics* 389
American Society of Human Genetics 468, 488
Andreev, Fedor 318
*Annales d'hygiène publique et de médecine légale* 164, 231
*Annals of Eugenics* 339, 345, 443, 635
*Annals of Human Genetics* 635

*Annals of the Fatherland* 34, 84, 93, 526
anthropogenetics 267, 271, 285, 319, 322, 324, 333–335, 337, 357, 365, 402, 418, 426
Anthropological Society of Paris 137, 163, 166–167, 541
anthropology 9, 16, 26, 73, 93, 106, 117, 195, 228–232, 245, 247, 256, 260, 262, 266, 273, 275, 288, 294–295, 344, 389, 434, 470, 478, 485–486, 488, 564–566, 573, 577, 584
anthropotechnique 246, 267, 271, 333, 357, 418, 426, 572
Antonovich, Maksim 112, 116
Anuchin, Dmitry 256–257, 260, 273, 275, 296, 381, 577, 583
*Archive of Legal Medicine and Social Hygiene* 195, 231–232, 547–548, 565
*Archiv für Rassen- und Gesellschafts-Biologie* 3, 251, 294, 396, 488, 587
Ardashnikov, Solomon 402
Arkadii, Archbishop 27–29, 32–33, 36, 508
artificial insemination 321–322, 331, 340, 365, 419, 425, 434, 441, 449–450, 457, 638
artificial selection. *See* selection, artificial
Astaurov, Boris 355, 357, 368, 610
Avdeev, Vladimir xx, 11, 19, 391–400, 406–407, 424, 427, 432, 464, 624–627

*Racial Essence of the Russian Idea, The* 392, 406, 627
*Raciology* 393–395, 406
*Russian Eugenics, A Reader* 11, 19, 391, 395, 397–400
*Russian Racial Theory before 1917* 392–393, 406

Babkov, Vasilii xi, 379, 398
Baer, Karl von 46, 130, 476–477, 534
Bakhtin, Mikhail 358
Balbiani, Konstantin 45, 51
Bateson, William 241, 253, 269
Batkis, Grigorii 309–310, 325–326, 328, 333, 594
Baur, Erwin 587, 599
Becquerel, Louis Alfred 154, 537
Bednyi, Dem'ian 322–323, 599–600
Bekhterev, Vladimir 249–250, 574
Beliaev, Dmitrii 353, 366, 608–609
*Bell, The* 77, 85
Bernard, Claude 54, 59
Bibikov, Petr 94
Bible 163, 288, 489
biology 16, 54, 73, 91, 93, 106, 196, 229, 262, 306, 309–310, 327, 336–337, 352–353, 358, 361–363, 366, 369, 419, 446, 457, 478, 499, 525, 527, 593
  evolutionary 95, 117, 326
  experimental 20, 262–263, 310, 359, 440–441, 450
  michurinist 346, 352–354
  molecular 20, 402, 444, 451, 453
  new 445–448, 450, 452, 454, 456
  racial 294
  "visionary" 441–442, 445
*Biometrika* 241, 569–570
Blagosvetlov, Grigorii xvii, 7, 15–17, 74–83, 87, 92, 97, 104–105, 109–110, 113, 116–124, 145, 150, 183–185, 187, 189–202, 225–226, 228–230, 233–234, 414–415, 464–465, 469–470, 473, 516–518, 523–524, 543, 545–550, 558
Blanter, Vladimir 356–357, 368, 610

Bochkov, Nikolai 355, 364, 373, 378, 385–386, 403, 410, 413, 422, 427, 458, 610, 619, 630
  *Clinical Genetics* 409–410
Bogoliubov, Vladimir 40, 42–43
Bolshevik Party 320, 334, 344
Bolshevik Revolution 13, 17, 255, 257–259, 263, 271, 277, 290, 300, 317, 327–328, 336, 376, 410, 416, 440, 464, 504, 579, 604
Bolsheviks 257–258, 260–261, 263, 265, 270, 275–277, 286, 288, 297, 327, 330, 375, 406, 416–420, 431–432, 584–586, 601
Borges, Jorge Luis 412–413, 493
Botkin, Sergei 53–54, 206, 554–555
Boudin, Jean-Marc 137, 163–164, 166–167, 169, 233, 536, 539–540
Bredov, Roman 206–207, 554–555
British Eugenics Society 342, 380, 448
Brock, J. F. 448, 453
Bronn, Heinrich Georg 101, 481, 483, 525
Bronovskii, Jacob 448, 453
Büchner, Ludwig 95, 98
Buckle, Henry T. 84, 104
Buffon, Georges-Louis 62, 362, 490, 492
Bunak, Viktor 260, 263–264, 266, 271, 275, 303, 333, 338, 379, 398, 577–579, 584, 589
Buniakovskii, Viktor 234, 567–568
Byron, George Gordon, Lord
  *Don Juan* 125, 162

Campanella, Tommaso 290, 436
Candolle, Alphonse de 236, 551, 569
Catherine the Great 28, 32
censorship 78, 80–82, 92–93, 120, 122–124, 184, 186, 189, 225, 288, 376, 469, 517–518, 526, 532, 543, 545, 550, 588, 618
  St. Petersburg Censorship Committee 87, 186, 225
Cervantes, Miguel de 412, 461
  *Don Quixote* 161, 413, 461

Chadwick, Edwin 438
Chaliapin, Fedor 279, 304, 317
Chernyshevskii, Nikolai 78, 84–85, 97, 118, 145, 179, 233, 507, 566, 601
  *What is to be Done?* 84–85, 227
Chistovich, Iakov 55, 57, 67, 120, 234, 467, 537, 556
Ciba Foundation 447, 449, 637
  *Man and his Future* 20, 363, 446–447, 613
Civic Code, Russian 227, 300–301
Civil War, American 107
Civil War, Russian 17, 255, 257–259, 263, 265, 330–331, 335, 359, 577
Claparède, Édouard 101, 481, 483, 525, 648
Clark, Colin 448, 453
Cold War 20, 293, 346, 352, 356, 362, 374, 402, 404, 411, 437, 443–444, 446, 457, 464
  Cuban missile crisis 448
Comfort, Alex 448, 450
Communist Academy 276, 309–310, 313–314, 319, 324, 595
  Society of Materialist-Biologists 314, 317, 596
conditional reflexes 334, 359, 361, 612
*Contemporary, The* 34–35, 78, 81, 84–85, 96–97, 107, 112, 118, 120, 126, 179, 185, 224, 226, 507, 517, 519–520, 526, 542
Council of People's Commissars (SNK) 265, 586
Crick, Francis 363, 448, 452, 454
Crimean War 9, 48, 51, 64, 73–74, 77, 89–90, 227, 410, 414, 464, 508
Crow, James F. 455

Dally, Eugène 163, 167, 541
Darwin, Charles 8, 15–16, 25, 74, 84, 92, 95, 100–110, 112, 114–117, 119, 121, 126, 134, 138, 146, 153, 156, 163, 166, 174–177, 193, 195–199, 201–202, 228, 232, 240, 243, 245, 251, 276, 279, 291, 297, 303, 306, 308, 326, 329, 337, 362, 369–371, 394, 413, 420, 428, 437–438, 440, 445, 450, 452, 470, 473, 481–485, 490–491, 493, 524–527, 530–531, 533, 551, 594, 648
  *Descent of Man, The* xviii, 17, 101, 198–199, 201, 229–230, 297, 415
  *On the Origin of Species* 8, 16, 26, 101–103, 105–106, 108, 128, 134, 138, 163, 176–177, 197, 199, 326, 371, 437, 475, 481–486, 490, 515, 527, 542, 590, 647–648, 652
  *Variation of Animals and Plants under Domestication, The* 196–197, 548
Darwinism 2, 17, 105, 125, 201, 276, 285, 295, 297, 317, 326, 337, 362, 369, 415, 419–420, 428, 468, 493, 525, 529, 585, 594, 612, 634
Darwin, Leonard 243, 332, 429
Davenport, Charles 241, 245, 269–270, 332, 336, 340, 433, 435, 581–582, 604
Davidenkov, Sergei 302, 324–325, 327, 333, 338–339, 341–342, 344–345, 353–356, 361, 366, 595, 607
Debay, Auguste 192, 490
*Deed* 17, 183–184, 190, 193–197, 200–202, 226, 229–230, 415, 465, 469, 517, 546–548, 558, 563–564, 583
degeneration 7–8, 10, 13, 16–18, 73, 121–122, 124–127, 136, 142, 150, 154, 156–157, 159–162, 164–165, 169–170, 172–173, 175–176, 178, 180, 183, 195–197, 202, 205, 224, 227–228, 232–236, 240, 245, 248–249, 251, 270, 280, 282, 286, 290, 294, 296–298, 309, 312–313, 316, 330, 356, 370, 373, 376, 378, 410, 413, 416, 420, 425, 431, 437–440, 443, 458, 487–488, 492–493, 566–568, 653
Delianov, Ivan 211–212, 218, 557
Devay, Francis 164, 166–167, 233, 490, 539–540
Dickens, Charles 111
Dickson, Sam 395

Dobroliubov, Nikolai 42, 78, 84, 145, 179, 507, 520, 601
Dobzhansky, Theodosius 358, 367, 370, 455, 596, 616
Dostoevsky, Fedor 34, 41, 75, 84, 87, 96, 108, 247, 266, 470, 644
Dubinin, Nikolai 352–353, 355, 364, 609, 614
Dubovitskii, Petr 51–53, 57, 70, 188–189, 204, 206, 510, 554–555
Dunn, Leslie C. 443, 468, 488

Edwards, William Frédéric 130
Efroimson, Vladimir 354–355, 364, 615
emancipation of the serfs. *See* serfdom
Engels, Friedrich 325–326, 438, 440, 601
eugamics 8–9, 328–332, 334, 351, 410, 412, 419–422, 499
Eugenics Record Office (ERO), United States 269, 282, 285, 433
Eugenics Record Office (ERO), University College London 4, 242–243
*Eugenics Review* 243, 251, 339–340, 583, 605, 633
*eugénnetique* 3–4, 17, 246, 581
euphenics 318–319, 323, 329, 419–420, 426, 450–452, 454, 456
euthenics 3, 244

Fando, Roman 379
fascism 2, 341, 346, 365
Filipchenko, Iurii xviii, xix, 249–250, 255, 267–270, 273, 278, 291, 299, 301, 303–304, 306–307, 309, 312–313, 315, 317, 320–321, 325, 327, 329, 334, 337, 352, 354, 357–359, 365–367, 370, 379, 396, 398, 580–582, 593, 596–597, 616–618
Florinskaia (née Fufaevskaia), Maria 70, 121, 187, 202–203, 223, 236, 465–467
Florinskii, Mark Iakovlev 27–30, 37, 65, 383

Florinskii, Vasilii
  childhood 26–31
  death 13, 17, 224, 236, 239, 245, 416, 465, 467
  establishes Tomsk University 214–215, 217–218, 220–223, 351, 370, 380, 382, 384, 387, 391, 409, 423, 458, 466–468, 559–560
  IMSA promotions 203, 212, 216
  letters 39, 60, 68, 465, 512
  marriage and children 71, 121, 187, 202–203, 207, 212, 215, 223, 465, 532
  memoirs 35, 46, 117, 215, 467, 504, 512
  moves from IMSA to Ministry of People's Enlightenment 212–213
  studies for priesthood 38–40
  studies medicine 42–58
  subject of scandal 185–187, 202
  travels to Europe 59–69, 465
  works in Siberia 214–223, 223, 384
*Fortpflanzungshygiene* 3, 244, 246, 426
Frank, Johann Peter 203, 438, 552
Freeze, Gregory L. 10, 227, 562
future, the 4–6, 9, 14, 16, 20, 48, 64, 67–68, 74–75, 94, 102, 119, 129, 174, 193, 201, 233, 243, 252–254

Galton, Francis 1–11, 13, 15–20, 23, 25–26, 139, 173–174, 176–178, 180, 183, 199–202, 224, 231, 235–236, 239–246, 249, 251, 265–267, 270, 273–274, 278–279, 284, 290–292, 294, 296–297, 299, 303–308, 312, 315, 318–321, 328–332, 345, 347, 351, 356–358, 361–362, 364–373, 378, 380, 384, 389, 398, 400–401, 403, 409–410, 415–417, 420–422, 424, 426–429, 432–433, 438–439, 442–443, 458, 474, 481, 491, 497, 499, 503–504, 550–551, 569–573, 575, 584, 592–593, 615
  *Essays on Eugenics* 243, 265, 303, 305, 307, 409, 497, 570–571, 622

*Hereditary Genius* 20, 180, 199–200, 245, 365, 368, 372, 378, 409–410, 422, 551, 569, 584, 607
Gamaleia, Nikolai 249–252, 254, 576
generative hygiene 249, 250, 251, 254, 436. *See also Fortpflanzungshygiene*
genetics xix, 11, 13, 18–20, 241, 249–250, 252–255, 263, 267, 269, 271, 288, 293, 298, 302–304, 310–312, 314–317, 319–322, 328–329, 334–335, 337, 339–340, 342–344, 346, 348, 351–355, 357, 359–363, 365–366, 369–371, 374, 379, 389–390, 401–402, 404, 409–410, 434, 440, 443, 445, 448–449, 453–454, 458, 470–471, 594–595, 602, 609–610, 613, 616, 631, 633, 635. *See also* medical genetics
*genofond* 316–317, 321, 331, 378, 396, 419–420, 445, 597
germinal choice 449–450, 452, 454, 456
GIZ, State Publishing House 303–307, 310, 592–593
*glasnost'* 48, 64, 75, 81, 227, 231, 375–376, 414, 422, 509
Glebov, Ivan 51–54, 61, 204, 384, 510, 555
Glikson, Artur 448, 453, 638
Goddard, Henry H. 418
Goethe, Johann Wolfgang 361–362, 367
Gogol, Nikolai 34, 40
Goncharov, Ivan 34, 87
Gorbachev, Mikhail 374–375, 422, 464
Gorky, Maxim 304, 317
Graham, Robert Clark 456
Great Break 293, 320–322, 324–325, 327, 339, 611
*Great Medical Encyclopedia* 299, 319, 353–354, 356, 373
Great Reforms 10, 48, 56, 64, 74, 84, 89, 94, 208, 210–211, 213, 218, 375, 410, 413–414, 463, 509, 556, 562

*Great Soviet Encyclopedia* 319, 325, 347, 364, 366
Great Terror 19, 293, 340
Grinberg, Lev 476, 481
Gruber, Ventseslav 63, 205
Gulag 352, 354, 376, 381
Gunn, John C. 9, 500

Haldane, J. B. S. 363, 426, 441, 448–449, 452–453, 614, 634
Hegel, Georg Wilhelm Friedrich 36
Heinlein, Robert A. 5, 446
hereditary diseases 19, 155, 167–170, 172, 283, 288, 302, 311, 321, 331, 341, 345, 356, 369, 373, 389, 404–405, 420, 422, 457, 490, 537
heredity 4–8, 13, 16–17, 19–20, 23, 110, 121, 123, 125–128, 133–138, 142–143, 147–148, 154, 165–166, 168–170, 172–174, 177–178, 196, 228–229, 234, 239–241, 243–244, 249, 251–253, 263, 266–267, 271–273, 278, 288, 295, 298, 301, 306, 308–311, 314–315, 318, 320–322, 324, 326–329, 332, 338, 343, 345, 347, 349, 351, 354, 356–357, 364, 366, 368–369, 371, 373, 386, 389–390, 400, 402, 409–410, 419, 424, 434, 437, 440–442, 458, 478, 490, 535, 594, 635
Herzen, Alexander 61, 77, 81, 83, 87, 516, 601–602
Hitler, Adolf 340, 342, 396
Hoagland, Hudson 448, 450
Hospital for Menial Workers 54, 91
human destiny 20, 435, 467. *See also* future, the
human evolution 8, 101, 198, 228, 270, 285, 291, 294, 297, 326–327, 363, 417, 467
  directed 8, 363
Human Genome Project 20, 389–390, 403, 437, 458
humaniculture 3, 244
human nature 5, 20, 100, 295, 298, 326, 328, 363, 365, 420–421, 424, 435–437, 458

Huxley, Aldous 436, 442, 446
Huxley, Julian 342, 363, 426, 445, 448–449, 452–453
Huxley, Thomas Henry 95, 101–102, 105, 114, 481
*Hygiene and Sanitation* 249–252

Iakobii, Pavel 193–195, 200–201, 228, 235–236, 465, 546, 569
Imperial Medical-Surgical Academy (IMSA) xvii, 7, 14–15, 40, 42–52, 55–56, 76, 91–92, 118–120, 163, 185, 187–189, 198, 202–204, 207–209, 211–213, 231, 373, 380, 390, 465–468, 476, 482, 507–510, 543–544, 565, 617
    council of 43–44, 46, 51, 56–57, 60, 69, 185, 202, 204, 206–207, 555
incest 4, 7, 127, 162, 165. *See also* marriage, consanguineous
inheritance of acquired characteristics 196, 278, 294, 298, 306, 310–313, 315, 318, 328, 360, 371, 420, 594–596. *See also* Lamarckism
Institute for the History of Natural Sciences and Technology (IIET) 361, 376, 379
Institute of Experimental Biology (IEB) 262–266, 270, 275–276, 304, 312, 324, 341, 345, 355, 580, 584, 600
Institute of Medical Genetics (IMG), Moscow 339, 351, 386
Institute of Medical Genetics (IMG), Tomsk 11, 19, 355, 385–386, 389, 403–404, 423
Institute of Physical Culture (IFK) 275, 277–278, 333, 470, 588
International Eugenics Congress
    First, London, 1912 2, 243, 251–252, 429, 433
    Second, New York City, 1921 2, 269–270, 282, 336, 417, 429
    Third, New York City, 1932 2, 336–337, 339, 429
International Federation of Eugenic Organizations 270, 321, 429

International Genetics Congress
    Fifth, Berlin, 1927 xix, 314, 335
    Seventh, Edinburgh 339, 342, 605
    Sixth, Ithaca, 1932 336
isolation 128, 132, 156, 269
Iudin, Tikhon 249–250, 252–254, 263–264, 271–272, 299, 302–303, 306–307, 309, 319, 325, 327, 333, 338–339, 355, 379, 396, 398, 574, 576
*Izvestiia* xxii, 271, 296, 300, 322–323, 347, 581, 599–600

James, William 412
Jordan, David Starr 440, 632

Kakushkin, Nikolai 268–269
Kaminskii, Grigorii 339, 341
Kammerer, Paul 310, 427, 594–595
Kanaev, Ivan 19, 268, 358–359, 361–362, 366–374, 378, 382, 384–386, 403, 413–414, 421–422, 427, 475, 611, 613, 615–617
    *Francis Galton* 592, 611, 616, 619
    *Twins and Genetics* 366–367, 370
Karakozov, Dmitrii 185, 188, 204, 208, 211, 226, 228, 414–415
Kashkin, Nikolai 96–97
Katkov, Mikhail 84, 87
Kazan University 50, 195, 214, 216–218, 249
Khrushchev, Nikita 19, 351–354
Kiter, Alexander 49–51, 53, 56, 205, 207, 509
Klein, Marc 448, 452–453, 638
*Knowledge* 198–200, 202, 550, 552
Kol'tsov, Nikolai xviii, 250, 255, 262–272, 276, 278–279, 291, 294, 297–299, 301–304, 306–308, 311–313, 315–319, 321–322, 324–325, 327–330, 333–334, 338, 341, 344–345, 352–355, 357, 360, 369–370, 379, 396, 398, 418, 428, 441, 451, 468, 488, 578–581, 589–590, 592–593, 598, 600, 610, 616, 650

Konstantin Nikolaevich, Grand Duke 210–211, 215, 463
Kozlov, Nikolai I. 205, 207
Kraevskii, Andrei 84, 87
Krasnogorskii, Nikolai 361, 612
Krasovskii, Anton 50, 53–54, 56, 69, 205–207, 510, 512
Kropotkin, Petr 34, 246–247, 573
Krzywicki, Ludwik 245, 247–248, 572
Kushelev-Bezborodko, Grigorii 75–78, 81, 463
Kuznetsova, Anna 210–211

Lamarckism 309–315, 317–318, 329, 420, 441, 594, 596, 612
Lamarck, Jean-Baptiste 249, 310–311, 318, 329, 332, 343, 360, 420, 424, 490
Lederberg, Joshua 363, 448–457, 641
Lenin, Vladimir 257, 265, 276, 294, 306, 346, 364, 386, 452, 585, 601
Lermontov, Mikhail 34, 69, 85
Levit, Solomon xix, 19, 311, 314, 319, 321–322, 325, 328, 333–342, 354–355, 379, 402, 521, 595, 604
Lewes, George H. 92, 95
Library of Racial Thought 391–392, 397, 626
Lipmann, Fritz A. 363, 448
Liublinskii, Pavel 248, 301, 379, 398
Lobashov, Mikhail 365–366
Lombroso, Cesare 247, 492
Lyell, Charles 95, 101, 105, 129, 533, 542
Lysenko, Trofim 19, 341, 343–346, 352–356, 401–402, 421, 435, 470–471, 607, 609, 612

MacKay, Donald M. 449, 453
MacKintosh, Elizabeth. *See* Tey, Josephine
Makarii, Archbishop 35–36, 216, 506
marriage 5, 7–8, 10, 16, 27, 29, 38, 120–121, 124–125, 127, 136, 150–152, 154, 156, 162–163, 169–173, 177–180, 195, 201, 224–225, 227–228, 234, 236, 239–240, 242, 253, 268, 272–273, 278–280, 283–284, 287, 293–294, 296–297, 299–303, 325, 331, 356, 369, 372, 389–390, 400, 403, 406, 415, 418–419, 425, 431, 436, 487, 490, 562
  consanguineous 4, 10, 124, 156, 163–170, 172, 177, 294, 301, 365, 373, 389, 404, 425, 541. *See also* incest
  hygienic 121, 126, 151, 154
  mixed 8, 10, 124, 136, 156, 228
  rational 8, 10, 16, 124, 127, 145, 150–151, 156, 159, 499
  regulation of 10, 18, 121, 156, 500, 542, 562
Marxism 18, 239, 275–279, 285, 295, 303–304, 306–307, 309–311, 313–320, 324, 326–329, 331–333, 338, 344, 351, 354, 363–364, 368, 372, 391, 400–401, 417, 419–422, 426, 428, 435, 464, 572, 585, 594
Marx, Karl 277–278, 306, 308–309, 325, 440, 585
Maudsley, Henry 242, 492
Medawar, Peter B. 363, 448, 453
Medical-Biological Institute (MBI) 319, 321, 323, 334, 337–339
medical genetics 11, 19, 333–334, 338–341, 344–347, 349, 351, 353–358, 365, 369–373, 378–379, 385–387, 389–390, 400–403, 405, 413, 420–422, 435, 457, 602
*Medical Herald* 57–58, 64, 67, 70, 73, 119–120, 163, 188–189, 536
Medical Institute, Leningrad 359, 361
Mendel, Gregor xix, 241, 249, 251–253, 266, 298, 303, 310, 314–315, 329, 335, 346–347, 356, 360, 366, 371, 389, 424, 442, 592, 594, 610
  *Experiments on Plant Hybrids* 303, 305, 354
Mendelism 253, 273, 295, 346, 354, 361, 441, 576

Mestergazi, Mikhail 317–318, 596, 598
Military-Medical Academy (VMA) 43, 380. *See also* Imperial Medical-Surgical Academy
*Military-Medical Journal* 49, 57, 69, 73, 185
Miliutin, Dmitrii 188, 207, 209, 211–212, 556
Mill, John Stuart 84–85, 566
Minaev, Dmitrii 78, 83, 125–126, 533
Ministry of Internal Affairs 82, 87, 543
  Medical Department 212, 231
Ministry of People's Enlightenment 82, 208–209, 211–213, 216, 218, 220, 260, 465, 467, 556
  Scientific Committee 208, 216
*Modern Psychiatry* 252, 254
Moleschott, Jacob 92, 95, 98, 194
Mol'kov, Al'fred 261, 263
Morel, Bénédict 169–170, 198, 492, 541–542, 653
Morgan, Thomas Hunt 298, 313–315, 335, 346, 360, 596
Morigerovskii, Alexander 36–38, 40–41, 118–119, 199, 506, 549
Moscow Society of Psychiatrists and Neurologists 302, 324
Moscow University xviii, 51, 92, 101, 118, 185, 230, 255, 259–260, 266, 273, 275, 296, 309, 311, 354, 381, 482, 583, 603
Motkov, Stanislav 377–378, 618
Muller, Herman J. xix, 62, 293, 335–343, 353, 363, 426, 435, 448–450, 452–457, 531, 603–607, 638–640
  *Out of the Night* 339, 449, 452, 605

Napoleonic wars 159, 438, 555
Narkompros (People's Commissariat of Enlightenment) 260, 263, 319, 577, 579, 586
Narkomzdrav (People's Commissariat of Health Protection) 260, 262–265, 269, 271, 275, 277, 300, 306, 319, 325, 327, 330, 338–339
  OMM (protection of maternity and infancy) Department 277, 283, 300
  OZDP (protection of children's and adolescents' health) Department 277, 283, 586
National Museum of the Tatarstan Republic 465, 504
natural selection. *See* selection, natural
*Nature* 241, 339, 454, 456
Neel, James V. 402–403
Nekrasov, Nikolai 78, 84, 87, 118
New Economic Policy (NEP) 265–266, 300, 320, 359, 410, 579–580
Niceforo, Alfredo 291, 588
Nicholas I 37, 48, 63, 89, 179, 218, 505, 543, 553
Nicholas II 220, 257
nihilists 88, 520, 562
NKVD (People's Commissariat of Internal Affairs) 324, 396
Norov, Avraam 89–90, 521
Nozhin, Nikolai 114–117, 531

Ol'khin, Pavel 92, 476
Orbeli, Leon 345, 361, 366
Orshanskii, Isaak 248–249
Orwell, George 436

Pargamin, Manus 235
Parkes, Alan S. 448, 450
Paul I 46, 210, 553
Pavlov, Ivan xix, 88, 339, 345, 359–361, 612
Pavlovskii, Evgenii 356, 593, 610
Pchelov, Evgenii 379, 398, 622
Pearson, Karl 3, 241–243, 245, 252, 278, 291, 332, 570, 635
Pelikan, Evgenii 54, 91, 231
Pelikan, Ventseslav 43–45, 51, 204
People's Commissariat of Enlightenment. *See* Narkompros

People's Commissariat of Health
    Protection. *See* Narkomzdrav
People's Commissariat of Internal
    Affairs. *See* NKVD
Peter-Paul Fortress 81–82, 190, 195,
    415, 526, 549
*Petersburg Leaf* 87, 185–187, 202, 545
Peter the Great 4, 141, 216, 273, 427,
    508, 583
Petrov, Osip 279–280
Pincus, Gregory 448, 450, 452
Pirie, N. W. 448, 452, 456
Pirogov, Nikolai 46, 49–50, 63, 477
Pisarev, Dmitrii 7, 15, 79, 81–85,
    88, 92, 97–99, 102–105, 122, 124,
    144–145, 176, 179, 184, 196, 224,
    475, 482, 491, 515, 517, 522–524,
    526, 528, 548–549, 601, 652
Plato 4, 274, 290, 436
Platonov, Petr 49
Ploetz, Alfred 2–3, 244, 426, 497
Polonskii, Iakov 75, 77, 97
Pomialovskii, Nikolai 31–32, 79, 506
Portugalov, Veniamin 183, 195–198,
    202, 229, 233, 235, 545–549
    *Issues in Social Hygiene* 198, 201, 415
*Pravda* xxii, 295, 336, 347
*Priroda* xxii, 250, 267, 296, 338, 341,
    356, 368, 370, 374, 393, 550, 552,
    626
Prokof'eva-Bel'govskaia, Alexandra
    353, 355, 366
*puériculture* 3, 244, 246, 426, 429
Pukirev, Vasilii 155
    *The Unequal Marriage* xvii, 154–155,
    227
Pushkin, Alexander 23, 31, 34, 69, 85
Putin, Vladimir 375, 464
Puzyrev, Valerii xx, 11, 19, 385–391,
    399, 403–406, 410, 422–423, 427,
    464, 621, 643

Quatrefages, Armand de 95, 105, 110
Quetelet, Adolphe 149, 233

race 3, 6, 8, 106–108, 111, 113–115,
    125, 129, 141, 160–161, 166, 169,
    177–178, 242, 246–247, 278, 282–
    283, 309, 317, 332, 356, 369, 371,
    392, 394, 398, 406–407, 431–432,
    444, 447–448, 453, 485–486, 488,
    491, 527, 530, 573, 652
Rachinskii, Sergei 101–103, 176,
    482–484, 490–491, 525–526, 533,
    648
racial hygiene 3, 244, 250–251, 254,
    262, 294, 343, 365, 395–396, 401,
    406, 421, 424, 436, 442–443, 575–
    576, 589. *See also Rassenhygiene*
raciology 11, 392–395, 398, 400, 406,
    623–624, 626
racism 2, 13, 270, 307, 341, 344, 346,
    393, 395, 402, 431, 444, 625
Radin, Evgenii 277–278, 327, 432,
    585–587
*Rassenhygiene* 3–4, 17, 244, 246, 295,
    340–342, 344, 401, 426, 428–429,
    442, 488, 497, 502, 575, 633
*raznochintsy* 179, 329, 542
Reclam, Karl H. 196–197, 201, 415,
    548–549
Reclus, Élie 79, 83, 110, 112, 114
Reich, Eduard 196–197, 415, 548–549
Richard III 462–463
Ritter, Carl 84, 95
Rklitskii, Ivan 49
Rokitansky, Carl 59, 62
Rokitskii, Petr 302, 320, 369
Rolle, Friedrich 104–105, 482, 529, 648
Royer, Clémence 101, 103, 106–110,
    160, 174, 481, 483, 491, 525–526,
    529–530, 652
Rushton, J. Philippe 394–395
Russell, Bertrand 441, 634
*Russian Eugenics Journal* (REZh) 275,
    306–308, 316, 357, 427–428, 581,
    583, 587, 616
Russian Eugenics Society (RES) 10,
    18, 239, 255, 264–267, 269–274, 277,
    294, 299–300, 302–303, 308, 316,

318, 333, 355, 357, 379, 398, 421, 470, 472, 488, 579–581
Russian Federation xxi, 374–375, 391, 406
*Russian Herald* 84, 87, 107, 224
Russian Orthodox Church 10, 36, 171, 562
*Russian Word* 7, 15–17, 72–85, 87–90, 92–95, 97, 100, 102, 105, 107, 110, 112–113, 116–118, 123–126, 133, 144–145, 163, 174–176, 179, 183–185, 187–194, 203, 224–226, 234, 273, 286–287, 370, 382, 387, 415, 463, 465, 469–470, 476, 487, 517, 519, 522–524, 532, 534, 563

Schleicher, August 104–105, 528
science fiction (SF) 5, 362, 436, 444, 446, 457
Sechenov, Ivan 53–54, 120, 196, 513, 532, 549–550
secret police 77, 81–82, 118, 185, 208, 469, 473, 547
selection
 artificial 16, 101, 147, 271, 378, 437, 484
 natural 8, 95, 106, 108, 110, 146, 177, 201, 251, 297, 335, 378, 437, 439, 483, 490–491, 526–527, 530, 594, 648
 sexual 16, 102, 146, 176, 297, 438, 484, 590
Semashko, Nikolai 261, 263, 266, 278, 300, 306, 308–309, 320, 325, 327, 329, 356, 432, 578–579, 593
Serebrovskii, Alexander xix, 239, 262, 271, 279, 313–317, 319, 321–325, 328–329, 331, 334–336, 338, 340–341, 344, 352, 355, 357, 368, 379, 396, 398, 419–420, 450, 582, 599–600, 603
serfdom 107, 157–158, 179, 232
 emancipation of the serfs 48, 56, 74, 84, 107, 109, 562
Serno-Solov'evich, Nikolai 94, 97
sexual selection. *See* selection, sexual
Shakespeare, William 25, 462

Sharp, Harry C. 264–266
Shchapov, Afanasii 79, 83, 92, 94, 117, 518–519, 534
Shelgunov, Nikolai 7, 15, 74, 79, 81, 83, 90, 92, 94, 99–100, 103, 105, 118, 120, 122, 124, 159, 190, 414, 469
Shklovskii, Isaak 247, 573
Shnirel'man, Viktor 393–394
Shorokhova, Antonina 269, 321
Shull, William 402–403
Siemens, Hermann W. 278, 284, 291, 308, 587
slavery 7, 107, 109–112, 114–116, 127, 150, 160, 163, 194, 232
Slavophiles 68, 75, 513
Slepkov, Vasilii 295–297, 299, 307, 328, 589–590
social hygiene 9, 14, 16, 18, 120, 126, 174, 176, 195–197, 201, 227–232, 251, 255, 261, 263, 285, 308–309, 325, 329, 333, 379, 415, 419–420, 438, 473, 485–487, 548, 565–566, 568, 578, 633, 649–650
 State Institute of 262, 300–301, 325, 591
 State Museum of xviii, 260–263, 578
Society for the Protection of People's Health 231, 249
Society of Enthusiasts for the Natural Sciences, Anthropology, and Ethnography 90, 230, 257
Society of Russian Physicians, St. Petersburg 55, 69, 73, 185, 189, 207, 234, 264, 553
Sokolov, Nikolai 79, 83, 579
Solov'ev, Zinovii 261, 306–309, 325, 327, 329, 356, 600–601
Soviet Union ii, xi, 11, 13, 18, 19, 293, 302, 312, 323, 326, 327, 328, 332, 333, 337, 339, 340, 341, 344, 345, 346, 347, 349, 351, 352, 356, 361, 363, 368, 374, 375, 376, 377, 379, 381, 384, 385, 391, 392, 396, 400, 401, 403, 404, 405, 406, 411, 418, 421, 422, 425, 435, 442, 443, 471,

472, 502, 503, 585, 586, 599, 605, 607, 614, 617, 622, 623, 624, 628, 632. *See also* USSR
*Spark* 87, 112–114
Spencer, Herbert 481
Speranskii, Mikhail 37, 213, 505
Stalin, Joseph 19, 320, 322, 324, 327, 333, 339–340, 342, 344, 346, 351–352, 354, 361, 364, 381, 398, 420–421, 442, 450, 605–606, 624
State Council 210–211, 214, 220
Steinach, Eugene 264, 266
sterilization 4, 246–248, 253, 264, 266, 271–273, 276, 283, 301, 325, 331, 369, 396, 403, 418, 425, 472, 579–580
Stowe, Harriet Beecher 107, 111, 114
St. Petersburg Foundling Home 57–58, 205
St. Petersburg Theological Academy. *See* Theological Academy, St. Petersburg
St. Petersburg University 76, 97, 209, 212, 428
Strakhov, Nikolai 87, 108–110, 482, 530
Suvorin, Aleksei 213–214, 558
Sysin, Aleksei 263, 300–301, 327, 379

Table of Ranks 44, 64, 72, 220, 468, 478, 508, 511
Tey, Josephine
  *Daughter of Time, The* 462, 472, 474
Thaw 19, 351–352, 355, 361, 363, 374, 411, 421–422
Theological Academy, St. Petersburg 14, 26, 34, 36–38, 41, 43–44, 507
Timiriazev, Kliment 245, 276, 482
Timiriazev State Scientific-Research Institute for the Study and Propaganda of the Natural-Science Foundations of Dialectical Materialism 275–278, 280, 286, 312, 316, 326, 585, 588, 645
Timofeev-Ressovskii, Nikolai 355, 366

Tkachev, Petr 7, 78–79, 83–84, 124, 184, 418, 522
Toll', Feliks 133, 476, 649
Tolstoi, Dmitrii 208–209, 213–215, 217–218
Tolstoi, Konstantin 163
Tomsk University xviii, 217–218, 220–222, 370, 382, 384–388, 391, 409, 458, 465–468, 513, 558–559
Trembley, Abraham 359, 362, 367
Trotsky, Leon 271, 327, 341, 417
Trowell, Hubert C. 449, 453
Turgenev, Ivan 34, 85, 370, 414
  *Fathers and Children* 84–85, 370
twins, studies of 240, 249, 334, 337, 339, 341, 345, 359, 361–362, 366, 373, 418, 616

*Under the Banner of Marxism* 309, 341, 344, 363, 597
University College London, Eugenics Record Office of. *See* Eugenics Record Office
Urals University 381–382, 620
USSR xxi, 11, 265, 322, 342, 344–345, 353, 378, 425, 502–503, 586, 590–591, 602, 605, 608–609, 614–615

Vaisenberg, Samuil 294–295, 488, 589
variability 4–8, 16, 85, 106, 121, 123, 126–134, 144, 157, 173–175, 177–178, 228, 234, 239–241, 288, 298, 309–310, 322, 324, 326–327, 349, 351, 364, 368, 434, 478, 533
Vavilov, Nikolai 269, 334, 337–338, 343, 345, 352, 355, 607
Vel'tishchev, Iurii 365
Virchow, Rudolf 54, 59–60, 62–63, 84, 95, 120, 134, 196, 438, 535, 548
Vishnevskii, Boris 296–297, 299, 307, 413
Vishniakov, Alexander 34–36, 49, 506
VOGiS (All-Union Society of Geneticists and Breeders) 355, 357

Vogt, Carl 92, 95, 98, 101–102, 105, 114–116, 129, 523–524, 526, 533–534
Voice 87, 188, 193, 224–225
Vol'f, Mavrikii 91, 93, 107, 522, 530
Volotskoi, Mikhail xviii, xix, 10–11, 17–18, 239, 255–257, 259–260, 262–266, 271–300, 302–304, 307, 309, 311–319, 321, 326, 328, 331–333, 338, 347, 355, 357, 368, 370–372, 379, 387, 389, 396, 398–401, 416, 418–419, 427, 464, 470–472, 475, 484, 488, 500, 576–577, 579–580, 583–584, 587–588, 592, 596–597, 616, 644

*Class Interests and Modern Eugenics* 278, 304, 593

*Elevating the Vital Forces of the Race* 272–273, 300, 471, 593

Vorob'eva, Anna 279–280

Wallace, Alfred Russel 8, 201, 415
*Contributions to the Theory of Natural Selection* 201, 415, 550, 603
War Ministry 46, 53, 87, 204, 209, 211–212, 465, 467

Military Medical Department 204, 206–207
Weismann, August 249, 251, 298, 346, 594, 603
Wells, H. G. 4–5, 242, 436, 440–441
Westernizers 68, 75, 513
Whewell, William 477
Wolstenholme, Gordon E. W. 447–449
World War I 17, 20, 255, 257–258, 269, 331, 425, 431, 433, 440
World War II 20, 345, 442–444, 446, 583
Wundt, Wilhelm 95, 194
Wylie, James 43, 204, 552–553

Yeltsin, Boris 374–375, 384

Zaitsev, Varfolomei 7, 15, 79, 83, 85, 92–95, 102, 105, 110–118, 124, 131, 145, 160–161, 174, 185, 193–194, 520, 522–523, 526, 531, 546
Zavadovskii, Mikhail 262, 441, 634
Zinin, Nikolai 46, 51–52, 204, 508, 553

# This book need not end here...

## Share

All our books—including the one you have just read—are free to access online so that students, researchers and members of the public who can't afford a printed edition will have access to the same ideas. This title will be accessed online by hundreds of readers each month across the globe: why not share the link so that someone you know is one of them?
This book and additional content is available at:
https://www.openbookpublishers.com/product/800

## Customise

Personalise your copy of this book or design new books using OBP and third-party material. Take chapters or whole books from our published list and make a special edition, a new anthology or an illuminating coursepack. Each customised edition will be produced as a paperback and a downloadable PDF. Find out more at:
https://www.openbookpublishers.com/section/59/1

Like Open Book Publishers

Follow @OpenBookPublish

Read more at the Open Book Publishers BLOG

www.ingramcontent.com/pod-product-compliance
Lightning Source LLC
Chambersburg PA
CBHW061701300426
44115CB00014B/2524